NEW ADVANCES IN CROP DISEASE MANAGEMENT

NEW ADVANCES IN CROP DISEASE MANAGEMENT

Edited by
Sangita Sahni, PhD
Ganesh Thapa, PhD
Bishun Deo Prasad, PhD

AAP | APPLE ACADEMIC PRESS

First edition published 2025

Apple Academic Press Inc.
1265 Goldenrod Circle, NE,
Palm Bay, FL 32905 USA

760 Laurentian Drive, Unit 19,
Burlington, ON L7N 0A4, CANADA

CRC Press
2385 NW Executive Center Drive,
Suite 320, Boca Raton FL 33431

4 Park Square, Milton Park,
Abingdon, Oxon, OX14 4RN UK

Library and Archives Canada Cataloguing in Publication

Title: New advances in crop disease management / edited by Sangita Sahni, PhD, Ganesh Thapa, PhD, Bishun Deo Prasad, PhD.
Names: Sahni, Sangita, editor. | Thapa, Ganesh, editor. | Prasad, Bishun Deo, editor.
Description: First edition. | Includes bibliographical references and index.
Identifiers: Canadiana (print) 20240350413 | Canadiana (ebook) 20240350456 | ISBN 9781774916070 (hardcover) | ISBN 9781774916063 (softcover) | ISBN 9781032667119 (ebook)
Subjects: LCSH: Food crops—Diseases and pests—Biological control. | LCSH: Plants, Protection of.
Classification: LCC SB608.F62 N49 2025 | DDC 632/.9—dc23

Library of Congress Cataloging-in-Publication Data

Names: Sahni, Sangita, editor. | Thapa, Ganesh, editor. | Prasad, Bishun Deo, editor.

Title: New advances in crop disease management / edited by Sangita Sahni, Ganesh Thapa, Bishun Deo Prasad

Description: First edition | Palm Bay, FL : Apple Academic Press, 2025 | Includes bibliographical references and index | Summary: "Crop protection is becoming increasingly important due to the need for food security and to meet growing environmental challenges. This new book, New Advances in Crop Disease Management,provides readers with an updated and comprehensive overview of diverse crop protection strategies with special emphasis on the recent advances in crop protection approaches, fundamental concepts, principles, methodology, relationships, and benefits for crop protection and human health. It covers a variety of disciplines such as plant pathology, biochemistry, cell, and molecular biology, plant physiology, host-pathogen interactions, precision agriculture, decision analysis, disease risk, new ways of exploiting host genetic diversity including host resistance deployment, biological control and microbial interactions, surveillance and detection technologies, and the consequences of climate change and signaling. The book contains twenty chapters and begins with the role of biotechnology and its implications in crop disease detection and protection. It then discusses the plant response to pathogen attack, molecular basis of host-pathogen interaction and induction of defense response, viz., R gene, phytoalexins, etc., and advancement in early detection methodologies. It also delves into induced mutation as new technology for resistant crop development. The last part details the recent approaches and applications of resistant crop lines development, including resistant pulses development approaches, mining and mapping of R genes, gene pyramiding, lactic acid bacteria as biocontrol agents, effecteromics and nanotechnology in crop disease management, and the applications of phytochemicals and CRISPR-Cas in the development of disease resistant crops for future. New Advances in Crop Disease Management will be a key reference in the field of crop science and plant pathology, agriculture for faculty and industry professionals as well as a valuable textbook/reference for undergraduate and graduate students"-- Provided by publisher

Identifiers: LCCN 2024013492 (print) | LCCN 2024013493 (ebook) | ISBN 9781774916070 (hardcover) | ISBN 9781774916063 (paperback) | ISBN 9781032667119 (ebook)

Subjects: LCSH: Food crops--Diseases and pests--Biological control. | Agricultural pests--Biological control. | Plants, Protection of.

Classification: LCC SB608.F62 N49 2025 (print) | LCC SB608.F62 (ebook) | DDC 632/.9--dc23/eng/20240425

LC record available at https://lccn.loc.gov/2024013492

LC ebook record available at https://lccn.loc.gov/2024013493

ISBN: 978-1-77491-607-0 (hbk)
ISBN: 978-1-77491-606-3 (pbk)
ISBN: 978-1-03266-711-9 (ebk)

About the Editors

Sangita Sahni, PhD
Associate Professor-cum-Sr. Scientist, Department of Plant Pathology, Tirhut College of Agriculture, Dholi, Dr. Rajendra Prasad Central Agricultural University, Pusa, Samastipur, Bihar, India

Sangita Sahni, PhD, is an Associate Professor-cum-Sr. Scientist in the Department of Plant Pathology, Tirhut College of Agriculture, Dholi, Dr. Rajendra Prasad Central Agricultural University, Pusa, Samastipur, Bihar, India. She acquired her PhD (Agriculture) in Mycology and Plant Pathology from the B.H.U., Varanasi. Subsequently, she worked as a post-doctoral research fellow at the University of Western Ontario, London, Ontario, Canada. She was a Visiting Scientist at John Innes Centre, Norwich, UK from 12th-29th February 12–29, 2024. She has published several research papers in reputed peer-reviewed national and international journals. She has published seven authored books, four edited and several book chapters. She has isolated several agriculturally important microbes from different sources and submitted their sequences to the National Center for Biotechnology Information (NCBI). Dr. Sahni has been awarded the Dr. Rajendra Prasad National Education Shikhar Awards for outstanding contribution to the field of education, the Young Scientist Award, Innovative Scientist of the Year award-2015, and other prestigious awards for outstanding contribution in the field of Plant Pathology. She is also an editor in international peer-reviewed journals (*Current Journal of Applied Science and Technology, International Journal of Plant and Soil Science, Asian Plant Research Journal,* and *Agriculture Letters*). She is involved in the All India Co-ordinated Research Program in MULLaRP, Chickpea, and Small Millets at T.C.A., Dholi, and PI of BRNS, a Government of India-funded project.

Ganesh Thapa, PhD

Biohazards Advisor, Trinity College of Dublin, Ireland

Ganesh Thapa, PhD, worked as Biohazard's Advisor at Trinity College of Dublin to provide expert advice, guidance, and CL3 training to all Trinity researchers and affiliated hospitals to work in SARS-CoV-2 and helped open 2 SARS-CoV-2 screening centers in Trinity College of Dublin during the first wave of Corona pandemic 2020-2021. He obtained his PhD degree in Biosciences and Bioengineering from the Indian Institute of Technology Guwahati, working on phytoremediation of heavy metal-contaminated sites. He worked as a leading scientist at the Center for Peptide Research, UKZN, Durban, to develop HIV-1 protease inhibitors for MDR patients suffering from both TB and HIV. He then joined the University of Dublin from 2014 to 2018 and developed TILLING wheat lines resources resistant to Fusarium head blight (FHB) and lodging tolerant through mutation of the activation loop of receptor kinase. During this period, his innovation was awarded a world patent on "Fusarium head blight disease resistance." He thereafter worked as a research fellow at the University of Limerick and discovered and validated anti-inflammatory peptides from blue whiting for therapeutic use against allergy and autoimmune diseases in humans.

Bishun Deo Prasad, PhD

Associate Professor, Department of Agricultural Biotechnology & Molecular Biology, College of Basic Sciences and Humanities, Dr. Rajendra Prasad Central Agricultural University, Pusa, Samastipur, Bihar, India

Bishun Deo Prasad, PhD, is an Associate Professor in the Department of Agricultural Biotechnology and Molecular Biology, College of Basic Sciences and Humanities, Dr. Rajendra Prasad Central Agricultural University, Pusa, Samastipur, Bihar, India. Dr. Prasad acquired his PhD from MS University, Baroda, Gujarat, India. He also worked at the John Innes Centre (JIC), Norwich, UK, during his PhD.

Subsequently, he worked as a post-doctoral research fellow at the University of Western Ontario, London, Ontario, Canada. He has published several research papers in reputed peer-reviewed international journals. He has also contributed to seven authored books, four edited books, several book chapters, and practical manuals (UG and PG courses). He has submitted several sequences (bacterial, fungal, and plant) to the National Center for Biotechnology Information (NCBI). He has also submitted bacterial isolates to the National Bureau of Agriculturally Important Microorganisms (NBAIM), Mau. He is also an editor in four international peer-reviewed journals. Dr. Prasad received the DAE – Young Scientist Research award in 2013 and the Fast Track Scheme for Young Scientists award from the Department of Science and Technology (DST), India, in 2012. He has also been awarded the Outstanding Achievement Award, Inventor of the Year Award (2015), Excellence in Teaching Award, and other prestigious awards. He has been awarded the ICAR-sponsored Best Teacher Award (2018) at BAU, Sabour. Dr. Prasad has received prestigious SERB SIRE, Govt. of India Award for the academic year 2022–23. As a SIRE awardee, he has received high-end training in frontiers area of Molecular Biology & Biotechnology at JIC, Norwich, UK.

Contents

Contributors

A. U. Akash
Department of Plant Pathology, College of Agriculture, Professor Jayashankar Telangana State Agricultural University, Hyderabad, Telangana, India

Mohammad Ansar
Department of Plant Pathology, Bihar Agricultural University, Sabour, Bhagalpur, Bihar, India

Suman Bakshi
Nuclear Agriculture and Biotechnology Division, Bhabha Atomic Research Center, Mumbai, Maharashtra, India

Bahadur Singh Bamaniya
Department of Horticulture, School of Life Sciences, Sikkim University, Tadong, Gangtok, East Sikkim, Sikkim, India

S. Ameer Basha
Department of Plant Pathology, College of Agriculture, Professor Jayashankar Telangana State Agricultural University, Hyderabad, Telangana, India

Ashok Bhattacharyya
Department of Plant Pathology, College of Agriculture, Assam Agricultural University, Jorhat, Assam, India

Shankar Bhujbal
Nuclear Agriculture and Biotechnology Division, Bhabha Atomic Research Center, Mumbai, Maharashtra, India

Michael B. Byrne
School of Agriculture and Food Science, University College, Dublin, Belfield Dublin, Ireland

G. Uma Devi
Department of Plant Pathology, College of Agriculture, Professor Jayashankar Telangana State Agricultural University, Hyderabad, Telangana, India

P. Dhanasekar
Nuclear Agriculture and Biotechnology Division, Bhabha Atomic Research Center, Mumbai, Maharashtra, India

Vinod Janardan Dhole
Nuclear Agriculture and Biotechnology Division, Bhabha Atomic Research Center, Mumbai, Maharashtra, India

Dinkar
Department of Plant Breeding and Genetics, Bihar Agricultural University, Sabour, Bhagalpur, Bihar, India

Pranab Dutta
School of Crop Protection, College of Post-Graduate Studies in Agricultural Sciences, Central Agricultural University, Umiam, Meghalaya, India

Padmanabh Dwivedi
Department of Plant Physiology, Institute of Agricultural Sciences, Banaras Hindu University, Varanasi, Uttar Pradesh, India

Khushbu Jain
Department of Molecular Biology and Genetic Engineering, Bihar Agricultural University, Sabour, Bhagalpur, Bihar, India

Sanjay J. Jambhulkar
Nuclear Agriculture and Biotechnology Division, Bhabha Atomic Research Center, Mumbai, Maharashtra, India

Pranjal Kr. Kaman
Department of Plant Pathology, College of Agriculture, Assam Agricultural University, Jorhat, Assam, India

Kshitiz
Division of Nematology, Indian Agricultural Research Institute, New Delhi, India

Anand Kumar
Department of Plant Breeding and Genetics, Bihar Agricultural University, Sabour, Bhagalpur, Bihar, India

Jitesh Kumar
Department of Molecular Biology and Genetic Engineering, Bihar Agricultural University, Sabour, Bhagalpur, Bihar, India

Mahesh Kumar
Department of Molecular Biology and Genetic Engineering, Bihar Agricultural University, Sabour, Bhagalpur, Bihar, India

Mankesh Kumar
Department of Plant Breeding and Genetics, Bihar Agricultural University, Sabour, Bhagalpur, Bihar, India

Pankaj Kumar
Department of Molecular Biology & Genetic Engineering, Bihar Agricultural University, Sabour, 813210 Bihar, India

Rajesh Kumar
Department of Horticulture, School of Life Sciences, Sikkim University, Tadong, Gangtok, East Sikkim, Sikkim, India

Ravi Ranjan Kumar
Department of Molecular Biology and Genetic Engineering, Bihar Agricultural University, Sabour, Bhagalpur, Bihar, India

Sanjay Kumar
Department of Plant Breeding and Genetics, Bihar Agricultural University, Sabour, Bhagalpur, Bihar, India

Saurav Kumar
Department of Molecular Biology and Genetic Engineering, Bihar Agricultural University, Sabour, Bihar, India

Vinod Kumar
Department of Molecular Biology and Genetic Engineering, Bihar Agricultural University, Sabour, Bhagalpur, Bihar, India

Diksha Kumari
Department of Plant Physiology, Institute of Agricultural Sciences, Banaras Hindu University, Varanasi, Uttar Pradesh, India

Rima Kumari
Division of Plant Biotechnology, College of Agricultural Biotechnology, Bihar Agricultural University, Sabour, 813210 Bihar, India

Gautam Kunal
Department of Plant Breeding and Genetics, Bihar Agricultural University, Sabour, Bhagalpur, Bihar, India

Jitendra Kumar Kushwah
Department of Agriculture, Government of Uttar Pradesh, India

Varsha C. Mohanan
St. Mary's College, Thrissur, Kerala, India

Suhail Muzaffar
UAB Research Center of Excellence in Arsenicals and Department of Dermatology, University of Alabama at Birmingham, Birmingham, Alabama, USA

Neelambari
Department of Botany, Purnea University, Purnia, Bihar, India

Rabiya Parveen
Department of Plant Breeding and Genetics, Bihar Agricultural University, Sabour, Bhagalpur, Bihar, India

Jay Prakash
Nuclear Agriculture and Biotechnology Division, Bhabha Atomic Research Center, Mumbai, Maharashtra, India

Bishun Deo Prasad
Department of Agricultural Biology and Molecular Biology, College of Basic Sciences and Humanities, Dr. Rajendra Prasad Central Agricultural University, Pusa, Samastipur, Bihar, India

Archana Rai
Nuclear Agriculture and Biotechnology Division, Bhabha Atomic Research Center, Mumbai, Maharashtra, India

Kumari Rajani
Department of Seed Science and Technology, Bihar Agricultural University, Sabour, Bhagalpur, Bihar, India

V. Ramya
Department of Plant Pathology, College of Agriculture, Professor Jayashankar Telangana State Agricultural University, Hyderabad, Telangana, India

Rakesh Deo Ranjan
Department of Plant Physiology and Biochemistry, Bihar Agricultural University, Sabour, Bhagalpur, Bihar, India

Tushar Ranjan
Department of Molecular Biology and Genetic Engineering, Bihar Agricultural University, Sabour, Bhagalpur, Bihar, India

Kumari Rashmi
Department of Plant Breeding and Genetics, Bihar Agricultural University, Sabour, Bihar, India

Sunayana Rathi
Department of Biochemistry and Agricultural Chemistry, Assam Agricultural University, Jorhat, Assam, India

Nand Kishore Sah
Fr. Vice Chancellor, Veer Kunwar Singh University, Ara, Bhojpur, Bihar, India

Sangita Sahni
Department of Plant Pathology, TCA, Dholi, Dr. Rajendra Prasad Central Agricultural University, Pusa, Samastipur, Bihar, India

Satyendra
Department of Plant Breeding and Genetics, Bihar Agricultural University, Sabour, Bhagalpur, Bihar, India

Surojit Sen
Department of Zoology, Mariani College, Mariani, Jorhat, Assam, India

Mohammad Shamim
Department of Molecular Biology and Genetic Engineering, Bihar Agricultural University, Sabour, Bhagalpur, Bihar, India

Lomash Sharma
Department of Horticulture, School of Life Sciences, Sikkim University, Tadong, Gangtok, East Sikkim, Sikkim, India

Birender Singh
Department of Plant Breeding and Genetics, Bihar Agricultural University, Sabour, Bhagalpur, Bihar, India

Digvijay Singh
Department of Plant Breeding and Genetics, Bihar Agricultural University, Sabour, Bihar, India

S. P. Singh
Department of Plant Breeding and Genetics, Bihar Agricultural University, Sabour, Bhagalpur, Bihar, India

Shishu Pal Singh
Department of Soil Science and Agricultural Chemistry, Institute of Agricultural Sciences, BHU, Varanasi, Uttar Pradesh, India

Sweta Sinha
Department of Molecular Biology and Genetic Engineering, Bihar Agricultural University, Sabour, Bihar, India

J. Souframanien
Nuclear Agriculture and Biotechnology Division, Bhabha Atomic Research Center, Mumbai, Maharashtra, India

Swapnil
Department of Plant Breeding and Genetics, Bihar Agricultural University, Sabour, Bihar, India

Ganesh Thapa
Biohazards and Biosafety, Trinity College of Dublin, The University of Dublin, College Green, Dublin, Ireland

Preface

Crop protection to sustain yield has become imminent to meet the food security and environmental challenges that will arise over the coming decades. The ever-increasing population, limited land resources, intensive farming systems, high doses of agrochemicals, increase in CO_2 concentration, and narrow genetic base have led to the evolution of pathogen and pest strains with increased susceptibility of plants. Hence, crop protection requires a multi-faceted, interdisciplinary approach to capture the complexity of interactions of any disease or syndrome to not only enhance yield but also be eco-friendly and sustainable.

This book provides readers with an updated and comprehensive overview of different crop protection strategies with special emphasis on the recent advances and state-of-the-art on crop protection approaches, fundamental concepts, principles, methodology, relationships, and benefits for crop protection and human health. It covers a variety of disciplines such as plant pathology, biochemistry, cell and molecular biology, plant physiology, host-pathogen interactions, precision agriculture, decision analysis, disease risk, new ways of exploiting host genetic diversity, including host resistance deployment, biological control, and microbial interactions, surveillance, and detection technologies, and the consequences of climate change, and signaling. Such a multidisciplinary approach, intrinsic to the field of crop protection, fills the gap between plant pathology and disease control while reviewing recent methodological progress in crop protection approaches and their effectiveness and applications in the real crop field and agricultural sectors. This book will act as a key reference for readers in the field of crop science, plant pathology, agriculture, functional foods/nutraceuticals, Plant health agency, biotechnology, industry, and marketing sectors, or as a textbook/reference for undergraduate and graduate students.

The book *New Advances in Crop Disease Management* contains 20 chapters, beginning with the role of biotechnology and its implications in crop disease detection and protection. Then, it discusses the plant response to pathogen attack, the molecular basis of host-pathogen interaction and induction of defense response, viz., R gene, phytoalexins, etc., and an advance in early detection methodologies. It also delves into induced mutation as a new technology for resistant crop development. The last part deals with the reporting of the recent approaches and applications of resistant crop line development in detail. This includes the resistant pulses development approach, mining, and mapping of R genes, gene pyramiding, lactic acid bacteria (LAB) as biocontrol agents, effector omics, and nanotechnology in crop disease management, and finally the applications of phytochemicals and CRISPR-Cas in the development of disease-resistant crops for the future.

CHAPTER 1

Glimpses of Biotechnology in Crop Disease Management

NAND KISHORE SAH

Fr. Vice Chancellor, Veer Kunwar Singh University, Ara, Bhojpur, Bihar, India

ABSTRACT

Global food production needs to keep pace with the world population growth. The Green Revolution, steered by several factors including crop disease management, has played a pivotal role in boosting food production to almost a peak level for the teeming millions in the world, albeit with numerous cautions to health and the environment. The Evergreen Revolution, with the principles, tools, and methods of biotechnology on the vanguard, has great promise in global food management. Every day, thousands of research papers appear with novel vistas to solve this challenging problem. Some of the latest cutting-edge technologies include the CRISPR-CAS system, RNAi, and RenSeq, which are capable of delivering crop plants with reduced threat from pathogens and other destructive agents, thus brightening the prospect of greater food production with some concerns on human and environmental health.

1.1 INTRODUCTION

Food management for the teeming millions living all over the world is one of the greatest challenges faced by modern scientists. By 2025, the world population is estimated to be over 8.1 billion, as per the current UN report (UN.org, 2019). The requirement of food grain by then may be over 3 billion

New Advances in Crop Disease Management. Sangita Sahni, Ganesh Thapa, and Bishun Deo Prasad (Eds)

tons. The world population may grow from 7.2 billion today to about 9.7 billion by 2050, and so will rise the requirement of food. Scientific research has contributed a great deal in boosting the production of food. These efforts ushered in the 'Green Revolution' that fulfilled the adequacy of food in the world till the last century. Several factors including innovated farm implements, protection, healthy treated seeds, chemical fertilizers, bio-fertilizers, disease management measures using insecticides and pesticides, development of disease-resistant, and climate-resilient plants, etc., contributed to achieving the goal. Associated with this also arrived numerous issues inimical to human health, soil fertility, ecosystem, flavor of food, particularly vegetables and so on. There is a plethora of microbes that reduce production. For example, bacterial and fungal pathogens have been estimated to dent crop production by over 15%. In some crops, as in potato, the level of reduction in the yield may be as high as 30% (Oliver & Pamela, 2019). The viral diseases have been evaluated to reduce the yield by about 3%. These figures may also depend upon several factors including the edaphic and other climatic conditions.

The focus has shifted now from just food production to food production without compromising the quality of human life and environment. Dr. M. S. Swaminathan, the architect of the first successful 'Green Revolution' in India, has put forward a novel concept of 'Evergreen Revolution' which takes care of most of the obnoxious issues that emerged from the first 'Green Revolution.' Biotechnology holds the key to solutions for achieving the goals of the 'evergreen revolution' (Swaminathan, 2010).

1.2 FACTORS FOR REALIZATION OF THE GOALS OF 'EVERGREEN REVOLUTION'

Numerous factors have been identified that may help overcome difficulties in crop management for accelerating food production, such as disease management, discouraging monocropping and soil management, appropriate use of green fertilizers, raising climate-resilient and resistant varieties, identification and use of additional crop plants from the biodiversity pool. The ancient Indian Rishis had reported the existence of 8.4 million living species on earth. The modern scientists have so far reported about 1.7 million of them. Still, some 6.7 million remain to be identified. There is little doubt that some in the biodiversity pool may fulfill the criteria of being listed as favorite crop plants (Sah, 2018).

Disease management of the crop plants is important in achieving the production targets. The first 'Green Revolution' succeeded primarily because of measures including disease management of the crop plants. Six major principles have been put forward for this purpose: exclusion, eradication, protection, resistance, treatment/therapy, control of vectors, such as insects and weeds.

1. **Exclusion:** This principle refers to prevention of entrance of pathogens and restricting them in the marked areas only. This aspect of the disease management holds great promise as it helps avoid the use of pesticides, insecticides, weedicides, nematicides, and so on that contribute to environmental pollution leading to health risk. Crop rotation and management is important in this pursuit.

2. **Eradication:** The principle of eradication applies to individual plant, seed lot in a limited area/field/region soon after introduction. Once established it is difficult to eradicate a pathogen. For a larger area, generally it does not work satisfactorily. Measures including sanitation, crop rotation, burning of infected plants, pasteurization, fumigation, and solarization of the soil, disinfection of irrigation water by ultraviolet (UV) irradiation, chlorination, bromination, and ozonation may help eradication (Ogle, 2021).

3. **Protection:** This practice of crop protection has been in practice ever since human civilization adopted agriculture as a primary means of livelihood. Protection then was largely from the herbivores and rodents. Later, certain diseases were also identified. Till 1940, most of these diseases were managed with the use of Bordeaux mixture made of inorganic components [copper sulfate ($CuSO_4$) and slaked lime, i.e., calcium hydroxide, $Ca(OH)_2$]. Several other agrochemicals including the organic chemicals were included in the pesti/fungicides regime, which enhanced crop production. Over the years, however, the pathogens developed resistance against several agrochemicals. They have now become a knotty challenge for the agricultural scientists to keep the pace of production vibrant while minimizing environmental pollution and health risk. One of the basic tenets of protection, therefore, is to avoid the use of pesticides and fungicides without compromising crop production. Multi-cropping and microbe-soil interface are also interesting areas where crop protection may be worked out in order to boost and sustain production (Boyer & Cohen 1973).

4. **Resistance:** Developing crop plants resistant to several types of vagaries including pathogens and climatic conditions is the need of the hour. Identification of the resistant ones from the natural pool and raising them to crop plants was followed when agriculture science was poorly developed. Success in these efforts was primarily based on luck and few and far between. Later, when the tools were innovated efforts got underway to artificially create and select the most useful ones. Around 10 to 20 years were still required to establish a viable resistant plant, that too, against a particular disease. Till the 20th century such efforts coupled with protection and control measures indeed brought about magnificent boost to production, albeit with serious concerns. Biotechnological tools and techniques have ushered in a new era of crop disease management. Precise knowledge of the disease and genetic engineering coupled with tissue culture-based resistance to the crop plants have greatly reduced the time of evolving new varieties. Several types of pathological conditions may now be overcome in one go if we like.

5. **Treatment/Therapy:** The crop plants that have already developed pathological conditions due to pathogens, deficiency of required elements in the soil, hostile climatic conditions and other reasons are treated with inorganic, organic, and biomolecules to render them healthy again. A large gamut of such chemical molecules has been investigated that help overcome the pathological conditions partially and sometimes fully. Under and over doses of such chemicals may yield unsatisfactory results. Some of these molecules enter food chain and have been found to disturb environmental harmony leading to unpleasant complications. The principle of crop rotation and cropping system may be followed to reduce the severity of pathogenicity.

6. **Control of Vectors:** The pathological conditions due to pathogens including bacteria, fungi, viruses, mycoplasma are brought in contact with the sensitive plants by several vectors which may be categorized as air, water, infected seeds and plants, birds, rodents, lepidodepterons insects, mollusks, nematodes, and farm implements, among others. Their complete elimination is not always possible. For a limited region it is possible, but for the vast tracts of cultivated lands it is extremely difficult. These efforts may support production but may also create ecological imbalance and other environmental problems associated with biodiversity.

1.3 MAJOR BIOTECHNOLOGICAL TOOLS AND TECHNIQUES

Three primary processes are followed in biotechnology, i.e., (i) genetic engineering or recombinant DNA (rDNA) technology; (ii) tissue culture; and (iii) transgenics.

1. **Genetic Engineering:** This involves synthesis of genes or part thereof, repair of genes, combining genes isolated from one organism/plant with others (rDNA), manipulation of the artificial genes or the parts thereof with the rDNA with an aim to improve microbes, plants, animals, and even humans.
   Genetic engineering works on two basic processes: (i) identification, synthesis or isolation/cutting of the desired gene or part thereof; and (ii) pasting them at suitable sites.
   The genes of interest may be cloned by the process of amplification (synthesis) or cutting them out from the host sites with the help of precise biochemical scissors termed as restriction endonucleases which cleave phosphodiester bonds at defined sites (DNA sequences). These are then inserted into suitable plasmid DNA which act as vehicles to transport the gene(s) to the target plants/organisms. These techniques have been perfected now. The credit for cloning a gene for the first time goes to Boyer and Cohen (1973), who succeeded inserting a foreign DNA fragment into a plasmid found in *Escherichia coli*. Plasmids are circular DNA molecules with replication ability independent of chromosomal DNA found in certain bacteria. Paul Berg in 1979 was possibly the first to introduce a gene of Simian virus 40 in bacteria through lambda phage, hence reckoned as 'Father of genetic engineering.' He got Nobel prize in 1980 with Gilbert and Stanley. Plant biotechnology got a shot in the arm since *Agrobacterium tumefaciens* was recognized as a sound vehicle to transfer foreign gene(s) into plants, particularly of the dicotyledonous group. This bacterium hosts versatile Ti-plasmids that transfer foreign genes into the plant cells during the process of transformation. *A. tumefaciens* is, therefore, sometimes referred to as a natural genetic engineer.

2. **Tissue Culture:** The transformed cells are raised as callus tissue which eventually gives rise to plants carrying the cloned foreign genes. Tissue culture is a versatile *in vitro* method of propagation of plants with desired traits. It has been developed and almost perfected

now over the years. Four major factors that influence tissue culture have been identified, i.e., genotype and physiological condition of plants, culture media, culture conditions and hormones. Significant progress has been made in agricultural science since the introduction of tissue culture. Not only with regard to food production, but also in the arena of horticulture, quality improvement, pharmaceutics, and environmental safety plant tissue culture has contributed immensely. In fact, plants and their parts are now being used as factories for mass production of necessary secondary metabolites, pharmaceuticals, nutraceuticals, and so on. Plant tissue culture has made significant inroads in the development of disease resistant, disease-free, and climate resilient crop plants (Twaij et al., 2020).

3. **Transgenics:** The swelling world population, climate change, genetic erosion and increasing need for nutritional supplements have necessitated the development of new cultivars adaptable to different biomes with better resistance to biotic and abiotic stresses for improved yield and quality that help us meet the future challenges (Arzani & Ashraf, 2017; Basso et al., 2020; Bigini et al., 2021). The new biotechnological tools and methods have brightened the prospect of such elite crop plants with smart features including quantitative and qualitative supremacy over the conventional counterparts. CRISPR-Cas9 mediated genome editing has gifted the scientists with a remarkable biochemical tool to develop choice crop plants bearing the desired traits. Maintenance and sustained propagation of the transgenics requires scientific skill, diligence, and perseverance.

1.3.1 CURRENT BIOTECHNOLOGICAL METHODS FOR BOOSTING VIRAL DISEASE RESISTANCE IN THE CROP PLANTS

Multilayered defenses against microbial pathogens have been tried recently through the versatile tools of genetic engineering, particularly against viral pathogens. These objectives have been realized either by cloning trans-resistance genes or by silencing the disease-sensitive genes by various ingenious methods. RNA interference (RNAi) system has been deployed in suitable plants for anti-viral defense through transgenic expression of the plus or minus strand of the viral genome, inverted repeats, or artificial microRNAs (amiRNAs). CRISPR-Cas [(clustered regularly interspaced short palindromic repeats)-(CRISPR-associated systems or proteins)] is being widely used these days to achieve resistance against DNA and RNA viruses. Significant broad resistance has been conferred on several crop plants by stacking several resistance genes. RenSeq (resistance gene enrichment sequence) is being employed these days to achieve a greater level of resistance against pathological attacks. RenSeq is a tool of comparative genomics which makes use of nucleotide-binding leucine-rich repeats (NLR) gene family members having a proven role in strengthening disease resistance. The NLR-gene specific DNA sequences are used as baits to fish out such sequences from the resistant germplasms. The enriched NLR-DNAs are sequenced and compared with the reference genomes to identify the polymorphisms in the NLR-genes aimed at jacking up resistance. Susceptibility-inducing genes are being inactivated to boost resistance (Oliver & Pamela, 2019).

1.3.2 DISEASES OVERCOME THROUGH BIOTECHNOLOGY

Bacillus thuringinensis toxin gene, commonly known as Bt gene, was one of the first to be cloned into cotton that effectively controls damage caused to the capsule of the cotton plant bearing tremendous pecuniary and industrial significance by the caterpillars of several pests, particularly of budworm, i.e., *Heliothis virescens* (F.), and bollworm, i.e., *Helicoverpa zea* (Boddie). Bt genes encode toxic protein that appear crystal-like in the carrier plant cells. The toxic proteins are commonly known as Cry-proteins. More than 60 types of Cry-proteins have been identified that may show varying degrees of protective effects against numerous pests. Actually, when the insects feed upon these plants, the crystals of Cry proteins enter the gut of insects and get activated by their own digestive enzymes. The active Cry proteins enter the lining cells of the alimentary canal through the specific receptors and rupture them. The insects then either die or stop feeding on the plant. Several crop plants have been protected by invoking this and other strategies (Table 1.1). The concerns about the human use of such crops still stay in the world. In India, except Bt-cotton many such crops are under scanners of the government machinery.

The first viral resistant crop plant was developed in 1986 by Abel and coworkers by transferring TMV viral coat protein (CP) to the tobacco plant. Ever since several crop plants have been produced that confer resistance against various viral diseases (Table 1.2). CRISPR-Cas mechanism is now applied often to raise the level of resistance against the viral diseases in the crop plants, although the results of the field trials are largely awaited.

TABLE 1.1 Biotechnological Interventions to Boost Resistance in the Crop Plants Against Bacterial/Fungal Pathogens

Sl. No.	Name of the Cereal	Pathogen(s)	Biotechnological Intervention	Genes Involved
1.	Wheat	*Pseudomonas syringae* pv. *oryzae*	Interspecies/interfamily. Transfer of known *PRRs	AtEFR
2.	Wheat	*Blumeria graminis f.* sp. *tritici*	Interspecies/interfamily. Transfer of known PRRs	TaRLK1 and TaRLK2
3.	Wheat and barley	*Blumeria graminis f.* sp. *hordei; Blumeria graminis f.* sp. *tritici*	Interspecies/interfamily. Transfer of known PRRs	HvLEMK1
4.	Wheat	*Blumeria graminis f.* sp. *Tritici*	Interspecies/interfamily. Transfer of known PRRs	HvLecRK-V
5.	Wheat	*Bipolaris sorokiniana*	Altered expression of transcription factors for better signaling	TaPIMP1
6.	Wheat	*Fusarium graminearum*	Host-induced gene. Silencing by RNAi	FgCh3b
7.	Wheat	*Fusarium culmorum*	Silencing of host. Genes by CRISPR-Cas9	FcGls
8.	Wheat	*Blumeria graminis f.* sp. *tritici*	Silencing of host. Genes by CRISPR-Cas9	TaMlo-A1
9.	Wheat	*Blumeria graminis f.* sp. *tritici*	Silencing of host. Genes by CRISPR-Cas9	TaEDR1
10.	Wheat	*Fusarium graminearum*	CRISPR-Cas9	TaLpx-1
11.	Wheat	*Fusarium graminearum*	CRISPR-Cas9	TaHRC
12.	Wheat	*Puccinia triticina, P. graminis*, and *P. striiformis*	Host-induced gene. Silencing by RNAi	PtMAPK1, PtCYC1, PtCNB
13.	Durum wheat, barely, rice, sorghum, maize	Multiple biotrophic pathogens	Transfer of APR alleles	TaLr34
14.	Barley	Multiple rusts and powdery mildew.	Transfer of APR alleles to boost resistance	TaLr67
15.	Barley	*Fusarium graminearum*	Host-induced gene. Silencing by RNAi	FgCYP51A, FgCYP51B, FgCYP51C
16.	Rice	*Xanthomonas oryzae pv. oryzae*-derived elf18	Interspecies/interfamily. Transfer of known PRRs	AtEFR
17.	Rice	*Acidovorax avenae* subsp. *avenae*	Interspecies/interfamily. Transfer of known PRRs	AtEFR

TABLE 1.1 *(Continued)*

Sl. No.	Name of the Cereal	Pathogen(s)	Biotechnological Intervention	Genes Involved
18.	Rice	*Pseudomonas syringae* pv. *tomato; Agrobacterium tumefaciens; Xanthomonas oryzae* pv. *oryzae*	Production of chimeric receptor kinases and R genes	AtEFR-OsXa21
19.	Rice	*Magnaporthe oryzae*	Production of chimeric receptor kinases and R genes	OsXa21-OsCEPiP
20.	Rice	*Xanthomonas oryzae* pv. *oryzae*	Deletion of effector binding sites	Os11N3/OsSWEET14
21.	Rice	*Xanthomonas oryzae* pv. *oryzae*	Addition of effector binding sites	OsXa27
22.	Rice	Broad-spectrum of pathogens.	Altered expression for signaling components	AtNPR1
23.	Rice	*Magnaporthe oryzae*	Transfer of APR alleles to boost resistance	OsIPA1/OsSPL14
24.	Rice	*Magnaporthe oryzae*	CRISPR-Cas9	OsSEC3A
25.	Rice	*Magnaporthe oryzae*	CRISPR-Cas9	OsERF922

**PPRs*: Pattern-recognition receptors.

TABLE 1.2 Biotechnological Manipulation to Enhance Resistance in Some Crop Plants Against the Viral Pathogens

Sl. No.	Name of the Crop	Type/Name of the Virus	Level of Biotechnological Intervention	Genes/Protein Involved
1.	Potato	Potato virus Y.	*A. tumefaciens* mediated viral coat protein transfer	Coat protein
2.	Potato	Potato leaf roll virus.	*A. tumefaciens* mediated viral gene transfer	Replicase and helicase
3.	Papaya	Papaya ring spot virus.	*A. tumefaciens* mediated viral coat protein transfer	Coat protein
4.	Squash	Watermelon mosaic virus 2, Zucchini yellow mosaic virus.	*A. tumefaciens* mediated viral coat protein transfer	Coat protein
5.	Sweet pepper	Cucumber mosaic virus.	*A. tumefaciens* mediated viral coat protein transfer	Coat protein
6.	Tomato	Cucumber mosaic virus.	*A. tumefaciens* mediated viral coat protein transfer	Coat protein
7.	Plum	Plum pox virus	*A. tumefaciens* mediated viral coat protein transfer	Coat protein
8.	Bean	Bean golden mosaic virus.	RNAi-mediated resistance	+, – RNA of viral replication protein

1.4 CONCLUSION

The population growth has been swelling over the years and is expected to exceed 9.7 billion by 2050, even though the land and other resources on earth remain limited. New strategies are needed henceforth for food security and the better future of mankind. Innovative efforts are required to exploit the 75% of the earth's surface spanned by water. Serious endeavor is required to delve deep into the treasure of biodiversity for identification of novel genes/species to meet the future challenges of food security. Feeding the ever-growing population and managing other requirements including clean water, health, quality life, and a tidy environment remains a great challenge. Magnificent progress has been made in achieving the goal of food adequacy in the world. Management of diseases of the crop plants is a proven step towards this mission. During the first green revolution, food production was optimized using chemical and biological control of diseases of the crop plants. These achievements also paved the way to numerous obnoxious issues, particularly related to health and the environment. In order to manage crop diseases for food security, the principles of 'Evergreen Revolution' must be followed with innovative ideas and the latest biotechnological tools and methods on the vanguard.

Six major points are taken into consideration for crop disease management: exclusion, eradication, protection, resistance, treatment/therapy, control of vectors, such as insects and weeds. Biotechnology contributes via three successive processes: (i) genetic engineering or rDNA technology; (ii) tissue culture; and (iii) transgenics. RNAi and RenSeq methods are employed to generate crop varieties that smother pathogenic attacks. CRISPR-Cas9 mediated genome editing has gifted the scientists with a remarkable biochemical tool to develop choice crop plants bearing the desired traits.

KEYWORDS

- **biotechnology**
- **climate resilient**
- **CRISPR-Cas**
- **evergreen revolution**
- **food management**
- **RNAi**
- **transgenics**

REFERENCES

Akanmu, A. O., Babalola, O. O., Venturi, V., et al., (2021). Plant disease management: Leveraging on the plant-microbe-soil interface in the biorational use of organic amendments. *Front. Plant Sci.* https://doi.org/10.3389/fpls.2021.700507.

Arzani, A., & Ashraf, M., (2017). Cultivated ancient wheats (*Triticum* spp.): A potential source of health-beneficial food products. *Comp. Rev. Food Sci. Food Saf., 16*, 477–488.

Basso, M. F., Monteiro, F. B. A., Grossi-De-Sa, M., et al., (2020). Insights into genetic and molecular elements for transgenic crop development. *Front. Plant Sci.* https://doi.org/10.3389/fpls.2020.00509.

Bigini, V., Camerlengo, F., Botticella, E., Sestili, F., & Savatin, D. V., (2021). Biotechnological resources to increase disease-resistance by improving plant immunity: A sustainable approach to save cereal crop production. *Plants, 10*, 1146. https://doi.org/10.3390/plants10061146.

Boyer & Cohen (1973). *GNN – Genetics and Genomics Timeline.*

Ogle, H., (2021). Disease management: Exclusion, eradication, and elimination. In: Brown, J. F., & Ogle, H. J., (eds.), Plant Pathogens and Plant Diseases: Australasian Plant Pathology Society (pp. 358–372).

Oliver, X. D., & Pamela, C. R., (2019). Genetic engineering for disease resistance in plants: Recent progress and future perspectives. *Plant Physiol., 180*, 26–38. doi: 10.1104/pp.18.01224.

Sah, N. K., (2018). Scope of plant biotechnology in the developing Countries. In: Prasad, B. D., Sahni, S., Kumar, P., & Siddqui, M. W., (eds.), Plant Biotech. (pp. 45–66).

Swaminathan, M. S., (2010). From Green to Evergreen Revolution. New Delhi: Academic Foundation.

Twaij, B. M., Jazar, Z. H., & Hasan, M. N., (2020). Trends in the use of tissue culture, applications, and future aspects. *Int. J. Plant Biol.* https://doi.org/10.4081/pb.2020.8385.

UN.org (2019). https://www.un.org/development/desa/en/news/population/world-population-prospects-2019.html.

CHAPTER 2

Importance and Role of Biotechnological Tools in Plant Pathology: Basic Concepts and Principles to Study Host-Pathogen Relationship

SUHAIL MUZAFFAR,[1] SANGITA SAHNI,[2] and BISHUN DEO PRASAD[3]

[1]*UAB Research Center of Excellence in Arsenicals and Department of Dermatology, University of Alabama at Birmingham, Birmingham, Alabama, USA*

[2]*Department of Agricultural Biology and Molecular Biology, College of Basic Sciences and Humanities, Dr. Rajendra Prasad Central Agricultural University, Pusa, Samastipur, Bihar, India*

[3]*Department of Plant Pathology, TCA, Dholi, Dr. Rajendra Prasad Central Agricultural University, Pusa, Samastipur, Bihar, India*

ABSTRACT

Biotechnology offers plant breeders access to a wide array of techniques for the development of crop yields with highly desirable traits. These traits include improved yield from crops, reduced susceptibility to environmental stresses, increased nutritional qualities, and reduced dependence on fertilizers or pesticides. Microbial diseases in food crops are one of the most problematic issues, and the development of disease-free plants is an important application of biotechnology. The understanding of the molecular mechanism underlying plant-pathogen interactions and genetic engineering tools have provided novel opportunities for engineering resistance to pathogens. Using biotechnology techniques, novel genes may be expressed into crops to provide disease resistance. Some of the biotechnology techniques, including

New Advances in Crop Disease Management. Sangita Sahni, Ganesh Thapa, and Bishun Deo Prasad (Eds)

genetic engineering, marker-assisted selection (MAS), plant tissue culture, bio-fertilizer technology, and vaccine production, have been widely used in crop improvement. Therefore, biotechnology has significantly amplified conventional crop improvement to meet the increased food demand in the 21st century.

2.1 INTRODUCTION

Biotechnology-based principles help farmers to grow more food at less cost in an environmentally sustainable manner. The principal goal of agricultural biotechnology is to improve the quality, quantity, and shelf life of crops thereby enabling farmers and consumers to obtain greater value at cheaper costs. The technological innovations in genetic engineering offer insights that are being implemented to address various agricultural challenges including drought resistance, disease resistance, and weed control. The benefits of biotechnology are more important considering the growing global population and increased demand for food. By 2050, the world population will grow to more than 9 billion, and feeding such a huge population will be an immense challenge. Biotechnology will be a major factor in controlling plant diseases caused by microbial pathogens, pests, and climatic conditions with minimal costs. For example, genetically engineered insect-resistant cotton has been an immense success story as it has reduced the use of synthetic pesticides that contaminate water sources and damage the environment.

Annually, an average of 26% of the worldwide crop yield is lost due to pre-harvest pests and pathogens. Identification of plant pathogens is an important step toward sustainable plant disease management. The development of fast and sensitive methods for the identification of plant pathogens is crucial for the decision-making process in disease management. Chemical control of plant diseases is mostly non-specific and it may harm beneficial organisms or may have undesirable health and environmental risks. The future of sustainable agriculture depends on the integration of traditional agricultural and advanced biotechnological techniques. Biotechnology improves our understanding of mechanistic regulation of plant-pathogen interactions as well as plant disease resistance. Advanced and cost-effective biotechnological advances have emerged as a control strategy against fungal and bacterial pathogens. Biotechnological tools enhance our ability to combat plant pathogens and improve crop productivity (Fagwalawa et al., 2013; Collinge, 2018).

Various biotechnology techniques used by scientists for crop improvement include plant tissue culture, genetic engineering, marker-assisted selection (MAS), artificial insemination, disease diagnostics, and vaccine production. Sequencing of 16s and 18s rDNA has also been critical in the accurate identification of microbial pathogens in plants. The development of rDNA technology involves the isolation of individual genes from their natural sources and incorporation into the new hosts to develop various desirable traits including disease resistance, higher productivity, and better grain quality (Kumar et al., 2020; Bonfim et al., 2007).

2.2 BIOTECHNOLOGY ADVANCES AND DISEASE RESISTANCE

The important achievement in plant sciences of the last century was the development of plant tissue culture and genetic engineering techniques, including transgenesis for gene transfer between distantly related species. Scientists in the late 1940s used the term "gene" as the unit of genetic information. However, they did not understand the chemical or structural aspects of the gene or how it was replicated from one generation to another. Watson and Crick recognized that a detailed understanding of the three-dimensional structure of the gene was the gateway to all biological processes. The discovery of the DNA structure by Watson and Crick and the deciphering of the genetic code laid the foundation of modern age biotechnology. Their discovery led to groundbreaking insights into the genetic code and protein synthesis, which later helped to produce new and powerful scientific techniques, specifically rDNA technology, genetic engineering, DNA fingerprinting, and the production of monoclonal antibodies. Later developments, including the discovery of restriction enzymes in 1970 and the establishment of rDNA technology by Cohen and Boyer, were fundamental for the rise of plant biotechnology. In 1973, the Cohen-Boyer team was able to cleave a plasmid from one bacterium and then insert a gene from a different bacterial species into the plasmid, thereby creating an rDNA molecule. Then they inserted the plasmid into the bacteria and demonstrated that the rDNA is functional inside the bacteria. This discovery created the first genetically modified (GM) organisms.

One of the important milestones in plant biotechnology was the fact that *Agrobacterium tumefaciens* Ti plasmid can be used to genetically engineer a plant genome with foreign DNA. *A tumefaciens* is a plant pathogen best known as nature's agent to create GM organisms. Before the advent of genome sequencing, the only known horizontal gene transfer process from bacteria to eukaryotes was mediated by *A. tumefaciens*. Plant biologists modified wild-type strains of *A. tumefaciens* in which DNA for transformation into a plant was placed between the left and right borders of the T-DNA. Later, a direct gene transfer technology such as particle bombardment or biolistic was established for the transformation of monocot species. In this method, foreign DNA is coated onto gold or tungsten particles are fired from a gene gun under high-pressure helium gas resulting in penetration into the host cell wall.

The development of rDNA technology and transgenic expression lead to a new era of transgenically engineered resistance in several crop species. The transgenic approach has been widely adopted in plant breeding programs to confer stress and disease resistance in crop plants. Numerous transgenic plants have been developed over the last few decades including cotton, rice, and corn. As of 2012, these GM transgenic crops were planted on 170 million hectares around the world. Recently, CRISPR (clustered regularly interspaced short palindromic repeats)-Cas9-based tools have transformed plant sciences due to their ability to edit the plant genome. Emmanuelle Charpentier and Jennifer Doudna share the award for developing this genome-editing technology and using this system as a gene-editing tool. CRISPR-Cas9 is a cost-effective and user-friendly approach compared with other techniques such as zinc-finger nucleases and transcription activator-like effector nuclease (TALEN). CRISPR-Cas9 has been used to successfully edit the genome of several plant species including monocot and dicots. The genome of model plants such as *Arabidopsis thaliana* as well as crops like maize, wheat, rice, tomato, potato, cotton, etc., have been edited using this technology despite their genome complexity. Various research groups developed resistance against tomato leaf curl virus, beet severe curly top virus, zucchini yellow mosaic virus (ZYMV) in different plants.

Plants are exposed to a wide range of pathogens, including insects, nematodes bacteria, fungi, and viruses. One of the biggest challenges for food security in the 21st century is to improve crop yield by the development of stress and disease-resistant plants. In traditional breeding, resistant crop varieties were selected and crossed to develop new plant varieties to attain desired physical and genetic traits. However, traditional breeding programs did not require understanding the mechanism of action of resistance (*R*) genes. Our initial understanding of plant disease resistance was developed by the work of H. Flor, who identified the gene-for-gene relationship during plant-pathogen interaction. During the process, the plant R gene product interacts with the pathogen Avr gene product to generate the hypersensitive response which leads to disease resistance. The first R gene identified to confer resistance in plants was tomato derived *Pto* which conferred resistance to *Pseudomonas syringae* strains carrying AvrPto. *Pto* encodes a serine-threonine protein kinase that specifically recognizes strains of *P. syringae*. Molecular characterization of plant *R* genes and pathogen *AVR* genes support the gene-for-gene receptor-ligand model. Direct interaction has been observed between the tomato Pto, and the corresponding avrPto avirulence gene product of *P. syringae*. The interaction between rice and the fungal pathogen *Magnaporthe oryzae* is a well-documented gene-for-gene system. Rice blast disease, caused by *M. oryzae*, is one of the most devasting plant diseases causing a major drop in rice production. Rice plants expressing the *Pi-ta* gene are resistant to strains *M. oryzae* expressing *AVR-Pita*. *Pi-ta* encodes a putative cytoplasmic receptor protein with a nucleotide-binding site and leucine-rich domain at the C-terminus. *AVR-Pita* encodes a metalloprotease with an N-terminal secretory signal and pro-protein sequences. Scientists are trying to exploit the R-Avr interaction between plants and pathogens to develop disease-resistant plants.

2.3 DETECTION OF PLANT PATHOGENS

The first and foremost step in plant disease management is to accurately identify the pathogen. Although some diseases can be identified by visual examination, others require laboratory tools for diagnosis. Diagnostic tools for the detection of plant pathogens with high multiplexing ability are critical to reducing the large agricultural losses caused by plant diseases. The essential features of pathogen diagnostic assays are specificity, sensitivity, reproducibility, and cost-effectiveness. The symptoms-based diagnostic methods are often time-consuming and the pathogen is often detected after a long time from the onset of the disease. Recent advances in biotechnology have presented new products and techniques that have replaced traditional time-consuming laboratory procedures. Molecular diagnosis of plant pathogens allows the accurate identification of morphologically similar species. Also, molecular techniques allow the detection of plant infection before symptom formation. The molecular biology-based method of disease diagnostics includes polymerase chain reaction (PCR), real-time PCR, DNA hybridization, and enzyme-linked immunosorbent assay (ELISA) (Collinge, 2018; Dong & Ronald, 2019).

2.3.1 POLYMERASE CHAIN REACTION (PCR)

The advent of the PCR in the 1980s allowed scientists to develop DNA-based approaches for plant pathogen detection. PCR in many ways has revolutionized plant pathology and diagnostic approaches. In PCR, a DNA template is exponentially amplified through repeated cycles using a thermostable DNA polymerase. PCR is a more sensitive and cost-effective technology compared to other techniques for the detection of plant pathogens such as bacterial, fungal, or viral pathogens. It can help farmers and researchers detect the presence of pathogens that have long latent periods between infection and symptom development. PCR is a highly sensitive technique that allows the detection of pathogens even in a small amount of DNA or pathogen isolate. In PCR, designing specific or universal primers to amplify pathogens followed by sequencing allows accurate detection of plant pathogens. The pathogens are detected by comparing nucleotide sequences of pathogen isolates with the GenBank database using basic local alignment search tool (BLAST) analysis.

2.3.2 REAL-TIME PCR

The improvement in DNA-based diagnostic methods came with the advent of real-time quantitative PCR that monitors the amplification of DNA in real-time based on fluorescence intensity. It not only allowed the detection of pathogens but also quantified them to determine the severity of the plant infection. In real-time PCR (qPCR) fluorescence intensity is measured after each cycle and this intensity reflects the amount of amplified DNA at that specific point intime. In earlier cycles, the fluorescence intensity is undetectable because of too little amplified DNA. During later cycles, the fluorescence intensity increases above the detectable threshold level (Ct value). There are two common technologies used for the real-time amplification of DNA such as non-specific fluorescent DNA dyes and fluorescently labeled oligonucleotide probes. Although the labeled probe-based approach is more specific and accurate, both approaches have been widely used for pathogen detection. The main advantages of qPCR-based diagnostics include fast and high-throughput detection as well as real-time quantification of target DNA. Another advantage is the multiplexing of amplification of different targets into a single reaction. For example, a target gene as well as an internal control like actin or GAPDH can be amplified in a single reaction.

2.3.3 DNA ARRAY HYBRIDIZATION

DNA array hybridization or DNA microarray is a technique based on the hybridization of amplified and labeled genome regions of interest to immobilized oligonucleotides spotted on a solid support platform. It is a powerful technique used for the identification of microbial pathogens from environmental samples without the need for isolation of pure culture. Most DNA microarray techniques are based on a single region for the detection of a specific taxonomic group. For example, the 16S ribosomal DNA region is widely for the detection of bacterial pathogens. For the detection of fungal pathogens, several regions of DNA, such as mitochondrial genes or ribosomal DNA spacers (ITS) are used. DNA arrays have been utilized for the detection of various bacterial and fungal plant pathogens in a wide range of environmental samples, such as greenhouse crops, potatoes, ginseng, and fruits.

2.3.4 ENZYME-LINKED IMMUNOSORBENT ASSAY (ELISA)

The enzyme-linked immunosorbent assay, also called ELISA is an immunological assay commonly used to measure proteins in biological samples including plant pathogens. ELISA is used to detect pathogens using a monoclonal antibody labeled with fluorescent probes. The main disadvantage of ELISA is that this process requires proper laboratory equipment. However, portable immunoassay methods have been developed for easier and quick detection of pathogens. Lateral flow devices are a simple paper-based dipstick assay to detect pathogens without the need for specialized laboratory settings. ELISA has been useful to detect and identify viruses responsible for plant diseases. In 1977, the publication of an ELISA-based detection of Plum pox virus and Arabis mosaic virus by Clark and Adams was a milestone in virus diagnostics. This technique revolutionized virus diagnosis by reducing the time required for the detection of pathogens. The advent of monoclonal antibodies also improved the assay sensitivity and specificity making ELISA a highly versatile assay for pathogen detection (Dong & Ronald, 2019).

2.3.5 GENETICALLY MODIFIED (GM) PLANTS FOR DISEASE RESISTANCE

Conventional breeding has been a major source of crop improvement over deceased; however, it is usually a very lengthy and labor-intensive process. Genetic engineering, which involves the ectopic expression of different genes in the plant genetic material, presents many advantages over conventional breeding. Genetically engineered plants are generated in a laboratory by altering the genetic makeup of plants by adding desirable genes. Most GM plants are generated by the biolistic method (gene gun method) or by the *Agrobacterium tumefaciens* mediated transformation (ATMT) method. The biolistic method has been a successful strategy for monocots, like wheat or maize, for which ATMT has been less successful. ATMT works comparatively better for the dicotyledonous plants such as tomatoes, potatoes, and tobacco plants. Genetically engineered plants show significant benefits over conventional plants by improving production yield, enhancing the nutritional content, and enhancing disease resistance. Genetic modification has mostly centered on conferring disease resistance towards insects, pests, and microbial pathogens. Disease resistance is indeed the best approach to control pathogen-derived plant diseases because no extra cost or management is required by the farmer once the resistant crop has been planted. Genetic resistance to diseases presents a highly economical approach to crop protection. For example, heterologous expression of pattern recognition receptors (PRRs) recognizes conserved molecules in pathogens and thereby trigger defense responses to a wide range of pathogens. Also, plants can be engineered to synthesize antimicrobial proteins (APs) that directly inhibit the growth and development of pathogens. The host RNAi machinery can be modified to confer viral immunity by degrading viral RNA (Chen, Lyngkjær, & Collinge, 2012; Punja, 2001). Different sources of resistance genes can be used to develop disease resistance in plants, the sources are discussed in subsections.

2.3.5.1 PLANT-DERIVED RESISTANCE GENES

Plants utilize various defense networks against invading pathogens including proteins and organic molecules produced before or during pathogen attack. Genetic engineering allows the expression of single dominant resistance genes derived from other plant species or overexpression of the already existing defense genes. The intrinsic plant responses that can be exploited to obtain durable disease resistance include the hypersensitive response (HR) and systemic acquired resistance (SAR). A plant could be engineered so that SAR is expressed constitutively even in the absence of activating pathogen. For example, the over-expression of the transcriptional regulator Npr1 enhances the disease resistance in *Arabidopsis thaliana* against a wide range of pathogens. Plants could also be genetically engineered to change a previously compatible reaction with a pathogen to an incompatible reaction to enable plants with the first level of defense.

2.3.5.2 ANTIMICROBIAL BIOMOLECULES

This involves antimicrobial peptides (AMPs) or proteins produced by transgenic plants to enhance plant resistance to microbial pathogens like bacteria and fungi. This approach utilizes genes found in microbes, insects, and animals to improve plant defense against a specific pathogen. APs like lysozyme that naturally occur in insects and animals have been used as a potential source of plant resistance. The production of reactive oxygen species (ROS) is known to play an important role in plant defense. Transgenic potatoes expressing glucose oxidase derived from the fungus *Aspergillus niger* continuously produce ROS and confer resistance to *Phytophtora infestans.* Lytic peptides are small proteins that cause cell lysis by making pores in the cell membrane. Cecropins are antibacterial lytic peptides derived from giant silk moth has been used for developing disease-resistant plants. Transgenic tobacco plants expressing cecropins confer resistance to *Pseudomonas syringae* that causes tobacco wildfire disease (Bonfim et al., 2007).

2.3.5.3 PATHOGEN-DERIVED RESISTANCE (PDR) GENES

It has been widely observed that transgenic plants expressing genes derived from some viral pathogens show resistance toward these viruses and other related strains. In 1987, it was demonstrated that tobacco plants transformed to express the coat protein (CP) gene of tobacco mosaic virus (TMV) were resistant to this viral pathogen. Later in 1990, it was reported that transgenic plants expressing the TMV sequence encoding for non-structural protein like RNA polymerase also conferred resistance against TMV. These results indicated that the ectopic expression of genes encoding viral proteins is interfering with the viral life cycle. More recent studies have demonstrated that this antiviral defense in transgenic plants is mediated by RNA interference (RNAi). Transgenic expression of viral RNA has been reported to generate dsRNA in plants, which in turn triggers RNAi. This process is referred to as cosuppression, a classical form of post-transcriptional gene silencing (PTGS). Particularly, transgenic squash (*Cucurbita* sp.) and papaya (*Carica papaya*) varieties created through an RNAi-mediated approach show durable resistance and have been grown commercially in the United States for more than 20 years (Punja, 2001; Bonfim et al., 2007).

2.4 CONCLUSION

Lack of arable land and insufficient supply of irrigation water have exerted pressure on farmers to produce more food with fewer resources. Conventional plant breeding approaches feel handicapped to meet the demands of food security in the 21st century. The plant breeding process by employing biotechnology techniques including genetic engineering and a transgenic approach remains the only viable option to meet such huge demands. The identification of potentially useful genes across the animal and plant kingdom is critical for the improvement of crops. When introduced into the desired crop species, these transgenic plants could be a boon to human development and welfare.

KEYWORDS

- **antimicrobial biomolecules**
- **biotechnology**
- **disease resistance**
- **DNA array hybridization**
- **enzyme-linked immunosorbent assay**
- **pathogen-derived resistance genes**
- **polymerase chain reaction**

REFERENCES

Bonfim, K., Faria, J. C., Nogueira, E. O., Mendes, É. A., & Aragão, F. J., (2007). RNAi-mediated resistance to bean golden mosaic virus in genetically engineered common bean (*Phaseolus vulgaris*). *Molecular Plant-Microbe Interactions, 20*, 717–726.

Chen, Y. J., Lyngkjær, M. F., & Collinge, D. B., (2012). Future prospects for genetically engineering disease-resistant plants. *Molecular Plant Immunity*, 251.

Collinge, D. B., (2018). Transgenic crops and beyond: How can biotechnology contribute to the sustainable control of plant diseases? *European Journal of Plant Pathology, 152*, 977–986.

Dong, O. X., & Ronald, P. C., (2019). Genetic engineering for disease resistance in plants: Recent progress and future perspectives. *Plant Physiology, 180*, 26–38.

Fagwalawa, L., Kutama, A., & Yakasai, M., (2013). Current issues in plant disease control: Biotechnology and plant disease. *Bayero Journal of Pure and Applied Sciences, 6*, 121–126.

Kumar, K., Gambhir, G., Dass, A., Tripathi, A. K., Singh, A., Jha, A. K., Yadava, P., et al., (2020). Genetically modified crops: Current status and future prospects. *Planta, 251*, 1–27.

Punja, Z. K., (2001). Genetic engineering of plants to enhance resistance to fungal pathogens: A review of progress and future prospects. *Canadian Journal of Plant Pathology, 23*, 216–235.

CHAPTER 3

Recent Advances in Early Detection of Plant Diseases Using Molecular Tools

SWAPNIL,[1] SWETA SINHA,[2] DIGVIJAY SINGH,[1] KUMARI RASHMI,[1] and SAURAV KUMAR[2]

[1]*Department of Plant Breeding and Genetics, Bihar Agricultural University, Sabour, Bihar, India*

[2]*Department of Molecular Biology and Genetic Engineering, Bihar Agricultural University, Sabour, Bihar, India*

ABSTRACT

Identification of plant pathogenic organisms is very important for appropriate disease management in plants. In this account, the readiness of quick, efficient, and precise techniques for the recognition of plant pathogenic organisms remains extremely necessary for the enhancement of disease control decision-making process. The diverse genomic approaches have made it easy and thus have converted investigation in the biological sciences subject. The approaches which are based on nucleic acid (NA) help to face the challenges in immunologic as well as microscopical recognition methods. For the precise and accurate target detection of the pathogens, DNA probes are utilized. A new emerging tool for the diagnosis of plant pathogens is the DNA microarray method linked to PCR, creating higher precision and sensitivity. Microarray, polymerase chain reaction (PCR)-based approaches, dot blot hybridization, PCR-RFLP, reverse-transcription-PCR, nested-PCR, multiplex-PCR, etc., are the approaches that are utilized for the pathogen detection. The operations of bioinformatics and proteomics as ultramodern technologies in the study of plant pathogenic organisms, detection of specific and precise DNA sequences has become probable, which eventually increases the precision of ultramodern approaches in detection of plant diseases.

New Advances in Crop Disease Management. Sangita Sahni, Ganesh Thapa, and Bishun Deo Prasad (Eds)

3.1 INTRODUCTION

Plant diseases are causing significant economic losses worldwide, which are due to different pathogens like fungi, bacteria, viruses, viroids, phytoplasma, etc. Plant diseases hamper crop production and may cause huge crop losses. It also poses a threat to food as well as nutritional security at the global level, which hampers the health of humans and animals (Miller et al., 2009).

The considerable limits of the prevailing approaches are the dependence on the capability of the organism to be raised, the time-consuming nature, the demand for widespread taxonomic know-how, and the tough samples with hidden symptoms. Thus, the prevalent approaches were substituted by immunological approaches. Pathogen detection based on antigen-antibody interaction is the quick approach to disease diagnosis, which involves monoclonal antibodies (Kohler et al., 1975) and enzyme-linked immunosorbent assays (ELISA) (Clark et al., 1977). These approaches are known as immunological approaches or serological approaches. Nevertheless, the approaches are generally utilized in the detection of viral diseases of plants. The introduction of antibody-based serological detections such as the monoclonal antibodies and the ELISA developed during the 1970s was a significant turning point in virology and bacteriology, which were followed by conventional approaches. The serological approaches like ELISA, which are important to detect and quantify the target pathogens, but have some boundaries like the demand for antibodies, specific antigens, etc., which are more costly. Still, the essential limitations of the conventional approaches are needed to look for approaches with improved levels of perfection and reliability (Narayanasamy et al., 2002, 2006). The failure to adequately identify and detect plant pathogenic organisms using culture-based morphological approaches and conventional methods has contributed to the development of nucleic acid (NA)-based molecular ways (Lievens et al., 2005; Lopez et al., 2003). Molecular approaches can supply accurate, consistent, and reproducible outcomes quickly (McCartney et al., 2008). Biotechnological approaches in the detection of plant pathogens can give significant information about the identification of specific plant pathogenic microorganisms related to the disease.

Some of the most important molecular approaches for the detection of plant pathogenic disease are fingerprinting (restriction fragment length polymorphism (RFLP), random amplified polymorphic DNA (RAPD), amplified fragment length polymorphism (AFLP), microsatellites), PCR isothermal amplification approaches, DNA hybridization technology, and sequencing (McCartney et al., 2003; Barnes & Szabo, 2007; Kang et al., 2010).

NA-based detection approaches overcome various challenges associated with conventional microscopic and immunological detection of plant pathogens. Furthermore, these can be utilized at any developmental stages of the plants since all living cells contain a whole set of the genome and are not affected by the environment. DNA probes are largely utilized for the precise detection of pathogens or pathogen propagules in infected tissues.

The successful management of plant pathogenic diseases is initially dependent on the precision and efficient identification of the plant pathogens, studying the quantity of genetic variability as well as the pathogenic variability present in a plant pathogenic population. The development of disease-resistant varieties and studying the disease resistance genes in various epidemiological regions is also an important measure. The diverse methods that are utilized for the detection of plant pathogenic organisms are listed in Figure 3.1.

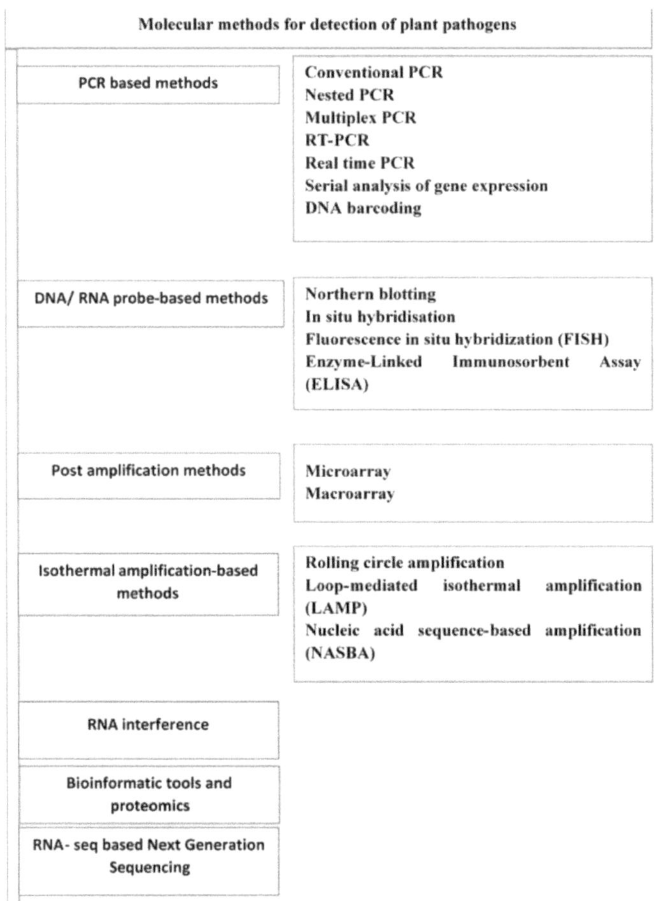

FIGURE 3.1 List of the molecular methods used for detection of plant pathogen.

Source: Adapted from Aslam et al., 2017.

3.2 MOLECULAR METHODS FOR DETECTION OF PLANT PATHOGENS

3.2.1 PCR-BASED METHODS

3.2.1.1 CONVENTIONAL PCR

PCR is utilized for synthesizing a particular DNA fragment with the help of an enzyme, DNA polymerase. Replication of the genetic material of cell (Figure 3.2) is done by this enzyme. In this technique, a million copies of DNA are produced using the three steps viz., denaturation, annealing, and extension with the use of specific primers. The plant pathology subject became revolutionized with the discovery of PCR due to its high specificity in detection of plant pathogenic organisms. It is a very sensitive technique, which requires a smaller quantity of pathogen DNA. The technique requires mixing of substances like dNTPs, DNA, primers, buffer, and DNA polymerase. Then, using the thermal cycler for amplification. *Mycosphaerella musicola* and *M. fijiensis* are identified with the help of PCR technology (Rahman et al., 2013).

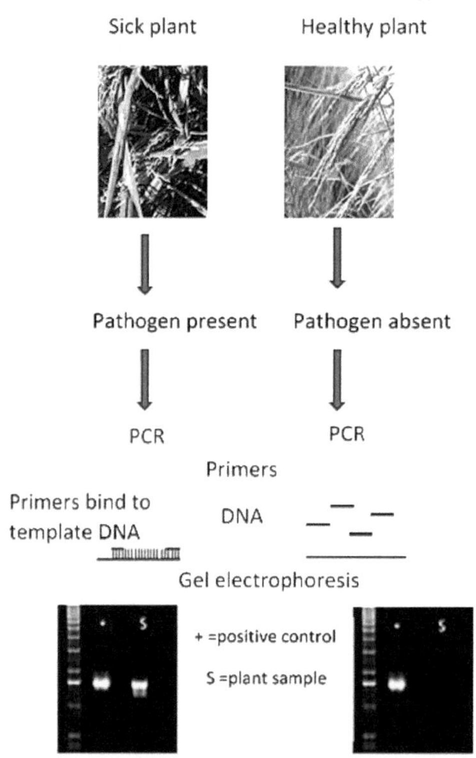

FIGURE 3.2 PCR-based diagnostic methods.

3.2.1.2 *NESTED PCR*

This PCR has an increased specificity due to lesser no specific binding because of the presence primers which are 2 in sets. This is a modification of PCR technique in which one primer binds to a specific target point while the other one binds away from the target site and thus multiplies/amplifies a much larger fragment. The next cycle of amplification has the second primer which amplifies the target segment of DNA. It is very specific and sensitive method. Its compassion is 1,000 times more than conventional PCR (Yeo & Wong, 2002). It is a useful method to study the phylogenetic relationships and in the identification of diverse group of pathogens. This method is laborious and sometimes gives false-positive results (Tsai et al., 2006). *P. palmivora* as well as *P. parasitica* were identified through the nested PCR (Chao et al., 2016).

3.2.1.3 *MULTIPLEX PCR*

This PCR is utilized for amplifying many targets in one run of PCR (Figure 3.3). For amplifying the DNA using the thermocycler, many primers and temperature dependent DNA polymerase enzyme is used. A particular annealing temperature has to be maintained as multiple primers are being used in the PCR. This is an additional benefit as many sequences are targeted at a time which saves reagents, time, and labor. This technology is used for pathogen or GMOs detection, genotyping, polymorphism analysis, mutation analysis, microsatellite STR analysis, etc. If diverse pathogens cause a same disease, then they can be detected using multiplex PCR. Multiplex PCR utilizes padlock probes for the identification pathogenic fungi. Detection of different pathogens in cactus like *P. nicotinae*, *P. cactorum*, *F. oxysporum*, and *B. cactivora* can be done using this PCR (Cho et al., 2016).

3.2.1.4 *RT-PCR*

Conventional PCR is modified to form RT-PCR in which the complementary DNA molecules are made using RNA molecules. The cDNA is used as a template in PCR for the purpose of amplification (Figure 3.4). The RT-PCR is done using one tube in one step or even few steps with multiple tubes. The first method is relatively better as it has lesser chances of contamination of the variations. This method distinguishes the alive and dead fungus diagnosis

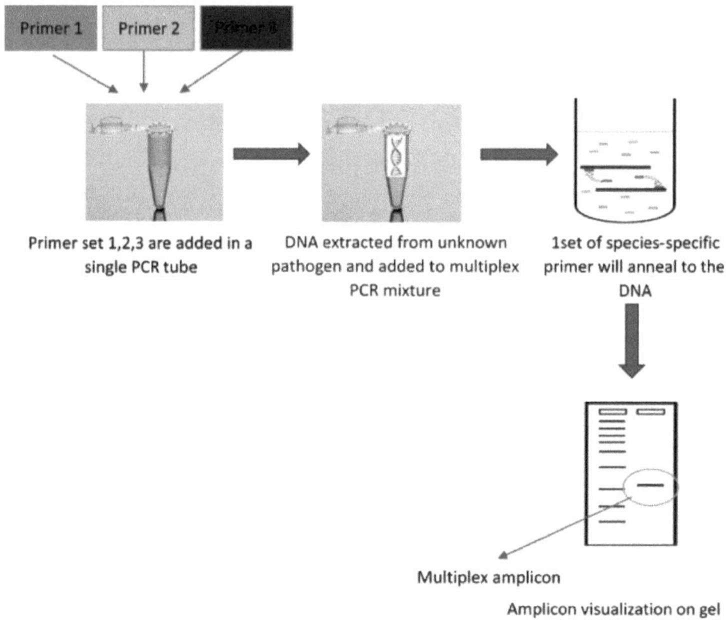

FIGURE 3.3 Pathogen identification through multiplex PCR.

of the RNA viruses such as retroviruses. The dead cells have their degraded mRNA and it can be detected with the help of RT-PCR to check the viability of the cell (Capote et al., 2012). Diagnosis of RNA viruses will help study the effectiveness of antimicrobial vaccines. The fungi (*Fusarium graminearum*) cause ear blight in crops like maize, barley wheat and oat (Brown et al., 2011).

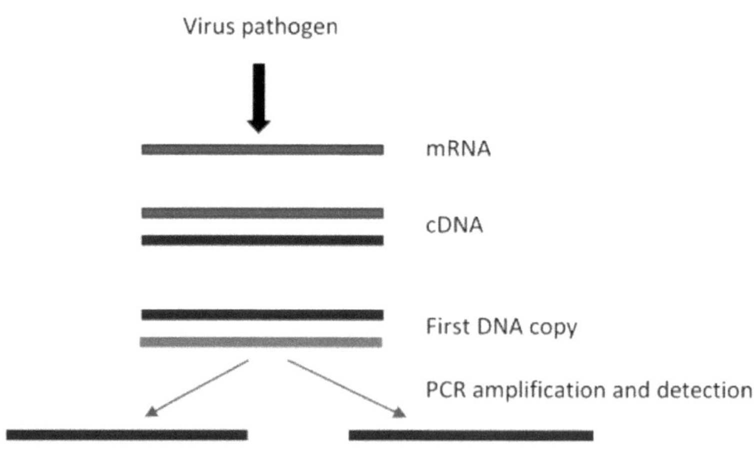

FIGURE 3.4 Pathogen detection through RT-PCR.

3.2.1.5 REAL-TIME PCR

It is also named Quantitative PCR (qPCR) or quantitative real-time PCR. It includes combination of amplification of a particular DNA segment (PCR-based) and quantifies the DNA in a particular reaction. The case of Conventional PCR takes a lot of time in which the products of PCR are analyzed using electrophoresis method. This PCR detects the PCR products in real time whenever the phase (exponential) is operating. The dye (fluorescent) which is used binds with the DNA and produces a fluorescent signal. As soon as the quantity of target DNA shoots up, the signal also shoots up after every amplification cycle. After each cycle of amplification when the amount of specific DNA segment increases, the signal also increases (McCartney et al., 2003). The fluorescent dye is used as a monitoring agent but these are not specific. The intercalating dye which binds to all the DNA forms the primer dimers which produces false results. Hence, the fluorescent probes were preferred because of their high specificity (Bu et al., 2005). *Aspergillus versicolor*, *Cladosporium cladosporioides*, *Stachybotry schartarum,* and *Alternaria alternate* have been identified and quantified by qPCR reported by Dawei et al., (2011).

3.2.1.6 SERIAL ANALYSIS OF GENE EXPRESSION

This is a PCR based technique for the identification of differences in different RNAs based on the levels of mRNAs. This method was developed by Ken Kinzler and Bert Vogelstein. The amount of expressed sequence tags (ESTs) which are specific to a target gene in a pool of cDNA gives the information regarding mRNA transcript levels. By determining the sequence of a statistically significant number of ESTs in one cDNA pool, relative to another, it is possible to identify gene transcripts that are differentially represented. Because only very short segments of cDNA are required to develop identity.

3.2.1.7 DNA BARCODING

Wherever there is variation in the DNA sequence of a standardized and short region of the genome (i.e., the barcode), this technique is utilized for the precise recognition of an organism species. This technique is based on the analysis of variation in the barcode segments of DNA that is needed to study the taxonomic relationships (Figure 3.5). For various species of fungi, bacteria, plant, and algae the method is useful for recognizing or identification. The use of barcodes for the identification of any plant species generally depends on the number of representatives which is present in the existing or available database. The advantages of this technique are species discovery based on the phylogenetic study of NA sequences, recognition of species at a specific life phase, providing knowledge about the diversity of life and helps in the making of NA specific sequencing tools (Savolainen et al., 2005). Pathogens that are difficult to be identified morphologically are recognized at the molecular level by using DNA barcoding. ITS marker is utilized worldwide as a marker for barcoding (Monteiro et al., 2015; Kim et al., 2010). ITS, SSU, RPB, and LSU markers are utilized for barcoding of DNA of fungal pathogens.

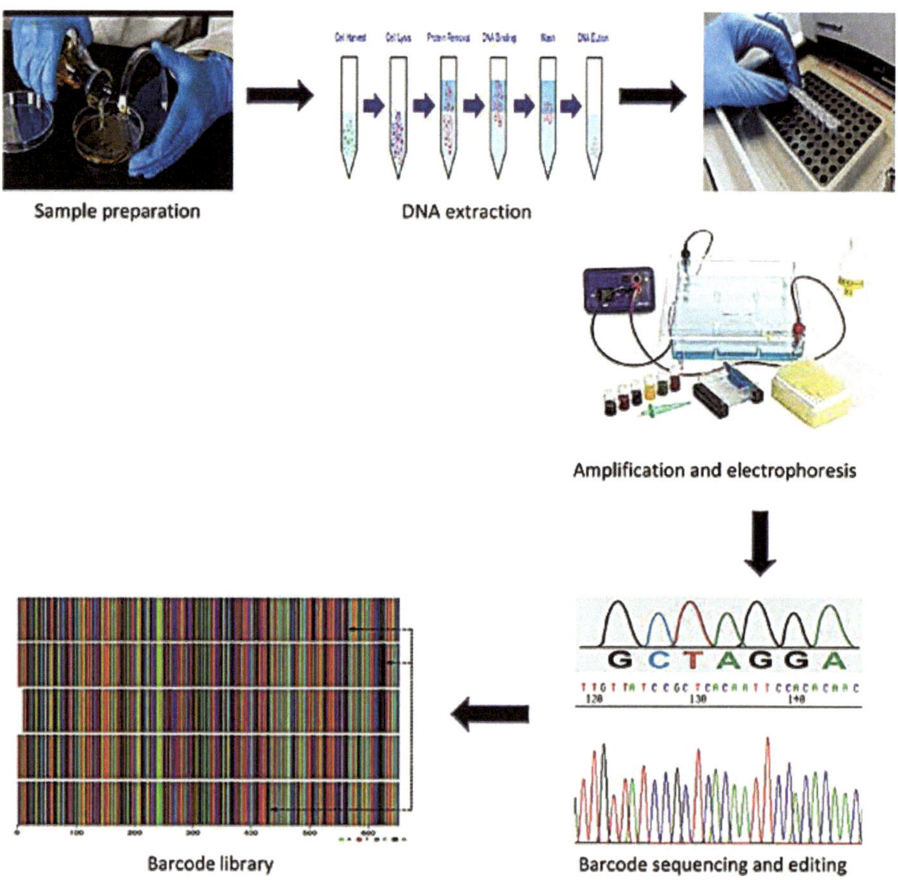

FIGURE 3.5 An outline of DNA barcoding method.

3.2.2 DNA OR RNA PROBE-BASED METHODS

These techniques are used to identify plant diseases that are caused by the pathogens like fungi using probes with high sensitivity and higher speed. The probe is utilized for analyzing NAs and amplification is not required. Probes are the short DNA segments that are single-stranded (ss) and labeling is done with isotopes like 35 S, 32P and 33P. Probes are utilized for identification of homologous sequences in the target DNA. Probes are generally used in PCRs combination (McCartney et al., 2003).

3.2.2.1 NORTHERN BLOTTING

This is a technique of RNA blotting using the electrophoresis method with the help of nitrocellulose membranes or nylon membrane. Then, the hybridization is done with NA sequences, either DNA or RNA. This process is alike Southern blotting but here RNA is used in place of DNA and blotting is done on the membrane. DNA is denatured before blotting in case of southern hybridization. This denaturation is not required in case of northern blotting. RNA is blotted on diazo benzyloxy methyl (DBM) cellulose paper. Further, o-amino phenyl thioether (APT) cellulose was better compared to DBM cellulose as it was easier to prepare and was more stable. Some disadvantages of northern blotting are the risks associated with mRNA degradation during the electrophoresis, difficulty in detection using multiple probes, and its lower sensitivity in comparison to the RT-PCR. In rice crop, *Magnaporthe grisea* is identified with the help of Northern Blotting and real-time PCR (Qi et al., 2002).

3.2.1.2 IN SITU HYBRIDIZATION

In this technique, riboprobes (single-stranded RNA probes) are used which are labeled with 35S. This technique is closely similar to the northern blot technique as both techniques are on the basis of the hybridization of probes (RNA or DNA) with their homologous mRNA sequences. In the case of the northern blot method, the digest of tissue is taken as preliminary material however, in this technique histological section is used. Hence this method is known as hybridization histochemistry. The labeling of probes is done either with radioactive or non-radioactive nucleotides. *In situ* hybridization probing can be done with cRNAs, cDNAs, and synthetic oligonucleotides. Riboprobes provide the most sensitive and efficient results. The riboprobes (35S) used for the identification of mRNA are highly sensitive one (Hayden et al., 2002). Some disadvantages associated with the method are that the utilization of radio-labeled probes which are expensive and unsafe to handle and dispose of. Also, due to low DNA and RNA copies, the targets are difficult to identify (Qian & Lloyd, 2003).

3.2.2.3 FLUORESCENCE IN SITU HYBRIDIZATION (FISH)

The method is utilized with microscopy for the identification and detection of bacteria and hybridization of fluorescent DNA probes. The probe binds to homology or complementation showing part of chromosome given by. FISH is applied to identify/detect plant fungi and virus's pathogens (Hijri, 2009; Kliot, 2016). Ribosomal RNA (rRNA) is specific in nature are already present in plants which are utilized by FISH for the detection of pathogen infections in plants. FISH is useful for the identification of several target sites at the same time quantitatively. This technique is safe, rapid, has a high resolution (Chao et al., 2016) and has single-cell sensitivity. This sensitivity is due to the binding of the probe with each ribosome in the sample and the detection is approximately 10^3 CFU/mL. Apart from the advantages, it has some problems like reduced specificity due to the use of autofluorescence materials which gives false-positive results.

3.2.2.4 ENZYME-LINKED IMMUNOSORBENT ASSAY TECHNIQUE (ELISA)

Antibodies and the color change form the basis of the ELISA method which is used for the detection of pathogens (Clark, 1977). More appropriately, ELISA is a very specific interaction technique of antibody with the antigens. In this technique, the antigen or specific and target macromolecule are immobilized on a surface (solid surface) and mixed with reporter enzyme linked antibodies. The antigens are taken from bacteria, viruses as well as fungi. For the detection, changes in color are studied. The activity of the reporter enzyme is measured for the detection of diseases using lateral flow devices. The sensitivity of ELISA depends on the organism, sample freshness, and titer.

3.2.3 POST AMPLIFICATION TECHNIQUE

3.2.3.1 MICROARRAY

It is used to study the gene expression of multiple genes simultaneously. DNA microarrays are slides with numerous tiny spots which are microscopic present in distinct positions in which each spot contains a sequence of DNA. They are also called DNA chips (Figure 3.6). Spot is the place where the probe is present on chip (Robinson et al., 2000). The probes are attached to the support matrix like nylon filters, glass chip and targeted cDNAs. In case of cDNA, radio-labeled probes or fluorescent tags are used. This method is easy to use as it does not need DNA sequencing. Numerous samples can be studied simultaneously in a single array. Some disadvantages associated with this method are expensive, requires a large quantity of mRNA, needs several replications for reducing the error. *Aspergillus candida* has been identified using this technique (Singh & Kumar, 2013) and mycobacterium (Gusev, 2001).

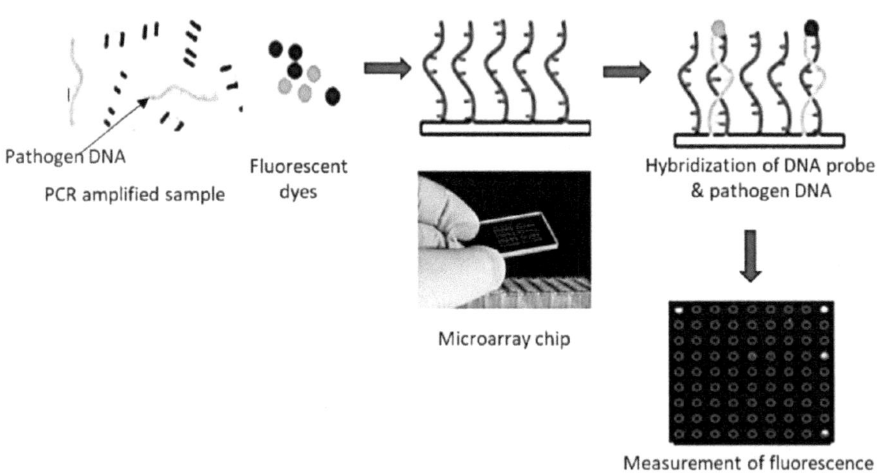

FIGURE 3.6 Microarray method for plant pathogen diagnosis.

3.2.3.2 MACROARRAY

It is also called reverse dot plot or DNA array hybridization. Small DNA spots that are present on a solid surface makes it a DNA microarray. This method does not involve radioisotopes (Singh & Kumar, 2013). PCR amplification as well as hybridization surges the sensitivity many folds as compared to only the PCR (Taoufik et al., 2004). It is used to detect mycobacterium species (Leinberger et al., 2005), *Fusarium solani, Alternaria alternata, Aspergillus fumigatus, Cladosporium herbarum,* and *Candida albicans* (Sato et al., 2010). It is also used to detect pathogens (oomycete and fungus) in case of solanaceous crops (Zhang et al., 2008).

3.2.4 ISOTHERMAL AMPLIFICATION METHODS

3.2.4.1 ROLLING CIRCLE AMPLIFICATION (RCA)

It is an isothermal as well as an enzymatic technique that makes single-stranded (ss) DNA molecules using DNA polymerases, circular ssDNA as well as DNA primer. In this method, probes that are oligonucleotide (PLPs), single-stranded DNA sequences of approx. 20-nucleotide. These are situated at 5′ and 3′ ends and are linked through linker sequences (40-nucleotide long) (Tsui et al., 2011). After hybridization, both the ends are ligated with ligase and hence, circularized. Now, this probe spans the whole target region and thus detects the DNA molecules with known sequences. RCA is a multipurpose DNA amplification tool with several applications viz., in the field of diagnosis, biosensing, nanotechnology, genomics, and proteomics. The fungi that cause Black grain eumycetoma have been detected by this technique.

3.2.4.2 LOOP-MEDIATED ISOTHERMAL AMPLIFICATION (LAMP)

Loop-mediated isothermal amplification (LAMP) is a PCR substitute which is used for amplification of nuclei acids (Figure 3.7). In this technique, a thermal cycler is not required but 4 primers and Taq DNA polymerase is used at a single temperature (isothermal conditions) for DNA amplification reported by Chao et al. (2016). There are Internal primer or forward and outer primer or backward primer given by Villari et al. (2013). There are inner primers which are two in number and they are known as forward inner primers and the other as backward inner primers (BIP), whereas the two outer primers are B3 and F3 (Fakruddin, 2011). LAMP is performed at 65°C for a duration of 1 hour in a water bath and the product of amplification is studied with the help of SYBR Green 1 dye and electrophoresis (Figure 3.3). This method is used for the detection of *Paracoccidioides brasiliensis*, a temperature-dependent fungus (Chang et al., 2012), and *Ophiostoma clavatum* (Chen et al., 2006). *Ochroconis gallopava* fungi can be detected (Tsui et al., 2011) and *Penicillium marneffei* (Sun et al., 2010) can be recognized through this method. Compared to normal PCR, this method is more accurate (10 times) and sensitive (Ren et al., 2009), no costly equipment, and has a smaller number of processes.

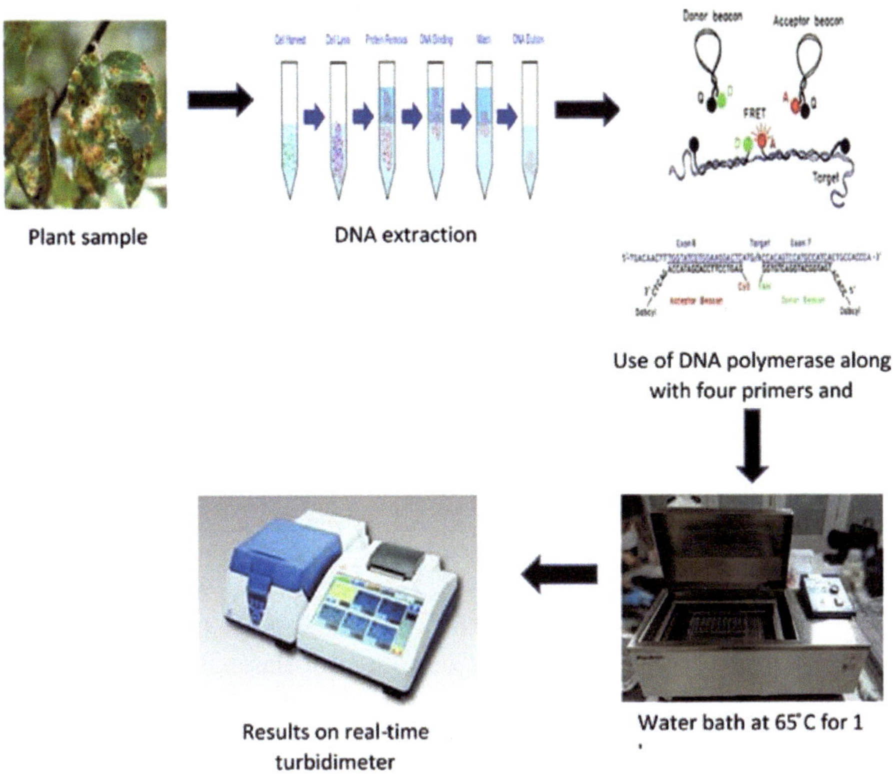

FIGURE 3.7 An outline of loop-mediated isothermal amplification (LAMP) method.

3.2.4.3 NUCLEIC ACID SEQUENCE-BASED AMPLIFICATION (NASBA)

This technique is utilized to make many copies of ss RNA. It is used to detect the RNA with high sensitivity using isothermal detection method in which the amplification is based on transcription. DNA amplification can also be done using this method. Some systems can also amplify the DNA. This technique has many advantages as 109 copies of nuclei acids could be amplified very fastly (approx. 1.5 hours) using three enzymes. Temperature is kept constant at 41°C and hence a thermal cycler is not required (Fakruddin et al., 2012). Compared to RT-PCR, the power of amplification of NASBA is better (Chang et al., 2012) because of the speedy amplification rates which is used for retrovirus detection. It also helps in the detection of mRNA sequences of humans without any risk of DNA contamination (Fakruddin et al., 2012). RNA is chemically and biologically more susceptible to RNase degradation than that of DNA. Hence, they should be properly handled and stored. This technique is useful in basic research, application-oriented fields like diagnosis of infectious diseases and development of medicine.

3.2.5 RNA INTERFERENCE (RNAI)

It is a system of gene regulation which occurs in a sequence specific manner with the help of silencing RNAs. It is being used to detect and control phyto-pathogenic fungi and acts in a very tissue-specific manner and kills the toxic disease-causing fungi (Panwar et al., 2012). MicroRNA and small interfering RNA are silencing RNAs which are required for RNA interference (RNAi). RNAi fights against the increasing fungicide resistant plant pathogens and thus controls the disease (Ishii & Holloman, 2015). This technique works by knocking down the genes with the help of the cell's own cellular defense. The detection of hairpin RNA and dsRNA by the fungal cells uses sequence homology which are important for silencing (Nakayashiki et al., 2005). A disadvantage associated with RNAi is the off-target effect, which tolerates a little bit of complementary relation to the sense strand. RNAi is used in maize for identifying *Aspergillus flavus* fungus (Masanga et al., 2015) and *Hordeum vulgare* for the identification of *Fusarium graminearum* (Chen et al., 2006).

3.2.6 RNA-SEQUENCING-BASED NEXT-GENERATION SEQUENCING

RNA sequencing (RNA-Seq) with the use of next-generation sequencing (NGS) is done to study the presence and amount of RNA in any plant sample at a particular time by studying the changes in cellular transcriptome levels. The RNA is made into cDNA using adaptors that are attached to single or both the sides. All the fragments are sequenced to get certain sequences from 1 end or 2 ends either with or without amplification. A library is developed by making the fragments of the RNA or cDNA which is required for next-generation sequencing. The priming of the resulting mRNA is done by the utilization of oligo primers or arbitrary primers.

The popular and most utilized platforms for next-generation sequencing of RNA like Ion Torrent, Illumina's HiSeq and SOLiD (Dawei & Peng, 2014). Finally, the reads are arranged according to the genome of reference. Various new software like Trinity and Rnnotator do not refer to the reference sequence and assembles the RNA-sequencing data. This technology is used for the identification of the *Magnaporthe oryzae* fungus, which is the disease-causing organism of blast in rice (Soanes et al., 2012), and the *Verticillium dahliae* fungus that forms vascular wilt in tomato (de Jonge et al., 2012).

Various methods for the detection of plant pathogens have been listed in Table 3.1.

3.2.7 ROLE OF BIOINFORMATICS IN PLANT DISEASE DIAGNOSIS

The appropriate use of communication and information technology in research of biology can be done effectively through bioinformatics. Bioinformatics has many uses in molecular and biotechnological research. It can many utilizations in agriculture discipline like management of generated data, data analysis, etc. (Jayaram & Dhingra, 2010). Diagnostic tools can be developed for traits like abiotic stress tolerance, disease resistance and pest resistance.

For the bacterial and viral identification, Next-Generation Sequencing (NGS) offers a reliable solution. It can be worked out even if there is no information about NA sequence (Quan et al., 2008). Through bioinformatics relationship between pathogen and disease can be easily established. The Carrot yellow leaf virus causes necrosis in carrots was found out by metagenomic approach (Adams et al., 2014). Diseased plant produced signals *via* short interfering RNAs (siRNA) and the levels of siRNA can be detected for the identification of concerned viruses (Mlotshwa et al., 2008). Some of the bioinformatics tools that are used for detection of plant viruses (Table 3.2) are Virfind, Taxonomer, VSD toolkit, and VirusDetect.

3.3 PROTEOMIC STUDIES AS A TOOL IN PATHOGEN IDENTIFICATION

Proteomic has a vital role in recognizing the changes in protein levels in host plants that are infected by pathogenic fungus, bacteria, and

TABLE 3.1 Molecular Methods Used in the Detection of the Various Pathogens in Some of the Crop Species (Majumder et al., 2012)

Crop	Pathogen Targeted	Molecular Method Used	References
Wheat	*Septoria tritici*	PCR, multiplex-PCR	Fraaije et al. (2001); Cartney et al. (2003)
	Pucciniastrii formis		
	P. recondite		
	Fusarium spp.	Competitive-PCR	Yabuta et al. (2010)
	Tilletia controversa, Tilletia caries	PCR	John et al. (1997)
	Phytophthora ramorum	Real-time PCR	Kim et al. (2008)
	Rust pathogens		Barnes et al. (2007)
Barley	Barley yellow dwarf virus.	Multiplex PCR	Deb et al. (2008)
	Blumeria graminis	SAGE	Thomas et al. (2002)
	Fusarium graminearum	RNA interference (RNAi)	Chen et al. (2006)
Rice	*Magnaporthe grisea*	Real-time PCR	Chadha et al. (2006)
		Northern blotting	
	Pseudomonas syringae pv. *oryzae*	Bioinformatic tools	Studholme et al. (2011)
	Magnaporthe oryzae	RNA-Seq-based next-generation sequencing.	Soanes et al. (2012)
	Rice tungro virus	Dot-blot, RT-PCR	Mangrauthia et al. (2010)
Maize	*Peronosclero sporasacchari*	Dot-blot	Yao et al. (1990)
	Ustilago maydis	PCR	Xu et al. (1999)
	Aspergillus flavus	RNA interference (RNAi)	Masanga et al. (2015)
Soybean	*Diaporthe phaseolorum, Phomopsis longicolla*	PCR-RFLP	Zhang et al. (1999)
	Phakopsora pachyrhizi, P. meibomiae	PCR *In situ* hybridization (ISH)	Frederick et al. (2002); Ellison et al. (2016)
Chickpea	*Ascochyta rabiei*	LAMP	Das et al. (2012)
Tobacco	Tobacco mosaic tobamovirus	Dot-blot	Ogras et al. (1994)
Cotton	Cotton leaf curl begomo virus	PCR	Sivalingam et al. (2007)
Potato	Potato spindle tuber viroid	Dot-blot	Owens et al. (1981)
	Phytophthora infestans	Multiplex PCR	Tooley et al. (1998)
	Verticillium spp.	Microarray	Boonham et al. (2007)
	Potato leaf roll virus.	Multiplex real-time PCR.	Boonham et al. (2000, 2007)
	Ralstonia solanacearum	Multiplex PCR	Pastrik et al. (2002)

TABLE 3.1 *(Continued)*

Crop	Pathogen Targeted	Molecular Method Used	References
	Phytoplasma associated potato diseases.	Multiplex PCR	Leyva-Lopez et al. (2002)
	Potato leaf roll virus.	RT PCR	Schoen et al. (2006); Singh et al. (1997)
	Potato poty virus	RT PCR	Mirmomeni et al. (2008)
	R. solanacearum	Real-time PCR	Ozakman et al. (2003)
	Clavibacter michiganensis	RFLP and RT-PCR	Lopez et al. (2006)
Tomato	Tomato spotted wilt virus.	RT PCR	Tsuda et al. (1994)
	Tomato yellow leaf curl virus.	Immunocapture PCR	Rampersad et al. (2003)
	Phytophthora, pythium	Real-time PCR	Lievens et al. (2006)
	Phytoplasma (little leaf pathogen).	Nested-PCR	Cervantes et al. (2008)
	Xanthomonas vesicatoria	PCR	Moretti et al. (2009)
	Verticillium dahliae	RNA-Seq-based next-generation sequencing.	de Jonge et al. (2012)
	X. campestris	PCR	Leite et al. (1995)
Pepper	*Xanthomonas vesicatoria*	PCR	Moretti et al. (2009)
Cucumber	Cucumber vein yellowing virus.	Real-time PCR	Pico et al. (2005)
Brinjal	Phytoplasma (little leaf pathogen).	PCR	Siddique et al. (2001)
Lettuce	*Fusarium oxysporum* fsp. *Lectucae*	Nested PCR	Mbofung et al. (2010)
Sugar beet	Beet necrotic yellow vein virus.	RT-PCR	Shahnejat et al. (2006)
Cauliflower	Cauliflower mosaic virus.	Real-time PCR	Cankar et al. (2005)
Cashew net	Cashew fungal-associated diseases.	Microarray ELISA DNA barcoding	Monteiro et al. (2015)
Solanaceous crops	Fungus, oomycete	Macroarray	Zhang et al. (2008)
Banana	Badnavirus	Rolling circle amplification (RCA)	James et al. (2011)
Apple	Apple scar skin viroid (ASSVd).	NASBA	Heo et al. (2019)
Grape	*Plasmopara viticola*	Fluorescence *in situ* hybridization (FISH)	Díez-Navajas et al. (2007)

TABLE 3.2 Bioinformatics Tools for the Detection of Plant Viruses from RNA-Sequences

Tool	Use	Availability	References
VirFind	Maps and provides taxonomic information.	http://virfind.org	Ho & Tzanetakis (2014)
Taxonomer	Metagenomics analysis tool.	https://www.taxonomer.com/	Flygare et al. (2016)
VSD toolkit	Identifies viral sequences in plants.	Source code can be used with Yabi. https://github.com/muccg/yabi	Barrero et al. (2017)
Virus detect	Analysis of sRNA data for the identification of viruses.	http://bioinfo.bti.cornell.edu/tool/VirusDetect	Zheng et al. (2017)

virus. Plant pathogens produces pathogenicity factors, inside cell virulence and outside cell virulence which can be described with the help of these tools (Lodha et al., 2013). In case of *X. axonopodis pv. citri*, differentially expressed proteins can be analyzed with the help of NH_2 terminal sequencing in *Citrus sinensis* (Mehta & Rosato, 2001). In case of *X. axonopodis,* the expression of proteins of was analyzed in *Passiflora edulis* (Tahara et al., 2003) and an upregulation of outer membrane protein was identified. In case of wild tomato, bacterial canker is caused by *Clavibacter michiganensis* (*Lycopersicon hirsutum*) were identified using proteomics. Around 26 proteins in tomato were recognized by utilizing 2DE and ESI-MS/MS.

3.4 CONCLUSION

Ultramodern molecular diagnostics can greatly enhance early disease detection. The rapidly growing databases generated by genomics give a unique chance to design precise and effective molecular detection systems which will help in the diagnosis of plant pathogens. Plant pathology along with biotechnological tools, bioinformatics, and proteomics has taken the pathogen detection and diagnosis to the next level.

KEYWORDS

- **bioinformatics**
- **disease**
- **genomics**
- **isothermal amplification methods**
- **molecular**
- **nucleic acid sequence-based amplification**
- **plant pathogens**
- **proteomics**

REFERENCES

Adams, I. P., Skelton, A., Macarthur, R., Hodges, T., Hinds, H., Flint, L., Nath, P. D., Boonham, N., & Fox, A., (2014). Carrot yellow leaf virus is associated with carrot internal necrosis. *Plos One*, e109125.

Ahmed, S. A., Van, D. E. B. H. G., Fahal, A. H., Van De, S. W. W., & De Hoog, G., (2014). Rapid identification of black grain eumycetoma causative agents using rolling circle amplification. *PLoS Negl. Trop. Dis.*, *8*(12), e3368.

Badali, H., & Nabili, M., (2012). Molecular tools in medical mycology; where we are! *Jundishapur J. Microbiol.*, *6*(1), 1–3.

Barnes, C. W., & Szabo, L. J., (2007). Detection and identification of four common rust pathogens of cereals and grasses using real-time PCR. *Phytopathol.*, *97*(6), 717–727.

Barrero, R. A., Napier, K. R., Cunnington, J., Liefting, L., Keenan, S., & Frampton, R. A., (2017). An internet-based bioinformatics toolkit for plant biosecurity diagnosis and surveillance of viruses and viroids. *BMC Bioinformatics, 18*, 26.

Boonham, N., Walsh, K., Mumford, R. A., & Barker, I., (2000). Use of multiplex real-time PCR for the detection of potato viruses. *EPPO Bull., 30*, 427–430.

Boonham, N., Walsh, K., Smith, P., Madagan, K., Graham, I., & Barker, I., (2003). Detection of potato viruses using microarray technology: Towards a generic method for plant viral disease diagnosis. *J. of Virol. Meth., 108*, 181–187.

Brown, N. A., Bass, C., Baldwin, T. K., Chen, H., Massot, F., Carion, P. W., Urban, M., et al., (2011). Characterization of the *Fusarium graminearum*-wheat floral interaction. *J. Pathogens, 2011*. 10.4061/2011/626345.

Bu, R., Sathiapalan, R. K., Ibrahim, M. M., Al-Mohsen, I., Almodavar, E., Gutierrez, M. I., & Bhatia, K., (2005). Monochrome light cycler PCR assay for detection and quantification of five common species of candida and aspergillus. *J. Med. Microbiol., 54*(3), 243–248.

Cankar, K., Ravnikar, M., Zel, J., Gruden, K., & Toplak, N., (2005). Real-time PCR detection of cauliflower mosaic virus to compliment the 35S screening assay for genetically modified organisms. *Journal of AOAC International., 88*, 814–822.

Capote, N., Aguado, A., Pastrana, A. M., & Sánchez-Torres, P., (2012). *Molecular Tools for Detection of Plant Pathogenic Fungi and Fungicide Resistance.* Valencia: INTECH Open Access Publisher.

Cartney, H. A., Foster, S. J., Fraaije, B. A., & Ward, E., (2003). Molecular diagnostics for fungal plant pathogens. *Pest Management Science, 59*, 129–142.

Cervantes, (2008). Detection and molecular characterization of two little leaf phytoplasma strains associated with pepper and tomato diseases in Guanajuato and Sinaloa, Mexico. *Plant Dis. 92*(7), 1007–1011.

Chadha, S., & Gopalakrishna, T., (2006). Detection of *Magnaporthe grisea* in infested rice seeds using polymerase chain reaction. *J. Appl. Microbiol., 100*, 1147–1153.

Chang, C. C., Chen, C. C., Wei, S. C., Lu, H. H., Liang, Y. H., & Lin, C. W., (2012). Diagnostic devices for isothermal nucleic acid amplification. *Sensors, 12*(6), 8319–8337.

Chao, H. J., Hong, S. W., Kim, H. J., & Kwak, Y. S., (2016). Development of a multiplex PCR method to detect fungal pathogens for quarantine on exported cacti. *Plant Pathol. J., 32*(1), 53–54.

Chen, X., Steed, A., Harden, C., & Nicholson, P., (2006). Characterization of *Arabidopsis thaliana–Fusarium graminearum* interactions and identification of variation in resistance among ecotypes. *Mol. Plant Pathol., 7*, 391–403.

Cho, H. J., Hong, S. W., Kim, H. J., & Kwak, Y. S., (2016). Development of a multiplex PCR method to detect fungal pathogens for quarantine on exported cacti. *Plant Pathol J., 32*(1), 53–54.

Clark, M. F., & Adams, A., (1977). Characteristics of the microplate method of enzyme-linked immunosorbent assay for the detection of plant viruses. *J. Gen. Virol., 34,* 475–483.

Das, A., Babiuk, S., & McIntosh, M. T., (2012). Development of a loop-mediated isothermal amplification assay for rapid detection of capripoxviruses. *J. Clin. Microbiol., 50*(5), 1613–1620.

Dawei, W., & Peng, Y., (2014). Assessing the impact of dominant sequencing-based gene expression profiling techniques (SGEPTs) on phytopathogenic fungi. *Chiang Mai. J. Sci., 41*, 922–9444.

De Jonge, R., Van, E. H. P., Maruthachalam, K., Bolton, M. D., Santhanam, P., Saber, M. K., Zhang, Z., et al., (2012). Tomato immune receptor Ve1 recognizes effector of multiple fungal pathogens uncovered by genome and RNA sequencing. *Proc. Natl. Acad. Sci., 109*, 5110–5115.

Deb, M., & Anderson, J. M., (2008). Development of a multiplexed PCR detection method for barley and cereal yellow dwarf viruses, Wheat spindle streak virus, Wheat streak mosaic virus. *J. Virol Methods, 148*, 17–24.

Díez-Navajas, A. M., Greif, C., Poutaraud, A., & Merdinoglu, D., (2007). Two simplified fluorescent staining techniques to observe infection structures of the oomycete *Plasmopara viticola* in grapevine leaf tissues. *Micron., 38*, 680–683.

Ellison, M. A., McMahon, M. B., Bonde, M. R., Palmer, C. L., & Luster, D. G., (2016). In situ hybridization for the detection of rust fungi in paraffin-embedded plant tissue sections. *Plant Methods, 12*(1), 37.

Fakruddin, M., (2011). Loop-mediated isothermal amplification (LAMP)-an alternative to a polymerase chain reaction (PCR). *Bangladesh Res. Pub J., 5*(4), 425–439.

Fakruddin, M., Mazumdar, R. M., Chowdhury, A., & Mannan, K., (2012). Nucleic acid sequence-based amplification (NASBA)-prospects and applications. *Int. J. Life Sci. Pharma. Res., 2*, 106–107.

Flygare, S., Simmon, K., Miller, C., Qiao, Y., Kennedy, B., & Di Sera, T., (2016). Taxonomer: An interactive metagenomics analysis portal for universal pathogen detection and host mRNA expression profiling. *Genome Biol., 17*, 111.

Fraaije, B., Lovell, D., Coelho, J. M., Baldwin, S., & Holloman, D. W., (2001). PCR-based assay to assess wheat varietal resistance to blotch (*Septoria tritici* and *Stagono sporanodorum*) and rust (*Puccinia striiformis* and *P. recondita*) diseases. *Eur. J. Plant Pathol., 107*, 905–917.

Frederick, R. D., Snyder, C. L., Peterson, G. L., & Bonde, M. R., (2002). Polymerase chain reaction assays for the detection and discrimination of the soybean rust pathogens *Phakopsora pachyrhizi* and *P. meibomiae*. *Phytopathol., 92*, 217–227.

Gusev, Y., Sparkowski, J., Raghunathan, A., Ferguson, H., Montano, J., & Bogdan, N., (2001). Rolling circle amplification: A new approach to increase sensitivity for immune histochemistry and flow cytometry. *Am. J. Pathol., 159*(1), 63–69.

Hayden, R., Qian, X., Roberts, G., & Lloyd, R., (2001). In situ hybridization for the identification of yeast-like organisms in the tissue section. *Diagn. Mol. Pathol., 10*(1), 15–23.

Heo, S., Kim, H. R., & Lee, H. J., (2019). Development of a quantitative real-time nucleic acid sequence-based amplification (NASBA) assay for early detection of apple scar skin viroid. *Plant Pathol. J., 35*(2), 164–171.

Hijri, M., (2009). The use of fluorescent in situ hybridization in plant-fungal identification and genotyping. In: *Plant Pathology* (pp. 131–145). Springer: Berlin, Germany.

Hirano, H., Islam, N., & Kawasaki, H., (2004). Technical aspects of functional proteomics in plants. *Phytochemistry, 65*, 1487–1498.

Ho, T., & Tzanetakis, I. E., (2014). Development of a virus detection and discovery pipeline using next generation sequencing. *Virology, 471–473*, 54–60.

Ishii, H., & Holloman, D. W., (2015). *Fungicide Resistance in Plant Pathogens*. Tokyo: Springer.

James, A. P., Geijskes, R. J., Dale, J. L., & Harding, R. M., (2011). *Molecular Characterization of Six Badnavirus Species Associated with Leaf Streak Disease of Banana in East Africa, 158*(3), 346–353.

Jayaram, B., & Dhingra, P., (2010). *Bioinformatics for a Better Tomorrow*. Indian Institute of Technology, Hauz Khas, New Delhi–110016, India.

John, I., Hackett, R., Cooper, W., Drake, R., Farrell, A., & Grierson, D., (1997). Cloning and characterization of tomato leaf senescence-related cDNAs. *Plant Mol. Biol., 33*, 641–651.

Kang, S., Mansfield, M. A., Park, B., Geiser, D. M., Ivors, K. L., Coffey, M. D., Grünwald, N. J., et al., (2010). The promise and pitfalls of sequence-based identification of plant-pathogenic fungi and oomycetes. *Phytopathology, 100*(8), 732–737.

Kim, S. G., & Park, C. M., (2008). Gibberellic acid-mediated salt signaling in seed germination. *Plant Sig. Behav., 3*, 877–879.

Kim, S. W., Li, Z., Moore, P. S., Monaghan, A. P., Chang, Y., & Nichols, M., (2010). A sensitive non-radioactive northern blot method to detect small RNAs. *Nucleic Acids Res., 38*, e98.

Kliot, A., Kontsedalov, S., Lebedev, G., Brumin, M., Cathrin, P. B., Marubayashi, J. M., Skaljac, M., et al., (2014). Fluorescence in situ hybridizations (FISH) for the localization of viruses and endosymbiotic bacteria in plant and insect tissues. *J. Vis. Exp., 84*, e51030.

Kohler, G., & Milstein, C., (1975). Continuous culture of fused cells secreting antibodies of predefined specificity. *Nature, 256*, 495–497.

Leinberger, D. M., Schumacher, U., Autenrieth, I. B., & Bachmann, T. T., (2005). Development of a DNA microarray for detection and identification of fungal pathogens involved in invasive mycoses. *J. Clin. Microbiol., 43*(10), 4943–4953.

Leite, R. P., Jones, J. B., Somodi, G. C., Minsavage, G. V., & Stall, R. E., (1995). Detection of *Xanthomonas campestris* pv. *vesicatoria* is associated with pepper and tomato seed by DNA amplification. *Plant Dis., 79*(9), 917–922.

Leyva-Lopez, N. E., Ochoa-Sanchez, J. C., Leal-Klevezas, D. S., & Martinez-Soriano, J. P., (2002). Multiplex phytoplasma associated potato diseases in Mexico. *Can. J. Microbiol., 48*(12), 1062–1068.

Lievens, B., Bart, P., & Thomma, P. H. J., (2005). Recent developments in pathogen detection arrays: Implications for fungal plant pathogens and use in practice. *Phytopathol., 95*, 1374–1380.

Lievens, B., Brouwer, M., Cammue, B. P. A., & Thomma, P. H. J., (2006). Real-time PCR for detection and quantification of oomycetes fungal pathogens of tomato in plant and soil samples. *Plant Science, 16*, 155–165.

Lodha, T. D., Hembram, P., & Basak, N. T. J., (2013). Proteomics: A successful approach to understand the molecular mechanism of plant-pathogen interaction. *Am. J. Plant Sci., 4*, 1212–1226.

Lodha, T. D., Hembram, P., Tep, N., & Basak, J., (2013). Proteomics: A successful approach to understand the molecular mechanism of plant-pathogen interaction. *American Journal of Plant Sciences, 4*, 1212–1226.

Lopez, M. L., Bertolini, E., Olmos, A., Caruso, P., Goris, M. T., Llop, P., Penyalver, R., & Cambra, M., (2003). Innovative tools for detection of plant pathogenic viruses and bacteria. *Int. Microbiol., 6*, 233–243.

Lopez, M. M., Bertolini, E., Marco-Noales, E., Llop, P., & Cambra, M., (2006). Update on molecular tools for detection of plant pathogenic bacteria and viruses. In: Rao, J. R., Fleming, C. C., & Moore, J. E., (eds.) *Molecular Diagnostics: Current Technology and Applications* (pp. 1–46). Horizon Bioscience, Wymondham, UK.

Mangrauthia, S. K., (2010). Rapid detection of rice tungro spherical virus by RT-PCR and dot blot hybridization. *J. Mycol. Pl. Pathol., 40*(3), 445–449.

Masanga, J. O., Matheka, J. M., Omer, R. A., Ommeh, S. C., Monda, E. O., & Alakonya, A. E., (2015). Downregulation of transcription factor aflR in *Aspergillus flavus* confers reduction to aflatoxin accumulation in transgenic maize with alteration of host plant architecture. *Plant Cell Rep., 34*, 1379–1387.

Mbofung, G., & Pryor, B. M., (2010). A PCR-based assay for detection of *Fusariumoxy sporum* f. Sp. *lactucae* in lettuce seeds. *J. Plant Pathol., 94*(7), 860–866.

McCartney, H. A., Foster, S. J., Fraaije, B. A., & Ward, E., (2003). Molecular diagnostics for fungal plant pathogens. *Pest Management Science, 59*(2), 129–142.

McCartney, H. A., Foster, S. J., Fraaije, B. A., & Ward, E., (2008). Molecular diagnostics for fungal plant pathogens. *Pest Management Science, 59*(2), 129–134.

Miller, S. A., Beed, F. D., & Harmon, C. L., (2009). Plant disease diagnostic capabilities and networks. *Annu. Rev. Phytopathol., 47*, 15–38.

Mirmomeni, M. H., Sharifi, A., & Sisakhtnezhad, S., (2008). Rapid detection of potato Y potyvirus in potato farms of Kermanshah using RT-PCR amplification of the P1-protease gene and its cloning. *Pak. J. Biol. Sci., 11*, 1482–1486.

Mlotshwa, S., Pruss, G. J., & Vance, V., (2008). Small RNAs in viral infection and host defense. *Trends Plant Sci., 13*, 375.

Monteiro, F., Romeiras, M. M., Figueiredo, A., Sebastiana, M., Balde, A., Catarino, L., & Batista, D., (2015). Tracking cashew economically important diseases in the West African region using metagenomics. *Frontiers in Plant Science, 6*(482), 1–6.

Monteiro, F., Romeiras, M. M., Figueiredo, A., Sebastiana, M., Baldé, A., & Catarino, L., (2015). Tracking cashew economically important diseases in the West African region using metagenomics. *Front Plant Sci., 6*, 482–483.

Moretti, C., Amatulli, M. T., & Buonaurio, R., (2008). PCR-based assay for the detection of *Xanthomonas euvesicatoria* causing pepper and tomato bacterial spot. *Letters in Applied Microbiol., 49*, 466–479.

Nakayashiki, T., Kurtzman, C. P., Edskes, H. K., & Wickner, R. B., (2005). Yeast prions [URE3] and [PSI+] are diseases. *Proc. Natl. Acad. Sci., 102*(30), 10575–10580.

Narayanasamy, P., (2002). *Microbial Plant Pathogens and Crop Disease Management.* Science Publishers, Enfield, USA.

Narayanasamy, P., (2006). *Postharvest Pathogens and Disease Management.* Wiley-Inter Science, John Wiley & Sons, Hoboken, NJ, USA.

Ogras, T. T., El-Fadly, G., Balaglu, S., Yilmaz, M. A., Cirakoglu, B., & Bermek, E., (1994). Development of a scientific detection technique for tobacco mosaic virus-based on dot-blot hybridization using an oligonucleotide probe. *Turkish J. Biol., 8*, 91–98.

Owens, R. A., & Diener, T. O., (1981). Sensitive and rapid diagnosis of potato spindle tuber viroid disease by nucleic acid hybridization. *Science, 213*, 670–672.

Ozakman, M., & Schaad, N. W., (2003). A real-time bio-PCR assay for detection of *Ralstonia solanacearum* race 3, biovar 2 in asymptomatic potato tubers. *Can. J. Plant Pathol., 25*, 232–239.

Panwar, V., McCallum, B., & Bakkeren, G., (2012). Endogenous silencing of *Puccinia triticina* pathogenicity genes in planta-expressed sequences leads to the suppression of rust diseases on wheat. *Plant J., 73,* 521–532.

Pastrik, K. H., Elphinstone, J. G., & Pukall, R., (2002). Sequence analysis and detection of *Ralstonia solanacearum* by multiplex PCR amplification of 16S-23S ribosomal intergenic spacer region with internal positive control. *Eur. J. Plant Pathol., 108*, 831–842.

Pico, B., Sifres, A., & Nuez, F., (2005). Quantitative detection of cucumber vein yellowing virus in susceptible and partially resistant plants using real-time PCR. *Journal of Virological Methods, 128*, 14–20.

Qi, M., & Yang, Y., (2002). Quantification of *Magnaporthe grisea* during infection of rice plants using real-time polymerase chain reaction and northern blot/phosphoimaging analyses. *Phytopathology, 92*(8), 870–876.

Qian, X., & Lloyd, R. V., (2003). Recent developments in signal amplification methods for in situ hybridization. *Diagn. Mol. Pathol., 12*(1), 1–13.

Quan, P. L., Briese, T., Palacios, G., & Ian, L. W., (2008). Rapid sequence-based diagnosis of viral infection. *Antiviral Res., 79*, 1–5.

Quirino, B. F., Candido, E. S., Campos, P. F., Franco, O. L., & Krüger, R. H., (2010). Proteomic approaches to study plant-pathogen interactions. *Phytochemistry, 71*, 351–362.

Rahman, M. T., Uddin, M. S., Sultana, R., Moue, A., & Setu, M., (2013). Polymerase chain reaction (PCR): A short review. *Anwer Khan. Mod. Med. Coll. J., 4*(1), 30–36.

Rampersad, S. N., & Umaheran, P., (2003). Detection of begomoviruses in clarified plant extracts: A comparison of standard, direct-binding, and immunocapture polymerase chain reaction techniques. *Phytopathol., 93*, 1153–1157.

Ren, C. H., Hu, C. Q., Luo, P., & Wang, Q. B., (2009). Sensitive and rapid identification of vibrio vulnificus by loop-mediated isothermal amplification. *Microbiol. Res., 164*(5), 514–521.

Robinson, B., Erle, D., Jones, D., Shapiro, S., Metzger, W., Albelda, S., Parks, W., & Boylan, A., (2000). Recent advances in molecular biological techniques and their relevance to pulmonary research. *Thorax, 55*(4), 329–339.

Sato, T., Takayanagi, A., Nagao, K., Tomatsu, N., Fukui, T., Kawaguchi, M., Kudoh, J., et al., (2010). Simple PCR-based DNA microarray system to identify human pathogenic fungi in the skin. *J. Clin. Microbiol., 48*(7), 2357–2364.

Savolainen, V., Cowan, R. S., Vogler, A. P., Roderick, G. K., & Lane, R., (2005). Towards writing the encyclopedia of life: An introduction to DNA barcoding. *Philos. Trans. R. Soc. B., 360*(1462), 1805–1811.

Schoen, C. D., Knorr, D., & Leone, G., (2006). Detection of potato leaf-roll virus in dormant potato tubers by immunocapture and a fluorogenic 5' nuclease RT-PCR assay. *Phytopathol., 86*, 993–999.

Shahnejat-Bushehri, A., Adel, J., & Dehkordi, M. K. H., (2006). Detection of beet necrotic yellow vein virus with a reverse transcription-polymerase chain reaction. *Int. J. Agri. Biol., 8*(2), 280–285.

Siddique, A. B., Agarwal, G. K., Alam, N., & Krishan, R. M., (2001). Electron microscopy and molecular characterization of a phytoplasma associated with the little leaf disease of brinjal (*Solanum melongena*) and periwinkle (*Cathrantus roseus*) in Bangladesh. *J. Phytopathol., 149*, 237–244.

Singh, A., & Kumar, N., (2013). A review on DNA microarray technology. *Int. J. Curr. Res. Rev., 5*(22), 1–5.

Singh, R. P., Kurz, J., Boiteau, G., & Moore, L. M., (1997). Potato leafroll virus detection by RT-PCR in field-collected aphids. *Amer. Potato. J., 74*, 305.

Sivalingam, P. N., Padmalatha, K. V., Monga, D., Ajmera, B. D., & Malathi, V. G., (2007). Detection of begomoviruses by PCR in weeds and crop plants in and around cotton field infected with cotton leaf curl disease. *Indian Phytopath., 60*(3), 356–361.

Soanes, D. M., Chakrabarti, A., Paszkiewicz, K. H., Dawe, A. L., & Talbot, N. J., (2012). Genome-wide transcriptional profiling of appressorium development by the rice blast fungus *Magnaporthe oryzae*. *PLoS Pathog., 8*, e1002514.

Studholme, D. J., Glover, R. H., & Boonham, N., (2011). Application of high-throughput DNA sequencing in phytopathology. *Annu. Rev. Phytopathol., 49*, 87–105.

Sun, J., Li, X., Zeng, H., Xie, Z., Lu, C., Xi, L., & De Hoog, G. S., (2010). Development and evaluation of loop-mediated isothermal amplification (LAMP) for the rapid diagnosis of *Penicillium marneffei* in archived tissue samples. *FEMS Immunol. Med. Microbiol., 58*(3), 381–388.

Tahara, S. T., Mehta, A., & Rosato, Y. B., (2003). Proteins induced by *Xanthomonas axonopodis* pv. passiflorae with leaf extract of the host plant (*Passiflora edulis*). *Proteomics, 3*, 95–102.

Taoufik, A., Nijhof, A., Hamidjaja, R., Jongejan, F., Pillay, V., Sonnevelt, M., & De Boer, M., (2004). *Reverse Line Blot Hybridisation in the Detection of Tick-Borne Diseases*. BTi.

Thomas, S. W., Glaring, M. A., Rasmussen, S. W., Kinane, J. T., & Oliver, R. P., (2002). Transcript profiling in the barley mildew pathogen *Blumeria graminis* by serial analysis of gene expression (SAGE). *Molecular Plant-Microbe Interactions, 15*(8), 847–856.

Tooley, P. W., Carras, M. M., & Lambert, D. H., (1998). Application of a PCR-based test for detection of potato late blight and pink rot in tubers. *Amer. J. Potato Res., 75*, 187–194.

Tsai, H. L., Huang, L. C., Ann, P. J., & Liou, R. F., (2006). Detection of orchid phytophthora disease by nested PCR. *Bot. Stud., 47*(4), 379–387.

Tsuda, S., Fujisawa, I., Hanada, K., Hidaka, S., Higo, K., Kameya-Iwaki, M., & Tomaru, K., (1994). Detection of tomato spotted wilt virus SRNA in individual thrips by reverse transcription and polymerase. *Ann. Phytopathol. Soc. Jpn., 60*, 99–103.

Tsui, C. K., Woodhall, J., Chen, W., Lévesque, C. A., Lau, A., Schoen, C. D., Baschien, C., et al., (2011). Molecular techniques for pathogen identification and fungus detection in the environment. *IMA Fungus, 2*(2), 177.

Villari, C., Tomlinson, J. A., Battisti, A., Boonham, N., Capretti, P., & Faccoli, M., (2013). Use of loop-mediated isothermal amplification for detection of *Ophiostoma clavatum*, the primary blue stain fungus associated with Ips acuminatus. *Appl. Environ. Microbiol., 79*(8), 2527–2533.

Wu, L., Han, Z., Wang, S., Wang, X., Sun, A., Zu, X., & Chen, Y., (2013). Comparative proteomic analysis of the plant-virus interaction in resistant and susceptible ecotypes of maize infected with sugarcane mosaic virus. *J. Proteomics., 89*, 124–140.

Xu, M. L., Melchinger, A. E., & L¨ubberstedt, T., (1999). Species-specific detection of the maize pathogens *Sporisorium reiliana* and *Ustilago maydis* by dot-blot hybridization and PCR-based assays. *Plant Dis. 83*, 390–395.

Yabuta, Y., Morishita, T., Kojima, Y., Maruta, T., Nishizawa-Yokoi, A., & Shigeoka, S., (2010). Identification of recognition sequence of ANAC078 protein by the cyclic amplification and selection of targets technique. *Plant Signal Behav., 5*, 1–3.

Yang, Y., Hu, M., Yu, K., Zeng, X., & Liu, X., (2015). Mass spectrometry-based proteomic approaches to study pathogenic bacteria-host interactions. *Protein Cell, 6*, 265–274.

Yao, C. L., Frederiksen, R. A., & Magill, C. W., (1990). Seed transmission of sorghum downy mildew: Detection by DNA hybridization. *Seed Sci. Technol., 18*, 201–207.

Yeo, S. F., & Wong, B., (2002). Current status of nonculture methods for diagnosis of invasive fungal infections. *Clin. Microbiol. Rev., 35*(3), 465–484.

Zhang, A. W., Hartman, G. L., Curio-Penny, B., Pedersen, W. L., & Becker, K. B., (1999). Molecular detection of *Diaporthe phaseolorum* and *Phomopsis longicolla* from soybean seeds. *Phytopathol., 89*(9), 797–804.

Zhang, N., McCarthy, M. L., & Smart, C. D., (2008). A macroarray system for the detection of fungal and oomycete pathogens of solanaceous crops. *Plant Dis., 92*(6), 953–960.

Zheng, Y., Gao, S., Padmanabhan, C., Li, R., Galvez, M., & Gutierrez, D., (2017). VirusDetect: An automated pipeline for efficient virus discovery using deep sequencing of small RNAs. *Virology, 500*, 130–138.

CHAPTER 4

Induction of Defense Responses: Pathogenesis-Related Proteins, HR, and Reactive Oxygen Species

SUHAIL MUZAFFAR,[1] SANGITA SAHNI,[2] and BISHUN DEO PRASAD[3]

[1]*UAB Research Center of Excellence in Arsenicals and Department of Dermatology, University of Alabama at Birmingham, Birmingham, Alabama, USA*

[2]*Department of Agricultural Biology and Molecular Biology, College of Basic Sciences and Humanities, Dr. Rajendra Prasad Central Agricultural University, Pusa, Samastipur, Bihar, India*

[3]*Department of Plant Pathology, TCA, Dholi, Dr. Rajendra Prasad Central Agricultural University, Pusa, Samastipur, Bihar, India*

ABSTRACT

Plants are continuously exposed to different types of microbial pathogens. However, successful infection is a rare event because plants generate a wide range of responses to protect themselves against pathogens. These plant responses include hypersensitive response, reactive oxygen species (ROS) production, and expression of pathogenesis-related (PR) genes. The hypersensitive response involves plant cell death at the site of pathogen ingress to prevent the spreading of the harmful pathogen from infected to uninfected tissues. Pathogen attack also results in a rapid oxidative burst (OB) that leads to ROS accumulation in infected tissues. ROS not only eliminates the pathogen directly but also activates signaling cascades involved in plant defense mechanisms. PR proteins are a class of proteins that accumulate in plants in response to biotic or abiotic stresses. The accumulated PR proteins

New Advances in Crop Disease Management. Sangita Sahni, Ganesh Thapa, and Bishun Deo Prasad (Eds)

also prevent plant reinfections thereby resulting in the development of systemic acquired resistance (SAR).

4.1 INTRODUCTION

Plants encounter various pathogens like bacteria, fungi, and viruses as they are a rich source of nutrients for these microbes. Plant-microbial interactions can be either beneficial or harmful resulting in symbiotic or pathogenic interactions, respectively. Plants have developed structural, chemical, and protein-based defense systems against invading pathogens. Depending on the type of invading pathogen, plants modulate their immune system so that an appropriate host response is expressed. They have developed highly efficacious constitutive and inducible defenses. Constitutive defenses include mechanical barriers such as plant wax, epidermal bark, cuticles, and cell walls. Induced plant defense specifically depends on the nature of the inducing agents and involves the production of endogenous signals and systemic expression of the defense molecules. Plants have evolved multi-faceted defense responses against pathogens including the cell death mechanism called hypersensitive response (HR), an oxidative burst (OB), and expression of pathogenesis-related (PR) genes.

Reactive oxygen species (ROS) play a crucial role in plant defense against pathogens, as the rapid accumulation of ROS at the pathogen attack site kills pathogens directly. It may also lead to HR that results in a zone of host cell death thereby preventing the further spread of biotrophic pathogens. The HR is a form of cell death associated with host resistance to plant pathogens. ROS and ion fluxes have been directly associated with HR, as OBs usually precede cell death. The gene-for-gene relationship between host plants and fungal or bacterial pathogens is well characterized based on genetic analysis of plants and pathogens. The link between HR and gene-for-gene theory was discovered in flax/flax rust system (*Linum margina*/*Melampspora lini*) by H. H. Flor during the 1930s. Flor discovered the specific plant-pathogen interaction that determined whether the pathogen attack results in successful infection or host resistance. Each dominant avirulence gene (*Avr* gene) in the pathogen interacts with the corresponding dominant resistance gene (*R*-gene) in the host. Resistance is conferred if the *R*-gene and the corresponding *Avr* gene were present during the interaction. The *R*-gene-mediated resistance is often associated with HR in most cases. This gene-for-gene relationship is broadly applicable in most higher plants and many *R*-gene and

Avr genes have already been characterized. Plants also react to the pathogen attack by inducing the expression of PR genes. PR proteins are structurally diverse toxic proteins that help eliminate invading fungal pathogens. Normally they are present in plants in trace amounts but their expression is drastically increased following the pathogen attack. Plants produce several phytohormones such as jasmonic acid (JA), salicylic acid (SA), and ethylene in response to pathogen attacks. The composition and quantity of these chemicals are different in different plant species and depend on the type of infecting pathogen.

4.2 PLANT DISEASE RESISTANCE

Plants are constantly challenged by various pathogenic microbial pathogens including bacteria, fungi, viruses, and oomycetes to compromise plant survival and fitness. These pathogens lead to a significant reduction in annual crop yield and pose a serious threat to our food security. Plants use structural as well as biochemical responses to combat pathogens. Cutin or lignin deposition on cell walls and production of antimicrobial molecules are the first line of defense to prevent the invasion of pathogens. However, many pathogens cross the first layer of defense and therefore an alternative defense mechanism has been developed by plants to counter the pathogens. One such defense mechanism is pathogen inducible defense response which includes HR followed by generation of ROS, and production of PR proteins. When a pathogen attacks a particular host plant, two outcomes are possible. A compatible interaction results in disease, while an incompatible interaction results in little or no disease. A plant species may be susceptible to a particular pathogen; however, some varieties may contain genes that help recognize the pathogen and activate host defenses. For example, some tomato cultivars get infected with the bacterial pathogen *Pseudomonas syringae* (a compatible response), while others like *Rio Grande* show resistance towards this pathogen (an incompatible response). Plant disease resistance is crucial for crop production and involves inherent structural plant defenses or infection-induced responses of the immune system. Plants naturally carry inherent disease resistance but may not be always sufficient for protection against every pathogen. Plants have developed multiple mechanisms to protect themselves against pathogens. Throughout evolution, plants have developed two types of defense mechanisms including non-inducible defense and inducible defense response. The inducible defense responses include:

- Hypersensitive response-defense response involving localized cell death;
- Defense chemicals like reactive oxygen species;
- Defense proteins such as defensins and PR proteins.

4.2.1 HYPERSENSITIVE RESPONSE

The hypersensitive response (HR) was described by Stakman in 1915. It is a critical event for successful pathogen recognition during host plant-microbe interactions. Pathogen recognition mechanisms due to the interaction of host resistance gene products and avirulence gene products from the pathogen lead to HR. The hypersensitive defense response is characterized by a rapid cell death at the site of pathogen entry and is found in almost all higher plants. HR is a disease resistance response induced by several classes of pathogens including bacteria, fungi, oomycetes, and viruses. HR is generally associated with biotrophic pathogens which derive their nutrition from living tissues and is less effective against necrotrophs that survive on dead host tissue (Figure 4.1). A similar cell death response has also been observed during resistant interactions between host and parasitic plants; however, it is not clear if it involves HR. The formation of necrotic lesions due to HR is a coordinated induction of an integrated defense response such as cell wall rigidification, phytoalexins production, and accumulation of PR proteins. The HR is not effective against necrotrophic pathogens as they feed on dead cells. Victoria blight of oats is caused by *C. victoriae* in which the necrotrophic pathogen secretes a host-selective peptide toxin called victorin. The cell death induced by the victorin is particularly interesting as it has evolved to use the host HR to eliminate host cells, which are then consumed by the pathogenic fungus. Plant cell death is beneficial for necrotrophic pathogens, such as *Botryotinia fuckeliana* and it leads to improved colonization. On the other hand, biotrophic pathogens are inhibited to some extent by HR, such as *Pyrenophora teres* and *Magnaporthe oryzae* (Morel & Dangl, 1997).

HR is a form of programmed cell death (PCD) that prevents the spreading of pathogens from the site of infection toward healthy tissue. It is well known that plant cells challenged by necrotrophic pathogens undergo PCD, as necrotrophic pathogens kill host cells to derive nutrients. HR-associated cell death induced by pathogen attack is a remarkable manifestation of PCD in plants. HR cell death is often accompanied by the enhanced growth of lytic vacuoles and tonoplast rupture. In addition to PCD, increased autophagic

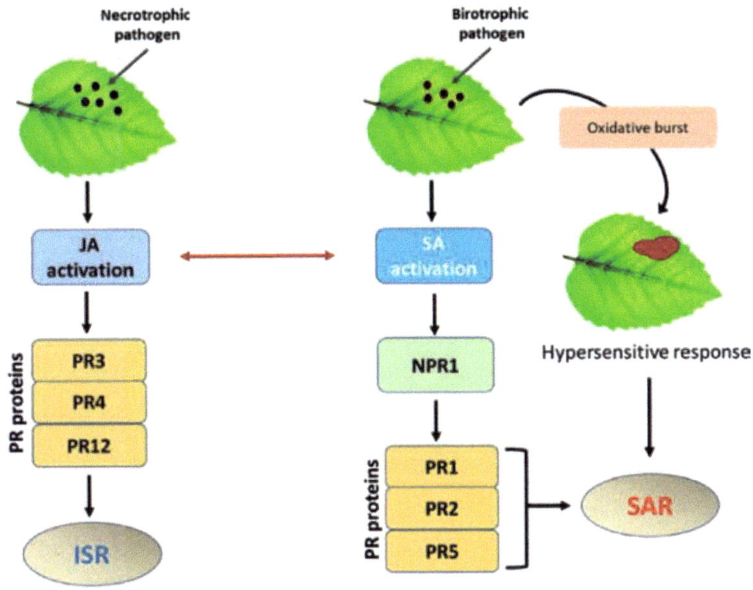

FIGURE 4.1 Salicylic acid (SA) induces disease resistance against biotrophic pathogens, whereas jasmonic acid (JA) activates resistance against necrotrophic pathogens. NPR1 (non-inducible pathogenesis-related 1) is a key regulator of SA signaling that leads to the activation of SA-dependent responses via the expression of PR proteins.

activity during HR cell death has been reported and the process is controlled by *ATG* genes. The morphology of plant tissue undergoing HR shares common features with apoptosis. The ultrastructural analysis of *P. syringae* infected lettuce cells shows mitochondrial swelling and other morphological features that are similar to animal cells undergoing apoptosis. The later stages of the plant infection displayed membrane dysfunction and vacuolization of the cellular cytoplasm. However, apoptosis-associated chromatin condensation and DNA disintegration were not observed. On the other hand, cowpea infected with rust fungus shows evidence of endonucleolytic DNA cleavage during HR. During several plant-pathogen interactions, the involvement of specific proteases in HR has been demonstrated. In plants, no clear homologs of caspases have been identified but caspase-like activity has been reported in plants based on biochemical and inhibitor studies.

The concept of HR is related to the gene-for-gene interaction model introduced by H. H. Flor. Studying the flax/flax rust system, Flor discovered the gene-for-gene relationship between the plant and pathogen resulting in either disease or resistance/HR. The dominant resistance gene (R-gene) in the host plant corresponded with a dominant avirulence gene (Avr gene) in the pathogen. R-Avr interaction is necessary to induce resistance against specific pathogens and this interaction is often exploited to control diseases in the field (Figure 4.2). The molecular basis of gene-for-gene interactions is a ligand and receptor model in which the R gene (receptor) recognizes the ligand produced by the AVR gene of the pathogen. Most *R*-genes encode NLR proteins characterized as intracellular multidomain receptor proteins carrying a typical nucleotide-binding site (NBS) and leucine-rich repeat (LRR) domain. These are called NLR immune receptors due to their analogous functionality with NOD-like receptors (NLRs) in the animal's innate immune system. NLR proteins belong to the signal transduction ATPases class of proteins where the ADP-bound form denotes the resting 'off' state. Following the pathogen recognition, a conformational change leads to ADP to ATP conversion resulting in the active or "ON" state which initiates downstream signaling. ATP hydrolysis is critical for the activated and inactivated states of downstream proteins. NLRs are among the most diverse protein families in plants that confer resistance to diverse pathogens including bacteria, fungi, viruses, and nematodes. More than 300 cell surface and intracellular *R* genes have been cloned and characterized in the model plant *Arabidopsis thaliana* as well as important crop plants such as *Triticum aestivum*, *Oryza sativa*, *Zea mays*, and *Solanum tuberosum*. Avr proteins are a class of pathogen effector proteins secreted by the pathogen into the host cells or apoplast region to enable pathogenesis. They are effectors whose presence can be detected by specific R-proteins resulting in the activation of the defense response. Plant NLRs can directly recognize specific effector proteins, or indirectly through effector modification on a host target. According to the specific pathogen recognition model, the first event in triggering the HR is the direct recognition of the pathogen Avr gene product by the corresponding plant R gene product. A similar interaction has been reported between the tomato Pto resistance gene product and the *Pseudomonas syringae* avirulence gene avrPto. The earliest response observed after pathogen recognition is OB, resulting due to production of ROS. The nature of the effectors of HR is comparatively less understood. Some of the elements of the defense response including ROS, phytoalexins, and SA are toxic to the plant cells and may directly result in cell death. ROS can cause loss of cell integrity as well as cell viability as their elevated lead to reactivity with membrane lipids, proteins, and DNA (Dey et al., 2020; Zurbriggen, Carrillo, & Hajirezaei, 2010; Morel & Dangl, 1997).

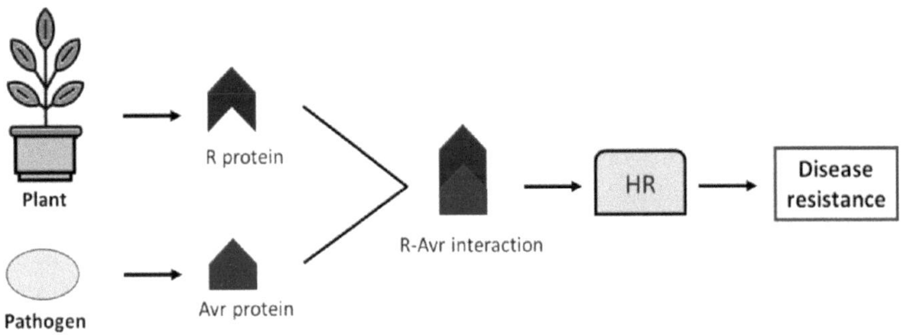

FIGURE 4.2 Plant R proteins recognize effectors or avirulence proteins from specific pathogens. This R-Avr interaction triggers defense responses such as the hypersensitive response and results in resistance of the host plant.

4.2.2 REACTIVE OXYGEN SPECIES (ROS)

One of the initial events associated with plant-pathogen interaction is the change in the redox status of the host cells due to the accumulation of ROS. ROS plays a crucial role in the survival and development of the plants as it is involved in limiting pathogen infection, induction of apoptosis, and signal transduction leading to plant defense processes. ROS is the first line of defense in plant cell walls against fungal cells. Plant cells require molecular oxygen in the ground state to carry out energy metabolism. However, 1 to 2% of molecular oxygen is converted to ROS in different cellular compartments whereas the rest of it is reduced to water. ROS signaling has been reported to propagate from plastids to the apoplast region and adjacent cells via activation of NADPH oxidases, leading to cell death in the region of infection. ROS are stored in different cellular compartments such as mitochondria, peroxisomes, chloroplasts, and recently identified apoplasts. The apoplastic OB is generated by the plasma membrane NADPH oxidases following sensing of pathogen attack by membrane receptor kinases.

During environmental stress or pathogen attack, ROS may generate from the cell surface, chloroplasts, mitochondria, or peroxisomes. In non-stressed plant cells, various antioxidants maintain redox homeostasis by neutralizing the detrimental effects of ROS. After a pathogen attack, different enzymes such as superoxide dismutases, NADPH oxidase, peroxidases, oxalates oxidases, lipoxygenases, and quinone reductase are up-regulated and lead to ROS generation. The accumulation of ROS has been implicated in various physiological processes including resistance to abiotic and biotic stress. The change in redox status induced by the pathogenic attack is a rapid plant response to infection as the increase in H_2O_2 can be detected within 2–3 minutes following the contact with the pathogen. H_2O_2 prevents pathogen growth by inducing the synthesis of phytoalexins and Pathogenesis Related (PR) proteins. Elevated ROS and phytoalexins levels in plant cells lead to the elimination of pathogens (Morel & Dangl, 1997; Zurbriggen et al., 2010). The increase in intracellular redox status of the infected tissue has been associated with different functions during defense response:

- Pathogen-induced burst of ROS drives peroxidase-mediated oxidative cross-linking of cell wall proteins thereby strengthening the physical barrier between plant cells and pathogen.
- The elevated concentration of ROS induces a signaling mechanism involved in the activation of inducible defense genes (Park, 2005; Dey et al., 2020).

Studies of different plant-pathogen interactions revealed a strong correlation between oxidative stress, nitrosative burst, and host defense response. The accumulation of ROS like apoplastic superoxide ions, hydrogen peroxide (H_2O_2), or nitric oxide (NO) is one of the early events during HR. The link between ROS and HR was first established when it was reported that superoxide was generated before HR in potato and tobacco plants infected with *Phytophthora infestans* (fungal pathogen) and tobacco mosaic virus (TMV), respectively.

4.2.3 PR-RELATED PROTEINS

Pathogenesis-related (PR) proteins are a structurally diverse group of plant proteins produced in response to biotic or abiotic stress. PR proteins are induced in response to pathogen attack as part of systemic acquired resistance (SAR). PR proteins are distributed in plants in minimal amounts, but the expression is induced in response to stress or invading pathogen attack. PR proteins and antimicrobial peptides (AMPs) are the key components of the plant's innate immune system that serve as diagnostic molecular markers of defense signaling. Overexpression of PR proteins such as glucanase, chitinase, and defensin augment the plant defense response against different pathogens. The induction and accumulation of PR proteins in plants in response to pathogens is an indispensable component of innate immunity in plants. Different types of PR proteins have been isolated from different plants and different plant organs like seeds, leaves, and roots may produce different forms of PR proteins. PR proteins show a strong antimicrobial activity including antifungal activity. These defense proteins are classified into 17 families based on molecular structure and enzyme activity. Among the 17 PR protein families, nine proteins show enzymatic activity including glucanases (PR-2), chitinases (PR-3, PR-4, PR-8, PR-11), osmotins and thaumatins (PR-5), protease inhibitors (PR-6), lysozymes (PR-8), peroxidase (PR-9), and ribonucleases (PR10). Some PR proteins inhibit the release of fungal spores or inhibit spore germination, while others like β-1,3-glucanase and chitinase degrade chitin-based cell walls of fungal pathogens. In response to pathogen attack, plants exhibit increased production of ROS SA and ethylene. Jasmonate serves as a signal molecule in the induction of PR proteins. Jasmonate also activates genes encoding defensins, protease inhibitors, and basic PR proteins (Singh et al., 2020; Ali et al., 2018).

SA plays a major role in plant disease resistance by inducing the production of PR proteins. PR proteins are also involved in HR and SAR responses against infections. Tobacco plants treated with SA lead to PR production and thereby attenuation of the disease symptoms caused by the TMV. These observations have been also confirmed in other plants infected with different fungal and bacterial pathogens. Several studies have demonstrated the role of the PR-4 family in plant defense regulated by SA, abscisic acid, and jasmonate. In rice, maize, and wheat, the *PR-4* gene expression is induced within the first hour following contact with the pathogenic fungus. However, the activation of *PR-4* genes in wheat depends on both SA and JA-dependent metabolic pathways. The presence of a conserved C-terminal Barwin domain containing 6 conserved cysteine residues forming disulfide bridges is a characteristic feature of the PR-4 proteins (Ali et al., 2018; Kitajima & Sato, 1999).

SAR is a form of induced resistance produced in plants when exposed to pathogens. Biochemical studies have revealed that the accumulation of SA is crucial for SAR in Arabidopsis and tobacco plants. Pathogen-induced mobile SAR signal generated in infected leaves transports to distant uninfected leaves resulting in SAR. SAR provides broad-spectrum and persistent resistance to secondary infections. This makes SAR a highly advantageous trait in crop protection. The first step in SAR is the recognition of a pathogen infection by host plants. After recognition, the signal is transmitted to trigger durable resistance in other organs of the plant. However, only compatible plant-pathogen interactions lead to SAR response. Plant immune systems depend on the ability to recognize pathogen molecules, signal transmission, and defense response. This hypothesis that SA may have a role in SAR was proposed in the 1990s since tobacco plants inoculated with TMV led to an increased endogenous level of SA. In addition to increased SA levels, SAR is often accompanied by elevated expression of SA-responsive genes in the uninfected parts of the plants. Overexpression of the SA-responsive PR1 gene is a molecular marker for SAR. In *Arabidopsis,* ics1 mutant is deficient in isochorismate synthase 1 activity, and SA synthesis is unable to show SAR response (Jain & Khurana, 2018; Kitajima & Sato, 1999; Sels et al., 2008).

4.3 CONCLUSIONS

The constant interactions between plants and pathogens are of critical importance for agriculture and development. Plant defense systems recognize pathogen molecules and respond by activating antimicrobial proteins (APs) and specific defense signaling. The most common type of plant resistance to pathogenic infections is HR which restricts pathogens from spreading within the plant tissue. HR is a combination of resistance and cell death, where cell death may be a prerequisite for pathogen resistance. The dynamic nature of plant disease resistance makes it a highly effective process that depends primarily on the timing and speed of the host response. PR proteins are induced by a wide range of pathogens as well as defense signaling molecules. These are the critical components of plant innate defense mechanism especially SAR that are often used as diagnostic molecular markers of plant defense. These attributes make PR genes highly promising candidates for the development of disease-resistant crop varieties.

KEYWORDS

- **defense responses**
- **disease-resistant crop varieties**
- **hypersensitive response**
- **pathogenesis**
- **plant disease resistance**
- **reactive oxygen species**
- **signal transmission**

REFERENCES

Ali, S., Ganai, B. A., Kamili, A. N., Bhat, A. A., Mir, Z. A., Bhat, J. A., Tyagi, A., et al., (2018). Pathogenesis-related proteins and peptides as promising tools for engineering plants with multiple stress tolerance. *Microbiological Research, 212*, 29–37.

Dey, N., Roy, U. K., Aditya, M., & Bhattacharjee, S., (2020). Defensive strategies of ROS in programmed cell death associated with hypertensive response in plant pathogenesis. *Annals of Systems Biology, 3*, 001–009.

Jain, D., & Khurana, J. P., (2018). Role of pathogenesis-related (PR) proteins in plant defense mechanism. In: *Molecular Aspects of Plant-Pathogen Interaction* (pp. 265–281). Springer.

Kitajima, S., & Sato, F., (1999). Plant pathogenesis-related proteins: Molecular mechanisms of gene expression and protein function. *The Journal of Biochemistry, 125*, 1–8.

Morel, J. B., & Dangl, J. L., (1997). The hypersensitive response and the induction of cell death in plants. *Cell Death & Differentiation, 4*, 671–683.

Park, J. M., (2005). The hypersensitive response. A cell death during disease resistance. *The Plant Pathology Journal, 21*, 99–101.

Sels, J., Mathys, J., De Coninck, B. M., Cammue, B. P., & De Bolle, M. F., (2008). Plant pathogenesis-related (PR) proteins: A focus on PR peptides. *Plant Physiology and Biochemistry, 46*, 941–950.

Singh, M., Srivastava, M., Kumar, A., Singh, A., & Pandey, K., (2020). Endophytic bacteria in plant disease management. In: *Microbial Endophytes* (pp. 61–89). Elsevier.

Zurbriggen, M. D., Carrillo, N., & Hajirezaei, M. R., (2010). ROS signaling in the hypersensitive response: When, where and what for? *Plant Signaling & Behavior, 5*, 393–396.

CHAPTER 5

Advances in Molecular Basis of Host-Pathogen Interaction

SANGITA SAHNI,[1] BISHUN DEO PRASAD,[2] VARSHA C. MOHANAN,[3] SUNAYANA RATHI,[4] MANKESH KUMAR,[5] and SUROJIT SEN[6]

[1]*Department of Plant Pathology, TCA, Dholi, Dr. Rajendra Prasad Central Agricultural University, Pusa, Bihar, India*

[2]*Department of Molecular Biology and Genetic Engineering, Bihar Agricultural University, Sabour, Bhagalpur, Bihar, India*

[3]*St. Mary's College, Thrissur, Kerala, India*

[4]*Department of Biochemistry and Agricultural Chemistry, Assam Agricultural University, Jorhat, Assam, India*

[5]*Department of Plant Breeding and Genetics, Bihar Agricultural University, Sabour, Bhagalpur, Bihar, India*

[6]*Department of Zoology, Mariani College, Mariani, Jorhat, Assam, India*

ABSTRACT

Plants are under the constant threat of various invading pathogens. However, the development of disease in plants is a rare event. This is due to the presence of sophisticated arsenals in host plants to fight against invading pathogens. Plants employ their survival strategies against biotic stresses, which are further fine-tuned by several lines of defense. The interaction of plants and pathogens is a complex and multifaceted process that involves both pathogens and plant-derived compounds. It is quite necessary to understand the molecular aspects of plant-pathogen interaction to formulate sustainable disease management strategies. The

New Advances in Crop Disease Management. Sangita Sahni, Ganesh Thapa, and Bishun Deo Prasad (Eds)

development of the RNA silencing-based approach has been successfully employed as a potential tool for crop protection. This chapter discusses the molecular basis of host-pathogen interactions as well as new disease management strategies.

5.1 INTRODUCTION

Plant disease is a malfunction of a plant's typical physiological functions brought on by pathogens. Pathogen and plant-derived chemicals are both involved in the complicated process known as plant-pathogen interaction (Boyd et al., 2013). The pathogens' secreted molecules are the fundamental variables that determine their pathogenicity and allow them to colonize the host successfully. Plant-derived chemicals, on the other hand, are involved in pathogen identification and eliciting the defense response. Based on the occurrence of diseases on the host, a pathogen which causes diseases is termed virulent and a pathogen that does not cause diseases is called avirulent. There are mainly three types of pathogens: necrotrophs (kill their host), biotrophs (plant cells remain alive), and hemibiotrophs (plant cells initially alive and killed later). Perfect sequential communication between the pathogen and the host is the key to a perfect plant pathogenic establishment. The pathogenic intervention starts with the first contact of the pathogen with the host surface. Resistance and susceptibility of a plant pathogen interaction depends upon several factors, genes, and their interactions. Many genes that function in or are connected with plant defensive mechanisms have been identified, characterized, and molecularly cloned as a result of decades of research on host-pathogen interactions in plants. These genes include Resistance genes that recognize the pathogen and initiate the defense responses in the plant (Gebhardt et al., 2000).

Being sessile in nature, plants are constantly threatened by pathogens such as bacteria, fungi, viruses, etc. These pathogens are causing a serious concern on the economic losses of plants. The arms race between host and pathogen determines the resistance and susceptibility reactions by the host against invading pathogens. The interactions leading to successful infection in host plants caused by pathogens are called "compatible interactions." However, pathogens not causing infection even if pathogens are growing happily in a plant is called incompatibility interaction.

5.2 HOST-BACTERIAL INTERACTION

Plant pathogenic bacteria are major plant pathogens that are found worldwide (Bar-On et al., 2018). During the life time plants are being infected by various bacterial pathogens causing huge losses in terms of yield as well as quality. They colonize plants through their entry from wounds or natural openings like as hydathodes or stomata, lenticels, etc., and reside in the apoplastic region of plant tissue. There are serious economic repercussions when disease-causing plant pathogenic bacteria invade plants and cause a variety of symptoms. Pathogenic bacteria attacked plant defenses using a variety of virulence strategies (Figure 5.1), which led to a rise in their population followed by more severe symptoms on infected plants. Plants, which are sessile in nature, have developed a complex cell wall as the first line of protection against pathogenic bacteria. On the other side, bacteria must evolve to conquer plant's cell wall to gain access to water and nutrients of plants. Being sessile in nature, plants have evolved a complex cell wall, which is the 1st line of plant defense to fight against invading pathogenic bacteria. Bacteria surpass this barrier by secreting cell wall degrading enzymes and avoid it by discharging cell wall-permeable toxins. In pathogenic bacteria, there are three different types of secretion systems exist. Using the type II secretion system (T2SS), gram-negative bacteria such as *Erwinia* spp., *Xanthomonas* spp., *Pseudomonas* spp., etc., secrete cell wall-degrading enzymes involved in the degradation of different constituents of cell walls in plants (Jha et al., 2005). Type II secreted proteins are also involved in inducing defense responses in host plants (Jha et al., 2005). In many Gram-negative plant pathogenic bacteria, the type III secretion system (T3SS) is a key element in pathogenicity through which pathogenic bacteria deliver effector molecules to the host cell (Lohou et al., 2013). Type IV secretion systems (T4SS), which secrete effector molecules or protein-DNA complexes into both eukaryotic and prokaryotic cells is thought to have been the main driver of bacterial evolution (Sgro et al., 2019). *Agrobacterium tumefaciens* that injects nucleoprotein complexes and protein factors into plant cells is a typical example of a type IV secretion system. Recently, *Xanthomonas citri* was reported to inject toxin as an effector molecule using T4SS into rival bacteria (Sgro et al., 2019).

5.2.1 BACTERIAL ELICITATION

Plants have evolved pattern recognition receptors (PRRs) to recognize specific microbe-associated molecular patterns (MAMPs) or pathogen-associated

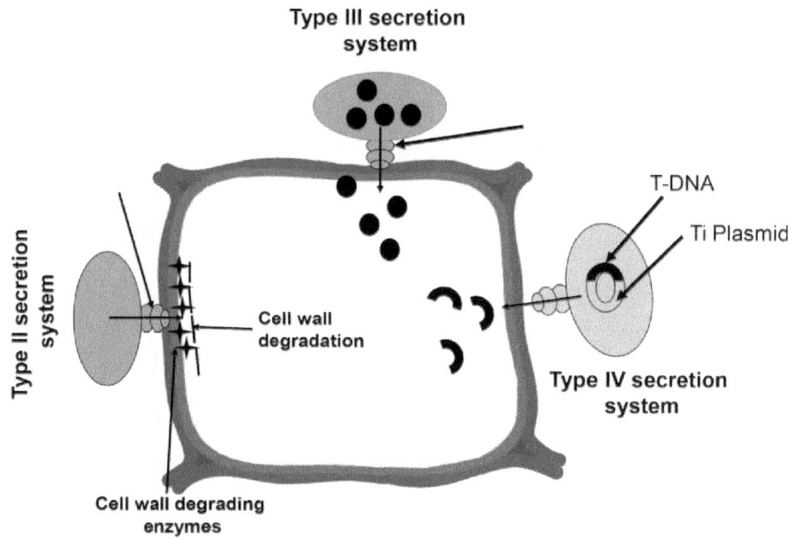

FIGURE 5.1 Virulence mechanisms of bacterial pathogens.

Source: Adapted from: Abramovitch et al. (2006); Sahni & Prasad (2020).

molecular patterns (PAMPs) (Newman et al., 2013). The recognition of a MAMP/PAMP activates numerous early, 'frontline' plant defenses to combat bacterial pathogens, which can be classified as 'basal' or 'resistant (R)-gene-mediated' defenses. Basal defenses occur early during plant–pathogen interaction in response to the perception by plant PRRs of extracellular MAMPs/PAMPs which is effective against nonspecific pathogens (Ingle et al., 2006).

5.2.2 FLAGELLIN/FLS2 PERCEPTION

Plants and other higher organisms have a highly sophisticated perception system for flagellin, a known bacterial MAMP/PAMP. A plethora of studies showed that FLAGELLIN SENSING2 (FLS2), a flagellin receptor of *Arabidopsis thaliana* and its orthologs in tomato and many other plants, is essential to trigger pattern-triggered immunity (PTI) (Robatzek et al., 2007; Mueller et al., 2010; Fürst et al., 2020) (Figure 5.2).

5.2.3 PLANT PERCEPTION OF ELONGATION FACTOR (EF)

A major obstacle to multicellular organisms' ability to defend themselves against microbial infections is their ability to distinguish between contagious

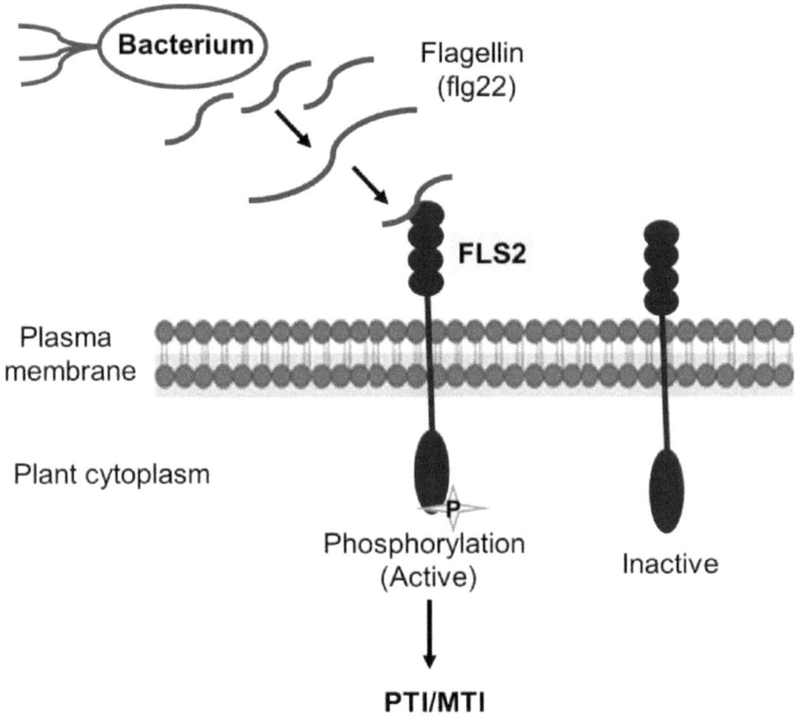

FIGURE 5.2 FLS2 recognizes the flg22 epitope, which causes phosphorylation of receptor kinases and the activation of defense responses, resulting in MAMP/PAMP-triggered immunity (PMI/MTI).

Source: Adapted from: Trdá et al. (2015); Sahni & Prasad (2021).

non-self and self (Kunze et al., 2004). However, the sophisticated perception systems of plants identify potential pathogens and trigger downstream defense signaling (Kunze et al., 2004). Elongation factor Tu (EF-Tu) is the most abundant bacterial protein. Several studies have shown that highly conserved bacterial EF-Tu acts as a PAMP in plant systems and induces defense responses (Kunze et al., 2004; Schoonbeek et al., 2015). A highly conserved N-acetylated peptide comprising of first 18 amino (elf18) is sufficient to trigger defense responses (Kunze et al., 2004). Transgenic wheat overexpressing Arabidopsis EFR confers responsiveness to elf18 and enhanced resistance to *Pst* (Kuwata, 1985). Rice, on the other hand, has been reported to identify bacterial EF-Tu via an epitope different than elf18 (Furukawa et al., 2014).

5.2.4 LYSM PERCEPTION SYSTEMS

Bacterial peptidoglycans (PGN), which are structurally related to chitin and plant symbiont-secreted lipochito-oligosaccharides (LCOs), are important constituents of bacterial cell walls and act as MAMPs or PAMPs to elicit immune responses (Newman et al., 2013). In Arabidopsis, two LysM-RLPs (AtLYM1 and AtLYM3) bind specifically to invading PGN and elicit defense responses (Gimenez-Ibanez et al., 2009; Willmann et al., 2011). Similarly, in rice, some LysM-RLKs, such as *Oryza sativa* CERK1 (*OsCERK1*) binds specifically to PGN and elicit defense responses (Liu et al., 2012; Ao et al., 2014; Kouzai et al., 2014). Additionally, *OsLYP4* or *OsLYP6* silencing impairs susceptibility to the bacterial rice blight brought on by *X. oryzae* (Liu et al., 2012). However, deletion of AtCERK1 makes Arabidopsis more vulnerable to bacterial infection brought on by Pto DC3000 (Gimenez-Ibanez et al., 2009).

5.2.5 BACTERIAL RESPONSES TO BASAL DEFENSES IN HOST

5.2.5.1 EVASION OF HOST BASAL DEFENSES

It makes sense for bacteria to change unrecognizably altered PAMPs or to completely lose them as a means of avoiding PRR-mediated detection. However, because of their high conservation and significance as bacterial traits, PAMPs appear to be the focus of PRRs. Bacteria are therefore restricted to changes that do not significantly reduce PAMP function. Subspecies-level PAMP changes were shown to be significant. The variations in flagellin proteins will intensely affect its capability to provoke *FLS2*-mediated responses in plants. The variations in flagellin allow *Xcc* to avoid PRR-mediated detection (Figure 5.3).

5.2.5.2 SUPPRESSION OF HOST BASAL DEFENSES USING EFFECTOR MOLECULES

Bacteria "inject" effector proteins into the host cell via the T3SS to counter defense responses of host. Such effectors include Avr proteins as well as those that potentially interfere with the early defense-signaling pathway. Pathogenic bacteria transfer effector proteins *via* T3SS into plant cells to defeat plant's PRR-mediated defenses. There are several effectors have been identified that suppress basal defenses in plant (e.g., AvrE1, AvrPto, AvrPtoB, AvrRpm1, HopAF1, XopX, etc.). Figure 5.4 showed the requirement of effector AvrPto for overpowering the recognition of flg22 and PAMPs.

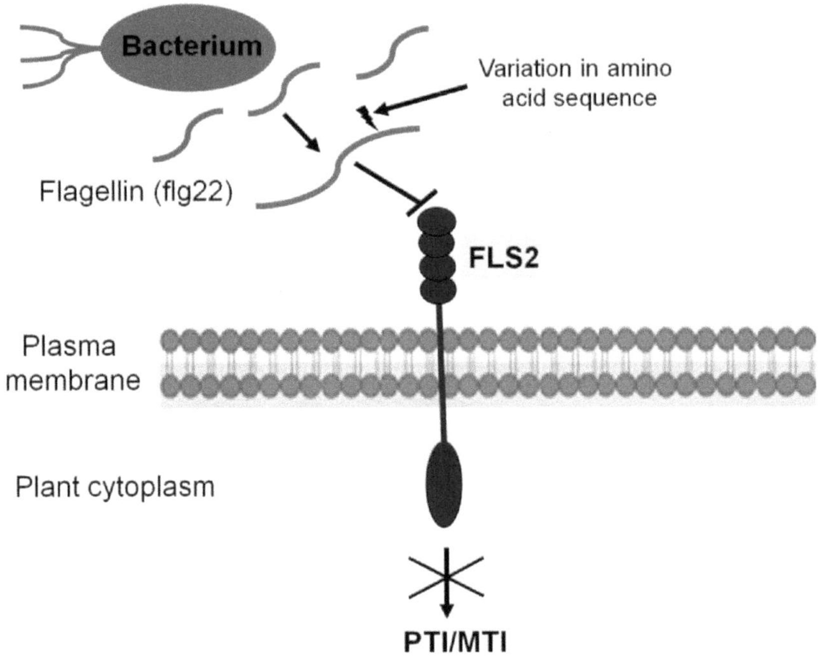

FIGURE 5.3 The alteration in the flg22 epitope to circumvent PRR-mediated detection in the host plant.

Source: Sahni & Prasad (2021).

5.2.5.3 BACTERIAL TYPE III EFFECTORS SUPPRESSES PAMP SIGNALING

The deployment of effector proteins by pathogenic bacteria via T3SS evades the plant's defenses. On the contrary, the evolution and diversification of the plant's R protein further accelerate the defense responses in plants either by direct or indirect interaction with effectors (Guard hypothesis). RIN4 from Arabidopsis is an important example of a host target of type III effectors that is "guarded" by R proteins. Two type III effectors, AvrRpt2 and AvrRpm1, have been shown to inhibit flg22-induced defenses in Arabidopsis. These type III effectors, AvrRpm1 and AvrB, interact with and induce phosphorylation of RIN4. RIN4 negatively regulates PAMP signaling. Bacterial type III effectors, AvrRpt2 and AvrRpm1 interacts with RIN4 and inhibit PAMP-induced defense responses in Arabidopsis (Figure 5.5).

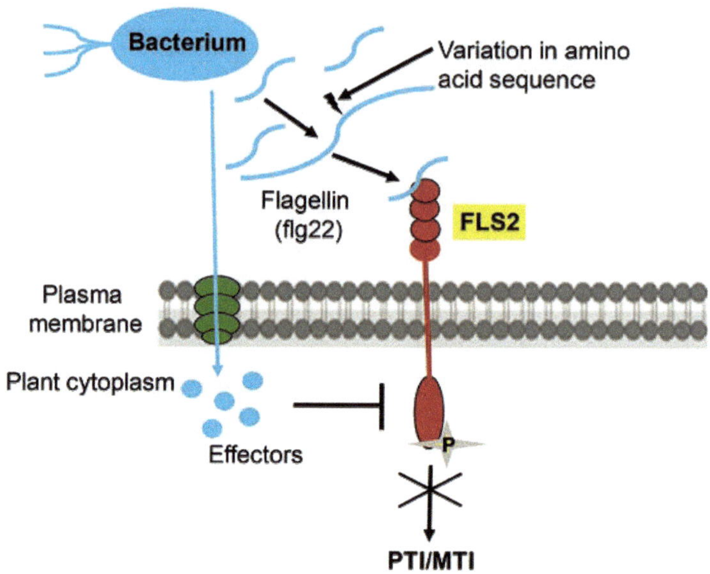

FIGURE 5.4 Suppression of host defense by bacterial effectors introduced into plant cells via the T3SS (Sahni & Prasad, 2021).

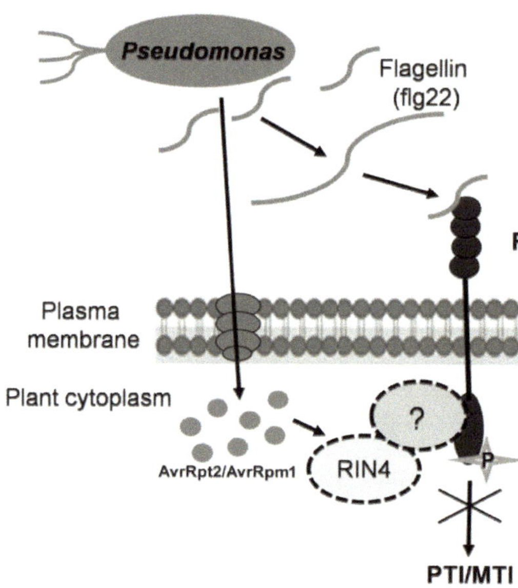

FIGURE 5.5 AvrRpt2/AvrRpm1 inhibit PAMP-induced defense responses by manipulating RIN4 and associated protein(s).

Source: Adapted from: Kim et al. (2005); Sahni & Prasad (2021).

5.3 HOST-FUNGI INTERACTION

Fungal pathogen-caused diseases pose a significant danger to crop yield and food security (Prell & Day, 2001). Phytopathogens enter the host cell either by direct penetration or through natural openings or wounds.

5.3.1 BIOTROPHIC VS. NECROTROPHIC FUNGAL PATHOGENS

Biotrophic and necrotrophic types of parasitism have traditionally been used to classify fungal plant diseases. Biotrophic infections rely entirely on the host to complete their life cycle, obtaining nutrition from living host cells through the development of specialized infection structures known as haustoria (Giraldo & Valent, 2013; Divon & Fluhr, 2006; Mendgen & Hahn, 2002). Necrotrophs, on the other hand, lack specialized infection structures and rely on sacrificed cells for nutrition (Lewis, 1973). The term 'hemibiotrophic' was coined to describe scenarios in which a pathogen renders its host mainly alive while establishing itself within the host tissue to support increasing diversification of lifestyles. It only switches to a necrotrophic lifestyle after this brief biotrophic-like phase (Divon & Fluhr, 2006; Perfect & Green, 2001). To obtain nutrients from living cells, biotrophs weaken the plant immune system and infect living plants. Necrotrophs, on the other hand, destroy the plant immediately or shortly after infection and feed on dead or dying tissues. Hemibiotrophs, like biotrophs, start by infecting the host plant before converting to necrotrophy and killing it (Giraldo & Valent, 2013; Divon & Fluhr, 2006).

Regardless of the differences in the lifestyle, all fungus that colonize plants are identified by the plant's innate immune system, which triggers a host defense response (Selin et al., 2016). The plant's innate immune sense generates both local and systemic responses, allowing it to respond to pathogen attack quickly and locally over a long period of time (Selin et al., 2016; Schwessinger & Ronald, 2012). As a result, the plant employs a two-tier innate immune response, which combines localized plasma membrane and intracellular sensors (Selin et al., 2016; Asai & Shirasu, 2015). The initial tier of the innate response in host plant is usually triggered by molecules essential to the pathogen which can't be modified without significant loss of viability (Selin et al., 2016). Numerous bacterial proteins, Ef-Tu, flagellin, and fungal cell wall components such as polysaccharides and chitin are among these compounds (Win et al., 2012; Newman et al., 2013). PAMPs and MAMPs are the terms for these components that are usually revealed (MAMPs) (Selin et al., 2016). PAMPs contained inside the plant apoplast are identified by membrane-localized PRRs after infection, triggering the first defense reaction known as PAMP-triggered immunity (PTI) (Selin et al., 2016; Koeck et al., 2011). The incorporation of pathogen's effector protein triggers ETS (effector-triggered susceptibility) in host (Sahni & Prasad, 2020). Further, the recognition of effector protein by host's R protein results in effector-triggered immunity (ETI) (Sahni & Prasad, 2020). This triggered ETI is an enlarged version of PTI that commonly causes HR (hypersensitive response), leading to disease resistance in the host cell against evading pathogens (Sahni & Prasad, 2020). Due to the natural selection pressure, pathogens modify effector proteins in order to evade ETI and cause susceptibility (Sonah et al., 2016; Sahni & Prasad, 2020). Similarly, the modifications and diversifying R protein again trigger ETI, which could be easily understandable using the "Zigzag model" (Jones & Dangl, 2006).

5.3.2 SPORE GERMINATION AND PENETRATION

Plant-pathogenic fungal interactions begin with some form of "communication" between the two. It starts with the landing of the inoculums such as spores, sclerotia, or fragments of mycelium of fungi on the plant surface and starts germinating or proliferating. The pathogen attaches itself to the host surface by producing adhesives, which are mucilaginous compounds usually made up of polysaccharides, glycoproteins, lipids, and fibrillary materials. This is followed by spore germination leading to the formation of germ tube formation. Once the proper environmental cues are received, the spore utilizes the reserved food materials, such as lipids, polyols, and carbohydrates and produces a germ tube. The germ tube senses the presence of particular signals like surface hardness, hydrophobicity, surface topography, and plant signals enabling further germ tube development and differentiation. Some fungal infections has evolved the ability to create specific infection like structures Appressoria or haustoria that allow them to penetrate the plant cuticle directly. A haustorium is formed when a specialized fungal hypha penetrates the mesophyll cells aiding in nutrient uptake from the host cells (Dean, 1997). It acts as an interface for exchange of various signals, as well as for nutrient uptake and considered to play a key role in disease establishment in host plant (Polonio et al., 2020). In rust fungus, the uredia which falls on the surface of the leaf forms a germ tube and develops an appressorial structure, but without penetrating the plant cuticle enters the host through the stomata.

The necrotrophic fungi are associated with other mechanisms like secretion of cell wall degrading enzymes and toxin which are mostly small peptides or secondary metabolites. Many fungi use PCD and the oxidative burst (OB) to fulfill their nutritional requirement (Mohanan, 2020). A hypersensitive response leading to OB is a defensive response by the host plant, whereas the induction of PDC around the affected areas in the plant restricts the growth of the pathogen (Mohanan, 2020). This mechanism of plant defense is particularly effective for biotrophs. But it has been seen that the necrotrophs has the ability to weaken this response to derive nutrients from dead host cells. Typical examples *are Botrytis cinerea* and *Sclerotinia sclerotiorum*. Hemibiotrophic fungi like *Cladosporium fulvum* produces GABA (γ-aminobutyric acid) to protect the host cells from oxidative stress thereby subverting the defense mechanism of the host and derive nutrients from it. Higher levels of these enzymes have been seen to produce in compatible interactions.

Some fungi especially those which are necrotrophic, produce toxins against the host plant. The majority of toxins produced by fungi are tiny peptides or secondary metabolites that can either be host- or non-host-specific (Mohanan, 2020). These are mainly low molecular weight molecules with varying structures. Many of these toxins elicit a PCD in the plant. Light has also shown to be an important factor of these toxins to elicit PCD (Mohanan, 2020).

5.3.3 ROLE OF EFFECTORS IN PHYTOPATHOGENICITY

Pathogens must overcome the plant's defense mechanism, specifically immunology, which is activated by pathogen components and proteins released by it, in order to successfully enter the plant (Dangl & Jones, 2001). Various cellular components of a host or the pathogen are identified by host machinery in an attempt to gain access, resulting in an immunological reaction against the infection. To get around the host's defense system, the pathogen secretes a collection of specialized proteins known as effectors (Jaswal et al., 2020). Researchers have found that those plants which are not capable of producing this hypersensitive response are resistant to these fungi. There is evidence that small molecule elicitors produced by the pathogen can induce a hypersensitive response in compatible interaction and lead to necrosis. These elicitors have been identified as pathogenicity factors in fungi like *Botrytis cinerea.*

Phytoathogens have to overcome a complex multistage host defense system for a successful infection. Chitin protects the fungi from direct host defense. But the host has also evolved efficiently to safeguard itself by producing chitinases. There are other such fungal effectors also like those with chitin binding domain as in *Cladosporium fulvum* with protect the fungus from degrading enzymes.

In the biotrophic phase immediately after penetration, in incompatible interaction fungi may express avirulence genes that prevent disease in those plants that possess a corresponding resistance gene. Examples of such avirulence genes comprise the *C. fulvum Avr4*, *Avr9*, and *Ecp2* genes, which encode small, secreted, cysteine rich proteins, *mig1* gene of *U maydis* which encode a cysteine-containing, secreted protein. *C. gloeosorioides CgDN3*, is induced by nitrogen starvation like the *C. fulvum* avirulence genes. Interestingly, *CgDN3* mutant strains elicited a localized hypersensitive-like response but, were nonpathogenic on intact leaves. Thus, CgDN3 appears to prevent a host response. Thus, the avirulence genes can be defined for their non-ability of a pathogen to infect the host, but also most of the avirulence genes are now known to be virulence factors or their subsets. There is evidence for the role of Avr proteins in infection of pathogens like *P. infestans*, *B. graminis, M. oryzae*, etc. Avr3a of *P. infestans* can inhibit necrotic response induced by its INF1 elicitor in *Nicotiana benthamiana.* This suggests its role in suppression of host resistant responses. Similarly, Avr-a10 and Avr-k1 expression enhances successful penetration of Bgh (*B. graminis* f.sp. hordei) haustoria suggesting its role in pathogen establishment. Avr4 of *C. flavum* is observed to protect the pathogen against chitinases. *Leptosphaeria maculans* AvrLm 4–7 is strongly required for virulence. Likewise, AvrPiz-t of *M. oryzae* is known to be involved in virulence processes apart from the conventional R-Avr interaction. All these instances put forward the probability of avr proteins to function as pathogen effectors as well.

Secondary metabolites that are either preformed or induced within the host system are the focus of fungal pathogen pathogenic intervention. This induces the expression of some detoxification compounds in the fungus. Instances of such mechanisms are seen in the following pathogenic fungi: (i) *Nectria haematococca* produces pisatin demethylase to detoxify phytoalexin pisatin; (ii) *Gaeumannomyces graminis* produces avenacinase to detoxify preformed triterpenoid oat root saponinavenacin A-1; and (iii) *Fusarium oxysporumf. sp. lycopersici* produces tomatinase which degrades alpha tomatine. Such genes are generally seen to lie in a gene cluster. For example, Pisatin demethylase *PDA1* is seen to lie in a gene cluster with three other genes *PEP1*, *PEP2*, and *PEP5* all of which is expressed during the pathogenic growth of the fungi. These pathogenic genes may or may not be essential for pathogenicity.

The importance of strategies to cope with toxic metabolites gained strong support from the analysis of a nonpathogenic *M. oryzae* strain harboring an insertion in the promoter of an ABC transporter. ABC1 was induced in presence of rice phytoalexin and was needed for the progression of the disease. ABC1 probably functions as an efflux pump to provide resistance to antimicrobial compounds produced by the rice plants. Another such an example are *Gloeocercospora sorghi*, as well as in *Leptosphaeria maculans*, inducible cyanide hydratases. This gene is implicated in degrading cyanide, which is released upon hydrolysis of cyanogenic compounds.

5.3.4 HOST DEFENSE SUPPRESSION AND MODULATION OF PLANT METABOLISM BY EFFECTORS

Effective pathogen establishment requires reprogramming of the host metabolism in favor of the pathogen. This objective is taken up by the effectors secreted by the pathogen. Effectors are small proteins secreted by the pathogen which are generally unique not similar to any known protein. They are expressed only during the biotrophic stage. The effector coding genes of the pathogen are expressed in response to the early plant signals. Effectors are mainly of two types – (i) those targeting host lytic enzymes to inhibit them; and (ii) those targeting plant immune response pathways. Plants secrete lytic enzymes like glucanases chitinases and proteases to check the pathogen. Effectors targeting these host lytic enzymes mostly inhibit or inactivate them. Effectors targeting plant immune system are of two types either apoplastic, which interact with apoplastic plant defense proteins and cytoplasmic effectors which enter the plant cell and interact with plant cytoplasmic proteins modulating its immune system. There are yet pathogenic effectors which have evolved to rise against the plant defense systems. For instance, during infection, plants produce chitin degrading enzymes which interact with fungal chitin and release GlcNAc oligomers which trigger plant immune response. To overcome this several pathogens, produce chitin binding effectors which either prevent the degradation of chitin to its oligomers or sequester the released oligomers to prevent the host immune response. Similarly, fungal pathogens may also produce peroxidize inhibitor protein effector to prevent oxidative burst (OB) of the pathogen caused by host ROS. Other effectors include fungal micro-RNAs which can suppress plant triggered immunity (PTI). There are yet other effectors which modulate the host metabolism to enhance certain biosynthetic pathways like aromatic amino acid biosynthesis, secondary metabolic biosynthesis like anthocyanin production, etc. All these processes together help the fungus to establish itself in the host.

5.4 HOST-VIRUS INTERACTION

Numerous viruses present a persistent threat to plants. Only a small percentage of these viruses may, however, successfully infect a specific host plant and cause illness (Sahni & Prasad, 2020). The rare occurrence of viral diseases in plant is due existence of defense mechanisms against viruses. These defense mechanisms comprises the pre-existing physical and chemical barriers, passive resistance, innate immunity and antiviral immune response (Cheng & Wang, 2017; Pallas & Garcia, 2011). The evolution and diversification of virus defense mechanisms that circumvent plant immunity and/or outwit plant defenses is a continuous process that occurs due to the co-evolutionary arms race (Sahni & Prasad, 2020; Anderson et al., 2010). Viruses and plants engage in active multifaceted defense and counter defense (Wang, 2018).

5.4.1 PROLIFERATION OF VIRUS IN PLANT

Plant viruses initiate their life cycle by penetrating into cell. A plethora of reports discussed regarding the passive entry of plant viruses and their transmission in plant tissues (Nazarov et al., 2020; Hipper et al., 2013). After entering, the host machinery is used for replication, transcription, translation, and assembly of the viral genome (Rampersad & Tennant, 2018). The DNA viruses localized to the nucleus undergo transcription, which enables them to access the cellular protein. In the cytoplasm, viral RNA is translated to create the viral proteins needed to complete the life cycle of the virus. Viral genome or virions were transported from cell to cell by plasmodesmata (Wang, 2021).

5.4.2 PLANT VIRUS SURVIVAL AND DISSEMINATION

Most plant viruses are spread from one plant to another by live creatures referred to as vectors (Whitfield et al., 2015). Arthropods that feed on plants, nematodes, fungi that parasitize plants, aphids, and whiteflies are a few of the main vectors that spread plant viruses (Whitfield et al., 2015). In vegetatively propagated plants, virus transmission occurs if mother plants are infected with virus. Transmission of viruses by a vector is a very specific process (Andret-Link & Fuchs, 2005). Infection of viruses in host plant depends upon the host range of individual virus. A virus having broad host range can infect a large number of plant families (e.g., cucumber mosaic virus), whereas narrow host range viruses only infect members of a single species and are often restricted to strains within a species (e.g., *Citrus tristeza).*

Genetic constituent of plants determines its resistance or susceptibility against viruses. The presence of active or passive defense in plants genotypes determines its resistance or susceptibility against infecting viruses. The inability of the plant to create one or more host components necessary for virus multiplication and propagation within the host is the cause of passive defense. Due to the activity of particular resistance genes in the plant, active defenses include identification and death of the virus-infected cells. Resistance genes often only function against a certain pathogen. Additionally, plants have a broad defense mechanism that resembles the immune system of animals in several ways (Sahni & Prasad, 2020).

5.4.3 POST-TRANSCRIPTIONAL GENE SILENCING (PTGS) IN PLANT

The production of double-stranded RNA (dsRNA) initiates the powerful sequence-specific RNA degradation mechanism known as PTGS (dsRNA). Viral dsRNA is primarily produced in virus-infected cells from replicative intermediates, intramolecular pairing of genomic RNA, and de novo products produced by cellular RNA-dependent RNA polymerases (RdRps) (Tenllado & Díaz-Ruíz, 2001). Cellular Dicer-like enzymes recognize and convert these viral dsRNAs into virus-derived small-interfering ribonucleic acids (vsiRNAs), which are 20–24 nucleotide (nt) RNA (Figure 5.6). These vsiRNAs are first incorporated into a multi-subunit ribonuclease complex, known as RISC (RNA-induced silencing complex). Sense strand (passenger strand) of vsiRNA gets degraded. The vsiRNAs guide the RISC to homologous RNA sequences, promoting a sequence-specific RNA degradation. One of the important components of RISC is the Argonaute (Ago) protein, which cleaves single-stranded (ss) RNA complementary to vsiRNAs into small RNA fragments. This gene inactivation process protects plants from invading viral RNA. PTGS also move systemically and protect plants from viral infection by generation of secondary vsiRNAs. vsiRNA spreads from one cell to another through plasmodesmata and over long distance via the phloem (Boualem et al., 2016).

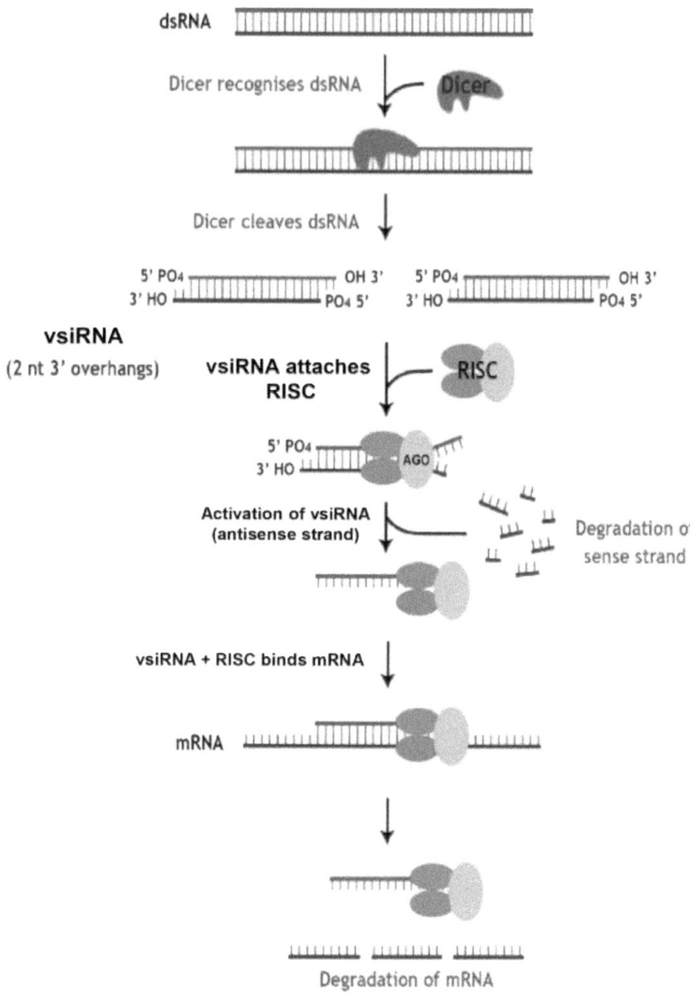

FIGURE 5.6 Overview of PTGS in plant (Sahni & Prasad, 2021).

5.4.4 SILENCING-BASED ANTIVIRAL RESPONSE SUPPRESSION AND SUPPRESSOR STRATEGIES

PTGS provide an effective measure in plant against viral infection. However, plant viruses are able to infect plant tissues and cause disease, which suggest the existence of effective counter defense strategies in viruses. This showed that viruses have evolved several mechanisms to suppress PTGS in plants. Viral suppressors of RNA silencing (VSRs) strategy are among the best studied strategies adopted by viruses to suppress PTGS in plants (Figure 5.7).

The VSRs, which are thought to be the product of recent evolutionary processes, are surprisingly diverse both within and between kingdoms and lack any clear sequence homology (Berry et al., 2009). Several VSR has been identified in plant viruses. VSRs have been divided into three categories based on how PTGS is suppressed in plants: (i) binding to dsRNA to prevent Dicer from producing vsiRNA; (ii) sequestering vsiRNA and preventing RISC assembly; (iii) cell-to-cell movement of vsiRNA; and (iv) direct targeting of effectors or processing factors that results in their inhibition or destabilization.

VSRs mediated silencing of natural PTGS in plants may leads to phenotypic variations. In soybean, seed coat color is determined by *CHS* gene, which encodes chalcone synthase (CHS) enzyme. A virus that infects soybean seeds is called the soybean mosaic virus (SMV). The mottling seed coat phenotype is brought on by VSR, which inhibits PTGS-mediated degradation of the CHS gene transcripts in soybean plants infected with SMV (Figure 5.8) (Wang, 2018; Senda et al., 2004).

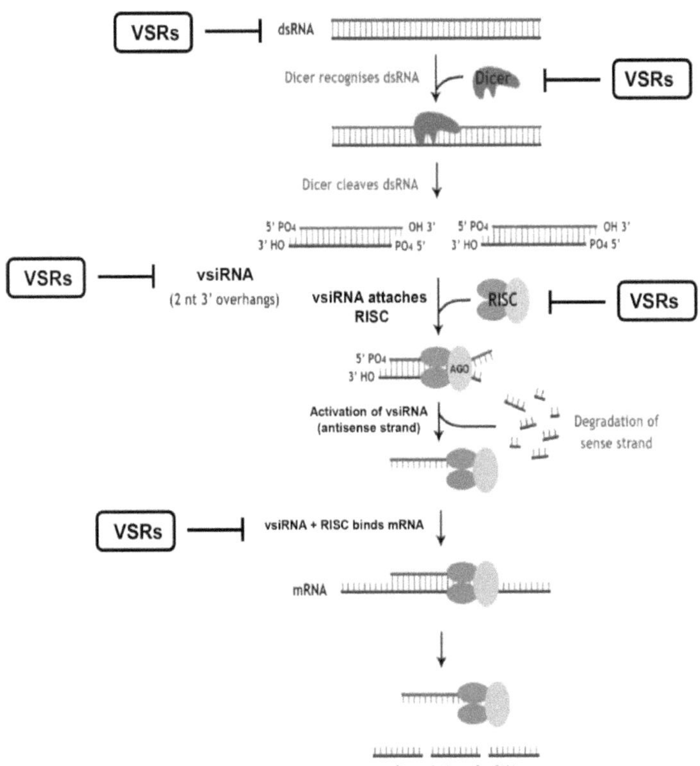

FIGURE 5.7 Suppression of PTGS in plants by VSRs (Sahni & Prasad, 2021).

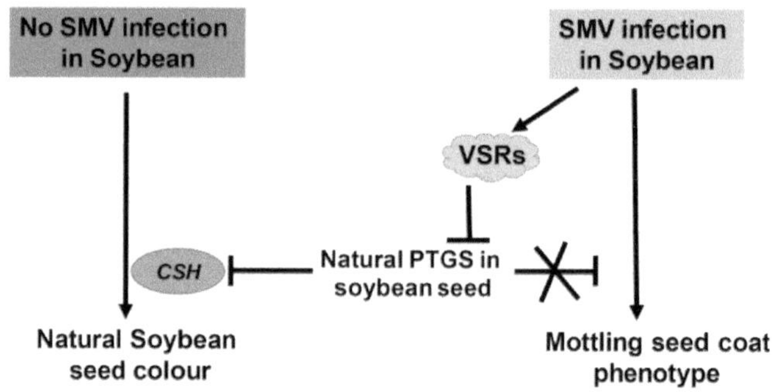

FIGURE 5.8 VSRs-mediated silencing of natural PTGS in soybean plants. The mottling phenotype of soybean seeds is observed due to suppression of natural RNA silencing-mediated degradation of the *CHS* gene transcripts by the SMV VSR (Sahni & Prasad, 2021).

5.4.5 VIRAL IMMUNITY AND COUNTER-IMMUNITY

Studies on how plants interact with nonviral pathogens led to the development of the zigzag model of plant immunity. Studies have shown that the plant immune system's ability to limit viral infection depends on both PTI and effector-triggered immunity (ETI). It has been demonstrated that the elicitor that causes PTI is viral dsRNA. Viruses also encode particular effector(s) to disrupt PTI as a counter-PTI mechanism. For instance, the plum pox virus's (PPV) coat protein (CP) works as an effector to stifle the PTI and promote PPV infection. Plants utilize PTGS mechanism to counter viral infection. VSRs counteract plant PTGS to overcome plant defense. Then plant antiviral R protein interact VSRs and cause effector-triggered immunity (ETI) (Figure 5.9). The majority of the known antiviral R proteins

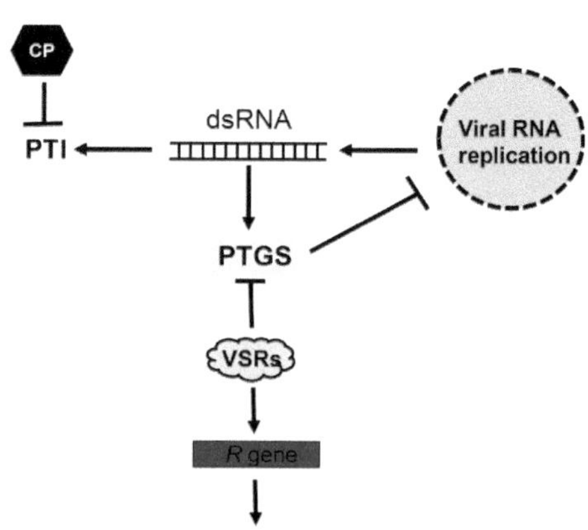

FIGURE 5.9 Viral immunity and counterimmunity.

Source: Wang (2018); Sahni & Prasad (2021).

contain nucleotide-binding and leucine-rich repeat (NB-LRR) domains. They share structural similarity to the antifungal and antibacterial R proteins.

KEYWORDS

- **effector-triggered immunity**
- **host defense suppression**
- **host-pathogen interaction**
- **microbe-associated molecular pattern**
- **PAMP-triggered immunity**
- **pathogen-associated molecular pattern**
- **post transcriptional silencing**

REFERENCES

Anderson, J. P., Gleason, C. A., Foley, R. C., Thrall, P. H., Burdon, J. B., & Singh, K. B., (2010). Plants versus pathogens: An evolutionary arms race. *Funct. Plant Biol., 37*(6), 499–512.

Andret-Link, P., & Fuchs, M., (2005). Transmission specificity of plant viruses by vectors. *Journal of Plant Pathology, 87*(3), 153–165.

Berry, B., Deddouche, S., Kirschner, D., Imler, J. L., & Antoniewski, C., (2009). Viral suppressors of RNA silencing hinder exogenous and endogenous small RNA pathways in *Drosophila. PLoS One, 4*(6), e5866.

Hipper, C., Brault, V., Ziegler-Graff, V., & Revers, F. (2012). Viral and cellular factors involved in Phloem transport of plant viruses. *Front. Plant Sci., 4*, 154.

Jha, G., Rajeshwari, R., & Sonti, R. V., (2005). Bacterial type two secretion system secreted proteins: Double-edged swords for plant pathogens. *Mol. Plant Microbe. Interact., 18*(9), 891–898.

Liu, B., Li, J. F., Ao, Y., et al., (2012). Lysin motif-containing proteins LYP4 and LYP6 play dual roles in peptidoglycan and chitin perception in rice innate immunity. *Plant Cell, 24*, 3406–3419.

Mendgen, K., & Hahn, M., (2002). Plant infection and the establishment of fungal biotrophy. *Trends in Plant Science, 7*(8), 352–356.

Mohanan, V. C., (2020). Molecular basis of host-pathogen (fungi) interaction. *Host-Pathogen Interaction: A Molecular Approach* (pp. 9–19). Todays & Tomorrow's Printers and Publishers.

Nazarov, P. A., Baleev, D. N., Ivanova, M. I., Sokolova, L. M., & Karakozova, M. V., (2020). Infectious plant diseases: Etiology, current status, problems and prospects in plant protection. *Acta Nature, 12*(3), 46.

Newman, M. A., Sundelin, T., Jon, T., Nielsen, J. T., & Erbs, G., (2013). MAMP (microbe-associated molecular pattern) triggered immunity in plants. *Front. Plant Sci., 4*, 139.

Pallas, V., & Garcia, J. A., (2011). How do plant viruses induce disease? Interactions and interference with host components. *Journal of General Virology, 92*, 2691–2705.

Perfect, E. S., & Green, J. R., (2001). Infection structures of biotrophic and hemibiotrophic fungal plant pathogens. *Mol. Plant Pathol., 2*, 101–108.

Polonio, Á., Pérez-García, A., Martínez-Cruz, J., Fernández-Ortuño, D., & De Vicente, A., (2020). The haustorium of phytopathogenic fungi: A short overview of a specialized cell of obligate biotrophic plant parasites. In: Cánovas, F. M., Lüttge, U., Risueño, M. C., & Pretzsch, H., (eds.), *Progress in Botany* (Vol. 82).

Prell & Day, (2001). Plant – fungal pathogen interaction: A classical and molecular. *Mycopathologia 154*(1), 55, 56.

Rampersad, S., & Tennant, P., (2018). Replication and expression strategies of viruses. *Viruses, 2018*, 55–82.

Richards, S., & Laura, E. R., (2019). The evolutionary history of LysM-RLKs (LYKs/LYRs) in wild tomatoes. *Richards and Rose BMC Evolutionary Biology, 19*, 141.

Sahni, S., & Prasad, B. D., (2020). Molecular basis of host-pathogen (bacteria) interaction. *Host-Pathogen Interaction: A Molecular Approach* (pp. 21–99). Todays & Tomorrow's Printers and Publishers.

Schoonbeek, H. J., Wang, H. H., Stefanato, F. L., Craze, M., Bowden, S., Wallington, E., Zipfel, C., & Ridout, C. J., (2015). Arabidopsis EF-Tu receptor enhances bacterial disease resistance in transgenic wheat. *New Phytol., 206*(2), 606–613.

Schwessinger, B., & Ronald, P. C., (2012). Plant innate immunity: Perception of conserved microbial signatures. *Annu. Rev. Plant Biol., 63*, 451–482.

Selin, C., de Kievit, T. R., Belmonte, M. F., & Fernando, W. G. D., (2016). Elucidating the Role of Effectors in Plant-Fungal Interactions: Progress and Challenges. *Front. Microbiol. 7*, 600. doi: 10.3389/fmicb.2016.00600.

Senda, M., Masuta, C., Ohnishi, S., et al., (2004). Patterning of virus-infected glycine max seed coat is associated with suppression of endogenous silencing of chalcone synthase genes. *Plant Cell 16*, 807–818.

Sgro, G. G., Oka, G. U., Souza, D. P., Cenens, W., Bayer-Santos, E., Matsuyama, B. Y., Bueno, N. F., et al., (2019). Bacteria-killing type IV secretion systems. *Front. Microbiol., 10*, 1078.

Sonah, H., Deshmukh, R. K., & Bélanger, R. R., (2016). Computational prediction of effector proteins in fungi: Opportunities and challenges. *Front. Plant Sci., 7*, 126.

Tenllado, F., & Díaz-Ruíz, J. R., (2001). Double-stranded RNA-mediated interference with plant virus infection. *J. Virol., 75*(24), 12288–12297.

Trdá, L., Boutrot, F., Claverie, J., Brulé, D., Dorey, S., & Poinssot, B., (2015). Perception of pathogenic or beneficial bacteria and their evasion of host immunity: Pattern recognition receptors in the frontline. *Front. Plant Sci., 6*, 219.

Wang, A., (2018). *Virus and Host Plant Interactions*. In: eLS. John Wiley & Sons, Ltd: Chichester.

Wang, A., (2021). Cell-to-cell movement of plant viruses via plasmodesmata: A current perspective on potyviruses. *Curr. Opin. Virol., 48*, 10–16.

Whitfield, A. W., Falk, B. W., & Rotenberga, D., (2015). Insect vector-mediated transmission of plant viruses. *Virology, 479, 480*, 278–289.

Willmann, R., Lajunen, H. M., & Erbs, G., (2011). Arabidopsis lysin-motif proteins LYM1 LYM3 CERK1 mediate bacterial peptidoglycan sensing and immunity to bacterial infection. *PNAS 108*(49), 19824–19829.

Willmann, R., Lajunen, H., Erbs, G., Newman, M., Kolb, D., Tsuda, K., et al., (2011). Arabidopsis lysin-motif proteins LYM1 LYM3 CERK1 mediate bacterial peptidoglycan sensing and immunity to bacterial infection. *Proc. Natl. Acad. Sci. U.S.A., 108*, 19824–19829.

Win, J., Chaparro-Garcia, A., Belhaj, K., Saunders, D. G., Yoshida, K., Dong, S., et al., (2012). Effector biology of plant-associated organisms: Concepts and perspectives. *Cold Spring Harb. Symp. Quant. Biol., 77*, 235–247.

CHAPTER 6

Recent Advancement in Tissue Culture Techniques for Crop Disease Management

RIMA KUMARI,[1] PANKAJ KUMAR,[2] SHISHU PAL SINGH,[3] and NEELAMBARI[4]

[1]*Division of Plant Biotechnology, College of Agricultural Biotechnology, Bihar Agricultural University, Sabour, 813210 Bihar, India*

[2]*Department of Molecular Biology & Genetic Engineering, Bihar Agricultural University, Sabour, 813210 Bihar, India*

[3]*Department of Soil Science and Agricultural Chemistry, Institute of Agricultural Sciences, BHU, Varanasi, Uttar Pradesh, India*

[4]*Department of Botany, Purnea University, Purnia, Bihar, India*

ABSTRACT

Plant diseases are a threat to world agriculture and general food security. Significant yield losses due to the attack of pathogens occur in most of the agricultural and horticultural crop species. Many conventional approaches such as biological control, chemical control, cultural control, and physical control are used for crop disease management. Furthermore, several biotechnological tools and approaches have been employed to overcome the hazardous effects of disease and pests in crop plants. But the majority of plant diseases can be successfully managed by tissue culture techniques that are products of biotechnology. Different *in vitro* techniques like shoot regeneration from cultured tissue by organogenesis and somatic embryogenesis for micropropagation, embryo rescue, somaclonal variation, eradication

New Advances in Crop Disease Management. Sangita Sahni, Ganesh Thapa, and Bishun Deo Prasad (Eds)

techniques, and gene transfer by somatic hybridization and transformation have been applied in crops for disease management. Thus, taking all the above into consideration, the present chapter is written to represent the recent advancements in tissue culture techniques for crop disease management.

6.1 INTRODUCTION

The primary danger to agricultural productivity is the emergence of new plant pathogens. Fungi, bacteria, viruses, and insect pests are the most common pathogens to inflict economic damage around the world. Pathogens and pests still destroy 20–30% of crops annually, despite widespread use of pesticides (Estruch et al., 1997; Peferoen, 1997). Crop protection technology is of the utmost importance in today's modern agriculture. Conventional crop disease management techniques include biological control, chemical control, cultural control, and physical control. Pesticide resistance has become a major issue, necessitating a method that is both focused and strategic in its search. Biotechnological solutions may be the only way to find a long-term solution to this pressing issue. Additionally, crop plants have been protected from disease and pests using a variety of biotechnological instruments and methodologies. It is now possible to create disease-resistant plants for future use because of advances in biotechnology techniques. There is a lot of potential in the development of transgenics, notably for improved resistance to a wide range of diseases and pests. Resistance to specific illnesses can be conferred by identifying and subsequently introducing tolerance genes from microbes or naturally resistant plant species into the susceptible host. It is possible to treat the majority of plant diseases successfully using biotechnological tissue culture techniques. As a result, tissue culture has become the primary platform for all genetic alterations in plant systems. The approach has helped change investigations leading to transgenics from simple callus induction to root and shoot regeneration. Furthermore, plant tissue culture techniques are based on totipotency (the ability to grow and regenerate) and the utilization of explants' nutritional condition. Conventional approaches, on the other hand, have a few disadvantages. Tissue culture has found new uses in the production of haploids, somaclonal variation, somatic hybrids, genetically engineered plants, and micropropagation, among other things. Disease-free plants might be produced using micropropagation and eradication procedures in the early stages of the development of these new tissue culture techniques. As a result, in vitro treatments are the most effective means of eradicating the virus from many crops. Morel and Martini proposed in 1952 that it is possible to extract the apical meristem of systemically infected plants and grow them in vitro to produce virus-free plants after the repeated subcultures of tomato roots in vitro by White (1934) and following work by Limasset & Cornuet (1949). Thus, this chapter is created to showcase recent advances in tissue culture techniques for agricultural disease control.

6.2 TISSUE CULTURE TECHNIQUES FOR CROP DISEASE MANAGEMENT

Almost all tissue culture techniques are used in plant pathology. Some of the important tissue culture techniques and their importance to plant pathology are briefly described.

6.2.1 SOMACLONAL VARIATION FOR CROP DISEASE MANAGEMENT

Regeneration of disease-resistant horticultural crops *in vitro* has been made possible by recent advances in tissue culture techniques (Krishna et al., 2016). Cellular variation is referred to as somaclonal variation in this context. This means that the genetics of plants regenerated from somatic cells utilizing tissue culture aren't exactly the same. When compared to the rate of naturally occurring mutation, this level of variation is astronomically high. Many plant species, including wheat, rice, oats, maize, tobacco, potatoes, sugarcane, and brassica, have shown somaclonal variation for numerous properties, such as resistance to fungus, virus, and bacterial illness. Cell cultures are grown for numerous cycles on nutritional medium without the addition of any selection agent, and then plants are regenerated.

Several plant species (Anil et al., 2018) have developed stable somaclonal variants, which are a means of creating variation and expanding the germplasm pool. Somaclonal variation is a simple alternative method that helps breeders quickly increase genetic variety in crops with low genetic variability or those are difficult to breed (Thakur et al., 2002; Jalaja et al., 2006; Penna et al., 2012). Many cultivars have been successfully released with improved features such as plant architecture, disease resistance, yield, appearance, and abiotic stress tolerance thanks to the somaclonal variety breeding strategy. There are several varieties of crops that have been improved in terms of quality and yield, such as Yidan No. 6 maize, CIMAP/ bio-13 aromatic grass, rice variant DAMA, and sugarcane cultivar 'Ono.'

'Ono' is a somaclonal variant that was created from the susceptible cultivar 'Pindar.' Sweet potato 'Scarlet,' a somaclonal variation cultivar, has a darker and more persistent skin color than the parent cultivar but has similar yields and disease resistance. During vegetative propagation, St. Augustine grass somaclonal variants demonstrated durable, beneficial modifications. For example, *in vitro* selection of desirable features has been used commercially to develop new horticulture varieties with somaclonal variation (2016).

The biotechnology strategy used to develop commercial variants in India has been somaclonal variation. Bio-13, a somaclonal variation of Citronella java (a medicinal plant), was just issued by CIMAP (Central Institute for Medicinal and Aromatic Plants) in Lucknow. In comparison to control types, this SV, bio-13, has 37% more oil and 39% more citronellol. 'Pusa Jai Kisan,' a somaclonal variation of the *B. juncea* variety 'Varuna,' has been released for commercial production. With bolder seeds and a higher yield, the new variety outperforms Varuna in every way. In 2002, Gupta & colleagues at CIMAP in India created an improved somaclonal version of Rose-scented Geranium. Immature embryos of two spring wheat types, HUW-206 and HUW-234, were used by Arun & collaborators (2003) to develop somaclonal variants (R2, R3, and R4 generations). Over the source types, these somaclonal modifications showed superior characteristics, including as resistance to spot blotch disease and increased yields. Sucrose content and disease resistance have been improved in a somaclonal sugarcane (CoC 671), Co 94012, which was launched as Phule Savitri in Maharashtra.

6.2.2 MICROPROPAGATION FOR CROP DISEASE MANAGEMENT

Material that has been infected by viruses and other systemic pathogens can be removed via propagation from meristematic tissue. For quick and true to type multiplication of plants on artificial nutrient media, micropropagation is a tissue culture (*in vitro*) approach. Vegetative propagated species benefit greatly from the widespread use of micropropagation for the creation of high-quality planting material. Micro-propagation's most significant advantages are the ability to rapidly produce large numbers of disease-free propagules from a single plant, propagation year-round and the ability to accommodate propagating material in a small space, reduction of labor costs for germplasm maintenance, avoidance of field inspections and environmental hazards, easy availability of material for micro-propagation, and rapid propagation.

6.2.3 ERADICATION TECHNIQUES

Almost all crops are infected with one or more viruses, which can be passed on to the next generation by asexual propagules. In the case of strawberry plants, this mandates the replacement of mother plants every year due to a variety of viruses and mycoplasmas. Seeds from virus-infected plants, on the other hand, are usually free of the virus. Aside from new infections, sexual progeny tends to be virus-free. This sense of relief, on the other hand, isn't quite accurate. Seed-transmitting viruses have been identified in a growing number of viruses as they are studied in greater detail. Viral infections have decimated many valuable clones of potato (*Solanum tuberosum*), sugarcane (*Saccharum officinarum*), and other clonal crops because they reproduce asexually. However, it should be noted that virus-free or extremely low-titer shoot apical meristems and some surrounding immature tissues tend to grow with distance away from the apical meristem. Because there is no vascular bundle or plasmodesmata link in the shoot apical meristem, there is no virus in the shoot apical meristem, and high amounts of phytohormones and metabolic activity are to blame for this. It is therefore useful in the recovery of virus-free plants from infected plants or clones using meristem culture.

Holmes was able to propagate virus-free Dahlia plants from virus-infected mother plants in 1948 via shoot-tip cuttings, thanks to his groundbreaking work. From virus-infected plants of dahlia in 1952, Morel & Martin extracted 100-micrometer-long stems and cultivated them *in vitro* to generate shoot-free shoots that failed to root. To obtain virus-free plants, they grafted the *in vitro* produced shoots onto healthy Dahlia root stocks. Many viruses have been eliminated from ornamentals, fruit crops, vegetables, and other food crops using tissue culture techniques since then. Therefore, three strategies have been found to be effective in eradicating and cultivating disease-free plants. In the first place, the apical meristem or the root tips of infected plants can be isolated and isolated. Use of anti-metabolites before bud excision or introduction into the nutritional medium are the other two options. Lastly, a temperature treatment (37°C) is applied to affected plants in order to slow or stop the spread of viruses. So, for example, several strategies such as meristem and shoot tip culture, thermotherapy, and chemotherapy combined with meristem and cryotherapy, as well as meristem and cryotherapy combined with plantlet regeneration from callus cultures, are employed in the fight against the virus. Heat-treated apical meristems have been found to be superior and even more effective at eliminating viruses than meristem culture alone (Quack, 1977).

6.2.4 MERISTEM CULTURE FOR CROP DISEASE MANAGEMENT

An infected germplasm can be eliminated by using meristem culture. Fast-growing meristems of plants have long been recognized to have a lower concentration of viruses compared to non-meristems. Meristem culture has been used to create virus-free plants in this situation. Cassava, potato, sweet potato, and ornamental plants are all popular places to find this chemical. The yield and quality of crops might be affected by the presence of viruses in the plants. The apical meristems of infected plants are usually free of the virus or have a very low quantity of it. They've come up with five basic theories to explain why viruses can't infect meristems:

- A lack of vascular or plasmodesmatal connections in the meristems prevents the virus from infecting them;
- The meristem cells' competition for important metabolites;
- The breakdown of the virus by the synthesis of chemicals in meristem cells;
- A lack of certain essential viral replication machinery components; and
- Presence of replication-inhibiting substances.

In order to produce pathogen-free plantlets through meristem tip culture, Tegen & Mohammed (2016) state that various steps include testing parent material for the presence of viruses and similar pathogens (viroids and phytoplasmas), thermotherapy/chemotherapy of parent material if disease-free material is not available, excision of meristem tip under aseptic conditions, and culture of apical dome plus one or two leaf primordia on suitable medium (Tegen & Mohammed, 2016). *In vitro* mass propagation of the virus-free plants is performed after indexing of plantlets for viral presence or absence, soil transfer, maintenance of pathogen free nuclear plant stocks, and meristem culture.

According to Al-Taleb et al. (2011), meristem culture, which has been successfully employed in potato, is a powerful and successful strategy for virus removal from plants that are infected. The Ministry of Agriculture provided them with 70 potato tubers (Solanum tuberosum L.), which they tested for virus infection using an enzyme-linked immunosorbent assay (ELISA). Spunta, Alaska, and Safrane cultivars had a PVY infection rate of 21.4%, 15.7%, and 12.8%, respectively. All three varieties of potato had their diseased tubers treated with meristem culture. Indole butyric acid (IBA) supplemented medium was shown to be the best for shoots and roots development, with a mean length of 7.71 cm, a root length of 9.41 cm, the number of shoots with a mean of 2.60, and the number of leaves with a mean of 2.60 (15.40). *In vitro*-grown plantlets examined by ELISA were shown to be devoid of PVY. The coat protein (CP) gene of PVY virus was successfully amplified by reverse transcription polymerase chain reaction (RT-PCR) in the infected samples and confirmed by ELISA. Around 90% of the Spunta cultivars and 80% of the Alaska and Safrane cultivars were effective in acclimatizing the plantlets developed from the *in vitro* shoot and root multiplication stage.

6.2.5 SHOOT TIP CULTURE

Shoot tip culture has been proposed as an alternative to the more traditional method of propagation, which involves planting tree seeds. Using shoot tip culture, an infection of seeds due to a variety of causes that could impede the propagation of healthy plants can be remedied (Roy et al., 1996). Genetic diversity and heterozygosity that spreads through the seeds would be eliminated by this procedure as well. A shoot tip culture has more advantages than a seed since it may produce a greater number of shoots than a seed. Cellular interactions between host cells and pathogens can be better understood using *in vitro* approaches (Takebe & Otsuki, 1974). All of the pathogens that cause disease have been successfully isolated into single cell clones. These clones have allowed us to better understand the interaction between host and virus thanks to their culture. The life cycle and pathogenesis of several pathogens have been studied in detail using tissue culture.

6.2.6 THERMOTHERAPY

Even the apical meristem of diseased plants has been found to be contaminated with tobacco mosaic virus (TMV), carnation mottle virus (CarMV), potato virus-X, cucumber mosaic virus (CMV), etc. (Hollings & Stone, 1968) and simple meristem-culture of such infected plant does not help to get disease-free plants. To combat the virus, it appears that thermotherapy-the treatment of affected parts/plants with heat-has been successful (Wang et al., 2018). Plants free of disease have been grown using a combination of thermotherapy and apical meristem culture in these situations. The mother plant is heated to a high temperature before the explants are removed. This method has worked well with a variety of potato cultivars (MacDonald, 1973;

Slack et al., 1987). The degree of sterilization is determined by the frequency and intensity of the heat treatment (being free from viruses). After an 8-week heat treatment, the percentage of virus-free plants in the white rose potato variety rose to 100% after 18 weeks. Shoot-tip culture isn't always necessary but heat-treated axillary buds have produced disease-free plants in some instances (Stace-Smith & Mellor, 1989). For 28–40 days, daily treatments of donor shoots were demonstrated to be efficient in eliminating grapevine fan leaf virus (GFLV), but not Arabis mosaic virus, when performed at 39°C and 22°C for six hours each day (Monette, 1986). *In vitro* virus-free plants of Chrysanthemum morifolium can be grown by exposing shoot tips to a temperature of –50°C for one month, according to Paludan (1985). Cold treatments between 5°C and 15°C have previously been used to remove potato spindle tuber virus from potatoes (Lizarraga et al., 1980).

6.2.7 CHEMOTHERAPY

To eradicate the virus from the plants, a wide variety of chemical groups were employed. Amino acids, antibiotics, growth regulators, purine, and pyrimidine analogs, and other such substances are among them. Although some progress has been made in eradicating, inhibiting, or inactivating the viruses, there is still much work to be done. In spite of a chemical treatment of the entire plant, the virus remained in a few plants (Tomlinson, 1982). When it comes to getting rid of a wide range of viruses, Virazol (l,2,4-triazole-3-corboxamide, 9-B-D-ribofuranosyl) has proven the most effective option (Shepard, 1977). Only a few viruses have been successfully treated with chemotherapy.

6.2.8 PLANTLET REGENERATION FROM CALLUS CULTURES

Plants grown from virus-infected Calli may actually be virus-free, according to multiple studies. Some of the cells in these Calli have been shown to be devoid of viral infection. There are a number of probable explanations for this: It is possible that the virus multiplies more slowly than the callus and that certain cells become virus resistant through mutation. Plants generated from these Calli were found to be virus-free in multiple experiments. This is consistent with the findings of previous research. Around 50% of the virus-free plants regenerated from the 1 mm explants of TMV-infected green tobacco leaves were one study's finding. Because of the potential of genetic variation during the callus phase, this method has not been preferred.

6.3 VIRUS-INDEXING

Meristem culture plants, regardless of whether they were grown with or without chemo or thermotherapy, are not free of viruses. As a result, these plants must undergo virus indexing, which is the process of testing for the existence of the virus(es). The following are a few of the more popular virus indexing techniques: Plants can be ranked based on the presence or absence of certain viral symptoms. However, obvious symptoms may take a long time to develop, and many viruses only show symptoms much later in plants grown from tissue culture. A bioassay test, also known as an infectivity test, involves injecting extremely sensitive and healthy indicator plants with sap from test plants. Indicator plants are those kinds or varieties of plants which are extremely vulnerable to the virus and quickly show symptoms of infection. The serological test is a highly sensitive and exact method of virus indexing that makes use of antibodies unique to the virus in question. ELISA may be the most convenient, quick, and efficient test system when dealing with a large number of samples. Antibodies against viral CPs are used in the ELISA test. ELISA cannot be used for viroids and viruses that have lost their CPs, and for infections that involve numerous related luteoviruses, such as potato leaf roll virus (PLRV), since the antiserum generated against the primary virus may not react with all the viruses. For viral identification, DNA and RNA probes can be employed, however they are time-consuming and cannot be applied on a wide scale like ELISA.

6.4 PROTOPLAST FUSION FOR CROP DISEASE MANAGEMENT

Closely related species or distantly related species can provide disease resistance in breeding programs. If an attempt is made to cross distantly related species, problems are almost always experienced. Protoplast fusion is a technique that can be utilized to avoid difficulties in the introduction of resistance genes into plants. Factors that contribute to species crossover may be avoided with this procedure, and even between distantly related species, viable hybrids (cybrids) can be retrieved (Harms, 2018; Tiwari, 2018).

During incubation, isolated protoplasts become sticky, collect in suspension, and spontaneously fuse. To increase the fusion rate, chemicals can be added to the mixture. In the presence of high Ca^{2+} and high pH (9–10) fusion can occur, although poly-ethylene glycol (Fusogen) is a regularly utilized chemical (Fusogen) (PEG). Protoplasts adhere to each other because to the presence of PEG, which can be seen under a microscope. The cytoplasm merges and mixes when stabilized PEG is diluted further, whether sequentially or all at once. Viruses can be inoculated into plant protoplasts in order to research their replication and physiological functions. To deliver foreign genetic material into plant cells, *Agrobacterium tumefaciens* or its modified Ti-plasmid and the cauliflower mosaic virus (CaMV) have been utilized.

6.5 ROLE OF BIOTECHNOLOGY AND GENETIC ENGINEERING IN PLANT DISEASE CONTROL

Transgenic plants can be created by using tissue culture and rDNA technology to insert foreign genes. Viral RNA must be stripped of its CP before it can be successfully infected. Naked viral RNA synthesis is practically nonexistent if the host is pushed to manufacture huge amounts of CPs. The virus's RNA will be encapsulated and prevented from replicating by the host CP. This will reduce and delay the onset of symptoms. Example: TMV CP expressed in transgenic plants shielded plants from this virus. Transgenic plants that produce viral genomes have a higher level of resistance to viral infection. Antisense RNA is used to CPs in these areas, which include a portion of the viral replication. Following inoculation with CMV, transgenic tobacco plants transformed with a DNA copy of the satellite RNA of CMV were found to produce high amounts of satellite RNA and symptoms were greatly decreased. Resveratrol (phytoalexin) is produced in transgenic tobacco plants when the groundnut stilbene synthase gene is expressed constitutively. The transgenic plants are resistant to *Botrytis cinerea*. *Bacillus subtilis* pv. tabaci is resistant to *Pseudomonas syringae* acetyltransferase-expressing transgenic tobacco plants that detoxify tabtoxin. Tobacco horn worm *Manduca sexta* chitinase gene was recently cloned into *P. fluorescens* to boost their antagonistic capability against *R. solani*.

Modern plant genetic engineering technologies for the control of bacterial illness are now available to the general plant population. Plant diseases can be controlled by tissue culture and genetic engineering, whereas biotechnology's function has been examined in depth (Fagwalawa et al., 2013; Mandahar & Khurana, 1998). Through rapid clonal multiplication, plant biotechnology impacts or aids plant pathology in numerous ways:

- To obtain pathogen-free mother plants through rapid clonal propagation.
- Genetically modified (GM) plants that have been created from seed.
- Plant pathogens, such as Agrobacterium tumefaciens and the CaMV, use this method as the primary means of transmitting genes from donor to recipient.

- Resistance genes are inserted into plants using genetic engineering techniques to control plant disease.
- Genetic engineering has already contributed significantly to the study of disease resistance genes in plants and pathogen virulence genes in pathogens.

6.6 CONCLUSION

Several plant tissue culture techniques and approaches have been emerging in recent years, which are becoming very effective in crop disease management. Among these successful tissue culture techniques, micropropagation, somaclonal variations, and eradication techniques were widely adopted by many countries in the early stages for producing disease-free plants. Later on, with the advent of new approaches of biotechnology and genetic engineering in disease management, it has become simpler and quicker in application. These technologies and approaches are economical, eco-friendly, and require less time in development compared to conventional breeding techniques. Thus, it is finally concluded that recent advancements in plant tissue culture techniques have offered remarkable and enormous opportunities to present-day plant breeders and pathologists to work in a consolidated and consorted way to overcome the issues posed by the new dynamics of disease and pest problems.

KEYWORDS

- **chemotherapy**
- **crop**
- **crop disease management**
- **diseases management**
- **eradication techniques**
- **genetic engineering**
- **protoplast fusion**
- **shoot tip culture**
- **thermotherapy**

REFERENCES

Al-Taleb, M. M., Hassawi, D. S., & Abu-Romman, S. M., (2011). Production of virus free potato plants using meristem culture from cultivars grown under Jordanian environment. *American-Eurasian J. Agric. & Environ. Sci.*, *11*(4), 467–472.

Anil, V. S., Bennur, S., & Lobo, S., (2018). Somaclonal variations for crop improvement: Selection for disease resistant variants *in vitro*. *Plant Science Today*, *5*(2), 44–54. https://dx.doi.org/10.14719/pst.2018.5.2.382.

Arun, B. A., Joshi, K., Chand, R., & Singh, B. D., (2003). Wheat somaclonal variants showing earliness, improved spot blotch resistance and higher yield. *Euphytica.*, *132*, 235–241. https://doi:10.1023/A:1025097224408.

Estruch, J. J., Carozzi, N. B., Desai, N., Duck, N. B., Warren, G. W., & Koziel, M. G., (1997). Transgenic plants: An emerging approach to pest control. *Nature Biotechnol.*, *15*, 137–141.

Fagwalawa, L., Kutama, A., & Yakasai, M., (2013). Current issues in plant disease control: Biotechnology and plant disease. *Bayero Journal of Pure and Applied Sciences*, *6*, 121–126.

Gupta, R., Banerjee, S., Mallavarapu, G. R., Sharma, S., Khanuja, S. P. S., Shasany, A. K., & Kumar, S., (2002). Development of a superior somaclone of rose-scented geranium and a protocol for inducing variants. *Hort Science*, *37*, 632–636.

Harms, C. T., (2018). Hybridization by somatic cell fusion. *Plant Protoplasts*. CRC Press.

Hollings, M., & Stone, O. M., (1968). Investigation of carnation viruses, I, carnation mottle. *Atm. Appl. Biol.*, *53*, 103–118.

Jalaja, N. C., Sreenivasan, T. V., Pawar, S. M., Bhoi, P. G., & Garker, R. M., (2006). co 94012– a new sugarcane variety through somaclonal variation. *Sugar Tech.*, 132–136. https://doi.org/10.1007/BF02943647.

Krishna, H., Alizadeh, M., Singh, D., Singh, U., Chauhan, N., Eftekhari, M., & Sadh, R. K., (2016). Somaclonal variations and their applications in horticultural crops improvement. *Biotech.*, *6*(1), 54.

Limasset, E. M., & Comuet, P., (1949). Organic growth factor requirements of tobacco tissue cultures. *Physiol. Plant*, *18*, 100–127.

Lizarraga, R. E., Salazar, L. F., Roca, W. M., & Schilde, R. L., (1980). Elimination of potato spindle tuber viroid by low temperature and meristem culture. *Phytopathology*, *70*, 754–755.

MacDonald, D. M., (1973). Heat treatment and meristem culture as a means of freeing potato varieties from virus X and S. *Potato Res.*, *16*, 263–269.

Mandahar, C., & Khurana, S., (1998). Role of biotechnology in controlling plant diseases. In: Khurana, S. M. P., (ed.), *Pathological Problems of Economic Crop Plants and Their Management* (637–648). Scientific Publishers Jodhpur (India).

Monette, P. L., (1986). Elimination *in vitro* of two grapevine *Nepoviruses* by an alternating temperature regime. *J. Phytopathol.*, *116*, 88, 89.

Morel, G., & Martin, C., (1952). Guerison de dahlias atteints d'une maladie a virus. *Compt. Rend. Acad. Sci. Ser. D.*, *235*, 1324–1325.

Paludan, N., (1985). Inactivation of viroids in chrysanthemum by low-temperature treatment and meristem-tip culture. *Acta Hortic.*, *164*, 181–186.

Peferoen, M., (1997). Progress and prospects for field use of Bt genes in crops. *Trends Biotechnol.*, *15*, 173–177.

Penna, S., Vitthal, S. B., & Yadav, P. V., (2012). *In vitro* mutagenesis and selection in plant cultures and their prospects for crop improvement. *Bioremediation, Biodiversity and Bioavailability*, *6*, 6–14.

Quack, F., (1977). Merstem culture and virus –free plants. In: Reinert, J., & Bajaj, Y. P. S., (eds.), *Plant Cell, Tissue, and Organ Culture* (pp. 598–615). Springer-Verlag, Berlin.

Roy, S. K., Islam, M. S., Sen, J., & Hadiuzzaman, S., (1996). Shoot tip an alternative to seed as a source of genetic material for conservation and propagation of forest trees. In: Islam, A. S., (ed.), *Plant Tissue Culture* (pp. 8–15). Oxford & mH Publishing Co. Pvt. Ltd., New Delhi.

Shepard, I. F., (1977). Regeneration of plants from protoplasts of potato virus X-infected tobacco leaves. *Virology*, *78*, 261–266.

Slack, S. A., Petrson, M. D., Rott, M., & Quinones, S. H., (1987). Effect of heat and chemical therapy treatments on *Potato virus S* elimination in Desiree. *Am. Potato J.*, *64*, 457.

Stace-Smith, R., & Mellor, F. C., (1989). Eradication of potato viruses X and S by thermotherapy and axillary bud culture. *Phytopathology*, *58*, 199–203.

Takebe, I., & Otsuki, Y., (1974). Multiple infection of tobacco mesophyll protoplasts by plant viruses. In: *3rd Intl. Congr. Plant Tiss. Cell Cult.* Leicester Univ., Leicester, Abstr. no 36.

Tegen, H., & Mohammed, W., (2016). The role of plant tissue culture to supply disease free planting materials of major horticultural crops in Ethiopia. *Journal of Biology, Agriculture and Healthcare*, *6*, 122–129.

Thakur, M., Sharma, D., & Sharma, S., (2002). *In vitro* selection and regeneration of carnation (*Dianthus cayophyllus* L.) plants resistant to culture filtrate of *Fusarium oxysporum* f. sp. dianthi. *Plant Cell Rep.*, *20*, 825–828. https://doi.org/10.1007/s00299-001-0412-1.

Tiwari, D. N., (2018). Current status and future prospective of use of biotechnology in plant disease management. *International Journal of Agricultural and Environmental Research*, *4*(4), 238–245.

Tomlinson, I. A., (1982). Chemotherapy of plant viruses and virus disease. In: Harris, K. F., & Maramorsch, K., (eds.), *Pathogens, Vectors and Plant Diseases: Approaches to Control* (pp. 23–44). Academic Press N.Y.

Wang, M. R., Cui, Z. H., Li, J. W., Hao, X. Y., Zhao, L., & Wang, Q. C., (2018). *In vitro* thermotherapy-based methods for plant virus eradication. *Plant Methods.*, *14*, 87. https://doi.org/10.1186/s13007-018-0355-y.

White, R. P., (1934). Multiplication of the viruses of tobacco and aucuba mosaic in growing excised tomato roots. *Phytopathology*, *24*, 1003–1011.

CHAPTER 7

Biotechnological Approaches for Bacterial Disease Management in Solanaceous Vegetable Crops

RAJESH KUMAR,[1] LOMASH SHARMA,[1] BAHADUR SINGH BAMANIYA,[1] and JITENDRA KUMAR KUSHWAH[2]

[1]*Department of Horticulture, School of Life Sciences, Sikkim University, Tadong, Gangtok, East Sikkim, Sikkim, India*

[2]*Department of Agriculture, Government of Uttar Pradesh, India*

ABSTRACT

Bacterial diseases are considered as major production constraints in different solanaceous vegetable crops such as potato, tomato, brinjal, pepper (sweet and hot). Among them, bacterial wilt, bacterial canker, bacterial spot and bacterial speck are the most dangerous diseases of solanaceous vegetables. They cause substantial yield losses throughout the world. Bacterial wilt alone causes 35–90% yield loss in tomato production in hot-humid tropical regions, between 50% and 100% yield loss in potato and 11.67 to 96.67% yield loss in brinjal, whereas bacterial canker causes 46 to 93% of tomato plant death. Over the past few decades, molecular diagnostics and management have become a viable alternative to traditional methodologies for infectious diseases and a valuable weapon in the battle against antibiotic resistance towards the diseases. Identification of activated hypersensitivity-related genes related to hypersensitive response and *in-vivo* immunogold localization in conjunction with isolation and purification of an anti-pathogen protein imparts the inhibitory activity to bacterial pathogens. Transfer of NPR1 protein involves the genes responsible for systemic acquired resistance (SAR) shows higher level of resistance towards bacterial diseases. Several

New Advances in Crop Disease Management. Sangita Sahni, Ganesh Thapa, and Bishun Deo Prasad (Eds)

genome editing techniques like TALENs, ZFNs, and CRISPR/Cas9 are the most effective tools to acquire broad spectrum resistance against different bacterial pathogens of solanaceous vegetable crops.

7.1 INTRODUCTION

Vegetables are fresh, tender, and edible portions of herbaceous plants, key constituents of a healthy diet for people. They are a significant source of dietary fibers, mineral components, phyto-vitamins, and potential antioxidants. Consumption of different vegetables regularly in sufficient quantity helps people maintain their health and also protects against the damages caused by free radicals. Besides, inadequate consumption of vegetables often leads to non-communicable diseases, including the deadliest cancer and cardiovascular diseases. Dietary fibers from green leafy vegetables reduce the danger of obesity, diabetes mellitus, and metabolic syndromes (Mursu et al., 2014). Many researchers have reported that consumption of vegetables protects people against different kinds of cancer (Hung et al., 2004; Wiseman, 2008). Solanaceae (Nightshade family) is considered one of the important families of flowering plants, including 75 genera with up to 2,000 species (Rubatzky & Yamaguchi, 1997). It includes different commercial fruit vegetables such as tomato (Solanum lycopersicum L.), brinjal (*Solanum melongena* L.), chili (*Capsicum annuum* L.), bell pepper (*Capsicum annuum* var. *grossum* L.), and one commercial tuber vegetable named potato (*Solanum tuberosum* L.). Solanaceous vegetable crops, including tomato, can be grown throughout the globe in a varied range of soil and climate, but as warm-season crops, the most favorable range of temperature for their record yield is 20°C to 24°C (Debnath et al., 2020). In India, tomato is the 3rd most important vegetable crop next to potato and onion with respect to area and production. The major phyto-constituents present in tomato are carotenoids, including 60–64% lycopene, 10–12% phytoene, 7–9% neurosporene, and 10–15% carotenes (Clinton, 1998). Quercetin and kaempferol are the important flavonoids in fresh tomato. Besides, tomato fruits are also an outstanding source of ascorbic acid, with 200 mg per kg of edible portion, and considered a major source of vitamin C next to citrus (Rao & Rao, 2007). Brinjal or eggplant is considered one of the most important, popular, and principal solanaceous vegetables (Sharma et al., 2020). It contains caffeic and chlorogenic acid as phenolic compounds and nasunin or delphinidin 3-(coumaroyl-rutinoside)-5-glucoside as flavonoids, predominantly found in the peel of eggplant (Noda et al., 2000). Another dominant phenolic compound found in brinjal fruit is chlorogenic acid, which acts as an anti-cancer, anti-microbial, anti-low-density lipoproteins, and antiviral activities (Matsuzoe et al., 1999). Peppers are considered one of the most important vegetables cum spices in the Indian subcontinent. Fresh peppers are remarkably an excellent source of ascorbic acid, vitamin K, flavonoids, and carotenoids (Bosland, 1996). Chilies are an important source of minerals like potassium, manganese, magnesium, and iron, where potassium is the major component of cell and body fluids that regulates the heart rate and blood pressure (Skrabal et al., 1981). Potato is an important staple food and considered the most chosen vegetable crop by growers in the world because of its capacity to provide more food energy per hectare on a per-day basis. There is an appreciable amount of phenolic acids, carotenoids, and anthocyanins reported in the skin and inner portion of potato (Ezekiel et al., 2013). Many researchers have demonstrated that these phytochemicals are correlated with antioxidants and actively involved in the prevention of atherosclerosis and cancer-like deadly diseases (McGill et al., 2013).

Bacterial infections cause significant productivity losses in crop plants all over the world. The major limiting factors of solanaceous vegetables in India are bacterial wilt, bacterial canker, bacterial spot, and bacterial speck (Singh et al., 2011). Bacteria are single-celled prokaryotic microorganisms with a definite nucleus, reproduce asexually through binary fission. Bacteria are categorized into two primary groups based on cell wall structure, which can be detected by using the gram staining process (Williams et al., 2017). Gram-positive bacteria stain purple, while gram-negative bacteria stain red or pink. The chemical content and structure of their cell walls are directly related to their color differences. Bacteria may be present in different shapes of the cells, i.e., rod-shaped, spherical, spiral-shaped, or filamentous. The majority of bacteria are motile with whip-like flagella that propel them through water films (Williams et al., 2017). Bacteria enter the plants through wounds or natural openings rather than directly into plant tissues, these openings may be caused by insects' attack, infections, and equipment during farm activities such as training and pruning (Williams et al., 2017). Bacteria become active and create infections only when conditions are favorable for their growth and multiplication. High humidity, dense planting, improper aeration, and stress caused by irregular irrigation, unhygienic soil, and nutritional imbalance are some of the conducive conditions that may lead to bacterial infection (AUSVEG, 2017). The pathogenic bacteria may survive in soil, crop residue, and plant propagules including seeds. Weeds may serve as collateral hosts and act as reservoirs for bacterial diseases. Bacteria may also spread the infections by diseased seed, propagules, and crop leftovers, as well as by water stream and contaminated farm equipment, and also agricultural workers' hands (AUSVEG, 2017). Some bacterial infections are encouraged by warm and humid conditions, whereas others are favored by cool and wet weather. Dry and hot conditions may arrest the disease development; symptoms might be increased if the plant is already infected (e.g., Bacterial wilt). Symptoms of bacterial infections are often confused with those of fungal diseases. To determine the type of pathogen causing the illness, it is critical to have diseased tissue analyzed at a plant diagnostics laboratory (AUSVEG, 2017).

In the last few decades, molecular diagnostics and management tools have become a reliable alternative to traditional techniques for the management of infectious diseases and a valuable weapon to combat antibiotic and chemical resistance towards the diseases. Diversity among individuals or population scan as well as determining the inheritance pattern of selective traits based upon DNA sequence polymorphism are not influenced by a set of environmental conditions (Kumar et al., 2018). Biotechnological techniques are important because of their rapid, less laborious, more sensitive, specific, and efficient nature, as compared to conventional management tools. Here, we are trying to give an overview of available major research work conducted in the recent past regarding new biotechnological techniques to manage the reviewed bacterial diseases of valuable solanaceous vegetable crops.

7.2 BACTERIAL WILT

Bacterial wilt is one of the most disastrous diseases of vegetable crops such as potato, chili, and tomato (Genin & Denny, 2012), reported to affect more than 450 species of 54 families. Among all crops, solanaceous vegetables are the most susceptible hosts (Lebeau et al., 2011; Kurabachew et al., 2017). The disease is caused by *Ralstonia solanacearum* bacteria, which was formerly known as *Pseudomonas solanacearum* (Cabi, 1999). This disease commonly found in wet tropical region, sub-tropical region and also in some temperate zones in different parts of the world (Boshou, 2005). *Ralstonia solanacearum*, is one of the dangerous phyto-pathogenic bacteria (Mansfield et al., 2012). It has been observed that bacterial wilt causes huge economic losses in solanaceous vegetable crops and tobacco as well (Elphinstone, 2005). The characteristic feature for bacterial wilt includes, initial wilting of upper leaves followed by wilting of entire plant including aerial parts (Kumar et al., 2021). *Ralstonia solanacearum* is a soil borne bacterium which first penetrates the roots of the plant and later invades the xylem tissues, then spreads further to the various aerial plant parts through the vascular bundles where its faster multiplication leads to the wilt and finally death of the plant (Genin, 2010). The spread of this bacterium is considered as a serious hazard to the crops and the pathogen is rated as a quarantine bacterium (Huet, 2014). The bacterium ability to remain dormant in the soil for several years and form latent infection within the collateral hosts makes the eradication of bacteria even more difficult (Hayward, 1991; Wenneker et al., 1999).

Pathogen is classified into five races based upon its ability to infect different plant species; it is also divided into six biovars based upon its ability to oxidize alcohols, hexoses, and sorbitol along with the disaccharides (Xue et al., 2013; Chandrashekara et al., 2012). Races 1 and 3 of *R. solanacearum* infect plant species belonging to the Solanaceae family, with race 1 being predominant in subtropical and tropical regions and race 3 is being restricted to temperate zones (Kim-Lee et al., 2005). These bacteria are classified into four phylotypes based on their geographical diversity: Phylotype I (Asia), Phylotype II (America), Phylotype III (Africa), and Phylotype IV (Indonesia). Phylotype II has been presented with two subgroups, IIA and IIB (Fegan & Prior, 2005), and only phylotype IIB strains are responsible for bacterial wilt of potatoes in temperate zones (Janse et al., 2004). In potato, tomato, peppers, and eggplant, all phylotypes are capable of causing disease (Janse et al., 2004; Cellier & Prior, 2010). The most significant difference between bacterial and fungal wilt is the presence of bacterial ooze on an exposed cut surface of plant organs such as roots, stems, flowers, fruits, and other vegetative parts in bacterial wilt, whereas such exudation is unlikely to be observed in fungal wilt. Apart from that, bacterial wilt develops from younger to older leaves; however reverse trend is followed in fungal wilts. Keeping the importance of bacterial wilts in mind, it becomes a pre-requisite factor to gain knowledge about the differences between different kinds of wilts occurring in solanaceous crops for accurate diagnosis and to formulate the management practices accordingly.

7.2.1 IMPACT OF BACTERIAL WILT ON SOLANACEOUS VEGETABLES

Ralstonia solanacearum is positioned as the second most disastrous pathogen amongst the 10 most lethal bacteria influencing the economic yield of crop plants (Mansfield et al., 2012). Significant loss of yield in different solanaceous vegetable crops are observed, with 88% yield loss in tomato observed in Uganda, and 70% yield loss of potato in India and considerable losses in different countries (Katafiire et al., 2005). Bacterial wilt is also accounted to influence 50–100% of yield loss of potatoes in Kenya (Muthoni et al., 2012). Bacterial wilt frequency is practically 55% on tomato, 63% on potato and 100% on pepper in Ethiopia (Assefa et al., 2015). Most wilted plants don't produce attractive or marketable tuber in potato decreases its consumer acceptance and considerable economic loss to the growers (Kurabachew et al., 2017). However, there is insufficient data on economic impact of the pathogen on solanaceous vegetable crops yield across the globe, considerable losses of roughly 75% in tomato and potato is reported because of their vulnerability to bacterial wilt (Elphinstone, 2005; Hayward, 1995). The disease is spreading at faster rate in light of the fact that agriculture is presently reaching across the nations, where vulnerable crops have not been grown earlier.

7.2.2 NOTABLE SYMPTOMS

In early stages, the first symptoms are typically seen on the leaves of plants. The symptoms may be observed a few days after infection and are recognized by younger leaves wilting at terminal end of the branches. These symptoms occur in the hottest part of the day (Elphinstone, 2005); however, it may recover if the temperature is cooled down especially in the night hours. The favorable temperature for symptom expression is high (85–95°F), which progresses immediately after infection (Ji et al., 2005). Entire plant may wilt and dry under the unfavorable conditions, however dried leaves may stay green, but plant dies in the end (Lambert, 2002). Stunting of the plants is the common symptom of bacterial wilt in the field which may appear at any growth stage of plant, it is not uncommon for the healthy plants that sudden wilting occurs when fruits are expanded rapidly (Jones et al., 1991). In young stems, vascular bundles are affected with visible appearance like dark brown streaks. However, in highly vulnerable cultivars, the young tender shoot may fall down (Pradhanang et al., 2005). It is observed that under favorable conditions, the plant which does not show symptoms may infected latently for longer period of time. After the infection, pathogen may continue to exist in infected plant and may be spread further (Gitaitis et al., 1992).

7.2.3 DISPERSAL OF PATHOGEN

Dispersal of the pathogen occurs through various methods; nevertheless, environmental factors like moisture and temperature substantially affect disease advancement and development by fungi and bacteria (Lopes & Rossato, 2018). The bacterium *Ralstonia solanacearum* may spread significantly at various locations through vegetative propagules, remaining due for around two to three years of survival within vegetative plant parts as an essential inoculum source (Coutinho, 2005). Weeds and contaminated soil, water, farm tools, waste of processing industry as well as latently infected plant parts, like tubers of potato and seeds of other solanaceous vegetables are at high risk to this lethal pathogen (van Elsas et al., 2001). Apart from it, contaminated crop residues in fields, also act as inoculums source in the adjoining region (Wang & Lin, 2005). Besides, several insects act as vectors that normally spread race 3 of *R. solanacearum* (Tahat & Sijam, 2010; Tomlinson et al., 2009). Thus, broad appropriation and long saprophytic survival of the pathogen makes the control more troublesome.

7.3 BACTERIAL SPOT

Doidge published the detailed account of the bacterial spot disease of tomato first time in 1920. Bacterium was with its rod-shape, aerobic, and gram-negative nature, with a single polar flagellum. *Bacterium vesicatorium* was the first name of the bacterium from the fact that vesicles grow in host cells during pathogenesis. Gardner & Kendrick (1921) also discovered the disease at the same time; however, they named causal organism as *B. exitiosum*. The term *B. exitiosum* was dropped as soon as it was realized that the organism was the same as reported by Doidge (1920) and her publication was at the priority. Gardner & Kendrick (1923) elaborated their first findings by demonstrating that *B. vesicatorium* may also cause a leaf spot on chili pepper (*Capsicum annuum*), however, at the same time, Higgins (1922) gave the detailed account on it. Cumulative efforts by several scientists discovered the fact that the bacterial pathogens utilized in the earlier investigations were pathogenic on both pepper and tomato, and for several years, it was assumed that cross infections may occur in the field. Although in the early 1940s, Burkholder & Li (1941) identified some host specificity of pepper and tomato strains. Cook (1973) revealed in the early 1970s that host specificity was correlated to hypersensitive resistance (HR) (Burkholder et al.,

1941; Canteros et al., 1991; Cook, 1973). The bacterial spot term has been used to characterize a specific set of symptoms on tomatoes and peppers, not to clearly designate the nomenclature of the causal organism. Currently, bacterial spot symptoms are caused by at least four taxonomically different *xanthomonads*, including *Xanthomonas vesicatoria*, *Xanthomonas euvesicatoria*, *Xanthomonas gardneri*, and *Xanthomonas perforans*. Jones et al. (2004) published the nomenclature and justification of the taxonomy of pathogens cause bacterial spot.

7.3.1 HOST RANGE AND RACE STRUCTURE

Bacterial spot affects a wide range of Solanaceous plants, but primarily tomato, cherry tomato, currant tomato, pepper, bird peppers, *C. anomalum, C. baccatum*, *C. chinensis* are being largely affected. Other solanaceous hosts from different genera also have been proposed (Dye et al., 1964; Elliott, 1951); however, one investigation has shown that some of these other genera are not considered as hosts for the bacterium (Laub & Stall, 1967). Plant resistance was recognized in pepper and tomato genotypes, which led to the identification of several pathogenic races on the basis of differential reactions (Potnis et al., 2015). Stall et al. (2009) identified four tomato races and 11 pepper races so far. Different pepper and tomato races like *X. vesicatoria*, *X. euvesicatori*, and *X. perforans* (tomato only) have been extensively reported among the four species (Horvath et al., 2012; Jones et al., 1995; Kebede et al., 2014). *Xanthomonas gardneri* has also been linked to pepper and tomato bacterial spots in different parts of the globe. The host range of species has been recently expanded as *X. perforans* strain was isolated from a pepper field in Florida in 2010. The strains of *X. vesicatoria*, *X. euvesicatoria*, and *X. gardneri* are identified, pathogenic exclusively either on tomato or on pepper, sometimes on both.

7.3.2 NOTABLE SYMPTOMS

The disease symptoms on tomato and pepper are identical, and lesions will emerge on all plant parts above the ground, however minor differences may also observe in lesions by different *xanthomonads*, including shot-hole symptoms on tomatoes caused by *Xanthomonas perforans*, against water-soaked lesions caused by *Xanthomonas gardneri*. Tomato and pepper's bacterial spot are predominant across the globe (Jones et al., 1998) and moreover arise concern in tropical and subtropical zones (Cox, 1966). Necrotic lesions on leaves, stem, flowers, and fruits of tomato and pepper are the characteristic feature of bacterial spot (Jones et al., 1991). Circular water-soaked lesions may appear during the early stages of disease development, subsequently dry and turn dark brown or black with a moist to greasy surface (Vallad & Goodman, 2004). When rainy weather exists and dew is prominent, the lesions may have a water-soaked appearance, which later on results in primary lesions coalesce, leads in severe necrosis with blighted appearance. When conditions favor rapid growth of bacteria, lesions of *X. perforans* on the foliage, often cause a shot-hole appearance and raised scabby lesions are developed on the surface of the fruit (Stall et al., 2009).

7.3.3 EPIDEMIOLOGY

Xanthomonads, like many other plant pathogenic bacteria, can survive between the crops on volunteer pepper or tomato plants or both, as well as for short periods in agricultural residues in the soil, which could be a source of inoculums for volunteer plants. After decomposition of the crop residues, generally the bacteria will no longer survive in the soil. Depending on the environmental conditions, decomposition may take a few weeks to the several months, and slow decomposition rates may lead to the inoculums for future crops (Jones et al., 1986). Although, *xanthomonads* have been found to survive epiphytically on weeds, but this appears to be of minor significance as a source of primary inoculums. The bacteria have been revealed to persist as an epiphyte in latent infections on host leaves also (McGuire et al., 1991), and furthermore to survive under these unfavorable circumstances (Timmer et al., 1987). Symptoms of bacterial spot *xanthomonads* on transplants may or may not be noticeable. Early research also indicated that bacterial spot infections may be transmitted through seeds (Gardner & Kendrick, 1923).

7.4 BACTERIAL CANKER

A serious threat to tomato cultivation is the bacterial canker of tomato, caused by *Clavibacter michiganensis* ssp. *michiganensis* (*Cmm*), caused terrible epidemics in the several cultivation pockets of tomato in the world (Blank et al., 2016; Kleitman et al., 2008; Smith, 1910; Volcani, 1985). In the year 1949, the disease was first discovered in Nagpur in India (Jain, 1951). The pathogen *C. michiganensis* subsp. *michiganensis* was reported to occur with an incidence up to 48% in Karnataka state (Umesha, 2006). In recent years, bacterial canker is also reported in hilly tomato-growing areas such as Himachal Pradesh's Solan and Sirmour districts (Singh et al., 2015). *Cmm*, the only recognized species of the genus *Clavibacter* is a gram-positive actinomycete and also non-motile. It has been categorized into a number of subspecies which cause disastrous diseases in agricultural crops (Davis et al., 1984), although dDDH (digital DNA–DNA hybridization) and ANI (average nucleotide identity) suggest that the subspecies meets with all essential criteria for a separate species (Tambong, 2017). *Cmm*-infected plants may exhibit a variety of symptoms depending on the host plant's age, cultivar vulnerability, and virulence of the pathogen, as well as certain environmental factors such as humidity and temperature (Gleason et al., 1993; De Leon et al., 2011). When plants infected at seeds or young seedlings stages, they obtain systemic infections that impact fruit quality and yield, and often lead to plant death. Older plants, on the other hand, are more likely to get secondary infections or foliar infections, which cause leaf chlorosis but may or may not interfere with crop quality and yield (Nandi et al., 2018). Plants with a systemic infection are first asymptomatic; symptoms arise after 6–8 weeks when the bacterium grows at a rapid rate (Gitaitis et al., 1991). *Cmm* infiltrates and proliferates in xylem vessels, causing internal vascular browning and tissue degradation. This interferes with water transport, causing wilt during initial stages of infection (Eichenlaub & Gartemann, 2011). Tomato plants infected in later stages of growth may be asymptomatic or show a delayed wilt (Gitaitis et al., 1991; Sharabani et al., 2013). They may produce marketable fruits, but they also serve as a source of infection for the next growing season (Sharabani et al., 2013).

7.4.1 TRANSMISSION

Low-quality infected and diseased seeds are primarily responsible for *Cmm's* long-distance transmission and spread; they may introduce the disease to the region where it was never been earlier (Fatmi et al., 1991). Genetic diversity of *Cmm* isolates from narrow geographical ranges suggested that the introduction of unidentified or foreign strains is not uncommon (Kleitman et al., 2008; Milijasevic-Marcic et al., 2012; Tancos et al., 2015; Thapa et al., 2017; Wassermann et al., 2017), because of their adaptability to the new ecosystems as a habitat (Jacques et al., 2012). *Cmm* infects the seed coat and endosperm when it enters through the vascular system of an infected female parent (Fatmi & Schaad, 1988) or through external contact with a bacterial source. *Cmm* also infect the other vegetative parts through wounds or natural openings, including stomata and hydathodes (Carlton et al., 1998; Tancos et al., 2013), allows the disease to spread sideways via dripping water or contaminated horticultural equipments (Carlton et al., 1998; Chang et al., 1991; Gitaitis et al., 1991; Ricker & Riedel, 1993; Xu et al., 2010).

7.5 MANAGEMENT THROUGH BIOTECHNOLOGICAL APPROACHES

A basic analysis of host-pathogen interaction by the utilization of classical and molecular approaches may efficiently manage these devastating bacterial diseases. For example, isolated genes may be placed on genetic maps in a population that segregates for susceptibility and resistance; these genes are involved in hypersensitivity reactions or resistance to pathogen (Prior et al., 1998). The use of biotechnological tools for the management of bacterial diseases with special reference to the solanaceous vegetables is explained here in different heads with some model examples.

7.5.1 LINKAGE OF GENETIC MARKERS TO STRAIN SPECIFIC RESISTANCE

Genetic analysis using molecular markers for resistance towards *R. solanacearum* has been undertaken so far and well documented in number of solanaceous vegetables; despite the modest degree of variability at DNA level available in the cultivars, it acquires more popularity in tomato due to its better genetic architecture than other crop species (Grimsley & Wang, 1998). Both general resistance and strain-specific resistance have been identified among the resistance loci. The utilization of highly polymorphic interspecific crosses permits for more extensive and detailed genetic analysis through multi-location testing of the RILs (recombinant inbred lines). DNA markers tightly linked to the resistance loci allow for molecular investigation of the resistance genes involved in longer turn, finally help in resistance breeding program.

7.5.2 IDENTIFICATION OF ACTIVATED HYPERSENSITIVITY-RELATED GENES IN RESPONSE TO BACTERIAL DISEASES

Any mechanism involves rapid, immediate, and localized cell death at the site of pathogen infection and therefore inhibits the pathogen dissemination and further spread, comes under hypersensitive response or HR (Klement et al., 1963), is considered as most efficient reaction against any plant pathogen. The fundamental reaction between the host and pathogen is cell death, which is mediated by microbial avirulence genes (*avr*) and plant resistance genes (R) (Flor, 1971; Godiard et al., 1994). Cell death is a common physiological response in plants that occurs to protect them from viral infections during the development (Vaux, 1993). This is ascertained through a genetic mechanism that has remained unchanged throughout the evolution process (Pontier et al., 1998). The study of "disease lesion mimics," mutants that acquired HR lesions spontaneously, provide evidence for a mechanism in plants (Greenberg et al., 1994; Dietrich et al., 1994). Certain genes, encoded by the host plant genome and controlled by the signals related to pathogen, are involved in cell death. Generally, these genes are active during the early stages of host pathogen interactions. *Hsr203J*, a tobacco gene with a similar activation profile, has been identified (Pontier et al., 1994). Its promoter is active, in the response to an HR-inducing bacterial isolate and is highly dependent on the pathogenic bacterial wilt's *hrp* (hypersensitive response and pathogenicity) genes. *Hsr203J* is suitable for identifying HR responsive *cis*-regulatory elements and characterizing the corresponding transacting factors. It was discovered that a 28-bp is required for transcriptional activation (Pontier et al., 1994), and it is involved in the creation of an HR inducible DNA protein complex. *hsr203J* was also functionally identified through expression in *E. coli*, where the gene product possessed esterase activity for the expression. Using sense and antisense techniques, the biological role of *HSR203* was determined, and the results suggest that this protein functions as a negative regulator of the HR (Pontier et al., 1994). *Hsr203J* acted to be a molecular marker for the hypersensitive response to *Ralstonia solanacearum* in tobacco, allowing for the identification of a regulatory element responsible for this specific activation. The characterization of this element facilitates the isolation of genes that may encode trans-acting factors: the characterization of the specific genes that encode HMG proteins. *HSR203's* enzymatic identification as an esterase suggests either a role in cell death control or a role in cell death establishment, as previously demonstrated for ICE family proteases in animal apoptosis or for endonucleases induced upon hypersensitive response in *Nicotiana* (Mittler et al., 1995). The genetic evidence is being established for an active role in the HR, indicating a regulatory role for this gene product. Similarly, in tomato HR resistance has been reported in for multiple races of *Xanthomonas*. *S. lycopersicum* accession Hawaii7998 (HI7998) is an important source of HR against Race 1 of *Xanthomonas* (Whalen et al., 1993). In *HI 7998*, the effect or *avrRxv*in T1 strain confer HR by interacting with the resistance proteins. Multiple non-dominant and independent genes influence HR resistance (Wang et al., 1994), as three genes *Rx1, Rx2*, and *Rx3* were identified in the Back cross 1 (BC_1) population resulting from the *HI 7998* and *LA 716* (*L. pennellii*) crosses (Yu et al., 1995). *Xanthomonas gardneri* causes bacterial spot in tomatoes through the *avrHah1* effector, which triggers *Bs3*-dependent HR in chili peppers (Schornack et al., 2008). HR has also been shown to confer resistance to *X. gardneri* in tomato under controlled environment and at multiple loci in the field condition. HR was observed in *Solanum pimpinellifolium LA1936, LA2533*, and *PI128216* in the greenhouse conditions, whereas resistance was observed in the field without HR (Liabeuf et al., 2015).

7.5.3 INVOLVEMENT OF NITRIC OXIDE (NO) IN INDUCED HYPERSENSITIVE RESPONSES

Nitric oxide (NO) is a reactive oxygen species (ROS), however meager information is available about its biological function in the higher plants (Leshem, 1996; Leshem & Haramaty, 1996). The accumulation of rishitin, a phytoalexin that imparts disease resistance to several solanaceous vegetables, is induced by NO treatment of the tissue of potato tuber, which releases the chemical 1-hydroxy-2-oxo-3, 3-bis (2-aminoethyl)-1-triazene (Noritake et al., 1996). Tobacco leaves were infiltrated with either 50 mM Tris-HCl buffer, pH 7.0, or a suspension of an avirulent mutant of *R. solanacearum* (5×10^8 cells/ml) in an experiment conducted by Huang & Knopp (1998). When bacteria-infiltrated leaves were in the early stages of hypersensitivity, i.e., 6 hours after bacterial infiltration, the leaves were ground to convert into powder in liquid nitrogen. The ground powder was homogenized in a buffer containing Tris-HCl, dithiothreitol, phenylmethyl-sulfonyl fluoride, and polyvinyl polypyrrolidone, the homogenate was further centrifuged, and supernatant was assayed for nitric oxide synthase (NOS) activity. According to the findings, there is a positive relationship between product formation and reaction time length. The activity of NOS in hypersensitive tissue extracts is substantially higher than in control tissue extracts. The presence of NOS in tobacco tissues may be detected by using an arginine/citrulline assay and anti-NOS antibodies can be detected by using a western blot analysis, indicating the existence of NOS in solanaceous vegetables too. It is also possible to determine that the NOS gene (*nos* gene) is preferentially activated in hypersensitive tissues, implying that NO is involved in bilaterally mediated hypersensitivity (Huang & Knopp, 1998).

7.5.4 ANTI-RALSTONIA SOLANACEARUM PROTEIN

Phytoalexins, phenols, lignins, tannis, and melanins are the inhibitory molecules with low molecular weight (Brokaert et al., 1995; Zalewskz & Sequeira, 1975); act as an antibacterial agent to protect themselves against the number of infections. In recent past decade, scientists have revealed that antibacterial proteins have a tremendous role in bacterial wilt resistance in potato crop. From the bacterial wilt resistant potato cultivars, isolation, purification, and *in-vivo* immunogold localization was done (Yuan & He, 1998). SDS-PAGE analysis of soluble proteins from MS-42.3 (bacterial wilt-resistant potato cultivar), revealed the isolation and purification of a 32 kD protein with the inhibitory activity against *R. solanacearum*. Antimicrobial protein (AP1) presence in potato lines was linked to the resistance towards bacterial wilt. *In-vitro* investigation of antimicrobial spectrum revealed that AP1 exhibited good efficacy against five different strains of *R. solanacearum* from various hosts, as well as two fungus *Alternaria solani* and *Rhizoctonia solani* (Yuan & He, 1998).

7.5.5 IDENTIFICATION AND MAPPING OF RESISTANT LOCUS TO BACTERIAL DISEASES

R. solanacearum is a soil-borne pathogen affects the vascular system of host plant, and its mode of invasion differs from that of other bacterial infections, and the mechanisms involved in this process is still unknown. Meagre information is available about the specificities of different bacterial races during the host/pathogen interactions. There is no convincing evidence of a gene-for-gene resistance to *R. solanacearum*, and subsequent investigations have indicated that this resistance is controlled by multiple loci in tomato (Aarons et al., 1993; Thoquet et al., 1996). *Arabidopsis* ecotypes were subjected to genetic diversity analysis to identify a resistant line to the *R. solanacearum* strain GMI1000. Accession Nd-1 shown the promising resistance to most strains tested in comparison to susceptible ecotype Col-5. After inoculation with some strains, plants of both ecotypes have experienced wilt disease, indicated heterogeneity in virulence (or avirulence) and resistance determinants in *R. solanacearum* and *Arabidopsis* (Deslandes et al., 1998).

The genetic basis of resistance to strain GMI1000 was investigated when *Arabidopsis* both sensitive and resistant lines were identified. In F_2 populations derived from cross between Col-5 and Nd-1, a 3: 1 segregation ratio of susceptible: resistant types revealed a single recessive locus specific resistance for *R. solanacearum*. The locus was mapped to chromosome V by using RILs derived from Nd-1 and Col-5. Through the testing of RFLP markers on the RILs, fine mapping of the *RRS1* locus is in progress (Deslandes et al., 1998).

7.5.6 GENETIC ENGINEERING FOR BACTERIAL SPOT RESISTANCE

The transgenic method includes transferring a resistant gene from one organism to another, which does not produce progeny when fertilized naturally (Sharma & Bhattarai, 2019). If transferred to tomato, the *Bs2* gene confers resistance to many races of bacterial spot may be of potent source of durable resistance. The co-expression of *avrBs2* confirmed the presence of *Bs2* in tomato (Tai et al., 1999). Hypersensitivity was conferred and growth of *X. campestris* was suppressed in an *avrBs2*-dependent manner through the transient expression of the Bs2 gene in tomato lines. *avrBs2* regulates the *X. campestris* fitness, and *Bs2* controls both virulence and avirulence in *X. campestris* (Kearney & Staskawicz, 1990). To manage the bacterial diseases in tomato, such as bacterial spot and bacterial wilt, the *EFR* gene from *Arabidopsis* either singly or combine with the *Bs2* gene from pepper have been transferred. In the field condition, the severity of bacterial spot was decreased from 22% to 98% through the expression of *EFR* and *Bs2/EFR* genes in tomato lines, while the commercial yield was increased by 43% to 170%, compared to non-transgenic control plants (Kunwar et al., 2018).

7.5.7 USE OF ANTIMICROBIAL PEPTIDES (AMPS) AGAINST BACTERIAL DISEASES

The genes integrations producing antimicrobial peptides (AMPs) from other plants may enhance the resistance of plants against the diseases (Mitsuhara et al., 2000; Oard & Enright, 2006). The AMPs such as cecropins, magainins, and sarcotoxin IA help in building biological defenses against pathogens (Ohshima et al., 1999; Smith et al., 1998; Hultmark et al., 1982). Cecropinsis a lytic peptide produced by Giant silk moth, *Hyalophora cecropia* (Hultmark et al., 1982; Jan et al., 2010). The use of cationic peptide cecropin B (CB) inhibits the growth of gram-negative and gram-positive bacteria (Sharma & Bhattarai, 2019). The fusion of the CB gene with the secretory signal peptide from the barley a-amylase gene results in the construct pBI121-spCB (Jan et al., 2010). After the transformation of construct into tomato, the increased resistance to bacterial spot was obtained, which was later validated by western blot analysis (Jan et al., 2010).

7.5.8 TRANSFER OF PR PROTEINS FOR BROAD SPECTRUM RESISTANCE AGAINST BACTERIAL DISEASES

Broad-spectrum disease resistance to different diseases was obtained when the non-expresser of PR genes (*NPR1*) from *Arabidopsis* is transferred to tomato (Lin et al., 2004). *NPR1* is a gene responsible in systemic acquired resistance (SAR). Transgenic tomato was effective against various diseases, with a moderate level of resistance against bacterial spot (Sharma & Bhattarai, 2019). The amount of *NPR1* protein accumulated by the transgenic lines was directly related to the level of resistance, with the fact that higher levels of *NPR1* proteins indicating a broader and higher level of disease resistance. Similarly, the *NPR1* protein imparted direct impact on inheritance pattern of the disease resistance (Lin et al., 2004).

7.5.9 GENOME EDITING FOR GENETIC MODIFICATION IN SOLANACEOUS VEGETABLES

Backcrossing in tomato is a traditional breeding process that involves crossing of resistant and susceptible cultivars to produce F_1 followed by several cycles of backcrosses to develop resistance towards bacterial spot (Sharma & Bhattarai, 2019). Several genome editing techniques, such as ZFNs (Zinc Finger Nucleases) and TALENs (Transcription Activator Like Effect or Nucleases) and are being explored in different crops to reduce the agony of classical breeding approaches which involves several breeding cycles. These genome-editing techniques make a cut in double-strand DNA at a recognized site with programmable nucleases, which aids in genetic alterations in a plant's gene or genome (Joung & Sander, 2013). Non-homologous end joining is used to insert or delete at DNA breaks, whereas homology directed repair utilizing a homologous repair template is used to repair double strand breaks (Liu & Fan, 2014; Li et al., 2013). However, this technology is not being adopted at large scale due high cost involved in the designing of these nucleases and specific technical skill is required to use them. The CRISPR/Cas9 system is a latest technology for genome editing that can be a substitute of TALENs and ZFN. CRISPR/Cas9 is a powerful tool to acquire broad-spectrum resistance to a variety of infectious bacterial diseases. Cas9 is a bacterial CRISPR-associated protein-9 nuclease from the bacteria *Streptococcus pyogenes*, and CRISPR/Cas9 is a modified bacterial immune system for genome editing (Li et al., 2013). Cas9 nuclease and Single guide RNA (sgRNA), which is specific to nucleic acid (NA) sequence, make up the system. As sgRNA target specificity is determined by the formation of a ribonucleotide-protein complex, designing target specific guide RNA for Cas9 recognition is quite simple (Li et al., 2013).

Through CRISPR/Cas9 mediated mutagenesis, the SlDMR6-1 ortholog Solyc03g080190.2 helps in the acquisition of broad range disease resistance in tomato (De Toledo Thomazella et al., 2016). Frame shift and premature truncation of the protein is caused by deletion of a few nucleotides from the SlDMR6-1 gene by the use of CRISPR/Cas9. After deletion, the progeny (T_1 plants) was infected with *X. perforans* (Xp4b) and *X. gardneri* (Xg153), and Sldmr6-1 mutants were found to be resistant without affecting plant growth and development compared to wild plants. During the infection in *Arabidopsis thaliana*, the up regulation of downy mildew resistance gene (DMR6) was observed (Zeilmaker et al., 2015). The increase in salicylic acid (SA) level during the mutagenesis of DMR6 was directly correlated with disease resistance in *Arabidopsis*, the ortholog of the same gene was implemented in tomato to acquire broad spectrum resistance against bacterial diseases.

7.5.10 PHAGE-BASED CONTROL OF CLAVIBACTER MICHIGANENSIS SSP. MICHIGANENSIS (CMM)

Because of the specificity, easy preparation, and low cost of production, phage-based control is also being investigated as a promising method of disease control (Jones et al., 2007). Bacteriophage *CMP1* is a member of the *Siphoviridae* virus family, which infects *Cmm* specifically, was first isolated from *Cmm*-infected tomato stems (Echandi & Sun, 1973). Its genome encodes an endolysin with peptidase activity that lyses *Cmm* but no other bacteria (Wittmann et al., 2011). Although the phage itself may have the ability to prevent *Cmm*, however its endolysin was expressed recombinantly in the tomato plants, providing total resistance to *Cmm* (Wittmann et al., 2016).

7.6 CONCLUSION

Bacterial wilt, bacterial canker, bacterial spot, and bacterial speck are the most important bacterial diseases among all the diseases affecting solanaceous vegetables. Bacterial diseases have been a major problem in the crops due to their complex nature, wider host range, and faster adaptability to changing environments, inviting huge challenges for the farming community. There are not sufficient management tactics to control these diseases till date; nevertheless, DNA markers tightly linked to resistance loci could help in breeding programs and allow molecular analysis of the resistance genes involved. Identification of active hypersensitivity-related genes linked to the hypersensitive response, which involves rapid and localized cell death at the infection site, limits the spread of the pathogen further. Isolation, purification, and *in-vivo* immunogold localization of anti-*Ralstonia solanacearum* protein from bacterial wilt resistant varieties of potato imparts the inhibitory activity to *R. solanacearum*. The resistance is controlled by several loci in the crop plants, the fine mapping of the single recessive locus (*RRS1* locus), which confers the bacterial wilt resistance by testing the RFLP markers on the RILs is of utmost preference. Resistance involves NPR1 protein shows broader level of resistance towards bacterial diseases, also correlates with the inheritance of the disease resistance. Genome editing tools like TALENs and ZFNs utilize programmable nucleases to create the cut in ds-DNA at specific site, helps in genetic alteration in the genome of plant. CRISPR/Cas9 is an effective tool to obtain broad spectrum resistance against various bacterial pathogens act as modified bacterial immune system for genome editing. Phage-based control also acts as a promising control of the disease due to its specificity, easy preparation, and cost-effective production. The integrative utilization of cultural, physical, biological, molecular, and chemical control may give the best possible outcome against these deadly bacterial diseases causing huge economic losses in solanaceous vegetables.

KEYWORDS

- **antimicrobial peptides**
- **bacterial canker**
- **bacterial spot**
- **bacterial spot resistance**
- **bacterial wilt**
- **biotechnology**
- **molecular techniques**

REFERENCES

Aarons, S. R., Danesh, D., & Young, N. D., (1993). DNA genetic marker mapping of genes for bacterial wilt resistance in tomato. In: Hartman, W. G. L., & Hayward, A. C., (eds.), *Bacterial. ACIAR Proceedings* (Vol. 45, pp. 170–175). Ferguson Publishing Company, Co. Brisbane, Australia.

Assefa, M., Dawit, W., Lencho, A., & Hunduma, T., (2015). Assessment of wilt intensity and identification of causal fungal and bacterial pathogens on hot pepper (*Capsicum annuum* L.) in Bako Tibbe and Nonno districts of west Shewa zone, Ethiopia. *Int. J. Plant Pathol.*, *4*(1), 21–28.

AUSVEG, (2017). *Bacterial Diseases in Vegetable Crops*. https://ausveg.com.au/biosecurity-agrichemical/crop-protection/solanaceous-vegetables/ (accessed on 12 December 2023).

Baker, R., Bragard, C., Candresse, T., & Van, D. W. J., (2014). Scientific opinion on the pest categorization of *Xanthomonas campestris pv.vesicatoria* (Doidge) Dye. *EFSA. J.*, *12*(6), 3720.

Blank, L., Cohen, Y., Borenstein, M., Shulhani, R., Lofthouse, M., Sofer, M., & Shtienberg, D., (2016). Variables associated with severity of bacterial canker and wilt caused by *Clavibacter michiganensis* subsp. *michiganensis* in tomato greenhouses. *Phytopathol.*, *106*(3), 254–261.

Boshou, L., (2005). A broad review and perspective on breeding for resistance to bacterial wilt. In: Allen, C., Prior, P., & Hayward, A. C., (eds.), *Bacterial Wilt Disease and the Ralstonia Solanacearum Species Complex* (pp. 225–238). American Phytopathological Society.

Bosland, P. W., (1996). Capsicums: Innovative uses of an ancient crop. *Progr. New Crops* (pp. 479–487). ASHS Press.

Broekaert, W. F., Terras, F. R., Cammue, B. P., & Osborn, R. W., (1995). Plant defensins: Novel antimicrobial peptides as components of the host defense system. *Plant Physiol.*, *108*(4), 1353–1358.

Burkholder, W. H., & Li, C. C., (1941). Variations in *Phytomonas vesicatoria*. *Phytopathol.*, *31*, 753–755.

Cabi, E., (1999). Distribution maps of plant diseases. *Ralstonia Solanacearum*, 783–785.

Canteros, B., Gerald, M. U. B., Pring, D., & Stall, R., (1991). A gene from *Xanthomonas campestris pv. vesicatoria* that determines avirulence in tomato is related to *avrBs3*. *Mol. Plant Microbe. Interact.*, *4*(6), 628–632.

Carlton, W. M., Braun, E. J., & Gleason, M. L., (1998). Ingress of *Clavibacter michiganensis* subsp. *michiganensis* into tomato leaves through hydathodes. *Phytopathol.*, *88*(6), 525–529.

Cellier, G., & Prior, P., (2010). Deciphering phenotypic diversity of *Ralstonia solanacearum* strains pathogenic to potato. *Phytopathol.*, *100*, 1250–1260.

Chandrashekara, K. N., Prasannakumar, M. K., Deepa, M., & Vani, A., (2012). A rapid, sensitive, and reliable method for detecting *Ralstonia solanacearum* using FTA (Whatman) card. *J. Plant Pathol.*, *94*, 219–221.

Chang, R. J., Ries, S., & Pataky, J., (1991). Dissemination of *Clavibacter michiganensis* subsp. *michiganensis* by practices used to produce tomato transplants. *Phytopathol.*, *81*(10), 1276–1281.

Clinton, S. K., (1998). Lycopene: Chemistry, biology, and implications for human health and disease. *Nutr. Rev.*, *56*, 35–51. https://doi.org/10.1111/j.1753-4887.1998.tb01691.x.

Cook, A. A., (1973). Characterization of hypersensitivity in *Capsicum annuum* induced by the tomato strain of *Xanthomonas vesicatoria*. *Phytopathol.*, *63*(7), 915–918.

Coutinho, T. A., (2005). Introduction and prospectus on the survival of *R. solanacearum*. *Bacterial wilt Disease and the Ralstonia Solanacearum Species Complex* (pp. 29–38). APS Publishing.

Cox, R. S., (1966). The role of bacterial spot in tomato production in south Florida. *Plant Dis. Rep.*, *50*, 699–700.

Davis, M. J., Gillaspie, A. G., Vidaver, A. K., & Harris, R. W., (1984). *Clavibacter*: A new genus containing some phytopathogenic coryneform bacteria, including *Clavibacter xyli* subsp. *xyli* sp. *Nov.,* subsp. *Nov.* and *Clavibacter xyli* subsp. *cynodontis* subsp. *Nov.,* pathogens that cause ratoon stunting disease of sugarcane and Bermuda grass stunting disease. *Int. J. Sys Bact.*, *34*(2), 107–117.

De Leon, L., Siverio, F., Lopez, M. M., & Rodriguez, A., (2011). *Clavibacter michiganesis* subsp. *michiganensis*, a seed borne tomato pathogen: Healthy seeds are still the goal. *Plant Dis.*, *95*(11), 1328–1338.

De Toledo, T. D. P., Brail, Q., Dahlbeck, D., & Staskawicz, B. J., (2016). *CRISPR-Cas9 Mediated Mutagenesis of a DMR6 Ortholog in Tomato Confers Broad Spectrum Disease Resistance*. bioRxiv., 064824.

Debnath, A., Kumar, R., Prasad, S. K., Sharma, N., & Kushwah, J. K., (2020). Multivariate analysis for genetic diversity estimation among tomato (*Solanum lycopersicum* L.) Genotypes. *Eco. Environ. & Conserv.*, *26*(3), 1208–1211.

Deslandes, L., Pileur, F., Liaubet, L., Camut, S., Beynon, J., Arlat, M., & Marco, Y., (1998). Identification and mapping of RRS1, a single recessive locus in *Arabidopsis thaliana* that confers resistance to *Ralstonia solanacearum*. *Bact. Wil. Dis.* (pp. 250–254). Springer.

Dietrich, R. A., Delaney, T. P., Uknes, S. J., Ward, E. R., Ryals, J. A., & Dangl, J. L., (1994). *Arabidopsis* mutants simulating disease resistance response. *Cell*, *77*(4), 565–577.

Doidge, E. M., (1920). A tomato canker. *J. Dep. Agric. Union S. Afr.*, *1*, 718–721.

Dye, D. W., Starr, M. P., & Stolp, H., (1964). Taxonomic clarification of *Xanthomonas vesicatoria* based upon host specificity, bacteriophage sensitivity, and cultural characteristics. *J. Phytopathol.*, *51*(4), 394–407.

Echandi, E., & Sun, M., (1973). Isolation and characterization of a bacteriophage for identification of *Corynebacterium michiganense*. *Phytopathol.*, *63*, 1398–1401.

Eichenlaub, R., & Gartemann, K. H., (2011). The *Clavibacter michiganensis* subspecies: Molecular investigation of gram-positive bacterial plant pathogens. *Annu. Revi. Phytopathol.*, *49*, 445–464.

Elliott, C., (1951). *Manual of Bacterial Plant Pathogens* (p. 186). Waltham, MA: Chronica Botanica.

Elphinstone, J. G., (2005). The current bacterial wilt situation: A global overview. *Bacterial Wilt Disease and the Ralstonia Solanacearum Species Complex*, 9–28.

Ezekiel, R., Singh, N., Sharma, S., & Kaur, A., (2013). Beneficial phytochemicals in potato-A review. *Food Res. Int.*, *50*(2), 487–496. https://doi.org/10.1016/j.foodres.2011.04.025.

Fatmi, M., & Schaad, N., (1988). Semi selective agar medium for isolation of *Clavibacter michiganense* subsp. *michiganense* from tomato seed. *Phytopathol.*, *78*(1), 121–126.

Fatmi, M., Schaad, N., & Bolkan, H., (1991). Seed treatments for eradicating *Clavibacter michiganensis* subsp. *michiganensis* from naturally infected tomato seeds. *Plant Dis.*, *75*(4), 383–385.

Fegan, M., & Prior, P., (2005). How complex is the *Ralstonia solanacearum* species complex. *Bacterial wilt Disease and the Ralstonia Solanacearum Species Complex*, *1*, 449–461.

Flor, H. H., (1971). Current status of the gene-for-gene concept. *Annu. Rev. Phytopathol., 9*(1), 275–296.

Gardner, M. W., & Kendrick, J. B., (1921). Bacterial spot of tomato. *J. Agric Res., 21*, 123–162.

Gardner, M. W., & Kendrick, J. B., (1923). Bacterial spot of tomato and pepper. *Phytopathol., 13*, 307–315.

Genin, S., & Denny, T. P., (2012). Pathogenomics of the *Ralstonia solanacearum* species complex. *Annu. Rev. Phytopathol., 50*, 67–89.

Genin, S., (2010). Molecular traits controlling host range and adaptation to plants in *Ralstonia solanacearum*. *New Phytologist, 187*(4), 920–928.

Gitaitis, R. D., Beaver, R., & Voloudakis, A., (1991). Detection of *Clavibacter michiganensis* subsp. *michiganensis* in symptomless tomato transplants. *Plant Dis., 75*(8), 834–838.

Gitaitis, R., McCarter, S., & Jones, J., (1992). Disease control in tomato transplants produced in Georgia and Florida. *Plant Dis., 76*(7), 651–656.

Gleason, M. L., Gitaitis, R. D., & Ricker, M. D., (1993). Recent progress in understanding and controlling bacterial canker of tomato in eastern North America. *Plant Dis., 77*(11), 1069–1076.

Godiard, L., Grant, M. R., Dietrich, R. A., Kiedrowski, S., & Dangl, J. L., (1994). Perception and response in plant disease resistance. *Curr. Opin. Genet. Dev., 4*(5), 662–671.

Greenberg, J. T., Guo, A., Klessig, D. F., & Ausubel, F. M., (1994). Programmed cell death in plants: A pathogen-triggered response activated co-ordinately with multiple defense functions. *Cell, 77*(4), 551–563.

Grimsley, N., & Wang, J. F., (1998). Chair's perspective: Host resistance. In: *Bacterial Wilt Dis.,* (pp. 197–199). Springer, Berlin, Heidelberg.

Hayward, A. C., (1991). Biology and epidemiology of bacterial wilt caused by *Pseudomonas solanacearum*. *Annu. Rev. Phytopathol., 29*, 65–87.

Hayward, A. C., (1995). Research on bacterial wilt: A perspective on international linkages and access to the literature. *Bacterial Wilt: The Disease and the Ralstonia Solanacearum Species Complex, EEUU* (pp. 1–8). American Phyto-pathological Society.

Higgins, B. B., (1922). The bacterial spot of pepper. *Phytopathol., 12*, 501–517.

Horvath, D. M., Stall, R. E., Jones, J. B., Pauly, M. H., Vallad, G. E., Dahlbeck, D., Staskawicz, B. J., & Scott, J. W., (2012). Transgenic resistance confers effective field level control of bacterial spot disease in tomato. *PLoS One, 7*(8), e42036.

Huang, J. S., & Knopp, J. A., (1998). Involvement of nitric oxide in *Ralstonia solanacearum*-induced hypersensitive reaction in tobacco. In: *Bact. Wilt Dis.* (pp. 218–224). Springer, Berlin, Heidelberg.

Huet, G., (2014). Breeding for resistances to *Ralstonia solanacearum*. *Fron. Plant Sci., 5*, 715.

Hultmark, D., Engstrom, A., Bennich, H., Kapur, R., & Boman, H. G., (1982). Insect immunity: Isolation and structure of cecropin D and four minor antibacterial components from *Cecropia pupae*. *Euro. J. Biochem., 127*(1), 207–217.

Hung, H. C., Joshipura, K. J., Jiang, R., Hu, F. B., Hunter, D., Smith-Warner, S. A., Colditz, G. A., et al., (2004). Fruit and vegetable intake and risk of major chronic disease. *J. Natl. Cancer Inst., 96*(21), 1577–1584. https://doi.org/10.1093/jnci/djh296.

Jacques, M. A., Durand, K., Orgeur, G., Balidas, S., Fricot, C., Bonneau, S., Quillevere, A., et al., (2012). Phylogenetic analysis and polyphasic characterization of *Clavibacter michiganensis* strains isolated from tomato seeds reveal that nonpathogenic strains are distinct from *C. michiganensis* subsp. *michiganensis*. *Appl. Environ. Microbiol., 78*, 8388–8402.

Jain, A. C., (1951). A bacterial disease (canker) of tomatoes new to India. *Sci. Cult., 17*, 46, 47.

Jan, P. S., Huang, H. Y., & Chen, H. M., (2010). Expression of a synthesized gene encoding cationic peptide cecropin B in transgenic tomato plants protects against bacterial diseases. *Appl. Environ. Microbiol., 76*(3), 769–775.

Janse, J., Beld, H. E., Elphinstone, J., Simpkins, S., Tjou-Tam-Sin, N. N. A., & Vaerenbergh, J., (2004). Introduction to Europe of *Ralstonia solanacearum* biovar 2, race 3 in *Pelargonium zonale* cuttings. *J. Plant Pathol., 86*, 147–155.

Ji, P., Momol, M. T., Olson, S. M., Pradhanang, P. M., & Jones, J. B., (2005). Evaluation of thymol as biofumigant for control of bacterial wilt of tomato under field conditions. *Plant Dis., 89*(5), 497–500.

Jones, J. B., (1986). Survival of *Xanthomonas campestris pv. vesicatoria* in Florida on tomato crop residue, weeds, seeds, and volunteer tomato plants. *Phytopathol., 76*(4), 430–434.

Jones, J. B., Jackson, L. E., Balogh, B., Obradovic, A., Iriarte, F. B., & Momol, M. T., (2007). Bacteriophages for plant disease control. *Annu. Rev. Phytopathol., 45*, 245–262.

Jones, J. B., Jones, J. P., Stall, R. E., & Zitter, T. A., (1991). S. M., Bacterial wilt. In: McCarter, M. N., (ed.), *Compendium of Tomato Diseases* (pp. 28, 29). APS Publishers.

Jones, J. B., Lacy, G. H., Bouzar, H., Stall, R. E., & Schaad, N. W., (2004). Reclassification of the *xanthomonads* associated with bacterial spot disease of tomato and pepper. *Sys. App. Micro., 27*(6), 755–762.

Jones, J. B., Stall, R. E., & Bouzar, H., (1998). Diversity among *xanthomonads* pathogenic on pepper and tomato. *Annu. Rev. Phytopathol., 36*, 41–58.

Jones, J. B., Stall, R., Scott, J., Somodi, G., Bouzar, H., & Hodge, N., (1995). A third tomato race of *Xanthomonas campestris* pv. *vesicatoria. Plant Dis., 79*(4), 395–398.

Joung, J. K., & Sander, J. D., (2013). TALENs: A widely applicable technology for targeted genome editing. *Nature Reviews. Mol. Cell Bio., 14*(1), 49–55.

Katafiire, M., Adipala, E., Lemaga, B., Olanya, M., El-Bedewy, R., & Ewell, P., (2005). Management of bacterial wilt of potato using one-season rotation crops in South western Uganda. *Bacterial Wilt Disease and the Ralstonia Solanacearum Species Complex*, 197–203.

Kearney, B., & Staskawicz, B. J., (1990). Widespread distribution and fitness contribution of *Xanthomonas campestris* avirulence gene *avrBs2*. *Nature, 346*(6282), 385, 386.

Kebede, M., Timilsina, S., Ayalew, A., Admassu, B., Potnis, N., Minsavage, G. V., Goss, E. M., et al., (2014). Molecular characterization of *Xanthomonas* strains responsible for bacterial spot of tomato in Ethiopia. *Eur. J. Plant Pathol., 140*(4), 677–688.

Kim-Lee, H., Moon, J. S., Hong, Y. J., Kim, M. S., & Cho, H. M., (2005). Bacterial wilt resistance in the progenies of the fusion hybrids between haploid of potato and *Solanum commersonii. Am. J. Potato Res., 82*(2), 129–137.

Kleitman, F., Barash, I., Burger, A., Iraki, N., Falah, Y., Sessa, G., Weinthal, D., et al., (2008). Characterization of a *Clavibacter michiganensis* subsp. *michiganensis* population in Israel. *Eur. J. Plant Pathol., 121*(4), 463–475.

Klement, Z., (1963). Rapid detection of the pathogenicity of phytopathogenic pseudomonads. *Nat., 199*, 299, 300.

Kumar, R., Chhetri, S., Bamaniya, B. S., Kushwah, J. K., & Verma, R. K., (2021). Molecular tools for bacterial wilt resistance in tomato. In: *Biotic Stress Management in Tomato*

(Biotechnological Approaches) in the Series- Innovation in Horticultural Science (pp. 25–42.). Apple Academic Press, CRC Press (Taylor & Francis Group), USA.

Kumar, R., Kumar, R., Prasad, B. D., Solankey, S. S., Kumar, J., & Bamaniya, B. S., (2018). Genetic variation study using morphological and DNA marker-based genotyping in bottle gourd (*Lagenaria siceraria* (Mol.) Standl.). *Curr. J. Appl. Sci. Tech., 31*(2), 1–10. https.//doi.10.9734/CJAST/2018/45876.

Kunwar, S., Iriarte, F., Fan, Q., Evaristo Da, S. E., Ritchie, L., Nguyen, N. S., Freeman, J. H., et al., (2018). Transgenic expression of EFR and Bs2 genes for field management of bacterial wilt and bacterial spot of tomato. *Phytopathol., 108*(12), 1402–1411.

Kurabachew, H., & Ayana, G., (2017). Bacterial wilt caused by *Ralstonia solanacearum* in Ethiopia: Status and management approaches: A review. *Int. J. Phytopath., 5*(3), 107–119.

Lambert, C. D., (2002). Agricultural bioterrorism protection Act of 2002: Possession, use and transfer of biological agents and toxins; Interim and Final Rule (7CRF part 331). *Federal Register, 67*, 76908–76938.

Laub, C. A., & Stall, R. E., (1967). An evaluation of *Solanum nigrum* and *Physalis minima* as suscepts of *Xanthomonas vesicatoria*. *Plant Dis. Rep., 51*, 659–661.

Lebeau, A., Daunay, M. C., Frary, A., Palloix, A., Wang, J. F., Dintinger, J., Chiroleu, F., et al., (2011). Bacterial wilt resistance in tomato, pepper, and eggplant: Genetic resources respond to diverse strains in the *Ralstonia solanacearum* species complex. *Phytopathol., 101*(1), 154–165.

Leshem, Y. Y., & Haramaty, E., (1996). The characterization and contrasting effects of the nitric oxide free radical in vegetative stress and senescence of *Pisum sativum* Linn. foliage. *J. Plant Physiol., 148*(3, 4), 258–263.

Leshem, Y. Y., (1996). Nitric oxide in biological systems. *Plant Growth Regulation, 18*(3), 155–159.

Li, J. F., Norville, J. E., Aach, J., McCormack, M., Zhang, D., Bush, J., Church, G. M., & Sheen, J., (2013). Multiplex and homologous recombination-mediated genome editing in *Arabidopsis* and *Nicotiana benthamiana* using guide RNA and Cas9. *Nat. Biotechnol., 31*(8), 688–691.

Liabeuf, D., Francis, D. M., & Sim, S. C., (2015). Screening cultivated and wild tomato germplasm for resistance to *Xanthomonas gardneri*. *Act. Horti., 1069*(1069), 65–70.

Lin, W. C., Lu, C. F., Wu, J. W., Cheng, M. L., Lin, Y. M., Yang, N. S., Black, L., et al., (2004). Transgenic tomato plants expressing the *Arabidopsis* NPR1 gene display enhanced resistance to a spectrum of fungal and bacterial diseases. *Transgenic Res., 13*(6), 567–581.

Liu, L., & Fan, X. D., (2014). CRISPR-Cas system: A powerful tool for genome engineering. *Plant Mol. Bio., 85*(3), 209–218.

Lopes, C. A., & Rossato, M., (2018). History and status of selected hosts of the *Ralstonia solanacearum* species complex causing bacterial wilt in Brazil. *Front. Microbiol., 9*, (1228).

Mansfield, J., Genin, S., Magori, S., Citovsky, V., Sriariyanum, M., Ronald, P., Dow, M., et al., (2012). Top 10 plant pathogenic bacteria in molecular plant pathology. *Mol. Plant Pathol., 13*(6), 614–629.

Matsuzoe, N., Yamaguchi, M., Kawanobu, S., Watanabe, Y., Higashi, H., & Sakata, Y., (1999). Effect of dark treatment of the eggplant on fruit skin color and its anthocyanin component. *Engei Gakkai Zasshi, 68*(1), 138–145. https://doi.org/10.2503/jjshs.68.138.

McGill, C. R., Kurilich, A. C., & Davignon, J., (2013). The role of potatoes and potato components in cardiometabolic health: A review. *Ann. Med., 45*(7), 467–473. https://doi.org/10.3109/07853890.2013.813633.

McGuire, R. G., Jones, J. B., Stanley, C. D., & Csizinszky, A. A., (1991). Epiphytic populations of *Xanthomonas campestris pv. vesicatoria* and bacterial spot of tomato as influenced by nitrogen and potassium fertilization. *Phytopathol.*, *81*(6), 656–660.

Milijasevic-Marcic, S., Gartemann, K. H., Frohwitter, J., Eichenlaub, R., Todorovic, B., Rekanovic, E., & Potocnik, I., (2012). Characterization of *Clavibacter michiganensis* subsp. *michiganensis* strains from recent outbreaks of bacterial wilt and canker in Serbia. *Euro. J. Plant Pathol.*, *134*(4), 697–711.

Mitsuhara, I., Matsufuru, H., Ohshima, M., Kaku, H., Nakajima, Y., Murai, N., Natori, S., & Ohashi, Y., (2000). Induced expression of sarcotoxin IA enhanced host resistance against both bacterial and fungal pathogens in transgenic tobacco. *Mol. Plant-Micro. Interact.*, *13*(8), 860–868.

Mittler, R., Shulaev, V., & Lam, E., (1995). Coordinated activation of programmed cell death and defense mechanisms in transgenic tobacco plants expressing a bacterial proton pump. *The Plant Cell*, *7*(1), 29–42.

Mursu, J., Virtanen, J. K., Tuomainen, T. P., Nurmi, T., & Voutilainen, S., (2014). Intake of fruit, berries, and vegetables and risk of type 2 diabetes in Finnish men: The Kuopio Ischaemic Heart Disease risk factor study. *Am. J. Clin. Nutr.*, *99*(2), 328–333. https://doi.org/10.3945/ajcn.113.069641.

Muthoni, J., Shimelis, H., & Melis, R., (2012). Management of bacterial wilt (*Ralstonia solanacearum*, Yabuuchi, et al., 1995) of potatoes: Opportunity for host resistance in Kenya. *J. Agric. Sci.*, *4*(9), 64.

Nandi, M., Macdonald, J., Liu, P., Weselowski, B., & Yuan, Z. C., (2018). *Clavibacter michiganensis* ssp. *michiganensis*: Bacterial canker of tomato, molecular interactions, and disease management. *Mol. Plant Pathol.*, *19*(8), 2036–2050.

Noda, Y., Kneyuki, T., Igarashi, K., Mori, A., & Packer, L., (2000). Antioxidant activity of nasunin, an anthocyanin in eggplant peels. *Toxicolog.*, *148*(2, 3), 119–123. https://doi.org/10.1016/s0300-483x(00)00202-x.

Noritake, T., Kawakita, K., & Doke, N., (1996). Nitric oxide induces phytoalexin accumulation in potato tuber tissues. *Plant Cell Physiol.*, *37*(1), 113–116.

Oard, S. V., & Enright, F. M., (2006). Expression of the antimicrobial peptides in plants to control phytopathogenic bacteria and fungi. *Plant Cell Rep.*, *25*(6), 561–572.

Ohshima, M., Mitsuhara, I., Okamoto, M., Sawano, S., Nishiyama, K., Kaku, H., Natori, S., & Ohashi, Y., (1999). Enhanced resistance to bacterial diseases of transgenic tobacco plants overexpressing sarcotoxin, IA, a bactericidal peptide of insect. *J. Biochem.*, *125*(3), 431–435.

Pontier, D., Baudouin, E., Czernic, P., Tronchet, M., Godiard, L., Bezombes, I., Deslandes, L., et al., (1998). Transcriptional regulation and function of hsr203j, an hypersensitivity-related gene of tobacco activated in response to *Ralstonia solanacearum*. In: *Bacterial Wilt Disease* (pp. 203–208). Springer, Berlin, Heidelberg.

Pontier, D., Godiard, L., Marco, Y., & Roby, D., (1994). hsr203J, a tobacco gene whose activation is rapid, highly localized, and specific for incompatible plant/pathogen interactions. *Plant Mol. Biol.*, *5*(4), 507–521.

Potnis, N., Timilsina, S., Strayer, A., Shantharaj, D., Barak, J. D., Paret, M. L., Vallad, G. E., & Jones, J. B., (2015). Bacterial spot of tomato and pepper: Diverse *Xanthomonas* species with a wide variety of virulence factors posing a worldwide challenge. *Mol. Plant Pathol.*, *16*(9), 907–920.

Pradhanang, P. M., Ji, P., Momol, M. T., Olson, S. M., Mayfield, J. L., & Jones, J. B., (2005). Application of acibenzolar-s-methyl enhances host resistance in tomato against *Ralstonia solanacearum*. *Plant Dis.*, *89*(9), 989–993.

Prior, P. H., Allen, C., & Elphinstone, J., (1998). *Bacterial Wilt Diseases: Molecular and Ecological Aspects* (pp. 197–200). Springer, Berlin, Heidelberg.

Rao, A. V., & Rao, L. G., (2007). Carotenoids and human health. *Pharmacol. Res.*, *55*(3), 207–216. https://doi.org/10.1016/j.phrs.2007.01.012.

Ricker, M. D., & Riedel, R., (1993). Effect of secondary spread of *Clavibacter michiganensis* subsp. *michiganensis* on yield of northern processing tomatoes. *Plant Dis.*, *77*(4), 364–366.

Rubatzky, V. E., & Yamaguchi, M., (1997). Tomatoes, peppers, eggplants, and other solanaceous vegetables. *World Veg.*, 532–576, Springer, Boston.

Sahin, F., & Miller, S. A., (1998). Resistance in *Capsicum pubescens* to *Xanthomonas campestrispv. vesicatoria* pepper race 6. *Plant Dis.*, *82*(7), 794–799.

Schornack, S., Minsavage, G. V., Stall, R. E., Jones, J. B., & Lahaye, T., (2008). Characterization of AvrHah1, a novel AvrBs3-like effector from *Xanthomonas gardneri* with virulence and avirulence activity. *New Phyto.*, *179*(2), 546–556.

Sharabani, G., Shtienberg, D., Borenstein, M., Shulhani, R., Lofthouse, M., Sofer, M., Chalupowicz, L., et al., (2013). Effect of plant age on disease development and virulence of *Clavibacter michiganensis* subsp. *michiganensis* on tomato. *Plant Pathol.*, *62*(5), 1114–1122.

Sharma, N., Bhutia, K. D., Kumar, R., Prasad, S. K., Debnath, A., & Sharma, M. M., (2020). Polygenic variation for morphological and biochemical traits of brinjal genotypes (*Solanum melongena* L.) and its wild relatives. *The Biosc.*, *15*(3), 303–309.

Sharma, S., & Bhattarai, K., (2019). Progress in developing bacterial spot resistance in tomato. *Agron.*, *9*(26), 1–11.

Singh, D., Mondal, K. K., & Kishun, R., (2011). Current status and future research strategies in plant bacteriology in India. *In: Plant Pathology in India: Vision, 2030* (pp. 115–120). Indian Phyto-pathological Society.

Singh, G., Bharat, N. K., & Sharma, M., (2015). Occurrence of bacterial canker of tomato in Himachal Pradesh, India: Identification and molecular characterization of the pathogen. *The Biosc.*, *10*(4), 1753–1757.

Skrabal, F., Aubock, J., & Hortnagl, H., (1981). Low sodium/high potassium diet for prevention of hypertension: Probable mechanisms of action. *Lancet*, *2*(8252), 895–900. https://doi.org/10.1016/s0140-6736(81)91392-1.

Smith, E., (1910). A new tomato disease of economic importance. *Sci.*, *31*, 794–796.

Smith, F. D., Gadoury, D. M., Van, E. J. M., Blowers, A., Sanford, J. C., Van, D. M. J., & Eisenreich, R., (1998). Enhanced resistance to powdery mildew in transgenic poinsettia conferred by antimicrobial peptides. *Phytopathol.*, *88*(83).

Stall, R. E., Jones, J. B., & Minsavage, G. V., (2009). Durability of resistance in tomato and pepper to *xanthomonads* causing bacterial spot. *Ann. Rev. Phytopathol.*, *47*, 265–284.

Tahat, M. M., & Sijam, K., (2010). *Ralstonia solanacearum*: The bacterial wilt causal agent. *Asian J. Plant Sci.*, *9*(7), 385–393.

Tai, T. H., Dahlbeck, D., Clark, E. T., Gajiwala, P., Pasion, R., Whalen, M. C., Stall, R. E., & Staskawicz, B. J., (1999). Expression of the *Bs2* pepper gene confers resistance to bacterial spot disease in tomato. *Proceed. Nation. Acad. Scienc. USA*, *96*(24), 14153–14158.

Tambong, J. T., (2017). Comparative genomics of *Clavibacter michiganensis* subspecies, pathogens of important agricultural crops. *PLOS One*, *12*(3), e0172295.

Tancos, M. A., Chalupowicz, L., Barash, I., Manulis-Sasson, S., & Smart, C. D., (2013). Tomato fruit and seed colonization by *Clavibacter michiganensis* subsp. *Michiganensis* through external and internal routes. *Appl. Environ. Microbiol.*, *79*(22), 6948–6957.

Tancos, M. A., Lange, H. W., & Smart, C. D., (2015). Characterizing the genetic diversity of the *Clavibacter michiganensis* subsp. *michiganensis* population in New York. *Phytopathol.*, *105*(2), 169–179.

Thapa, S. P., Pattathil, S., Hahn, M. G., Jacques, M. A., Gilbertson, R. L., & Coaker, G., (2017). Genomic analysis of *Clavibacter michiganensis* reveals insight into virulence strategies and genetic diversity of a gram-positive bacterial pathogen. *Mol. Plant-Micro. Inter.*, *30*(10), 786–802.

Thoquet, P., Olivier, J., Sperisen, C., Rogowsky, P., Laterrot, H., & Grimsley, N., (1996). Quantitative trait loci determining resistance to bacterial wilt in tomato cultivar Hawaii 7996. *Mol. Plant-Micro. Inter.*, *9*, 826–836.

Timmer, L. W., (1987). Growth and survival of *Xanthomonads* under conditions non-conducive to disease development. *Phytopathol.*, *77*(9), 1341–1345.

Tomlinson, D. L., Elphinstone, J. G., Soliman, M. Y., Hanafy, M. S., Shoala, T. M., Abd El-Fatah, H., Agag, S. H., et al., (2009). Recovery of *Ralstonia solanacearum* from canal water in traditional potato-growing areas of Egypt but not from designated pest-free areas (PFAs). *Eur. J. Plant Pathol.*, *125*(4), 589–601.

Umesha, S., (2006). Occurrence of bacterial canker in tomato fields of Karnataka and effect of biological seed treatment on disease incidence. *Crop Prot.*, *25*(4), 375–381.

Vallad, G. E., & Goodman, R. M., (2004). Systemic acquired resistance and induced systemic resistance in conventional agriculture. *Crop Sci.*, *44*(6), 1920–1934.

Van, E. J. D. V., Kastelein, P., De Vries, P. M., & Van, O. L. S., (2001). Effects of ecological factors on the survival and physiology of *Ralstonia solanacearum* bv.2 in irrigation water. *Can. J. Microbiol.*, *47*(9), 842–854.

Vaux, D. L., (1993). Toward an understanding of the molecular mechanisms of physiological cell death. *Proceed. Nation. Acad. Scienc. USA, 90*(3), 786–789.

Volcani, Z., (1985). *Bacterial Diseases of Plants in Israel*. Agricultural Research Organization.

Wang, J. F., & Lin, C. H., (2005). *Integrated Management of Tomato Bacterial Wilt (No. Other)*. AVRDC – The World Vegetable Center.

Wang, J. F., Stall, R. E., & Vallejos, C. E., (1994). Genetic analysis of a complex hypersensitive reaction to bacterial spot in tomato. *Phytopathol.*, *84*(2), 126–132.

Wassermann, E., Montecchia, M. S., Correa, O. S., Damian, V., & Romero, A. M., (2017). *Clavibacter michiganensis* subsp. *michiganens* is strains virulence and genetic diversity. A first study in Argentina. *Europ. J. Plant Pathol.*, *149*(1), 35–42.

Wenneker, M., Verdel, M. S. W., Groeneveld, R. M. W., Kempenaar, C., Van, B. A. R., & Janse, J. D., (1999). *Ralstonia* (*Pseudomonas*) *solanacearum* race 3 (biovar 2) in surface water and natural weed hosts: First report on stinging nettle (*Urtica dioica*). *Eur. J. Plant Pathol.*, *105*(3), 307–315.

Whalen, M. C., Wang, J. F., Carland, F. M., Heiskell, M. E., Dahlbeck, D., Minsavage, G. V., Jones, J. B., et al., (1993). Avirulence gene *avrRxv* from *Xanthomonas campestris pv. vesicatoria* specifies resistance on tomato line Hawaii 7998. *Mol. Plant–Microbe Interact.*, *6*(5), 616–627.

Williams, S. D., Boehm, M. J., & Hand, F. P., (2017). *Bacterial Diseases of Plants*. Ohioline. https://ohioline.osu.edu/factsheet/plpath-gen-6 (accessed on 12 December 2023).

Wiseman, M., (2008). The second World Cancer Research Fund/American Institute for Cancer Research expert report. Food, nutrition, physical activity, and the prevention of cancer: A global perspective. *Proceed. Nutrit. Soci.*, *67*(3), 253–256.

Wittmann, J., Brancato, C., Berendzen, K. W., & Dreiseikelmann, B., (2016). Development of a tomato plant resistant to *Clavibacter michiganensis* using the endolysin gene of bacteriophage CMP1 as a transgene. *Plant Pathol.*, *65*(3), 496–502.

Wittmann, J., Gartemann, K. H., Eichenlaub, R., & Dreiseikelmann, B., (2011). Genomic and molecular analysis of phage CMP1 from *Clavibacter michiganensis* subspecies *michiganensis*. *Bacteriop.*, *1*(1), 6–14.

Xu, X., Miller, S. A., Baysal-Gurel, F., Gartemann, K. H., Eichenlaub, R., & Rajashekara, G., (2010). Bioluminescence imaging of *Clavibacter michiganensis* subsp. *michiganensis* infection of tomato seeds and plants. *Appl. Env. Micro.*, *76*(12), 3978–3988.

Xue, Q. Y., Ding, G. C., Li, S. M., Yang, Y., Lan, C. Z., & Guo, J. H., (2013). Rhizocompetence and antagonistic activity towards genetically diverse *Ralstonia solanacearum* strains-an improved strategy for selecting biocontrol agents. *Appl. Microbio. Biotech.*, *97*(3), 1361–1371.

Yu, Z. H., Wang, J. F., Stall, R. E., & Vallejos, C. E., (1995). Genomic localization of tomato genes that control a hypersensitive reaction to *Xanthomonas compestris pv. vesicatoria* (Doidge) dye. *Genet., 141,* 675–682.

Yuan, F. H., & He, L. Y., (1998). An anti-*Ralstonia solanacearum* protein from potato and its immunogold localization *in vivo*. *Bact. Wilt Dis.* (pp. 209–217). Springer.

Zalewskz, J. C., & Sequeira, L., (1975). An inhibitor of bacterial growth from potato and its role in resistance to bacterial wilt. *Phytopathol.*, *65*, 1336–1341.

Zeilmaker, T., Ludwig, N. R., Elberse, J., Seidl, M. F., Berke, L., Van, D. A., Schuurink, R. C., et al., (2015). Downy mildew resistant 6 and DMR 6-like oxygenase 1 are partially redundant but distinct suppressors of immunity in *Arabidopsis*. *Cell. Mol. Biol.*, *81*(2), 210–222.

CHAPTER 8

Recent Advances in Development of Diagnostic Kits for Early Detection of Viral Diseases in Crop Plants

KHUSHBU JAIN,[1] JITESH KUMAR,[1] TUSHAR RANJAN,[1] RAVI RANJAN KUMAR,[1] ANAND KUMAR,[2] MOHAMMAD ANSAR,[3] MAHESH KUMAR,[1] MOHAMMAD SHAMIM,[1] KUMARI RAJANI,[4] VINOD KUMAR,[1] BIRENDER SINGH,[2] SANGITA SAHNI,[5] and BISHUN DEO PRASAD[6]

[1]*Department of Molecular Biology and Genetic Engineering, Bihar Agricultural University, Sabour, Bhagalpur, Bihar, India*

[2]*Department of Plant Breeding and Genetics, Bihar Agricultural University, Sabour, Bhagalpur, Bihar, India*

[3]*Department of Plant Pathology, Bihar Agricultural University, Sabour, Bhagalpur, Bihar, India*

[4]*Department of Seed Science and Technology, Bihar Agricultural University, Sabour, Bhagalpur, Bihar, India*

[5]*Department of Plant Pathology, TCA, Dholi, Dr. Rajendra Prasad Central Agricultural University, Pusa, Samastipur, Bihar, India*

[6]*Department of AB&MB, CBS&H, Dr. Rajendra Prasad Central Agricultural University, Pusa, Samastipur, Bihar, India*

ABSTRACT

Plant viral infections are the greatest serious threat to agriculture's long-term viability, resulting in enormous losses every year. New techniques that can

New Advances in Crop Disease Management. Sangita Sahni, Ganesh Thapa, and Bishun Deo Prasad (Eds)

enable reliable identification of the causal agents are in high demand for the successful treatment and prevention of viral infections. Several diagnostic methods with enhanced acuteness and precision for the identification of common and unidentified plant viruses are continuously being developed, thanks to improvements in biochemical and molecular biology techniques. Currently, the most extensively used approaches for plant virus diagnostics are serological and nucleic acid (NA) assays. On the other hand, interest in creating real-time approaches that will make in-field diagnostics easier. Innovative technologies based on next-generation sequencing (NGS) have shown significant promise in detecting numerous viruses at the same time; nevertheless, such techniques are still in their early phases in plant viral disease diagnostics. This chapter addresses current advances in plant viral disease detection, determination, and diagnosis. In-field diagnostics require new portable instruments and technology that can give real-time assessments in a reasonably short period of time. It is also thoroughly described how these tools and procedures are now being developed and used in plant virology, as well as any potential drawbacks.

8.1 INTRODUCTION

Infections in plants, animals, and humans are caused by viruses, which are the most genetically diverse organisms. Viruses, in general, have a little genome that only produces a few proteins, making it challenging to regulate them using a variety of methods (Rojas et al., 2008). Viruses, among all disease-producing pathogens in plants, present a threat to agricultural production, having been reported to cause half of emergent infectious disease among a range of crop plants, with viral infection causing over 40% of total crop losses (Ranjan et al., 2021). Viruses, as a result, are one of the biggest restrictions in agricultural output around the world, as they reduce crops' yields and quality as well as number (Kumar et al., 2020; Mehetre et al., 2021). Worldwide, a variety of crop species are known to be infected by more than 25 plant virus families, causing considerable economic losses. Vertical and horizontal modes of transmission are the most common mechanisms of transmission for plant virus species. In the vertical method, the infectious agent passes occurs either by vegetative propagation or through the parent plants or sexual reproduction via contaminated seeds (Kumari et al., 2020). In the horizontal method, diseases are mostly transmitted by insect vectors, agricultural equipment, and other direct or indirect types of contamination. Various signs, such as mosaic damage, yellowing, chlorosis, stunting, and necrosis, are found in infected plants after infection by a virus species, resulting in the loss of proper growth and reproductive capabilities of the plants (Hancinsky et al., 2020).

There are now just a few treatments to treat or prevent plant viral infectious diseases in the field. Therefore, early detection of the underlying causes is crucial to stopping the spread of infectious diseases. The adoption of precise and sensitive technology for precise detection is required for effective and successful plant disease control. In this context, new techniques and approaches for viral diagnostics are constantly being developed and implemented (Rubio et al., 2020). However, the majority of viral detection and identification tests are costly, time-consuming, and inappropriate for collecting new viruses. Nonetheless, there is a necessity for these diagnostic procedures to be updated in order to capture new and emerging viral species as well as associated infections. The fact that plants commonly have concurrent infections with different virus species or variants of the same virus species poses the biggest problem for diagnosing plant-based viral diseases. The primary problem with viral diagnostics in this situation is the simultaneous identification of many viruses. In this context, the fundamental bottleneck in viral diagnostics is the simultaneous identification of numerous viruses. Furthermore, because plant viruses have such a wide genetic variety, using a single approach for effective identification becomes more difficult (Jaurej et al., 2013). Viruses are made up of genetic material, such as RNA or DNA, which has a high rate of mutation. To accomplish speedy and reliable detection, preferred procedures and techniques such as electron microscopy (EM), immune-serological methods (ELISA), molecular approaches, and biosensor-based methods are currently frequently used. To improve sensitivity and specificity in disease diagnosis, a variety of approaches are frequently used. The study of plant viral dynamics is gaining popularity; yet efficient methodologies for detecting viruses in infected plants have been observed to be limited when compared to viruses that affect humans and animals.

Because of recent advances in molecular biology and genomics methodologies, we have been able to construct methods for detecting plant viruses by doing away with viral culture (Jaurez et al., 2013; Rubio et al., 2020). Invariably, the introduction of next-generation sequencing (NGS) technologies has resulted in a revolution in the detection of plant viruses. These methods allow for the simultaneous sequencing of millions or billions of nucleotides and do not require prior knowledge of viral sequences, making it possible to detect any viruses, including previously unreported ones. Furthermore, the use of such a developing field of study uses contemporary methods to comprehend the ecology of disease risk, emergence, and dynamics of viral diseases in addition to earlier knowledge of the virus and its diagnostics (Villamor et al., 2016).

8.2 METHODS FOR PLANT VIRAL DIAGNOSTIC

Plant virus diagnostics have been developed and commercialized in a variety of ways. However, the practical implementation of any common approach is dependent on a variety of parameters, including cost, sensitivity, speed, equipment availability, and illness stage. When plants become infected with a virus, they might exhibit a variety of physiological signs (Martinelli et al., 2015). Visual inspections like these have historically been used to identify viral infections using indicator plants. There have also been developed a number of additional quantitative high-throughput imaging-based methods. Serological and nucleic acid (NA)-based technologies are becoming crucial tools for diagnosing plant diseases as molecular biology techniques. In recent decades, advances in NGS technology have provided a system for diagnosing viruses, and these technologies have been used to detect and identify a variety of viral species (Barba et al., 2014). Some of the most popular techniques for the accurate detection of plant viruses are shown in Figure 8.1.

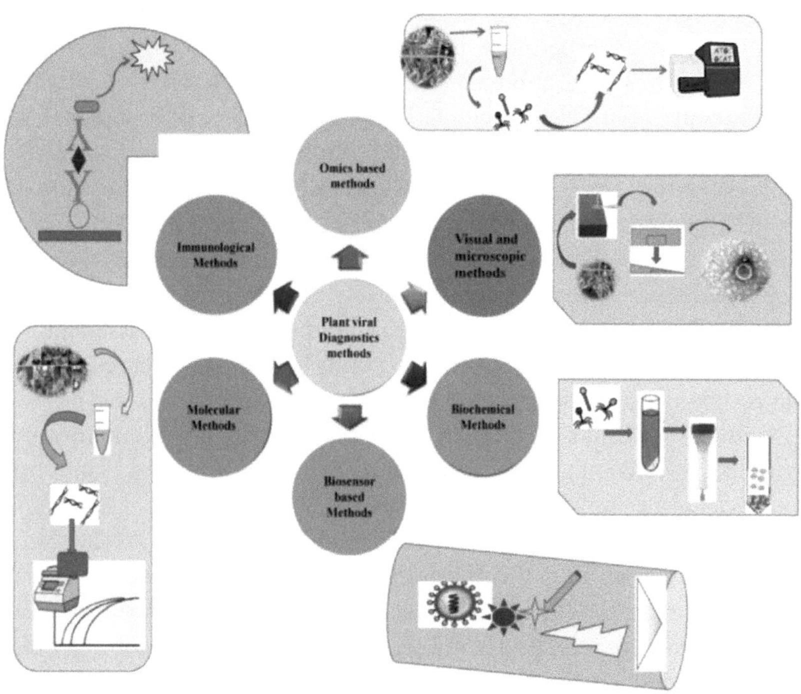

FIGURE 8.1 Common diagnostic techniques for identifying plant viruses and diagnosing diseases of diverse crops.

Source: Reprinted with permission from Mehetre et al., 2021.

A glimpse of the key diagnostic techniques is listed in subsections, ranging from conventional visual examinations of the indicator plants to cutting-edge third-generation sequencing technologies.

8.2.1 VISUAL INSPECTIONS AND INDICATOR PLANTS

Visual observation of infected seeds and plants were once the most prevalent approach for detecting plant pathogens. At the turn of the 20th century, plant viral diseases' symptoms were proposed as a basis for virus taxonomy. The symptoms produced in the host plants are usually used to characterize a plant viral disease with a known origin (Webster et al., 2004). This is due to the simplicity with which symptoms can be identified, particularly if they seem to be disease specific. Mosaics and chloroses are two different plant viral infection symptoms that are closely associated to particular physiological problems. While chlorosis comprises systemic diseases that impact the entire plant, mosaics involve variations in the colors and shapes of the leaves. To determine the titer infected plants, contain infectious viral particles, it is advised to use the plant infection method and count the mosaic (necrotic) spots that grow on the leaves of indicator plants. The number of spots that arise will vary depending on the viral particle concentration. Different approaches, such as visible light imaging, chlorophyll fluorescence imaging, hyperspectral imaging, and thermal imaging, can be used to conduct a visual examination in most cases. These approaches, on the other hand, are heavily influenced by environmental and biological factors. Visual checks are also ineffective in asymptomatic situations (Balodi et al., 2017). Fluorescent proteins have thus been widely employed as indicators of infection and to measure the amount of viral particles in host plants and have remained efficient investigative tools for interpreting biological processes. One of the most effective strategies for viral diagnosis was the imaging of fluorescence tagged proteins in host plants using microscopic techniques (Mukta & Bart, 2015).

8.2.2 MICROSCOPIC METHODS

One of the most popular methods for spotting viruses in plant tissues is microscopic detection, which makes use of contemporary light and high-resolution electron microscopes. Under a microscope, inclusions or Ivanovsky crystals viral particle aggregates that are produced after infection by some viruses can be observed in plant cells. Virus inclusions take different forms depending on the type of virus (Kitajima, 2004). Tobacco mosaic virus (TMV) produces needle-like and hexagonal crystals, whereas Potato virus X (PVX) and wheat mosaic virus produce spherical amorphous entities. The characterization of viral particles' morphology in both unprocessed and processed samples has been improved using transmission electron microscopy (TEM). One of the first ideas to categorize viruses into distinctive groups based on their morphological and serological links, as well as some of their biological characteristics, has emerged from TEM research (Richert-Pöggeler et al., 2019). Direct detection is a possibility with TEM, where the diseased tissue is initially homogenized and then stained negatively. Various plant-infecting viruses were observed and researched using EM techniques, including tomato yellow leaf curl disease (TYLCD), potato virus S (PVS), rice stripe mosaic virus (RSMV), tomato brown rugose fruit virus (ToBRFV), pepino mosaic virus (PepMV), and potato virus M (PVM) (Zhang et al., 2020). Using EM techniques, researchers examined and investigated many plant-infecting viruses such as TYLCD, PVS, RSMV, ToBRFV, PepMV, and PVM (Zhang et al., 2020). Other techniques, like as serology, are frequently used in conjunction with EM to improve detection sensitivity. Immune electron microscopy (IEM), which essentially utilizes particular antiserum to combat viral antigen and has been created in this area. Several IEM variations are predominantly employed in viral diagnostics, such as solid-phase immune electron microscopy (SPIEM) and immunosorbent electron microscopy (ISEM) (Kitajima, 2004). IEM can also be performed directly on raw serum, obviating the need for immunoglobulin purification. Thus, microscopy techniques are combined with serological methods to detect localized viral infections within infected plants. An important seed-borne pathogen, bean common mosaic potyvirus (BCMV), was discovered in immature infected leaves of French beans using the ISEM assay (Klap et al., 2020). Cryo-electron microscopy (cryo-EM), which can disclose virtually the atomic-level structures of viruses, is currently the most advanced microscopy technology utilized in the study of plant-virus interactions. Cryo-EM technology has aided in the understanding of virus structural conformations, such as the identification of specific nucleoproteins from negative-stranded RNA viruses, as well as the by shedding light on the components and processes involved in viral replication, we can better comprehend viral infections (Zamora et al., 2017). Thanks to improvements in microscopic techniques, the identification of viral indicator molecules has helped us interpret the localization of viral infections (Ertel et al., 2017). Future improvements in these methods could have a significant effect on plant virus surveillance and detection.

8.2.3 SEROLOGICAL METHODS

Serological approaches, such as ELISA (enzyme-linked immunosorbent assay), are commonly employed in plant virus diagnostics. They are based on using polyclonal or monoclonal antibodies to precisely detect a protein molecule. The ELISA approach was originally used to diagnose viral plant diseases by Clark and Adams, but it had previously been used to determine whether viruses are present in plant tissues via serological methods (Clark et al., 1977). In viral identification, ELISA-based procedures such as direct tissue blot immunoassay (DTBIA), double antibody sandwich (DAS) ELISA, and tissue-print (TP) ELISA are the most common (Al-Aziz et al., 2019). Various ELISA modifications have been created and used to detect various plant virus infections in a variety of host plants. The ACP-ELISA and antigen-coated plate ELISA were utilized for RSMV identification in rice plants, among the several variations of ELISA-based methodologies produced. Rice dwarf disease is caused by the rice dwarf virus (RDV), and monoclonal antibodies were developed utilizing pure RDV virions as the immunogen. A dot enzyme-linked immunosorbent assay (dot-ELISA) was developed for the fast detection of RDV in rice using a plate-trapped antigen enzyme-linked immunosorbent assay (PTA-ELISA).

PTA-ELISA has been discovered to be helpful for quickly assessing a large number of samples for the identification of viruses in a variety of settings. Squash mosaic virus (SQMV), cowpea severe mosaic virus (CPSMV), cucumber mosaic virus (CMV), cowpea aphid borne mosaic virus (CABMV), zucchini yellow mosaic virus (ZYMV), and papaya fatal yellowing virus (PLYV) were found in squash, cowpea, and papaya plants, according to a study (Wu et al., 2014). A TAS-ELISA (triple antibody sandwich enzyme-linked immunosorbent assays) was used to detect Tomato yellow leaf curl Thailand virus (TYLCTHV) in tomato, pepper, eggplant, okra, and cucurbit plants. Various Badnaviruses have been detected in West African yam plants using the Protein A sandwich enzyme-linked immunosorbent assay (PAS-ELISA) (Seal et al., 2014).

8.2.4 NUCLEIC ACID (NA)-BASED METHODS

Many domains of technology based on NAs are used in several types of diagnostics, including as clinical, food safety, and environmental analysis. The scope of molecular and genomics approaches has expanded in the diagnosis of infectious pathogens as a result of their rapid growth. NA (DNA or RNA)-based approaches have been generally recognized in viral diagnostics due to their rising prevalence sensitivity. The three essential phases in NA-based analysis are isolation of the NA (either DNA or RNA), amplification, and product analysis. The most important stage is the last one since it shows the outcome right away.

8.2.5 PCR-BASED METHODS

One of the most important techniques is called the polymerase chain reaction (PCR) for molecular detection of different infections and is regarded as the benchmark. The analysis of viral detection using PCR has undergone extensive changes. PCR-based approaches are among the most extensively used methods for viral detection because they use molecular primers to amplify a specific section of the viral genetic material to detect the virus's DNA or RNA signatures (Li et al., 2008). It is usual practice to use PCR-based methods and specialized sets of degenerate primers for viral identification in a variety of host plants. The tomato plant, which is grown all over the world and afflicted with viruses such as tomato mosaic virus (ToMV), Tomato leaf curl New Delhi virus, Tomato leaf curl Gujarat virus, Tomato leaf curl Palampur virus, and others, was subjected to viral detection using primer sets designed for these viruses. Typically, conventional PCR is thought to be the most important tool for diagnosing plant viruses; however, due to the difficulties in adjusting conditions and the time required for detection, these techniques are less widely used in plant viral diagnostics (Mirmajlessi et al., 2015). Furthermore, because most plant viruses have an RNA-based genome, standard PCR is limited Given that it is based on just accurate information, in plant virology on pure DNA. Reverse transcriptase PCR (RT-PCR), which operates on the principle of cDNA synthesis from RNA, is one of the most promising technologies in the diagnosis of plant viruses.

The multiplex RT-PCR method also employed to detect numerous viruses at the same time. In lily plants with mixed viral infections, a study employing simplex and multiplex RT-PCR using virus-specific primers revealed the presence of three distinct viruses, CMV, lily mottle virus (LMoV), and Lily symptomless virus (LSV) (Kwon et al., 2013). Similarly, researchers discovered five potato viruses at the same time: the potato aucuba mosaic virus (PAMV), the Potato leafroll virus (PLRV), the PVM, the PVS, and the Potato virus X (PVX). Multiplex RT-PCR was used to detect PVM, PepMV, ToMV, and PVS in pepino plants (Kumar et al., 2020; Kumari et al., 2020).

These methods, however, have limitations when it comes to identifying distinct variations of known viruses. As a result, PCR and RT-PCR-based techniques need to be improved for results that are significantly more sensitive. During the preceding few decades, numerous groups simultaneously developed such PCR-bound virus-specific antibody strategies. Among these, immunocapture PCR or immunocapture RT-PCR (IC-PCR, IC-RT-CPR) technologies based on the principle of linking the sensitivity of ELISA and PCR were used in the detection of plant viral infections. The viral agents are caught by virus-specific antibodies bound to PCR tubes or micro-well plates, and then the viral particles are PCR or RT-PCR of the viral particles (Figure 8.2).

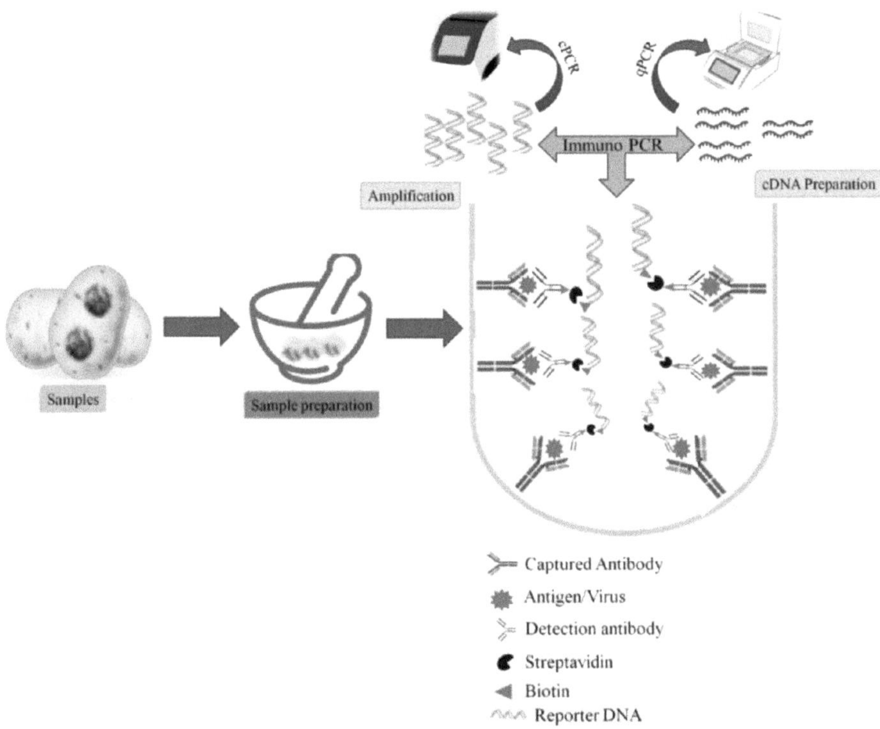

FIGURE 8.2 Schematic representation of immunocapture PCR (IC-PCR) used in plant viral diagnosis.

Source: Adopted from: Mehetre et al. (2021).

8.2.6 *DIAGNOSTIC KIT FOR VIRUS DETECTION*

Effective ELISA and PCR-based techniques have been devised for the identification of most commercially relevant phytopathogenic viruses in laboratories. With the broad adoption of the DNA microarray diagnostics technique in recent years, it is now possible to simultaneously detect all diseases in a single sample of the crop culture being studied. These instrumental analysis techniques all have high sensitivity and productivity, making them useful for monitoring viral infections, managing quarantines, and certifying plant materials. These approaches, on the other hand, are sophisticated and time-consuming; they necessitate experienced personnel and costly equipment, limiting their application to well-equipped facilities and laboratories. As a result, phytosanitary and quarantine control does not apply to the majority of cultivated and imported vegetables, berries, ornamental, and horticultural crops, and potatoes. In order to effectively monitor viral infections, rapid, and sensitive detection technologies must be accessible in both laboratory and field situations. Immunochromatographic assay (ICA), which is based on the interaction between the target virus and immunoreagents (antibodies and their conjugates with colored colloidal particles) put on membrane carriers, is one of the potential options for overcoming this difficulty (test strips). The sample liquid flows across membranes when the test strip is dipped into the sample being examined, triggering immunochemical reactions that result in visible coloring in the test and reference lines (Byzova et al., 2009). Spot Check LF (Adgen Ltd., UK), Pocket Diagnostic (Forsite Diagnostics Ltd., UK), and Immunostrips are just a few of the overseas companies that make plant virus test strips (Agdia, United States). The results of these strips show that virus infection can be identified in just a few minutes. The great sensitivity of this method, as well as the convenience of sample preparation and analysis, are all significant advantages. The goal of this research was to create immunochromatographic test systems for the express detection of plant viruses with different shapes and sizes of virions, such as spherical bean mild mosaic virus (BMMV) and carnation mottle virus (CarMV), rod-shaped TMV, and filamentous potato viruses X and Y (PVX, PVY). The detailed principle of ICA is depicted in Figure 8.3.

8.2.7 NEXT-GENERATION SEQUENCING (OMICS)-BASED METHODS

Research in both basic and practical plant virology has evolved as a result of NGS technology. Total NA (DNA or RNA) extraction from samples (environments, infected host plants, etc.), followed by DNA fragmentation for library formation, are typical processes in NGS-based techniques. Different sequencing chemistries and platforms can be used to analyze fragmented

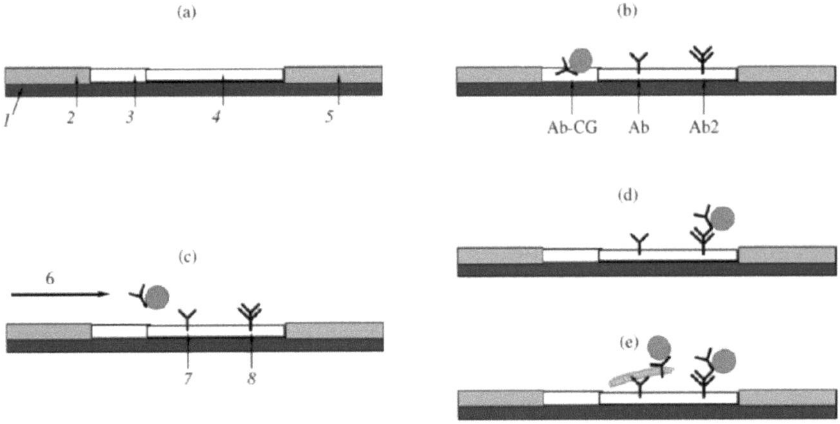

FIGURE 8.3 The principle of immunochromatographic assay. (a) test-system composition: 1 – plastic mount; 2 – absorption membrane; 3 – conjugate mount; 4 – working membrane; and 5 – absorption pad. (b) Reagents introduced: Ab – virus-specific antibodies; Ab2 – antibodies against rabbit IgG; and Ab-CG – colloidal gold conjugated to virus-specific antibodies. (c) A schematic representation of analysis: 6 – spreading of liquid front; 7 – test line; and 8 – control line. (d) Test result in the absence of the virus: one colored band in the control line area. And (e) test result in the presence of the virus: two colored bands in both test and control line areas.

DNA by using a collection of primers and adapters for synthesized DNA. The term omics refers to a number of biological fields, including metagenomics, genomics, proteomics, and metabolomics, all of which focus on the investigation of diverse cellular molecules using next-generation sequencing (NGS) technology. Metagenomics, for example, is the study of a collection of entire DNA or RNA from a diverse group of organisms (Riesenfeld et al., 2004). For the detection of viruses, metagenomics is being developed. Total NA (DNA or RNA) is isolated from infected plant samples and sequenced using next-generation sequencing (NGS) technologies, with viral sequences discovered using bioinformatics methods. Blawid et al. detailed the sample preparation and analytical methods required for routine sequencing, which is currently utilized to find plant virus genomes in metagenomics NGS data. In crops and wild plants, viral metagenomics has been used to study the biodiversity of plant viruses over the past 10 years. Numerous of these studies have used the "ecogenomics" method, which looks at viral populations in individuals rather than the environment as a whole (Blawid et al., 2017). Because viral metagenomics does not require prior knowledge of the virus or virus-specific primers, it has been utilized to find novel viruses in plants without bias.

8.2.7.1 ANALYSIS USING SECOND-GENERATION SEQUENCING TECHNOLOGIES

Second-generation sequencing technologies arose from the combination of high-throughput sequencing and low-cost results delivery. Commercially available platforms include 454 Life Sciences/Roche Diagnostics' platforms; Illumina, Inc.'s HiSeq, MiSeq, and NextSeq; ABI's SOLiD; and Ion Torrent PGM, Proton, S5, and BGISeq-500 (Ambardar et al., 2017). In comparison to traditional methodologies, these sequencing platforms were proven to be more accurate and sensitive when diagnosing viruses. The use of second-generation sequencing technology to detect new and re-emerging viruses has emerged as the most promising technique in plant viral disease diagnostics. Surprisingly, these techniques have also been proven to be useful for identifying normal viral infections in plants using small RNAs (sRNA). Next-generation sequencing systems can also be used to detect plant viruses or viroids in an indirect manner (Xing et al., 2020). In plants with viral infections, a specific type of RNA molecule known as a short interfering RNA is produced by the host plant (siRNA). A cytoplasmic cell surveillance mechanism called RNA silencing (RNAi) employs short interfering RNAs (siRNA) as a guide to recognize dsRNA and specifically target and eliminate single- and double-stranded RNA (dsRNA) molecules that are the same as the inducer. These technologies, which include the following, provide good opportunities to detect viruses or viroids infecting plants, even at extremely low titers and in asymptomatic infections viruses or viroids that have never been identified before (Maliogka et al., 2018).

Deep sequencing of the sweet potato viral disease (SPVD) infecting sweet potatoes using the Illumina MiSeq technology revealed the genetic variety of the viruses in South African plant samples, both symptomatic and asymptomatic. The authors demonstrated that using Illumina sequencing platforms, mixed virus infections in individual plants might be examined for begomoviruses, potyviruses, criniviruses, DNA, and RNA viruses without harming the plant. Prior knowledge. These findings also showed that different viruses, such as SPCSV and SPFMV, can cause a disease, such as SPVD, and begomoviruses significantly contribute to the development of the disease (Nhlapo et al., 2018).

8.2.7.2 ANALYSIS USING THIRD-GENERATION SEQUENCING TECHNOLOGIES

Second-generation sequencing has revolutionized biological research, and it is now widely utilized in studies of microbial diversity in the environment, the human-associated microbiome, plant genomes, and other fields.

Third-generation sequencing technologies, also known as single-molecule sequencing technologies, have received a lot of interest in the previous decade (Edwards et al., 2010). These methods have a number of advantages over second-generation sequencing methods. They can, for example, significantly generate each RNA or DNA sample length reads, obviating the requirement to build contigs from short sequence reads from scratch. These approaches also have extremely quick runtimes, and most crucially, need very little input template (DNA or RNA), as well as providing real-time analysis. The Helicos™ Genetic Analysis System from SeqLL, which is currently available, is a third-generation sequencing platform, pacific biosciences' single-molecular real-time (SMRT) sequencing, and Oxford Nanopore Technologies' MinIONnanopore sequencing (Xiao et al., 2019).

With a sequence length of greater than 15 kb on average, Oxford NanoporeMinION has proven to be useful for real-time analysis. MinION has been used to identify and determine the entire genome sequences of a variety of viruses that affect both humans and animals, such as the Ross River viruses, ebola33, dengue34, zika35, influenza31, cowpox36, but its use to plant viruses is still in the early stages (Quick et al., 2017). Using MinIONnanopore sequencing, a few related investigations have shown encouraging findings. Several studies have proven that this methodology is accurate and cuts down on analysis time; because of its mobility, it is being actively exploited to build modern diagnostic procedures for pests and illnesses in agriculture. The use of MinIONnanopore sequencing in crop plants has been expanded to microbiome characterization and pathogen identification. Currently, only a few research have used MinIONnanopore sequencing to detect or analyze plant viruses such as Maize streak virus, maize yellow mosaic virus (YMV), and Maize totivirus in maize crops, or Plum pox virus in plum trees (among others).

The portable nanopore sequencer is reportedly capable of promptly differentiating and accurately detecting widely disseminating plant viruses in a variety of agricultural plants. The most economically significant virus to infect cultivated potatoes (Solanum tuberosum L.) is the potato virus Y (PVY), which is challenging to identify and classify due to its extreme genetic variety. Della Bartola and colleagues (Della et al., 2020) used traditional genetic and serological approaches to analyze Irish PVY isolates. In eight potato plants in part, nanopore sequencing was also utilized to detect and reconstruct the entire five PVY genotypes and the genomic sequences of four viruses (PVY, PVX, PVS, and PLRV). Nanopore technologies have significant potato production systems' potential for viral detection and for studying the genetic diversity of highly diverse viruses like PVY, according to genome reconstructions of RNA viruses like PVY and other viruses. This study demonstrated how portable nanopore technology can be used to detect RNA viruses in potato plants. The implementation of a unique platform to monitor optimal crop hygiene procedures in commercial potato production systems demonstrates the wide application of nanopore sequencing in plant virology. Third-generation sequencers generate sequence reads of hitherto unheard-of lengths, which will greatly improve the quality of genome assemblies. Furthermore, sequencing in the field will be facilitated by the speed of sequencing and the convenience of sample preparation. Because the ONT device is small, portable, with laptop control, it is ideal for in-field diagnostics. During the 2014 Liberia Ebola outbreak, ONT was used for the first time in the field to treat human infections (Hoenen et al., 2016). As a result, these approaches offer a straightforward path to diagnosing plant viral diseases. Figure 8.4 shows a possible pathway for diagnosing plant viral infections quickly and on-site utilizing portable sequencing equipment (MinION) developed by ONT.

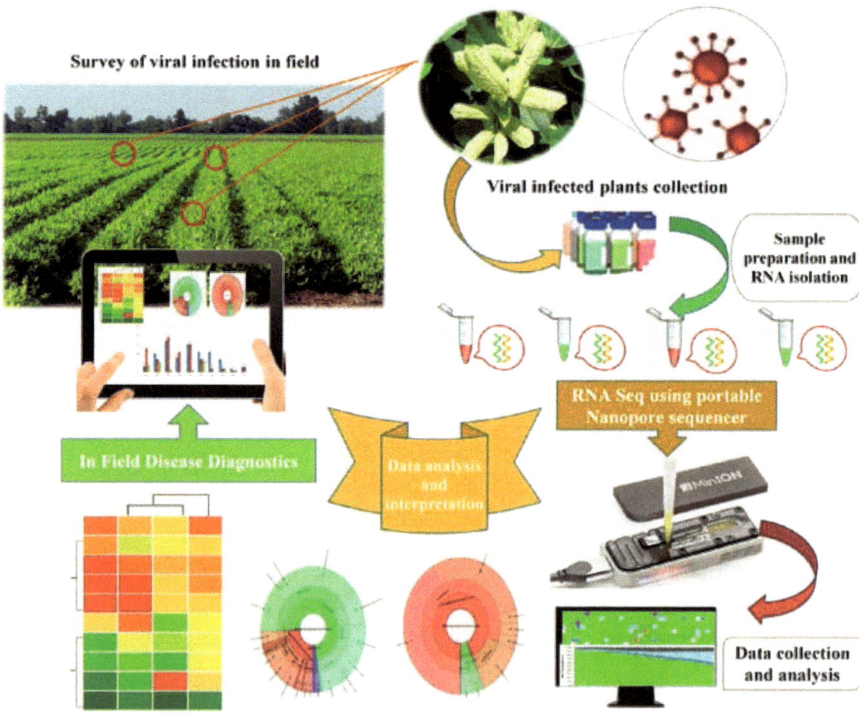

FIGURE 8.4 A putative workflow for rapid and on-site diagnosis of plant viral infections using portable sequencing device (MinION) developed by ONT, UK.

Source: Reprinted with permission from Mehetre et al., 2021. © 2021 by the authors. Licensee MDPI, Basel, Switzerland. http://creativecommons.org/licenses/by/4.0/

8.3 CONCLUSIONS

Plant health must be monitored, and pathogens must be quickly identified in order to stop viral disease proliferation and support efficient management techniques. Drawbacks of diagnostic methods for identifying plant diseases include low sensitivity, the requirement for prior knowledge of the genome sequence, and the inability to simultaneously identify a large number of pathogens. ICA, which is based on the interaction between the target virus and immunoreagents (antibodies and their conjugates with colored colloidal particles) put on membrane carriers, is one of the potential options for overcoming this difficulty. Total NA content in biological samples may now be determined thanks to the advancement of advanced DNA sequencing methods. The use of omics approaches, such as third-generation sequencing platforms like Oxford Nanopores, as a generic tool for diagnosing plant diseases is most easily applied to viral diagnostics. However, much more has to be learned in order to improve the current level of viral diagnostics in plant diseases. Despite significant progress in the field of pathogen diagnosis tools, there are still not enough universally applicable, low-cost, and easy procedures. Furthermore, mixed infections continue to be a stumbling block in disease diagnosis. All approaches, such as cell culture techniques, molecular techniques, immunological techniques, and other modern methods, must be improved. Discovering the proper target location for molecular detection is essential for detection success, yet locating such targets is difficult. Recent discoveries of numerous novel viral organisms suggest that metagenomics is a promising method. Further advances in bioinformatics tools based on NGS technologies could pave the way for disease diagnosis technology developments that will help reduce crop losses from newly developing diseases.

KEYWORDS

- **diagnostics**
- **in-field diagnosis**
- **next-generation sequencing**
- **nucleic acid**
- **plant viruses**
- **serological and nucleic acid methods**
- **third-generation sequencing**

REFERENCES

Abd El-Aziz, M. H., (2019). Three modern serological methods to detect plant viruses. *J. Plant. Sci. Phytopathol., 3,* 101–106.

Ambardar, S., Gupta, R., Trakroo, D., Lal, R., & Vakhlu, J., (2016). High throughput sequencing: An overview of sequencing chemistry. *Ind. J. Microbiol., 56,* 394–404.

Balodi, R., Bisht, S., Ghatak, A., & Rao, K. H., (2017). Plant disease diagnosis: Technological advancements and challenges. *Indian Phytopathol., 70,* 275–281.

Barba, M., Czosnek, H., & Hadidi, A., (2014). Historical perspective, development, and applications of next-generation sequencing in plant virology. *Viruses, 6,* 106–136.

Blawid, R., Silva, J. M., & Nagata, T., (2017). Discovering and sequencing new plant viral genomes by next-generation sequencing: Description of a practical pipeline. *Ann. Appl. Biol., 170,* 301–314.

Bos, L., (1970). *Symptoms of Virus Diseases in Plants* (2nd edn., p. 132). Pudoc: Wageningen, The Netherlands.

Byzova, N. A., Safenkova, I. V., Chirkov, S. N., Zherdev, A. V., Blintsov, A. N., Dzantiev, B. B., & Atabekov, I. G., (2009). Development of immunochromatographic test systems for express detection of plant viruses. *Applied Biochemistry and Microbiology, 45*(2), 204–209.

Clark, M. F., & Adams, A. N., (1977). Characteristics of the microplate method of enzyme-linked immunosorbent assay for the detection of plant viruses. *J. Gen. Virol., 34,* 475–483.

Della, B. M., Byrne, S., & Mullins, E., (2020). Characterization of potato virus Y isolates and assessment of nanopore sequencing to detect and genotype potato viruses. *Viruses, 12,* 478.

Ertel, K. J., Benefield, D., Castano-Diez, D., Pennington, J. G., Horswill, M., Den, B. J. A., Otegui, M. S., & Ahlquist, P., (2017). Cryo-electron tomography reveals novel features of a viral RNA replication compartment. *eLife, 6,* 25940.

Hancinsky, R., Mihálik, D., Mrkvová, M., Candresse, T., & Glasa, M., (2020). Plant viruses infecting *Solanaceae* family members in the cultivated and wild environments: A review. *Plants, 9,* 667.

Hoenen, T., Groseth, A., Rosenke, K., Fischer, R. J., Hoenen, A., Judson, S. D., Martellaro, C., Falzarano, D., Marzi, A., Squires, R. B., et al., (2016). Nanopore sequencing as a rapidly deployable Ebola outbreak tool. *Emerg. Infect. Dis., 22,* 331.

Juarez, M., Legua, P., Mengual, C. M., Kassem, M. A., Sempere, R. N., Gomez, P., Truniger, V., & Aranda, M. A., (2013). Relative incidence, spatial distribution, and genetic diversity of cucurbit viruses in eastern Spain. *Ann. Appl. Biol., 162,* 362–370.

Kitajima, E. W., (2004). Electron microscopy in plant virology: Past, present, and future. *Microsc. Microanal., 10,* 212–213.

Klap, C., Luria, N., Smith, E., Bakelman, E., Belausov, E., Laskar, O., Lachman, O., et al., (2020). The potential risk of plant-virus disease initiation by infected tomatoes. *Plants, 9,* 623.

Kumar, R. R., Ansar, M., Rajani, K., Kumar, J., & Ranjan, T., (2020). First report on molecular basis of potato leaf roll virus (PLRV) aggravation by combined effect of tuber and prevailing aphid. *BMC Research Notes, 13*(1), 1–4.

Kumari, P., Kumar, J., Kumar, R. R., Ansar, M., Rajani, K., Kumar, S., & Ranjan, T., (2020). Inhibition of potato leafroll virus multiplication and systemic translocation by siRNA constructs against putative ATPase fold of movement protein. *Scientific Reports, 10*(1), 1–1.

Kwon, J. Y., Ryu, K. H., & Choi, S. H., (2013). Reverse transcription polymerase chain reaction-based system for simultaneous detection of multiple lily-infecting viruses. *Plant Pathol. J., 29*, 338.

Li, R., Mock, R., Huang, Q., Abad, J., Hartung, J., & Kinard, G., (2008). A reliable and inexpensive method of nucleic acid extraction for the PCR-based detection of diverse plant pathogens. *J. Virol. Methods, 154,* 48–55.

Maliogka, V. I., Minafra, A., Saldarelli, P., Ruiz-García, A. B., Glasa, M., Katis, N., & Olmos, A., (2018). Recent advances on detection and characterization of fruit tree viruses using high-throughput sequencing technologies. *Viruses, 10*, 436.

Martinelli, F., Scalenghe, R., Davino, S., Panno, S., Scuderi, G., Ruisi, P., Villa, P., Stroppiana, D., Boschetti, M., Goulart, L. R., et al., (2015). Advanced methods of plant disease detection. A review. *Agron. Sustain. Dev., 35*, 1–25.

Mehetre, G. T., Leo, V. V., Singh, G., Sorokan, A., Maksimov, I., Yadav, M. K., Upadhyaya, K., et al., (2021). Current developments and challenges in plant viral diagnostics: A systematic review. *Viruses, 13*(3), 412.

Mirmajlessi, S. M., Loit, E., Maend, M., & Mansouripour, S. M., (2015). Real-time PCR applied to study on plant pathogens: Potential applications in diagnosis: A review. *Plant. Prot. Sci., 51*, 177–190.

Mutka, A. M., & Bart, R. S., (2015). Image-based phenotyping of plant disease symptoms. *Front. Plant Sci., 5*, 734.

Naidu, R. A., & Hughes, J. D., (2001). A. Methods for the detection of plant virus diseases. In: *Proceedings of the Conference on Plant Virology in Sub Saharan Africa Organized by IITA* (pp. 233–253). Ibadan, Nigeria.

Nhlapo, T. F., Rees, D. J., Odeny, D. A., Mulabisana, J. M., & Rey, M. E., (2018). Viral metagenomics reveals sweet potato virus diversity in the Eastern and Western Cape provinces of South Africa. *S. Afr. J. Bot., 117*, 256–267.

Quick, J., Grubaugh, N. D., Pullan, S. T., Claro, I. M., Smith, A. D., Gangavarapu, K., Oliveira, G., Robles-Sikisaka, R., Rogers, T. F., Beutler, N. A., et al., (2017). Multiplex PCR method for MinION and Illumina sequencing of Zika and other virus genomes directly from clinical samples. *Nat. Protoc., 12*, 1261–1276.

Ranjan, T., Pal, A. K., Prasad, B. D., Kumar, R. R., Kumar, M., Shamim, M., & Jambhulkar, S., (2021). Reassessing the mechanism of genome packaging in plant viruses with lessons from ATPase fold. *Australasian Plant Pathology, 18*, 1–4.

Richert-Pöggeler, K. R., Franzke, K., Hipp, K., & Kleespies, R. G., (2019). Electron microscopy methods for virus diagnosis and high-resolution analysis of viruses. *Front. Microbiol., 9*, 3255.

Riesenfeld, C. S., Schloss, P. D., & Handelsman, J., (2004). Metagenomics: Genomic analysis of microbial communities. *Annu. Rev. Genet., 38*, 525–552.

Rojas, M. R., & Gilbertson, R. L., (2008). Emerging plant viruses: A diversity of mechanisms and opportunities. In: Roossinck, M. J., (ed.), *Plant Virus Evolution* (pp. 27–51). Springer: Berlin/Heidelberg, Germany.

Rubio, L., Galipienso, L., & Ferriol, I., (2020). Detection of plant viruses and disease management: Relevance of genetic diversity and evolution. *Front. Plant. Sci., 11*, 1092.

Seal, S., Turaki, A., Muller, E., Kumar, P. L., Kenyon, L., Filloux, D., Galzi, S., et al., (2014). The prevalence of badnaviruses in West African yams (*Dioscorea cayenensis*-rotundata) and evidence of endogenous pararetrovirus sequences in their genomes. *Virus Res., 186*, 144–154.

Villamor, D. E., Mekuria, T. A., Pillai, S. S., & Eastwell, K. C., (2016). High-throughput sequencing identifies novel viruses in nectarine: Insights to the etiology of stem-pitting disease. *Phytopathology, 106*, 519–527.

Webster, C. G., Wylie, S. J., & Jones, M. G., (2004). Diagnosis of plant viral pathogens. *Curr. Sci., 86*, 1604–1607.

Wu, J., Ni, Y., Liu, H., Ding, M., & Zhou, X., (2014). Monoclonal antibody-based serological assays and immunocapture-RT-PCR for detecting rice dwarf virus in field rice plants and leafhopper vectors. *J. Virol. Methods, 195*, 134–140.

Xiao, H., Li, C., Al Rwahnih, M., Dolja, V., & Meng, B., (2019). Metagenomic analysis of Riesling grapevine reveals a complex virome including two new and divergent variants of *Grapevine leafroll-associated virus 3. Plant Dis., 103*, 1275–1285.

Xing, F., Gao, D., Liu, H., Wang, H., Habili, N., & Li, S., (2020). Molecular characterization and pathogenicity analysis of prunus necrotic ringspot virus isolates from China rose (*Rosa chinensis* Jacq.). *Arch. Virol., 165*, 2479–2486.

Zamora, M., Méndez-López, E., Agirrezabala, X., Cuesta, R., Lavin, J. L., Sánchez-Pina, M. A., Aranda, M. A., & Valle, M., (2017). Potyvirus virion structure shows conserved protein fold and RNA binding site in ssRNA viruses. *Sci. Adv., 3*, 1–7.

Zhang, M., Ye, J., He, J. S., Zhang, F., Ping, J., Qian, C., & Wu, J., (2020). Visual detection for nucleic acid-based techniques as potential on-site detection methods: A review. *Anal. Chim. Acta, 1099*, 1–15.

CHAPTER 9

Development of Disease Resistant Crop Plants Through Induced Mutagenesis

SANJAY J. JAMBHULKAR,[1] SUMAN BAKSHI,[1] ARCHANA RAI,[1] SHANKAR BHUJBAL,[1] JAY PRAKASH,[1] SANGITA SAHNI,[2] and BISHUN DEO PRASAD[3]

[1]*Nuclear Agriculture and Biotechnology Division, Bhabha Atomic Research Center, Mumbai, Maharashtra, India*

[2]*Department of Plant Pathology, TCA, Dholi, Dr. Rajendra Prasad Central Agricultural University, Pusa, Samastipur, Bihar, India*

[3]*Department of AB&MB, CBS&H, Dr. Rajendra Prasad Central Agricultural University, Pusa, Samastipur, Bihar, India*

ABSTRACT

Major threat to food security is the losses incurred by the infestation of pathogens and subsequent disease outbreaks across the globe. Development of disease resistant varieties is a daunting task due to continuously evolving pathogens. However, mutation breeding has been found as promising methods to evolve disease resistant mutant varieties all over the world. Induced mutagenesis has helped to develop 474 disease resistant varieties comprising 334 for fungal, 85 for bacterial and 55 for viral pathogens. So far 264 disease resistant varieties have been developed in cereals, 81 in pulse and 71 in oilseeds crops.

9.1 INTRODUCTION

Sustainable food security needs sufficient food production by conquering the challenges of climate change, population growth, shrinking arable

New Advances in Crop Disease Management. Sangita Sahni, Ganesh Thapa, and Bishun Deo Prasad (Eds)

land, biotic, and abiotic stresses, and quality food. The only option is to evolve climate resilient stress tolerant high yielding varieties. Generation and exploitation of novel variability could help extensively to achieve the targets. Mutation breeding is one of the laser techniques (Chopra, 2005) among the crop improvement methodologies to isolate desirable variability keeping parental characters intact and translating that mutation into variety in the shortest possible time (Bhatia, 1991; Jambhulkar, 2007). Worldwide, 3365 varieties have been developed using induced mutagenesis (IAEA/MVD, 2021). Some of the varieties have notably contributed to the country's economy of the agriculture sector (Konzak, 1984; Maluszynski, 1990; Micke, 1991; Bhatia, 1991; Ahloowalia et al., 2004).

Disease pathogens chiefly fungi, bacteria, and viruses are main threats to plant health and known to cause global yield loss of 16% (Oerke, 2006). Constant challenge in plant breeding is developing disease-resistant varieties to continuously changing pathogens. Among the crop improvement techniques, mutation breeding has played an important role in the isolation of variability for disease resistance (IAEA, 1975) and to tackle the most destructive diseases in the event when no source of resistance is available. The best-cited example in the resistance breeding of wheat for tackling deadly wheat stem rust race Ug99 caused by *Puccinia graminis f. sp. Tritici (Pgt)*. The global initiative by IAEA resulted in the development of stem rust-resistant mutant wheat varieties Eldo Ngano1 and Eldo Mavuno1 in Kenya (Forster, 2014) and the mlo mutations against powdery mildew in barley (Miklis et al., 2007) which is ever-lasting resistance to all races of fungus due to loss of functional mutation occurred at the locus (Panstruga & Schulze-Lefert, 2002). The Mutant Variety Database, International Atomic Energy Agency (IAEA/MVD, 2021) documented a total of 474 disease-resistant mutant varieties which includes 334 varieties resistant to fungal, 85 to bacterial, and 55 to viral pathogens in all crop species. A total of 264 disease-resistant mutant varieties released in cereal, 81 in pulse, and 71 in oilseed crops worldwide. The scope of induced mutagenesis shall be strengthened with the integration of the modern mutation breeding techniques with robust selection and speed breeding tools. The gene-specific mutagenesis like TILLING and CRISPR in the era of genomics are likely to enhance the speed of development of disease-resistant varieties. An overview of the above aspects is highlighted in this review.

9.2 INDUCED MUTAGENESIS AND SCREENING FOR DISEASE RESISTANCE

Climate change has been responsible for creating newer pathogenic races of existing diseases as well as novel diseases for which resistant sources may not be available in the germplasm. In such situations, the immediate answer could be sought from a simpler but effective technology called induced mutagenesis. The success of developing disease-resistant varieties is through the exploitation of major genes. However, polygenic, durable, and acquired (Ribeiro et al., 2001) resistance also opens up wide opportunities to develop disease-resistant varieties through induced mutagenesis. Induced mutagenesis helps to generate variability by evolving novel genes that impart disease resistance. Commonly used mutagens for the treatment of seeds are chemical and physical. Physical mutagens such as X-rays, Gamma rays, alpha particles, UV, fast neutrons, and ion beams usually cause deletions, while chemical mutagens such as sodium azide (SA), ethyl methane sulfonate (EMS), methyl methane sulfonate (MMS), hydrogen fluoride (HF), diethyl sulfate (DES), hydroxylamine, and N-methyl-N-nitrosourea (MNU) usually cause point mutations (Bhatia, 1963). In addition to physical and chemical agents, transposable elements such as mPing are also responsible for the induction of mutations in plants (Hancock et al., 2011).

Mutation breeding for disease resistance has been reported in various IAEA programs (IAEA, 1975). A mutation breeding program is successful when it rectifies one or two shortfalls in an otherwise high-yielding and popular variety. The selection of such a variety and the isolation of a high-yielding desirable mutant without alteration of any other characters shall lead to the development of the mutant variety in the shortest possible time. The second most important objective is the development of a mutant population. In comparison to spontaneous mutation rates, i.e., 10–5–10–8 (Zhong-Hua et al., 2014), the induced mutation rate is 1000–1–10000–1. Looking into this rare probability, raising a large number of plants in the M2 generation (1,00,000–2,00,000) shall help to identify the desirable mutant. The third most important criterion is screening methods. In general, mutations are recessive in nature and appear in the M_2 generation. Therefore, stringent screening in the M_2 generation would strengthen the case of isolating the desirable mutant. Mutant screening can be carried out *in vivo* as well as *in vitro* (Maluszynski et al., 1995). An important point in screening is the contact of the pathogen to the host plant, proper temperature, and humidity for disease development. Screening for disease resistance could be carried out using various methods such as field screening, sick plot and disease nursery screening, single plant and leaf cut methods, *in vitro* cultures, culture filtrate methods, phytotoxins, and pathotoxins methods. Direct laboratory-based techniques are also useful for disease identification, such as polymerase chain reaction (PCR), immunofluorescence (IF), fluorescence *in situ* hybridization (FISH), enzyme-linked immunosorbent assay (ELISA), flow cytometry (FCM), and gas chromatography-mass spectrometry (GC-MS). Indirect methods include thermography, fluorescence imaging, hyperspectral techniques, biosensors based on highly selective bio-recognition elements such as enzymes, antibodies, DNA/RNA, and bacteriophage as a new tool for the early identification of crop diseases (Fang & Ramasamy, 2015).

9.3 MUTATION BREEDING FOR DEVELOPMENT OF DISEASE RESISTANT VARIETIES IN CEREAL CROPS

Cereal crops, which include rice, wheat, maize, and millets, are the most predominant part of food, contributing 75% of calories and 67% of protein intake all over the world. These crops are attacked by fungal, bacterial, and viral pathogens and cause significant yield loss. However, fungal diseases are the most yield-diminishing factor in cereal crops and cause 15-20% yield loss and may reach up to 50% depending on the severity of the disease (Róz˙ewicz et al., 2021). Developing resistant varieties is the most economical and long-term solution to curb crop yield losses due to disease pathogens.

Wheat rusts, namely stripe, stem, and leaf rust, affect grain yield and quality in all wheat-growing areas in the world. The stripe and stem rust infection may cause 100% crop loss, and leaf rust can cause damage up to 50% (Prasad et al., 2021). Mutation breeding has been effectively used in wheat to induce broad-spectrum resistance to all three rust pathogens and powdery mildew fungus using chemical mutagen (Campbell et al., 2012). Induced mutants using fast neutrons have been isolated against powdery mildew disease (Duggal et al., 2002) and yellow rust (Smith et al., 2004) in wheat. Stripe rust-resistant mutants have been isolated in wheat using electron beam and gamma-ray irradiation (Bakshi et al., 2021).

Rice is staple food for more than half of world population (Khush, 2005). The most destructive fungal disease is rice blast disease caused by *Magnaporthe oryzae* and a total of 102 rice blast R genes have been known in rice (Xiao et al., 2018). Zhang et al. (2003) developed a japonica rice mutant (R917) after irradiating F_1 seeds of cross between Chengte 232 and Xiushui 37 with broad-spectrum of resistance to blast. The resistance in R917 is controlled by single dominant genes which is non-allelic to those of its parent and other known genes. Jung et al. (2005) reported a blast lesion mimic (*blm*) mutant in rice which is isolated from an MNU-mutagenized population of cultivar Hwacheong (wild type) which confers broad-spectrum resistance to blast pathogens by activating multiple defense-associated signaling pathways. Among bacterial diseases, bacterial blight caused by the *Xanthomonas oryzae pv. oryzae* is the most destructive pathogen and generally cause yield loss by 10–20% however, it resulted in total crop loss if infection starts at tillering stage (Mew et al., 1993). Around 40 genes known to contribute resistance to various races. Taura et al. (1991b, 1992) identified two rice mutant lines using MNU and these lines carry mutant genes *xa19* and *xa20 and found* resistant to all Japanese and Philippines races of bacteria. Busungu et al. (2016) identified new mutant 'XM14,' and found resistant to all Japanese Xoo races. Genetic analysis showed that resistance is controlled by a single recessive gene *Xa42* and precisely located around the centromeric region of chromosome 3 through linkage analysis. Mutation induction studies in cultivar IR64 using gamma ray irradiation resulted into isolation of rice mutant that are resistant to blast disease, bacterial blight and tungro viral disease (Wu et al., 2005). Rice mutants with enhanced bacterial blight resistance have been identified in Nagina22 mutagenized population using EMS (Mohapatra et al., 2014). Mutants showing increased resistance to biotrophic fungal pathogens have been isolated in other cereals including barley (Jorgensen, 1992), maize (Simmons et al., 1998) and rice (Takahashi et al., 1999; Zeng et al., 2002).

Development of high yielding disease resistant varieties through induced mutagenesis has been successful in cereal crops. A total of 264 disease resistant mutant varieties have been released worldwide, which comprise of 43 bacterial (Table 9.1 and Figure 9.1(A)), 212 fungal (Table 9.2 and Figure 9.1(B)), and 9 viral disease (Table 9.3 and Figure 9.1(C)) resistant (Figure 9.1 and Tables 9.1–9.3). The crop-wise distribution of cereal mutant varieties shows that the maximum number of varieties were released in rice (119), followed by wheat (69), maize (29), barley (29), and 13 varieties of millet. Among the diseases, the maximum number of resistant varieties for fungal diseases have been released in rice (88), followed by wheat (69), barley (28), maize (16), and millets (11). For bacterial resistance, the maximum number of resistant varieties were released in rice (24), followed by maize (12), wheat (4), and millet (2). For viral diseases, seven resistant varieties were released in rice and one each in maize and barley. The data on disease resistance of cereal crop mutant varieties shows that both physical and chemical mutagens were successfully used to induce desired traits.

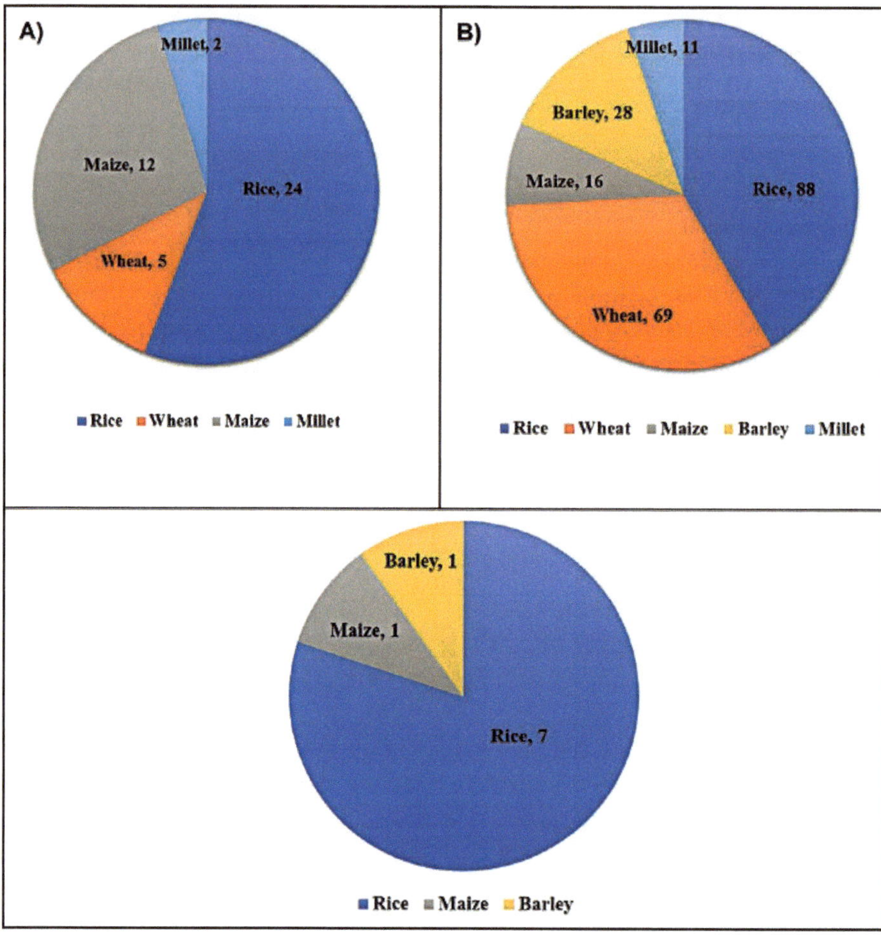

FIGURE 9.1 Cereal mutant varieties resistant to bacterial (A); fungal (B); and viral (C) diseases.

The most frequently used mutagen is gamma rays and 66% of resistant varieties are gamma ray induced mutants, followed by X-rays (6%), neutrons (5%), aerospace (3%), laser (2.2%), ion beam (1.5%) and other physical mutagens (7%). However, chemically induced resistant mutant varieties were only 18. The most successful chemical mutagen for developing resistant

TABLE 9.1 Mutant Varieties of Cereal Crops Resistant to Bacterial Diseases

Sl. No.	Variety Name	Year	Country	Mutagen	Mutant	Resistance to Disease	Other Improved Traits
				Crop: Wheat [5]			
1.	Longfumai 4	1988	China	Gamma rays	Direct	Bacterial diseases	Altered maturity.
2.	Longfumai 7	1996	China	Gamma rays	Direct	Bacterial diseases	Early maturing and quality.
3.	Wanmai 32	1997	China	Ion beam	Direct	Bacterial diseases	–
4.	Longfumai 8	1998	China	Gamma rays	Direct	Bacterial diseases	–
5.	Yannong 5158	2007	China	Aerospace	Direct	Bacterial diseases	Stem rust sheath blight resistance.
				Crop: Rice [24]			
1.	Fulianai	1966	China	Gamma rays	Direct	Resistance to blast.	Short culm, resistance to blast, early.
2.	Yangfuxian 2	1991	China	Gamma rays	Direct	Bacterial diseases	High grain yield and good quality.
3.	Fuchuerai	1978	China	Gamma rays	Indirect	Bacterial leaf blight.	Shorter culm.
4.	Fuxian 6	1989	China	Gamma rays	Indirect	Resistance to BLB.	Early and good cooking quality.
5.	Zhe 852	1989	China	Gamma rays	Direct	Bacterial diseases	Wide adaptability and grain quality.
6.	Zhefu 9	1990	China	Gamma rays	Direct	Bacterial diseases	Grain quality and altered maturity.
7.	Yangfuxian 3	1993	China	Gamma rays	Direct	Bacterial diseases	High grain yield, good quality.
8.	Xiangzaoxian 21	1996	China	Gamma rays and He-Ne laser.	Direct	Bacterial diseases	High grain yield, early maturity.
9.	Yuanjing 7	1999	China	Gamma rays	Direct	Bacterial diseases	High grain yield.
10.	Xiangzaoxian 25	1997	China	Gamma rays	Direct	Bacterial diseases	High grain yield.
11.	Atomita 3	1990	Indonesia	Gamma rays	Direct	Bacterial leaf blight, and bacterial leaf stripe	Resistance to brown plant hopper and good cooking.
12.	DB 250	1987	Vietnam	Gamma rays and MNH	Direct	Bacterial blight	Non-lodging and resistant to *Pyricularia oryzae*.

TABLE 9.1 *(Continued)*

Sl. No.	Variety Name	Year	Country	Mutagen	Mutant	Resistance to Disease	Other Improved Traits
13.	DT-10	1989	Vietnam	Gamma rays and MNH.	Direct	Bacterial leaf blight	Non-lodging and resistance to insects.
14.	Yangfuxian 9850	2004	China	Gamma rays	–	Bacterial leaf and sheath blight.	Good quality, resistance to blast, non-lodging.
15.	Yangfujing 4298	2004	China	Gamma rays	Direct	Bacterial diseases.	Improved agronomic traits.
16.	Yangfujing 4901	2004	China	Gamma rays	Direct	Bacterial leaf blight.	Resistance to blast, non-lodging, good quality.
17.	Zhenuo #3	2003	China	Gamma rays	Direct	Bacterial diseases	High yield.
18.	Chiyou S162	2005	China	Gamma rays	Direct	Bacterial blight	Plant type and resistance to blast.
19.	Nanhua 11	1987	China	CO_2 laser	Direct	Bacterial diseases	Wide adaptability.
20.	Yangfuxian 5	2000	China	Gamma rays	Direct	Bacterial diseases	High quality and wide adaptability.
21.	Kahayan	2002	Indonesia	Gamma rays	Direct	Bacterial diseases	Amylose content (19–20%).
22.	Winongo	2002	Indonesia	Gamma rays	Direct	Leaf blight	Amylose content (19–20%).
23.	DiahSuci	2003	Indonesia	Gamma rays	Direct	Leaf blight	Amylose content (19–20%).
24.	Mira 1	2006	Indonesia	Gamma rays	Direct	Bacterial diseases	High yield.
				Crop: Maize [12]			
1.	Luyuandan 7	1981	China	Gamma rays	Indirect	Leaf spot	Long ear.
2.	Yuanqi 123	1978	China	Gamma rays	Indirect	Bacterial diseases	Early maturity, flint kernel.
3.	Xinnongfuyu 1	1987	China	–	Indirect	Bacterial diseases	Improved vigor.
4.	Longfuyu 2	1987	China	Gamma rays	Direct	Bacterial diseases	–
5.	Liaoyuan 1	1988	China	–	Direct	Large and small spot, bacterial wilt.	Quality, resistant to corn bug weevil and borer, drought, and lodging resist.

TABLE 9.1 *(Continued)*

Sl. No.	Variety Name	Year	Country	Mutagen	Mutant	Resistance to Disease	Other Improved Traits
6.	Longbaoyu 1	1990	China	–	Direct	Bacterial diseases	High yield.
7.	Longfuyu 3	2007	China	–	Direct	Bacterial diseases	–
8.	Luyuandan 16	1995	China	–	Direct	Bacterial wilt	Wider adaptability.
9.	Luyuandan 14	1997	China	–	Direct	Bacterial diseases	Non-lodging and early.
10.	Zhongyuandan 32	1997	China	–	Direct	Large and small spot.	Early and good quality.
11.	Ludan 50	1998	China	–	Direct	Large and small leaf spot.	Non-lodging.
12.	Luyuandan 22	2002	China	Gamma rays	Direct	Bacterial diseases.	Better quality.
				Crop: Millets 92)			
1.	Lugu 2	1991	China	Gamma rays	Direct	Bacterial diseases	High yield.
2.	Longgu 29	1992	China	Fast neutron	Direct	Bacterial diseases	High yield.

TABLE 9.2 Mutant Varieties of Cereal Crops Resistant to Fungal Diseases

Sl. No.	Variety Name	Year	Country	Mutagen	Mutant	Resistance to Disease	Other Improved Traits
						Crop: Wheat [69]	
1.	Sinvalocho Gama	1962	Argentina	Gamma rays	Direct	Black stem and leaf rust.	–
2.	BR4	1979	Brazil	Gamma rays	Indirect	Stem rust	High yield.
3.	Changwei 19	1978	China	Gamma rays	Direct	Stripe rust and mildew.	Tolerance to salt and alkaline.
4.	Changwei 20	1979	China	Gamma rays	Direct	Stripe rust	Fertility tolerance and good quality.
5.	Emai 6	1966	China	Gamma rays	Direct	Rust	Wide adaptability, high temperature tolerant.
6.	Luten 1	1968	China	Gamma rays	Direct	Resistance to stripe rust.	Semi-dwarf straw, non-lodging.
7.	Nanyang 75-6	1979	China	Gamma and dES	Direct	Resistance to stripe rust.	Plant uniformity.
8.	Ningmai 3	1973	China	Gamma rays	Direct	Stem and leaf rust PM.	Bigger spike, late maturity.
9.	Taifu 1	1966	China	Gamma rays	Direct	Stripe rust.	Early maturity, resistance to drought.
10.	Xinshukuang 1	1971	China	Gamma rays	Direct	Diseases	Strong stalk and large grains.
11.	Yannoun 685	1974	China	Physical	Indirect	Rust	Good grain color at maturity.
12.	Yuandong 1	1979	China	Gamma rays	Direct	Stripe rust.	Early maturity and good quality.
13.	Yuandong 7848	1978	China	Gamma rays	Direct	Stripe and leaf rust.	Early and white grain color.
14.	Yuangnong 53	1971	China	Gamma rays	Direct	Rust	Suitable for North China.
15.	Yuangnong 61	1971	China	Gamma rays	Direct	Rust	High yield, plant height (90 cm) and suitable for North China (spring wheat region).
16.	Yunfuzao	1980	China	Gamma rays	Direct	Diseases	Early maturity, fits well for cotton/wheat rotation.
17.	Zhengliufu	1979	China	Gamma rays	Direct	Resistance to stripe rust.	Resistance to drought.
18.	092	1966	China	Gamma rays	Direct	Stripe and stem rust resistance.	Early maturity, higher yield, resistance to lodging.
19.	1161	1966	China	Gamma ray	Direct	Resistance to stripe rust.	Resistance to low temperature.

TABLE 9.2 *(Continued)*

Sl. No.	Variety Name	Year	Country	Mutagen	Mutant	Resistance to Disease	Other Improved Traits
20.	Yuandong 3	1985	China	Gamma rays	Direct	Resistance to rust and powdery mildew.	Tolerance to dry hot wind and high stable yield.
21.	62-8	1985	China	Fast neutron	Direct	Resistance to stripe rust and early maturity.	Early maturity.
22.	62-10	1985	China	Fast neutron	Direct	Resistance to stripe rust and shorter culm.	Shorter culm.
23.	Emai 9	1980	China	Gamma rays	Direct	Tolerance to *Gibberella zeae*, powdery mildew, salinity.	Tolerance to salinity, alkalinity, and high temperature at late stage, high, and stable yield.
24.	Fuer	1977	China	Gamma rays	Direct	Resistance to rust, tolerance to powdery mildew.	Tolerance to stress.
25.	Heichun 2	1979	China	Gamma rays	Indirect	Resistance to leaf rust.	Early maturity, strong culm, tolerance to drought at seedling stage and to moisture at late stage.
26.	Kexing 15	1972	China	Gamma rays	Direct	Resistance to rust and powdery mildew.	High TGW.
27.	Neimai 5	1979	China	Gamma rays	Direct	Early maturity, escapes *Meromysa saltatrix*, high resistance to rust.	Intermediate early maturity, tight stature, tolerance to hot and dry wind.
28.	Taifu 10	1968	China	Gamma rays	Direct	Resistance to stripe rust.	Drought tolerance, wide adaptability.
29.	Taifu 15	1968	China	Gamma rays	Direct	Resistance to stripe rust.	Early maturity, tolerant to drought.
30.	Xiaoyan 6	1979	China	Laser	Direct	Resistance to stripe rust, tolerance to *Gibberella zeae*.	Resistance to stripe rust, tolerance to *Gibberella zeae*.
31.	Xifu 3	1977	China	Gamma rays	Direct	Resistance to diseases.	Good growth and tolerance to stress.

TABLE 9.2 *(Continued)*

Sl. No.	Variety Name	Year	Country	Mutagen	Mutant	Resistance to Disease	Other Improved Traits
32.	77 L15	1983	China	Laser	Direct	Resistance to stripe, leaf, and stem rust, tolerance to *Gibberella zeae.*	Strong culm, resistance to lodging, large spike, good appearance at maturity stage.
33.	352	1983	China	Laser	Direct	Early maturity, tolerance to *Gibberella zeae.*	Early maturity and high yield.
34.	Lumai 5	1984	China	Physical	Indirect	Strong resistance to stripe rust and powdery mildew.	Short stem, resistance to lodging, dry hot wind tolerance.
35.	Weifu 6757	1986	China	Gamma rays	Direct	Resistance to rust.	Resistance to soil conditions (salinity, alkalinity) and better adaptability.
36.	Zhemai 3	1983	China	Laser	Direct	Resistance to diseases.	Early maturity, high yield, vegetation period (180 days) and plant height (95 cm).
37.	Chuanfu 2	1989	China	Gamma rays	Direct	High resistance to stripe rust.	High yield and early maturity.
38.	Chuanfu 3	1989	China	Gamma rays	Direct	High resistance to stripe rust.	High yield
39.	Fusheabo 1	1987	China	Fast neutron	Direct	Resistance to stripe rust.	–
40.	Fuou 1	1974	China	Gamma rays	Direct	Resistance to fungal diseases.	High grain yield.
41.	Jinmai 23	1980	China	Gamma rays	Direct	Resistance to fungal diseases.	Altered maturity.
42.	Qinmai 6	1983	China	Laser	Direct	Resistance to fungal diseases.	Good quality.
43.	Yumai 4	1984	China	Gamma rays.	Direct	Altered maturity.	Altered maturity.
44.	Xinchun 3	1986	China	Gamma rays	Direct	Resistance to fungal diseases.	High grain yield.
45.	Jimai 28	1988	China	Gamma rays	Direct	Resistance to fungal diseases.	Tolerance to low temperatures.
46.	Xifu 7	1989	China	Gamma rays	Direct	Resistance to fungal diseases.	Altered maturity.
47.	Guifu 12	1986	China	Physical	Direct	Resistance to fungal diseases.	High grain yield.
48.	Qinchun 415	1993	China	Gamma rays	Direct	Resistance to fungal diseases.	–

TABLE 9.2 *(Continued)*

Sl. No.	Variety Name	Year	Country	Mutagen	Mutant	Resistance to Disease	Other Improved Traits
49.	Wei 9133	1993	China	Fast neutron	Direct	Resistance to fungal diseases.	–
50.	Mv 8	1978	Hungary	Gamma rays	Indirect	Resistance to powdery mildew.	Short culm, strong straw, resistance to lodging, good quality.
51.	NI-5643	1975	India	Gamma rays	Indirect	Resistance to rust.	Early maturity, amber grain color, high yield.
52.	Soghat 90	1991	Pakistan	Sodium azide	Direct	Resistance to leaf rust.	Improved protein content.
53.	Tatara	1996	Pakistan	Gamma rays	Direct	Resistance to diseases.	Drought tolerance and semi-dwarfness.
54.	Payne	1981	United States	X-rays	Indirect	Resistance to leaf rust.	Resistance to leaf rust, resistance to lodging, good baking quality and middle early maturity.
55.	Polukarlykovaja-49	1979	Russian Federation	N-nitroso-N-ethyl urea (NMU).	Indirect	Resistance to brown rust and mildew.	Resistance to lodging, short culm, high yield.
56.	Pitikul	1982	Russian Federation	N-nitroso-N-ethyl urea (NMU).	Indirect	Resistance to lodging, middle early maturity, resistance to brown rust and mildew.	Resistance to lodging, medium maturity, large grain size and high yield.
57.	Yunnatodesskii	1989	Russian Federation	N-nitroso-N-ethyl urea (NMU).	Indirect	Resistance to brown rust, resistance to mildew, resistance to bunt.	Resistance to lodging and large grain size.
58.	Meshenskaya	1989	Russian Federation	N-nitroso-methyl urea (NMU).	Direct	Resistance to brown rust, resistance to mildew, resistance to bunt.	Resistance to low temperature and good baking quality.
59.	Kazanskaya 84	1992	Russian Federation	N-nitroso-methyl urea (NMU).	Direct	Resistance to mildew.	Resistance to low temperature, resistance to Sweden fly and soft grains.
60.	Skifyanka	1992	Russian Federation	Chemical	Direct	Resistance to mildew.	Resistance to lodging, large grain size and good baking quality.

TABLE 9.2 *(Continued)*

Sl. No.	Variety Name	Year	Country	Mutagen	Mutant	Resistance to Disease	Other Improved Traits
61.	Jubileinaya 75	1992	Russian Federation	N-nitroso-N-ethyl urea (NMU).	Indirect	Resistance to brown rust, resistance to mildew.	Large size grains and resistance to insects.
62.	Longfumai 8	1998	China	Gamma rays	Direct	Resistance to fungal diseases.	Good quality.
63.	Longfu 2	2003	China	Ion beam	Direct	Resistance to fungal diseases.	Drought tolerant.
64.	Shannong 12	2005	China	Gamma rays	Direct	High disease resistance.	Strong gluten, high disease resistance, early maturity, resistance to late sowing, and high yield and stable yield.
65.	Longfumai 16	2006	China	Gamma rays	Direct	Resistant to diseases.	High yield.
66.	Jingdong 23	2008	China	Gamma rays	Indirect	Immune to stripe rust.	High yield, good quality and medium tillering ability.
67.	Xinchun No. 30	2009	China	Gamma rays	Direct	Rust and powdery mildew.	High and stable yield, wide adaptability, lodging resistance, good growth potential, early maturity, good quality.
68.	Fermer	2009	Bulgaria	Gamma rays	Direct	Brown rust	High yield, good quality, excellent drought, and cold tolerance.
69.	Xinchun 33	2010	China	Gamma rays	Direct	Rust	Good bread quality and high yield.
				Crop: Rice [88]			
1.	Aifu 9	1966	China	Gamma rays	Direct	Blast	Short culm, higher yield.
2.	Fuwan 23	1978	China	Gamma rays	Direct	Yellow stunt and *Xantomonas.*	Bigger spike, large grain size and good quality.
3.	Fuxuan 3	1970	China	Gamma rays	Direct	Blast	Good tillering.
4.	Fuxuan 124	1972	China	Gamma rays	Direct	Blast	Intermediate maturity.
5.	Jinfu 1	1969	China	Gamma rays	Direct	Blast	Early maturity (7 days earlier).

TABLE 9.2 *(Continued)*

Sl. No.	Variety Name	Year	Country	Mutagen	Mutant	Resistance to Disease	Other Improved Traits
6.	Kefuhong 2	1981	China	Gamma rays	Indirect	Blast	Early maturity.
7.	Wanfu 33	1978	China	Gamma rays	Direct	Blast	Early maturity, resistance to low temperatures.
8.	Wangeng 257	1975	China	Gamma rays	Direct	Blast	Higher yield and fertilizer tolerant.
9.	Xiangfudao	1976	China	Gamma rays	Direct	Blast and *Xantomonas.*	Resistance to low temperature.
10.	Xiongyue 613	1965	China	Gamma rays	Direct	Blast	Higher yield and good quality.
11.	Yifunuo 1	1973	China	Gamma rays	Direct	Blast	Bigger spike and higher grain number.
12.	Fulianzao 3	1968	China	Gamma rays	Direct	Disease	Early maturity and short culm.
13.	Fushe 410	1974	China	Gamma rays	Direct	Blast	–
14.	Fu 769	1976	China	Gamma rays	Direct	Diseases	High yield and suitable for mountain areas.
15.	Fu 756	1975	China	Gamma rays	Direct	Blast	Higher yield and short culm.
16.	M 112	1981	China	Gamma rays	Direct	*Sogatella furcifera*	Resistance to low temperature and high yield.
17.	Wanhongfu	1980	China	Gamma rays	Direct	Diseases	Resistance to low temperature and suitable for mountain areas.
18.	Zhuqin 40	1978	China	Gamma rays	Direct	Blast	Suitable for mountain areas.
19.	240	1980	China	Gamma rays	Direct	Diseases	Early maturity.
20.	Fushenongken 58	1973	China	Gamma rays	Direct	Fungal diseases	High grain yield.
21.	Ejingnuo 6	1986	China	Gamma rays	Direct	Blast and blight.	Good grain quality and higher grain yield.
22.	Erjiufeng	1982	China	Gamma rays	Indirect	Fungal diseases	Higher yield, large spike, more grains per spike, early maturity.
23.	Taifu 4	1979	China	Gamma rays	Direct	Diseases	Fertilizer efficient.
24.	652	1979	China	Gamma rays	Direct	Fungal diseases	Improved growth habits.
25.	Xiushui 48	1981	China	Gamma rays	Indirect	Blast	Tolerance to low temperature, tolerance to day length and high yield.

TABLE 9.2 *(Continued)*

Sl. No.	Variety Name	Year	Country	Mutagen	Mutant	Resistance to Disease	Other Improved Traits
26.	Xianghu 24	1983	China	Gamma rays	Indirect	Blast and blight.	Tolerance to day length and glutinous grain.
27.	Ailiutiaohong	1989	China	Gamma rays	Direct	*Pyricularia oryzae*	Dwarfness (88 cm), resistance to insects and low temperatures, high yield.
28.	Qingwei 1	1985	China	Gamma rays	Direct	Diseases	High yield and late maturity.
29.	Fu 8-1	1988	China	Gamma rays	Direct	Resistance to fungal diseases.	High grain yield.
30.	Tangernian	1985	China	Gamma rays	Direct	Resistance to diseases.	High yield and late season variety.
31.	Wanhua	1983	China	Gamma rays	Indirect	Resistance to low temperature and suitable for mountain areas.	Resistance to low temperature and high yield.
32.	Fuwan 81-548	1989	China	Gamma rays	Direct	Resistance to fungal diseases.	Good quality.
33.	Meisanwu 2	1990	China	Gamma rays	Direct	Resistance to fungal diseases.	Resistance to insects.
34.	Xiuxui 117	1984	China	Gamma rays	Direct	Resistance to fungal diseases.	Altered maturity.
35.	Xianghu 93	1984	China	Gamma rays	Direct	Resistance to fungal diseases.	Altered maturity
36.	Zijiangnuo	1984	China	Gamma rays	Direct	Resistance to fungal diseases.	High grain yield.
37.	Xiushui 04	1985	China	Gamma rays	Direct	Resistance to fungal diseases, resistance to blast and bacterial blight.	High yield, good grain quality and altered maturity.
38.	Ganwannuo	1993	China	Gamma rays	Indirect	Resistance to fungal diseases.	High grain yield.
39.	Wandao 20	1994	China	Ion beam	Direct	Resistance to fungal diseases.	Altered maturity.
40.	Wandao 45	1994	China	Ion beam	Direct	Resistance to fungal diseases.	Altered maturity.
41.	Shenxiangjing	1994	China	–	Direct	Resistance to fungal diseases.	Improved plant structure.
42.	Zhefu 762	1993	China	Physical	Direct	High resistance to blast and bacterial blight.	High grain yield, good cold tolerance, seedling age flexibility, good eating quality, improved grain quality.

TABLE 9.2 *(Continued)*

Sl. No.	Variety Name	Year	Country	Mutagen	Mutant	Resistance to Disease	Other Improved Traits
43.	Ganwanxian 23	1994	China	–	Indirect	Resistance to fungal diseases.	High quality
44.	Fuxuan 8	1998	China	Gamma rays	Direct	Resistance to fungal diseases.	Tolerance to low temperatures.
45.	Camago-8	1996	Costa Rica	Gamma rays	Direct	Resistance to blast and resistance to viruses.	–
46.	UNP 9027	1994	Costa Rica	Gamma rays	Direct	Resistance to *Pyricularia oryzae*.	Better nitrogen response.
47.	IRAT 216	1985	Cote D'Ivoire	Gamma rays	Indirect	Resistance to *Pyricularia*.	Good adaptability to wetland rice culture and tolerance to flooding.
48.	Calendal	1979	France	Fast neutrons	Direct	Resistance to *Sclerotinium oryzae*.	Longer grains, improved threshability.
49.	Marathon	1985	France	Gamma rays	Direct	Resistance to *Pyricularia*.	–
50.	Nucleoryza	1972	Hungary	Fast neutrons	Direct	Blast resistance.	Early maturity and improved yield.
51.	Mutashali	1980	Hungary	Fast neutrons	Direct	Resistance to *Pyricularia oryzae* (blast).	Resistance to lodging, resistance to shattering of grains and high yield.
52.	Pusa-NR-381	1989	India	Gamma rays	Indirect	Resistance to blast.	–
53.	CRM 49	1999	India	Sodium azide (NaN_3)	Direct	Resistance to blast disease.	–
54.	CRM 51	1999	India	Sodium azide (NaN_3)	Direct	Resistance to blast disease.	–
55.	CRM 53	1999	India	Ethyl methane sulfonate	Direct	Resistance to blast disease.	–
56.	Atomita 1	1982	Indonesia	Gamma rays	Direct	Resistance to blast disease.	Early maturity, resistance to BPH (biotype 1), GLH.

TABLE 9.2 *(Continued)*

Sl. No.	Variety Name	Year	Country	Mutagen	Mutant	Resistance to Disease	Other Improved Traits
57.	Danauatas	1988	Indonesia	Gamma rays	Direct	Resistance to blast disease.	High yield, resistance to drought and low pH.
58.	Fulgente	1973	Italy	X-rays	Direct	Resistance to blast disease.	High productivity.
59.	Sachiminori	1978	Japan	Gamma rays	Indirect	Resistance to blast disease.	Stiff culm
60.	ITA 123	1980	Nigeria	Gamma rays	Direct	Resistance to blast disease.	Semi-dwarfness
61.	RD 6	1977	Thailand	Gamma rays	Direct	Resistance to blast disease.	Glutinous endosperm and improved high-quality rice.
62.	Pooya	2004	Iran, Islamic Republic of	Gamma rays	Direct	Resistance to blast disease.	Resistance to lodging and higher yield.
63.	Tabesh	2004	Iran, Islamic Republic of	Gamma rays	Direct	Resistance to blast disease.	Resistance to lodging, short culm, and higher yield.
64.	Minnuo 706	1988	China	Gamma rays	Indirect	Resistance to blast disease.	Good tillering, higher yield, glutinous, and good quality.
65.	Jinhang-Simiao	2006	China	Aerospace	Indirect	Resistance to fungal diseases.	Good quality.
66.	Huahang-simiao	2006	China	Aerospace	Direct	Resistance to fungal diseases (rice blast) and good quality.	Good quality.
67.	Peiza 130	2008	China	Aerospace	Direct	Resistance to fungal diseases.	High yield, resistance to fungal diseases and early maturity.
68.	Liangyouhang 2	2008	China	Aerospace	Direct	Resistance to fungal diseases, resistance to resistance to blast and bacterial blight.	High yield and good grain quality.
69.	Hangxiang 18	2008	China	Aerospace	Indirect	Resistance to fungal diseases.	Late maturity.
70.	Yuanjing 41	2004	China	Gamma rays	Direct	Resistance to blast disease.	Dwarf stalk, lodging resistance, large ears and multiple tillers, and high yield, stable production.
71.	Zhenuo 5	2004	China	Gamma rays	Direct	Resistance to blast disease.	High yield, high quality, and wider adaptability.

TABLE 9.2 *(Continued)*

Sl. No.	Variety Name	Year	Country	Mutagen	Mutant	Resistance to Disease	Other Improved Traits
72.	Yuanjing 35	2005	China	Gamma rays	Direct	Moderately resistant to rice blast and bacterial blight.	Moderate plant height, compact plant, more effective panicle is, the panicle type is medium, more grains per panicle, improved yield.
73.	Guangyinruanzhan	2008	China	Gamma rays	Indirect	Resistance to blast.	Early-maturity.
74.	Early Samba	2000	India	Gamma rays	Indirect	Resistance to leaf and panicle blast.	Early-maturity.
75.	IACuba 28	2001	Cuba	Gamma rays	Indirect	Resistance to fungal diseases.	–
76.	Michinoku-wase	1988	Japan	Gamma rays	Indirect	High resistance to rice stripe disease and panicle blast and moderately resistance to bacterial leaf blight.	–
77.	Okini-Iri	1996	Japan	Gamma rays	Indirect	High resistance to the panicle blast.	Suitable for sake brewing, high cold resistance.
78.	Hayatsukushi	1997	Japan	Gamma rays	Indirect	Field resistance to blast.	Early-maturity.
79.	Hiroshima No. 21	1998	Japan	Gamma rays	Indirect	Resistance to leaf and panicle blast.	Early-maturity.
80.	Koshihikari Toyama BL No. 2	1998	Japan	Gamma rays	Indirect	Resistance to fungal diseases.	–
81.	Aichi-no-kaori SBL	1999	Japan	Gamma rays	Indirect	High resistance to rice stripe disease and panicle blast and moderately resistance to bacterial leaf blight.	–
82.	Fusa-no-mai	2000	Japan	Gamma rays	Indirect	High resistance to the panicle blast.	Suitable for sake brewing and high cold resistance.

TABLE 9.2 *(Continued)*

Sl. No.	Variety Name	Year	Country	Mutagen	Mutant	Resistance to Disease	Other Improved Traits
83.	Koshihikari Niigata BL No. 4	2002	Japan	Gamma rays	Indirect	Resistance to blast.	–
84.	Koimusubi blast resistance and	2002	Japan	Gamma rays	Indirect	Resistance to blast.	Low protein content, highly cold tolerance, excellent eating quality.
85.	Churahikari	2003	Japan	Gamma rays	Indirect	Resistance to blast.	Medium-late maturity, shorter culm, and lodging resistance.
86.	Sai-no-Kagayaki	2002	Japan	Gamma rays	Indirect	Field resistance to blast and resistance to stripe disease.	Resistance green rice leaf hopper, tolerant to lodging with short culm and stiffness, good eating quality.
87.	Zhejing 41	2009	China	–	–	Resistance to blast and bacterial leaf blight and resistance to brown planthopper.	Medium maturity, Medium tillering ability, large panicle, high yield, good grain quality.
88.	Moretsu	–	Japan	MNU	Direct	Resistance to stripe rust.	Resistance to lodging.
Crop: Maize [16]							
1.	Jidan 101	1974	China	–	Indirect	Resistance to leaf spot.	Good root structure, resistance to lodging and good quality.
2.	Luyu 3	1980	China	–	Indirect	Good resistance to diseases.	Higher yield.
3.	Luyuandan 1	1976	China	–	Indirect	Resistance to leaf spot.	Bigger grains and good quality.
4.	Luyuanshan 2	1981	China	–	Indirect	Resistance to diseases.	Higher yield.
5.	Yuanlian 5	1980	China	–	Indirect	Resistance to leaf spot.	Early maturity, suitable for close planting and good quality.
6.	Yuan 74-751 (inbred)	1974	China	Gamma rays and microwave (8 mm/min).	Direct	Resistance to leaf spot.	Good plant type, erect leaf, good combining ability.

TABLE 9.2 *(Continued)*

Sl. No.	Variety Name	Year	Country	Mutagen	Mutant	Resistance to Disease	Other Improved Traits
7.	Yuan 79-418 (inbred)	1979	China	Fast neutrons	Direct	Resistance to leaf spot.	Growth period (95 days), tolerant to waterlog, good combining ability, suitable for Beijing region.
8.	Zhongyuandan 4	1982	China	–	Indirect	Resistance to leaf spot.	Early maturity, resistance to leaf spot and higher yield.
9.	Jidan 1	1967	China	Gamma rays	Indirect	Resistance to leaf blight *Sphacelotheca reliana*.	–
10.	Luyuandan 3	1976	China	Gamma rays	Indirect	Good stature	Resistance to diseases.
11.	Xiangsan 1	1980	China	Gamma rays	Indirect	Resistance to diseases.	High yield
12.	Yuanqi 722	1978	China	Gamma rays	Indirect	Resistance to diseases.	Medium maturity, semi-dent kernel type, good combining ability.
13.	Mudan 7	1983	China	Physical mutagen	Indirect	The mutant variety Mudan 7 was officially approved in 1983. It was developed by treatment with physical mutagen.	Main improved attributes of mutant variety are resistance to fungal diseases and altered maturity.
14.	Keduo 6	1991	China	–	Direct	Resistance to fungal diseases.	High yield
15.	Xinongdanjiao 1	1991	China	–	Direct	Resistance to fungal diseases.	Medium maturity, growth period of 120 days, high yield, high quality, strong stress resistance, coordinated flowering period, large pollen quantity, good seedling emergence, vigorous growth, plant height 230–250 cm, lodging resistant.
16.	Hybrid ChKG 280 MV	1992	Russian Federation	Chemical	Indirect	Main improved attributes of variety are resistance to bunt.	Resistance to insects and high yield.

TABLE 9.2 *(Continued)*

Sl. No.	Variety Name	Year	Country	Mutagen	Mutant	Resistance to Disease	Other Improved Traits
Crop: Barley [28]							
1.	Alexis	1986	Germany	Gamma rays	Indirect	Resistance to mildew.	–
2.	Vienna	1959	Austria	X-rays	Direct	Resistance to mildew.	High yield, high 1000-kernel weight, and resistance to lodging.
3.	Lupidamai 1	1987	China	Gamma rays	Direct	Resistance to fungal diseases.	Early maturity and good quality.
4.	Qianlu 1	1995	China	Gamma rays	Direct	Resistance to fungal diseases.	High-quality and high-yielding malting barley variety.
5.	Keti	1982	Denmark	X-rays	Indirect	Resistance to mildew.	High yield.
6.	Betina	1970	France	EMS	Direct	Excellent mildew tolerance.	Short stiff straw, resistance to lodging, high yield.
7.	Nadja	1975	Germany	X-rays	Indirect	Resistance to mildew, stripe rust and leaf rust.	Short stem, lodging resistance, top brewing quality, very good malting quality and high yield.
8.	Trumpf	1973	Germany	X-rays	Indirect	Resistance to mildew, stripe rust and leaf rust.	Short stem, lodging resistance, top brewing quality, very good malting quality and high yield.
9.	DL-253	1981	India	Gamma rays and EMS (0.30%).	Direct	Disease resistance to smut and yellow rust.	Seed yield, high tillering.
10.	Amil	1994	Iraq	Gamma rays	Direct	Resistance to stripe rust, resistance to powdery mildew.	High yield.
11.	Baraka	1994	Iraq	Gamma rays	Direct	Resistance to powdery mildew.	High yield and malting quality.
12.	Samir	1993	Iraq	Gamma rays	Direct	Resistance to leaf rust.	High yield and malting quality.
13.	Tone-Nijo	1989	Japan	Gamma rays	Indirect	Resistance to leaf stripe and resistance to mildew.	–

TABLE 9.2 *(Continued)*

Sl. No.	Variety Name	Year	Country	Mutagen	Mutant	Resistance to Disease	Other Improved Traits
14.	Atlas	1976	Czech Republic	X-rays	Indirect	Resistance to mildew (Mla 70).	Good tillering, resistance to lodging, high yield and good grain weight.
15.	Rapid	1976	Czech Republic	X-rays	Indirect	Resistance to mildew (genes Mla 6 and Mlg).	Higher grain production, resistance to lodging, shorter vegetation period and early maturity.
16.	Karat	1981	Slovakia	X-rays	Indirect	Resistance to powdery mildew.	High yield, suitable for intense conditions.
17.	Koral	1978	Czech Republic	X-rays	Indirect	Resistance to powdery mildew.	High yield, resistance to lodging.
18.	Krystal	1981	Czech Republic	X-rays	Indirect	Resistance to powdery mildew.	High yield and very good brewing quality.
19.	Jenny	1980	Sweden	X-rays	Indirect	Resistance to powdery mildew.	High yield
20.	Mona	1970	Sweden	X-rays	Indirect	Resistance to powdery mildew.	High yield and good straw.
21.	Rupal	1972	Sweden	X-rays	Indirect	Resistance to powdery mildew.	Short straw and sprouting resistant.
22.	Salve	1974	Sweden	Gamma rays	Indirect	Field tolerance to powdery mildew.	Large plump grains, stiff straw, high seed production.
23.	Visir	1970	Sweden	X-rays	Indirect	Resistance to powdery mildew.	High yield.
24.	Midas	1970	United Kingdom	Gamma rays	Indirect	Resistance to powdery mildew.	Short stiff straw and erectoides habit.
25.	Heriot	1983	United Kingdom	X-rays	Indirect	Resistance to mildew, resistance to yellow and brown rust.	Semi-prostrate type, short stiff straw, late maturity, high yield.

TABLE 9.2 *(Continued)*

Sl. No.	Variety Name	Year	Country	Mutagen	Mutant	Resistance to Disease	Other Improved Traits
26.	Bonneville 70	1969	United States	Thermal neutrons	Direct	Field resistance to loose smut.	White-kernelled, six-rowed, smooth-awned spring barley and has brittle awns with improved threshability.
27.	Temp	1978	Russian Federation	NMU	Direct	Resistance to powdery mildew.	Early maturity, short stem, high yield, and mildew resistant.
28.	IZ Bori	2009	Bulgaria	Sodium azide	Direct	Resistance to powdery mildew, brown, black, and stem rust.	Tolerance to low temperatures, high grain yield, high grain protein and lysine content and wider adaptability.
Crop: *Pennisetum* spp. [5]							
1.	New Hybrid Bajra 5	1974	India	Gamma rays	Indirect	Powdery mildew.	–
2.	Pusa 46	1982	India	Gamma rays	Direct	Fungal diseases.	–
3.	ICMH 451	1986	India	Gamma rays	Indirect	Powdery mildew.	–
4.	NHB 3 (hybrid)	1975	India	Gamma rays	Indirect	*Sclerospora graminicola*	–
5.	NHB 4 (hybrid)	1975	India	Gamma rays	Indirect	*Sclerospora graminicola*	–
Crop: Millet [6]							
1.	Changwei 75	1975	China	Gamma rays	Direct	*Pyricularia setaria*	Good quality.
2.	Nunxuan 12	1986	China	Fast neutrons	Direct	Fungal diseases	Drought tolerance.
3.	Longgu 27	1988	China	Fast neutrons	Direct	Fungal diseases	Resistance to fungal diseases.
4.	Longgu 28	1989	China	Fast neutrons	Direct	Fungal diseases	Drought tolerance.
5.	Fugu 3	1989	China	Gamma rays	Direct	Fungal diseases	High yield.
6.	Cheget	1993	Russian Federation	Chemo mutants	Indirect	Resistance to smut.	Drought tolerance.

TABLE 9.3 Mutant Varieties of Cereal Crops Resistant to Viral Diseases

Sl. No.	Variety Name	Year	Country	Mutagen	Mutant	Resistance to Disease	Other Improved Traits
Crop: Rice [7]							
1.	Zhengguang 1	1978	China	Gamma rays	Direct	Yellow stunt virus.	–
2.	Camago-8	1996	Costa Rica	Gamma rays	Direct	Viruses	Resistance to blast.
3.	Heiseimochi	1988	Japan	Gamma rays	Indirect	Rice stripe virus.	Resistance to lodging.
4.	ITA 235	1988	Nigeria	Chemical	Indirect	Viruses (RYMV).	Semi-dwarfness.
5.	Zhongzao 21	2003	China	–	–	Viruses	Large spike and more grains, good grain quality, blast resistance.
6.	Shengba-simiao	2005	China	Aerospace	–	Viruses	Better cooking quality.
7.	Zhe 101	2005	China	Aerospace	–	Viruses	Blast and bacterial blight resist.
Crop: Maize [1]							
1.	Kneja435	2005	Bulgaria	NMU and Colchicine	Indirect	Viruses	Resistance to fungal diseases.
Crop: Barley [1]							
1.	Masakado mugi	1989	Japan	Gamma rays	Indirect	Soil born barley yellow mosaic virus.	Early maturity.

varieties is N-methyl-N-nitrosourea (NMU) followed by sodium azide (SA), EMS and NMU.

9.4 MUTATION BREEDING FOR DEVELOPMENT OF DISEASE RESISTANT VARIETIES IN OILSEED CROPS

Edible oils are composed of triglycerides which are esters of one molecule of glycerol and three molecules of fatty acids. It mainly includes soybean, groundnut, rapeseed and mustard, sesame, sunflower, sunflower, flax/linseed, etc. In 2020/21, 209.14 million metric tons (MMT) of vegetable oil production was recorded worldwide (https://www.statista.com/statistics/267271/worldwide-oilseed-production-since-2008/) whereas oilseed production in India is forecasted to 38.4 MMT with productivity of one metric ton per hectare. To meet the increasing need of population for edible oil there is a need to reduce the production losses caused by various biotic (pathogen, insect-pest) and abiotic stresses (drought, high temperature, salinity) and weeds (Raza et al., 2019). Globally yield losses caused by plant disease are estimated up to 16% (Oerke, 2006). For overall vegetable oil crops, about 14.00 million tons/year loss was estimated worldwide due to diseases (Chattopadhyay et al., 2016). Further, the variation in yield loss depends upon the severity of pathogen attack and unfavorable climatic conditions.

Most of the oilseed crops are affected by foliage diseases such as the rusts, downy mildews, leaf spots and blights or soil-borne root diseases like charcoal rot, Sclerotinia rots, Verticillium wilts, and *Fusarium wilt*s. Most of the time, these crops are being grown under marginal agricultural practices. Therefore, least expensive measures like host resistance will help more oil seed production in sustainable manner. However, narrow genetic variability among oilseed crops is a major bottleneck.

Mutation breeding has proven to be an important tool for the generation of resistance against diseases (Oladosu et al., 2016). By using mutagenesis technique, disease tolerant mutants have been identified in various oilseed crops. In *B. juncea*, using γ-rays and EMS mutagen, *Alternaria* tolerant lines (Agnihotri et al., 2009; Meena et al., 2020) and white rust tolerant lines (Agnihotri et al., 2009) have been developed. Using gamma irradiation, tolerant lines against *Alternaria* blight have also been developed in *B. campestris* (Das & Rahman, 1988) and *B. napus* (Sharma et al., 2012). In groundnut, late leaf spot and rust diseases tolerant mutant have been developed using gamma irradiation (Mothilal et al., 2010). Soybean mutants showed less severe disease symptoms in response to virulent *Pseudomonas syringae* pv. glycinea and *Phytophthora sojae* isolated using EMS, NMU or nitrosoguanidine treatment (Hoffman et al., 1999). Using mutation breeding technique, 71 oilseed crop varieties have been released till date (Figure 9.2(A) for tolerance to different bacterial, fungal, and viral diseases (MVD database; https://mvd.iaea.org) (Figure 9.2(B)). Details of these disease tolerant oilseed mutant varieties have been complied in Table 9.4 for bacterial disease, Table 9.5 for fungal disease and Table 9.6 for viral disease (Figures 9.2 and Tables 9.4–9.6).

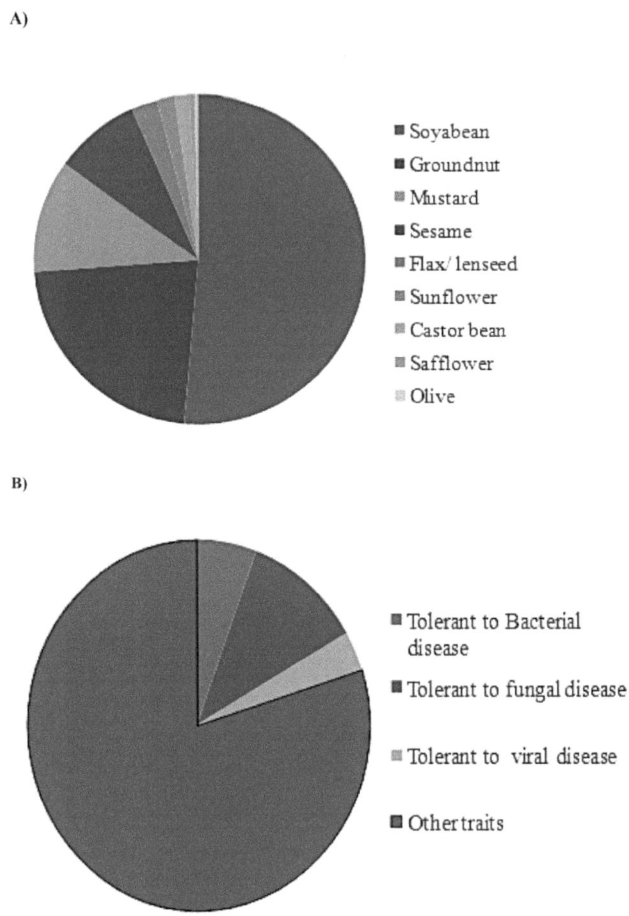

FIGURE 9.2 Oilseed crop varieties developed through mutation breeding (numbers) (A). And oilseed crop varieties for disease tolerance developed through mutation breeding (B).

TABLE 9.4 Mutant Varieties of Cereal Crops Resistant to Viral Diseases

Sl. No.	Variety Name	Year	Country	Mutagen	Mutant	Resistance to Disease	Other Improved Traits
					Crop: Rice [7]		
1.	Zhengguang 1	1978	China	Gamma rays	Direct	Yellow stunt virus.	–
2.	Camago-8	1996	Costa Rica	Gamma rays	Direct	Viruses	Resistance to blast.
3.	Heiseimochi	1988	Japan	Gamma rays	Indirect	Rice stripe virus.	Resistance to lodging.
4.	ITA 235	1988	Nigeria	Chemical	Indirect	Viruses (RYMV).	Semi-dwarfness.
5.	Zhongzao 21	2003	China	–	–	Viruses	Large spike and more grains, good grain quality, blast resistance.
6.	Shengba-simiao	2005	China	Aerospace	–	Viruses	Better cooking quality.
7.	Zhe 101	2005	China	Aerospace	–	Viruses	Blast and bacterial blight resist.
					Crop: Maize [1]		
1.	Kneja435	2005	Bulgaria	NMU and Colchicine	Indirect	Viruses	Resistance to fungal diseases.
					Crop: Barley [1]		
1.	Masakadomugi	1989	Japan	Gamma rays	Indirect	Soil born barley yellow mosaic virus.	Early maturity.

TABLE 9.5 Mutant Varieties of Oilseed Crops Resistant to Bacterial Diseases

Sl. No.	Crop	Variety Name	Year	Country	Mutagen	Mutant	Resistance to Disease	Special Character
1.	Soybean	Hefeng 33	1992	China	TN	Direct	Bacterial	Resistant to virus diseases.
2.	Soybean	Heinong 31	1987	China	TN	Direct	Bacterial diseases	Nutritional quality. and drought tole.
3.	Soybean	Yedadou 2	1990	China	Gamma rays	Direct	Bacterial diseases	Nutritional quality.
4.	Soybean	Yudou 4	1987	China	Gamma rays	Direct	Bacterial diseases	Good nutritional quality.
5.	Soybean	Kefeng 34	1993	China	–	Direct	Bacterial diseases	High yield.
6.	Soybean	Kexin 3	1995	China	EMS	Direct	Bacterial diseases	Nutritional quality.
7.	Soybean	Kexin 5	2000	China	EMS	Direct	Bacterial diseases	–
8.	Soybean	Liaodou 7	1992	China	Gamma rays	Hybrid	Bacterial diseases	Tolerance to drought.
9.	Soybean	Mufeng 6	1987	China	Gamma rays	Direct	Bacterial diseases	–
10.	Soybean	Prikarpatskaya 81	1991	RF	NMU	Direct	Bacterial	Fungal diseases
11.	Soybean	TAMS 98-21	2007	India	Gamma rays	Direct	Bacterial pustules	Myrothesium leaf spot and SMV.
12.	Soybean	Mageva	1991	RF	Chemical	Direct	Bacteria spots	Early maturity.
13.	Groundnut	Xianghuasheng 4	1996	China	Gamma rays	Direct	Bacterial diseases	Early maturity.
14.	Groundnut	Fenghua 1	2001	China	Gamma rays	Direct	Bacterial disease	–
15.	Groundnut	KARISMA seren	2002	Malaysia	Gamma rays	Direct	Cercospora	High vitamin A, protein, and fat.
16.	Groundnut	KARISMA sweet	2002	Malaysia	Gamma rays	Direct	Bacterial diseases	High vitamin A, high invert sugar.
17.	Groundnut	Katantan 1	2017	Indonesia	Gamma rays	Direct	Bacterial diseases	High yield.
18.	Groundnut	Kehua 1	2008	China	EMS	Direct	Bacterial diseases	–
19.	Sesame	Taka 2	1996	Egypt	Gamma rays	Direct	Wilt and root rot	–
20.	Sesame	NIAB-Sesame-2016	2016	Pakistan	–	–	Bacterial blight	Phyllody and charcoal-rot.

Note: TN: Thermal neutrons; and RF: Russian Federation.

TABLE 9.6 Mutant Varieties of Oilseed Crops Resistant to Fungal Diseases

Sl. No.	Crop	Variety Name	Year	Country	Mutagen	Mutant	Resistance to Disease	Special Character
1.	Soybean	Chiang Mai 5	2006	Thailand	Gamma rays	Direct	Resistant to rust with reddish brown.	–
2.	Soybean	Doi Kham	1986	Thailand	Gamma rays	Direct	High resistance to rust.	High yield and larger seed size.
3.	Soybean	Heihe 36	2004	China	–	–	–	–
4.	Soybean	Heinongxiaolidou 1	1989	China	Fast neutrons	Direct	Resistance to fungal diseases.	–
5.	Soybean	Mageva	1991	Russian Federation	Chemical	Direct	Resistance to brown rust.	Early maturity
6.	Soybean	Meratus	1998	Indonesia	Gamma rays	Direct	Resistance to leaf rust.	High yield.
7.	Soybean	Prikarpatskaya 81	1991	Russian Federation	Chemical (0.025% NMU).	Direct	Resistance to bacterial and fungal diseases.	–
8.	Soybean	Rajabasa	2004	Indonesia	Gamma rays	Mutant in crossing.	Resistance to leaf rust.	High yield.
9.	Soybean	Tengger	1991	Indonesia	Gamma rays	Direct	Medium-resistance to leaf rust.	Early maturity, yellow color of seeds.
10.	Soybean	Wei 7610-13	1983	China	Gamma rays and fast neutron.	Direct	Resistance to sporotrichosis.	Early maturity, high yield.
11.	Groundnut	Binachinabadam-1	2000	Bangladesh	Gamma rays	Direct	Moderately resistant to collar rot, CLS, and rust.	High yield
12.	Groundnut	Fu 22	1985	China	Gamma rays	Direct	Resistance to *Aspergillus flavus* (causing aflatoxin development).	High yield

TABLE 9.6 *(Continued)*

Sl. No.	Crop	Variety Name	Year	Country	Mutagen	Mutant	Resistance to Disease	Special Character
13.	Groundnut	Golden	2002	Pakistan	Gamma rays	Direct	Resistance against cercospora leaf spot (tikka disease).	High yield, reddish seed coat color.
14.	Groundnut	GPBD 5	2010	India	Gamma rays	Direct	Resistance to late leaf spot and rust diseases.	–
15.	Groundnut	Mutant 28-2	2003	India	Chemical mutagen (0.5% EMS)	Direct	Resistance to late leaf spot.	Resistance to *Spodoptera litura* and thrips.
16.	Groundnut	Shanyou 27	1985	China	Beta rays	Mutant in crossing.	Resistance to rust.	–
17.	Sesame	ANK-S2	1995	Sri Lanka	Gamma rays	Direct	Resistance to *Fusarium*, Phytophthora, and other diseases.	High yield.
18.	Sesame	Ansanggae	1984	Korea	Gamma rays	Direct	Resistance to diseases (leave blight, *Phytophtora stemrot* and *fusarium wilt*.	Resistance to shattering, high yield.
19.	Sesame	Taka 1	1996	Egypt	Physical	Direct	Resistance to wilt.	–
20.	Sesame	Taka 2	1996	Egypt	Physical	Direct	Resistance to wilt and root rot.	–
21.	Sesame	Binatil-3	2013	Bangladesh	Gamma rays	Direct	Moderately resistant to stem rot.	–
22.	Sesame	Binatil-4	2016	Bangladesh	Gamma rays	Direct	Moderately resistant to stem rot.	Tolerance waterlogged.

TABLE 9.6 *(Continued)*

Sl. No.	Crop	Variety Name	Year	Country	Mutagen	Mutant	Resistance to Disease	Special Character
23.	Sesame	Cairo White 8	1992	Egypt	Gamma rays	Direct	Tolerance to wilt pathogens.	–
24.	Sesame	Sinai White 48	1992	Egypt	Gamma rays	Direct	Wilt resistance.	–
25.	Sesame	Zhongzhi 11	2003	China	Aerospace	Direct	Resistance to stem blight and *fusarium wilt*.	Lodging resistance.
26.	Rapeseed (*Brassica napus* L.)	Ganyou 5	1984	China	Gamma rays	Direct	Resistance to white rust and Sclerotinia.	Early maturity, higher yield.
27.	Rapeseed (*Brassica napus* L.)	Tismenitskii	1989	Russian Federation	Chemical (NMU, 0.0025%).	Direct	Resistance to mildew and resistance to snow mold.	Good quality, oil content.
28.	Mustard (*Brassica juncea*)	TPM-1	2007	India	β-rays	Direct	Tolerance to powdery mildew.	Yellow seed coat.
29.	Mustard (*Brassica juncea*)	NIFA-mustard canola	2003	Pakistan	Gamma rays	Direct	Moderately resistant to *Alternaria*, *Sclerotinia* stem rot and downy mildew.	–
30.	Mustard (*Brassica juncea*)	Binasarisha-7	2011	Bangladesh	Gamma rays	Direct	Moderately resistant to stem rot, tolerant to *Alternaria* blight.	–
31.	Mustard (*Brassica juncea*)	Binasarisha-8	2011	Bangladesh	Gamma rays	Direct	Moderately resistant to stem rot, tolerant to *Alternaria* blight.	–

TABLE 9.6 *(Continued)*

Sl. No.	Crop	Variety Name	Year	Country	Mutagen	Mutant	Resistance to Disease	Special Character
32.	Rapeseed (*Brassica napus*)	Binasarisha-9	2013	Bangladesh	Gamma rays	Direct	Moderately resistant to stem rot, tolerant to *Alternaria* blight.	–
33.	Rapeseed (*Brassica napus*)	Binasharisha-3	1997	Bangladesh	Gamma rays	Direct	Tolerance to *Alternaria* disease.	–
34.	Rapeseed (*Brassica napus*)	Ivanna	1990	Russian Federation	Chemical 0.0025% NMU	Direct	Resistance to mildew.	Resistance to insects.
35.	Oriental mustard (*B. juncea*)	Agrani	1991	Bangladesh	Gamma rays	Direct	Tolerant to *Alternaria* disease.	–
36.	Oriental mustard (*B. juncea*)	Safal	1991	Bangladesh	Gamma rays	Direct	Tolerant to *Alternaria* disease.	–
37.	Safflower	Inshas 11	2011	Egypt	Gamma rays	Crossing with one mutant.	Resistant for leaf spots and smut.	Resistant for tunnel insects and main pests.
38.	Flax/ linseed	Dufferin	1979	Canada	X-ray	Crossing with one mutant.	Better resistance to *Melampsora lini* (Ehrenb.) Lev.	–
39.	Flax/ linseed	Baltyuchai	1991	Russian Federation	Chemical (NMU, 0.012%)	Direct	Resistance to Fusariosis.	Resistance to lodging.

9.5 MUTATION BREEDING FOR DEVELOPMENT OF DISEASE RESISTANT VARIETIES IN PULSE CROPS

Pulses are the best source of proteins after animal-based food. Pulse crop accounts for 6–7% of total food grain production in India. Globally, pulses are grown in an area of about 81 mha with 73 million tons production. India ranks first both in area and production of pulses with 35% of global acreage and 25% of world production. Variety of pulse crops suffers due to plant pathogens. Crop loss due to wilt and root rot, Ascochyta blight and botrytis-gray mold ranged from 5–25% individually in chickpea (Chandrashekhar et al., 2014). The major reasons for the crop losses are damage due to insect infestation and plant diseases like *fusarium wilt*, root rots, Ascochyta blight, powdery mildew, rusts, and viral diseases.

Increase in pulse seed production and simultaneous mitigation of the crop losses due to the diseases necessitates the high yielding and disease resistant crop varieties. Through induced mutagenesis it is possible to overcome the saturating genetic variability and create newer mutants. Mabrouk et al. (2018) used pulverization and inoculation of detached leaf methods to isolate the gamma radiation induced Ascochyta blight resistant genotype. Naturally occurring mutations were the earliest source of disease resistance. The spontaneous mutation in pea was reported way back in 1948. It was resistant to powdery mildew. The gene was later determined to be *Er*, Erysiphe resistance (*PsMLO1*) having a transversion mutation, C→G at 680 position resulting into premature termination. Later on, two more mutations were induced chemically in *Er* in order to achieve the PMR (Sharma, 2016; Humphry et al., 2011). Using gamma radiation, a mutant for yellow mosaic virus (YMV) has been isolated in urd bean. Along with YMV resistance, the mutant also had purple pigmentation on different plant parts and upwardly protruding pods with bold seeds (Vanniarajan et al., 2019). Resistance against MYMV was induced using a chemical mutagen, EMS with improved yield capacity in mung bean (Raihan et al., 2018). Bhatia et al. (2001) has extensively reviewed the development of pulse crop varieties developed through induced mutagenesis.

Development of high yielding varieties with disease resistance is a continuous breeding process. Induced mutagenesis has contributed significantly to generate disease resistance mutant and to develop high yielding varieties. Among total 3,365 mutant varieties, 466 varieties have been released in pulse crops and 81 have been bred for diseases resistance (Figure 9.3).

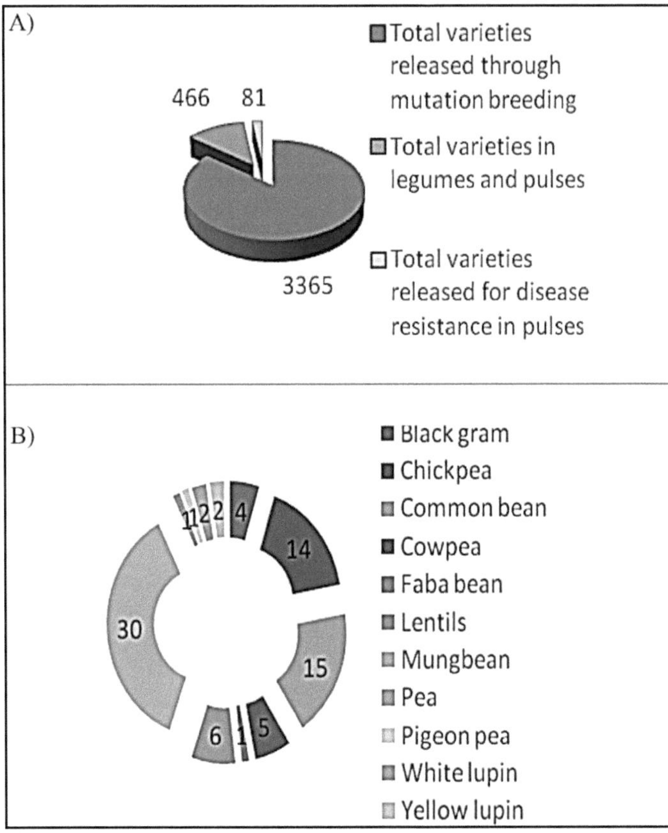

FIGURE 9.3 Total pulse crop varieties and disease resistant varieties developed through mutation breeding (A). Crop wise distribution of the disease resistant crop varieties in legumes and pulses (B).

9.6 MUTATION BREEDING FOR DEVELOPMENT OF DISEASE RESISTAN VARIETIES IN FRUIT AND VEGETABLE CROPS

Besides the crop plants, induced mutagenesis has played an important role in development of disease resistant fruit trees and well as vegetable. Though it is beyond the scope of this chapter, we would like to make a mention that mutation breeding has played an important role in strengthening variability in fruit and vegetable crops and limited examples of the same has been mentioned in Table 9.10. Most prominent example is development of resistant mutant for black spot disease caused by *Alternaria alternata* in Japanese pear. One of these mutants was registered as 'Gold Nijisseiki' (Tables 9.7–9.10).

TABLE 9.7 Mutant Varieties of Pulse Crops Resistant to Bacterial Diseases

Sl. No.	Crop	Variety Name	Year	Country	Mutagen	Mutant	Resistance to Disease	Other Improved Traits
1.	Chickpea	CM-2008	2008	Pakistan	EMS	Direct	Resistant to wilt and tolerant to blight disease.	Seed size and high yield.
2.	Chickpea	NIFA-95	1995	Pakistan	Gamma rays	Direct	Resistant to bacterial blight.	–
3.	Chickpea	Pusa 413 (Atul)	1985	India	Gamma rays	Direct	Moderately blight resistant, resistant to stunt virus, root rot.	High number of branching.
4.	Common bean	Saparke	1967	Russian federation	Gamma rays	Direct	Bacterial diseases.	High yield.

TABLE 9.8 Mutant Varieties of Pulse Crops Resistant to Fungal Diseases

Sl. No.	Crop	Variety Name	Year	Country	Mutagen	Mutant	Resistance to Disease	Other Improved Traits
1.	Black gram	TAU 1	1985	India	Gamma rays	Indirect	Powdery mildew.	High yield.
2.	Chickpea	BGM 547	2005	India	Gamma rays	Direct	Wilt, root rot, stunt, and *Helicoverpa armigera.*	High yield, bold grain size, attractive golden-brown color.
3.	Chickpea	CM 72	1983	Pakistan	Gamma rays	Direct	Blight (*Ascochyta rabiei*).	High yield.
4.	Chickpea	CM-88	1994	Pakistan	Gamma rays	Direct	*Ascochyta*, and *Fusarium.*	High yield.
5.	Chickpea	CM-98	1998	Pakistan	Gamma rays	Direct	*Ascochyta* blight, and *Fusarium wilt.*	–
6.	Chickpea	Hassan-2K	2000	Pakistan	Gamma rays	Direct	Blight and wilt	High yield, higher protein content.
7.	Chickpea	NIFA-88 (CM1918)	1990	Pakistan	Gamma rays	Direct	Moderately resistant to *Ascochyta rabiei.*	Early maturity, high yield, and nitrogen fixation.
8.	Chickpea	Pusa 408 (Ajay)	1985	India	Gamma rays	Direct	Blight resistance.	High yield, semi erect plant architecture.

TABLE 9.8 *(Continued)*

Sl. No.	Crop	Variety Name	Year	Country	Mutagen	Mutant	Resistance to Disease	Other Improved Traits
9.	Chickpea	Pusa 417 (Girnar)	1985	India	Gamma rays	Direct	Wilt moderate resistance to stunt virus, collar root, foot root, root rot, low pod borer and nematode damage.	High yield, short, profused branched, high pod number, etc.
10.	Chickpea	TAEK-SAGEL	2006	Turkey	Gamma rays	Direct	Resistant to *Ascocyta* blight.	Early maturity, high yield, high seed protein, high 1,000 seed weight.
11.	Chickpea	THAL-2006	2006	Pakistan	–	Indirect	Tolerance to blight.	Tolerance to moisture and bold seed size.
12.	Common bean	Alfa	1972	Slovakia	EMS	Direct	Resistant to *Colletotricum lindemuthianum.*	White seed color, improved seed yield, high protein content, early maturity.
13.	Common bean	FT-Paulistinha	1992	Brazil	Gamma rays	Indirect	Resistant to *Anthracnose*, resistance to leaf spot.	Altered plant architecture and high yield.
14.	Common bean	Fu-22	1980	Giza 80	Gamma rays	Direct	Resistant to rust.	High yield, high TGW, white seed coat, higher protein content.
15.	Common bean	Longyundou 4	1994	China	–	Direct	Resistance to fungal diseases.	High yield.
16.	Common bean	Sanilac	1956	United States	X-rays	Direct	Resistant to alpha, beta, and gamma races of *Colletotricum lindemuthianum* and bean common mosaic virus 1 and 123, resistant to *Sclerotinia sclerotinium.*	Bush type, early maturity.

TABLE 9.8 *(Continued)*

Sl. No.	Crop	Variety Name	Year	Country	Mutagen	Mutant	Resistance to Disease	Other Improved Traits
17.	Common bean	Seafarer	1967	United States	X-rays	Direct	Resistant to alpha, beta, and gamma races of *Colletotricum lindemuthianum* and bean common mosaic virus 1, 15, and 123.	Very early maturity, bush type.
18.	Common bean	Unima	1957	Germany	X-rays	Indirect	Immune to *Colletotricum lindemuthium* and resistant to *Psuedomonas phaseolicola.*	–
19.	Common bean	Universal	1950	Germany	X-rays	Direct	Good resistance to *Colletotricum lindemuthianum.*	Early maturity, high yield.
20.	Cowpea	Lukusuzi	2018	Zambia	Gamma rays	Direct	Tolerant to resistant against *Ascochyta* and *Cercospora*, medium tolerant to cowpea aphid borne mosaic virus.	Determinant and upright plant type, increased number of pods/plant and number of seeds per pod.
21.	Cowpea	V16 (Amba)	1981	India	DMS	Direct	Resistant to fungal and bacterial diseases.	High yield.
22.	Cowpea	V240	1984	India	DMS	Direct	Resistant to fungal, bacterial, and viral diseases.	High yield.
23.	Lentil	Binamasur-1	2001	Bangladesh	Extract of *Datura* seeds.	Direct	Tolerant to rust and blight.	High yield and black seed coat.
24.	Lentil	Binamasur-3	2005	Bangladesh	EMS	Direct	Rust and blight tolerance.	High yield, early maturity.
25.	Lentil	Djudje	2008	Bulgaria	Gamma rays	Direct	Resistance to *Fusarium* and botrytis.	High yield, dwarf bushy habit, non-shattering, high protein content.

TABLE 9.8 *(Continued)*

Sl. No.	Crop	Variety Name	Year	Country	Mutagen	Mutant	Resistance to Disease	Other Improved Traits
26.	Lentil	Mutant 17 MM	1999	Bulgaria	–	–	Resistant to anthracnose, *Sternohylium*, and viruses.	Vigorous growth habit, large leaflet, pods, and seeds, etc.
27.	Lentil	NIAB MASOOR-2006	2006	Pakistan	Gamma rays	Direct	Resistant to rust and blight.	High pod number, lodging resistance.
28.	Lentil	Zornits	2000	Bulgaria	EMS	Direct	Resistance to anthracnose, *Ascochyta* blight.	High yield, high protein content, good culinary and organoleptic qualities.
29.	Mung bean	Camar	1991	Indonesia	Gamma rays	Direct	Resistant to *Cerospora* leaf spot, resistant to *Uromyces* sp., medium resistance to scrap diseases.	High yield and tolerance to salinity and acid soil.
30.	Mung bean	Chai Nat 72	1999	Thailand	Gamma rays	Direct	Resistant to fungal diseases.	High yield, large seed size.
31.	Mung bean	TAP-7	1983	India	Gamma rays	Direct	Resistance to mildew and leaf spot.	Early maturity and high yield.
32.	Mung bean	TARM-1	1997	India	Gamma rays	Direct	Powdery mildew resistant.	–
33.	Mung bean	TARM-18	1996	India	Gamma rays	Indirect	Resistant to powdery mildew.	High yield.
34.	Mung bean	TARM-2	1994	India	Gamma rays	Direct	Resistant to powdery mildew.	High yield.
35.	Mung bean	TM 2000-2	2010	India	–	–	Resistance to powdery mildew.	Suitable for rice fallows.
36.	Mung bean	TM-96-2	2007	India	Gamma rays	Indirect	Powdery mildew resistant, resistance to *Corynespora* leaf spot.	Early maturity, suitable for rice fallows.

TABLE 9.8 *(Continued)*

Sl. No.	Crop	Variety Name	Year	Country	Mutagen	Mutant	Resistance to Disease	Other Improved Traits
37.	Pea	Sredetz	–	Bulgaria	Gamma rays and EMS.	Direct	Resistance to *Ascochyta pisi.*	High yield.
38.	Pigeon Pea	TJT-501	2009	India	–	–	Tolerant to *Phytopthora* blight.	Early maturity, high yield.
39.	White lupin	Kopylovskii	1985	Russian Federation	–	Indirect	*Fusarium* resistance.	–
40.	White lupin	Martin 2	1984	Russian Federation	–	Indirect	*Fusarium* resistance.	–
41.	Yellow lupin	Aga	1981	Poland	X-rays	Direct	Resistant to *Fusarium.*	Early maturity, high yield potential.
42.	Yellow lupin	Narochanskii	1983	Russian Federation	Gamma rays	Direct	Resistance to *Fusarium.*	Early maturity, high yield potential and good fodder quality.

TABLE 9.9 Mutant Varieties of Pulse Crops Resistant to Viral Diseases

Sl. No.	Crop	Variety Name	Year	Country	Mutagen	Mutant	Resistant to	Other Improved Traits
1.	Black gram	Binamash-1	1994	Bangladesh	Gamma rays	Direct	YMV and *cercospora* leaf spot).	High yield and early synchronous maturity.
2.	Black gram	TU-94-2	1999	India	Gamma rays	Indirect	YMV	High yield.
3.	Black gram	Vamban 2	1997	India	–	Direct	YMV	High yield, early maturity.
4.	Chickpea	Pusa 547	2006	India	Gamma rays	Direct	Stunt virus *Fusarium wilt*, and root rot.	High yield, good cooking quality.
5.	Common bean	Gratiot	1962	United States	X-rays	Direct	Common bean mosaic virus 15.	High protein content.
6.	Common bean	IAPAR 57	1992	Brazil	EMS	Indirect	Golgen mosaic virus.	–
7.	Common bean	IAPAR 65	1993	Brazil	EMS	Indirect	Golgen mosaic virus.	–
8.	Common bean	Mogano	1985	Italy	EMS	Direct	Viruses	Dwarf, early maturity, and high yield.
9.	Common bean	Ouray	1982	United States	X-rays	Indirect	Common bean mosaic virus.	Bush habit.
10.	Common bean	Seaway	1960	United States	X-rays	Direct	Common bean mosaic virus.	Short season, upright bush type.
11.	Cowpea	Cowpea-88	1990	India	Irradiation	Direct	YMV	High grain and fodder yield.
12.	Cowpea	TC-901	2018	India	Gamma rays	Direct	Cowpea mosaic virus and root-rot diseases.	Semi-determinate, increased seed size, pod number and pod size.
13.	Faba bean	Tuwaitha	1994	Iraque	Gamma rays	Direct	Wilt	High protein content.
14.	Mung bean	BM 4	1992	India	–	–	YMV	High yield, early maturity.

TABLE 9.9 *(Continued)*

Sl. No.	Crop	Variety Name	Year	Country	Mutagen	Mutant	Resistant to	Other Improved Traits
15.	Mung bean	Binamoog-2	1994	Bangladesh	Gamma rays	Indirect	YMV and *Cercospora* leaf spot.	Large seed size, early, and synchronous maturity.
16.	Mung bean	Binamoog-3	1997	Bangladesh	Gamma rays	Direct	Leaf yellow mosaic virus and *Cercospora* leaf spot.	–
17.	Mung bean	Binamoog-4	1997	Bangladesh	Gamma rays	Direct	Leaf yellow mosaic virus and *Cercospora* leaf spot.	Early maturing, dwarf.
18.	Mung bean	Binamoog-5	1998	Bangladesh	Gamma rays	Direct	Leaf yellow mosaic virus and *Cercospora* leaf spot.	High yield.
19.	Mung bean	Binamoog-6	2005	Bangladesh	Gamma rays	Direct	Yellow mosaic virus.	Early maturity.
20.	Mung bean	Binamoog-7	2005	Bangladesh	EMS	Direct	Leaf yellow mosaic virus and *Cercospora* leaf spot.	High yield.
21.	Mung bean	Binamoog-8	2010	Bangladesh	Gamma rays	Direct	Yellow mosaic virus disease.	Early maturing, 22–23% protein content.
22.	Mung bean	Binamoog-9	2017	Bangladesh	Gamma rays	Direct	Yellow mosaic viruses and *Cercospora* leaf spot.	Early and synchronous maturity, non-shattering pods, etc.
23.	Mung bean	Dhauli (TT9E)	1979	India	–	Indirect	Resistant to yellow mosaic virus.	High yield, early maturity.
24.	Mung bean	LGG 450	1993	India	–	–	Tolerance or resistance to yellow mosaic virus.	High yield, early maturity.
25.	Mung bean	LGG-407	1993	India	–	–	Yellow mosaic virus.	High yield, early maturity.

TABLE 9.9 *(Continued)*

Sl. No.	Crop	Variety Name	Year	Country	Mutagen	Mutant	Resistant to	Other Improved Traits
26.	Mung bean	ML 26-10-3	1983	India	Gamma rays	Direct	Yellow mosaic virus.	High yield.
27.	Mung bean	NIAB Mung 19-19	1985	Pakistan	Gamma rays	Direct	Mung bean yellow mosaic virus.	Early maturity, determinate type, high yield, etc.
28.	Mung bean	NIAB Mung 20-21	1986	Pakistan	Gamma rays	Direct	Yellow mosaic virus and *Cercospora* leaf spot.	Early maturity, shortness, etc.
29.	Mung bean	NIAB Mung 51	1990	Pakistan	Gamma rays	Direct	Yellow mosaic virus and *Cercospora* leaf spot.	Early maturity, shattering resistant large seed size.
30.	Mung bean	NIAB Mung 54	1990	Pakistan	Gamma rays	Direct	Yellow mosaic virus and *Cercospora* leaf spot.	Early maturity, shattering resistant large seed size.
31.	Mung bean	NIAB Mung 92	1992	Pakistan	–	Indirect	Yellow mosaic virus.	Early maturity, shattering resistant large seed size.
32.	Mung bean	NIAB Mung 98	1998	Pakistan	Gamma rays	Indirect	Yellow mosaic virus and *Cercospora* leaf spot.	High yield.
33.	Mung bean	Pant Moong 2	1982	India	Gamma rays	Direct	MYMV	More pods, high yield.
34.	Mung bean	TJM-3	2007	India	Gamma rays	Indirect	YMV, PM, R*hizoctonia* root-rot.	Early maturity, large seeds.
35.	Mung bean	TMB-37	2005	India	–	Indirect	Resistant to yellow mosaic virus.	Early maturing, high yield, large seed.

TABLE 9.10 Disease Resistant Varieties in Fruit and Vegetable Crops

Crop	Developed Resistant Mutant for Disease	Mutagen	References
Potato	*Rhizoctonia solani*	Gamma rays	Kara & Arici (2019)
Japanese pear	*Alternaria alternata*	Gamma rays	Masuda et al. (1997)
Highgate (*Musa aciminata*)	*Fusarium oxysperum*	Gamma rays	Bhagwat & Duncan (1998)
Tomato	Virus resistance	EMS	Piron et al. (2010)
Tomato	Bacterial wilt (*Raestoni solanacearum*)	Gamma ray	Peiris et al. (2009)
Banana	Black sigatoka (*Mycosphaerella fijiensis Morelet*)	Ion beam	Reyes-Boria et al. (2007)
Malus domestica	Alternaria blotch	X-ray	Saito et al. (2001)
Black pepper	Fusariosis	Gamma rays	Hidalgo et al. (1998)
Mentha	*Verticillium wilt*	Gamma rays	Todd (1990)
Musk melon	*Fusarium oxysporum*	Gamma rays	Burger et al. (2013)
Morus spp.	Dogare disease (*Diaporte nomurai*)	Gamma rays	Nakajima (1973)
Zinnia elegance	Leaf curl virus	X-ray	Swarup & Raghava (1974)
Corchorus capsularis	*Macrophomina phaseoli*	X-ray	Singh et al. (1973)
Vitis vinifera	*Plasmophora viticola*	X-ray	Coutinho (1977)
Peppermint	*Wilt verticillium*	Gamma rays	Murray (1977)
Pirus mallus	Apple powdery mildew	Gamma rays	Lapins (1973)

9.7 TILLING IN REVERSE GENETICS FOR DISEASE RESISTANCE

TILLING is one of the reverse genetics approaches which involved inducing mutations in gene of interest followed by identifying point mutation. Vast literature has been generated since the inception of the idea of TILLING. Here, we have mentioned limited examples to show that how TILLING was used to induce mutation for specific gene responsible for disease resistance. An EMS mutagenized population of genotype UC1041+*Yr36* has been screened with primers of Wheat Kinase Start 1 and 2 and START domains (Fu et al., 2009). Mutants obtained through TILLING were found to carry change in conserved amino acid in WKS-1 and premature stop codons for WKS-2 domain. The WKS-1 mutants showed susceptible reaction similar to control whereas none of WKS-2 mutants were lost their resistance indicating WKS-1 is Yr36 gene. Krattinger et al. (2009) applied TILLING method to examine the multi-pathogen resistance locus *Lr34* on short arm of chromosome 7D in wheat. The gene, Lr34 is known for providing durable resistance to leaf rust and was shown co-segregate with partial resistance to adult plant stripe rust (*Yr18*), powdery mildew (*Pm38*) and leaf tip necrosis (*Ltn1*). Single chromosome substitution line Lalbahadur carrying *Lr34*, *Yr18*, *Pm38*, and *Ltn1* were mutagenized with SA and γ rays. The susceptible plants obtained from these populations showed single base substitution and single based deletion in SA and γ rays developed population, respectively. Induced mutations in the ABC transporter were suitable to confer loss of the leaf rust, stripe rust and powdery mildew resistances encoded by *Lr34/Yr18/Pm38* and leaf tip necrosis. TILLING of single gene contributing multiple pathogen resistance in wheat which included *Lr34*, *Yr18*, *Pm38*, and morphological marker *Ltn1*, is controlled by the same gene indicted that presence of single locus could provide durable resistance against multiple diseases. Acevedo-Garcia et al. (2017) established the use of this approach for inducing and identifying mutation at multiple loci for powdery mildew in wheat. Induced missense mutations at three wheat *TaMlo* homoeologs, *TaMlo-A1, TaMlo-B1*, and *TaMlo-D1* cause single amino acid substitutions. Wheat mutants with all three loci mutated and homozygous double mutants showed enhanced resistance to *Blumeria graminis f. sp. tritici* without any pleiotropic effect on phenotype. TILLING in maize to decipher the involvement of genetic loci for resistance against northern corn leaf blight (NCLB) hemibiotrophic fungus *Exserohilum turcicum. Htn1* locus is known for quantitative and partial NCLB resistance and an important source of genetic resistance. This *Htn1* locus is located through map-based cloning in a region of 131.7-kb on chromosome 8 that has three candidate genes encoding a wall-associated receptor-like protein (*ZmWAKRLP1*) and receptor-like kinases (*ZmWAK-RLK1* and *ZmWAKRLK2*). However, analysis of TILLING mutants in *ZmWAK-RLK1* showed susceptibility to NCLB and an alternate splice variant of this locus appeared to be associated with susceptible mutants. TILLING approach assisted to find out quantitative nature of *Htn1* resistance in maize is encoded by an innate immune receptor with an extracellular wall-associated kinase domain (Hurni et al., 2015). TILLING is considered as a novel approach to create new resistance alleles against plant viruses. In tomato, TILING population was screened for mutations on the five translation initiation factors, *eIF4E1, eIF4E2*, *eIF(iso)4E*, *eIF4G*, and *eIF(iso)4G* and found that a splicing mutation in factor eIF4E1 led to immune response against two strains of potato virus Y (PVY) and pepper mottle virus (Piron et al., 2010). Talamè et al. (2008) utilized SA mutagenized barley population of cultivar 'Morex,' and population is screened for barley stem rust resistance protein gene 1 (*Rpg1*) and eukaryotic translation initiation factor 4E (*eIF4E*) genes. In the *Rpg 1* locus, three mis-sense mutations and a single mutation in non-coding region was detected whereas in *eIF4E* genes, three missense and two mutations in non-coding region was found.

9.8 ECOTILLING FOR DISEASE RESISTANCE

EcoTILLING approach is used in barley to analyze allelic variation in race nonspecific (*mlo*) and race specific (*Mla*) resistance against barley powdery mildew and several new variants were induced in both genes that can be exploited to develop cultivars with enhanced resistance level (Mejlhede et al., 2006; Seeholzer et al., 2010). Nieto et al. (2007) employed EcoTILLING to screen allelic variation for *eIF4E* locus in a collection of *Cucumis species* for resistance to Melon Necrotic Spot Virus (MNSV) and Cucumber Vein Yellowing Virus (CVYV). The exonic regions of *eIF4E* was found to be highly conserved and only six polymorphic alleles found from EcoTILLING of 113 accessions. Several studies in germplasm collection of crop plants especially wheat and barley were undertaken for allele mining and resulted in identification additional functional alleles contributing resistance to diseases (Bhullar et al., 2010). EcoTILLING using cDNA was carried for 233 cultivated accessions of the genus *Capsicum*. Sequencing of variants resulted in 21 haplotypes for *eIF4E* and 15 haplotypes *eIF (iso)4E* locus. Functional relevance these alleles showed that five new *eIF4E* variants (*pvr2*^{10}, *pvr2*^{11}, *pvr2*^{12}, *pvr2*^{13}, and *pvr2*^{14}) showed PVY-resistance reaction.

9.9 CLUSTERED REGULARLY INTERSPACED SHORT PALINDROMIC REPEATS (CRISPR) FOR DISEASE RESISTANCE

Clustered regularly interspaced short palindromic repeats associated protein 9 [CRISPR (Cas9)] is an RNA based genome editing system which has been extensively used to improve disease resistance traits in crop plant. Enhanced resistance to bacterial blight (*Xanthomonas oryzae pv. Oryzae*:*Xoo*) of rice was found by editing *Os8N3* (Kim et al., 2019) and *OsSWEET11*, *OsSWEET13*, and *OsSWEET14 genes* (Olivia et al., 2019). Knockout in negative regulator *Oserf922* induced rice null mutants (Wang et al., 2016) exhibiting enhanced resistance to rice blast. Toledo et al. (2016) generated null mutants *SlDMR6-1* via the CRISPR/Cas9 system showed resistance to *P. syringae, P. capsici and Xanthomonas spp.* without unfavorable effects. Wang et al. (2014) edited wheat *Mlo* genes and observed resistance to fungus *Blumeria graminis f. sp. tritici (Bgt).* In tomato, resistance to powdery mildew fungus *Oidiumneo lycopersici* was generated by targeting *SlMlo1* with two sgRNAs simultaneously (Nekrasov et al., 2017). *Taedr1* locus in wheat was mutated by targeting all three homoeologs of Enhanced Disease Resistance (*EDR1*) simultaneously (Zhang et al., 2017) for Bgt resistance. Resistance to citrus canker caused by *Xanthomonas citri ssp. citri* (*Xcc*) was achieved by editing of *CsLOB1* promoter (Peng et al., 2017) and coding region of *CsLOB1* (Jia et al., 2017). Lellis et al. (2002) observed that Arabidopsis mutants which harbor a loss-of function mutation in the *eIF(iso)4E* gene exhibited enhanced resistance to Turnip mosaic virus (TuMV) (Pyott et al., 2016). The precise modification in *eIF4E* locus showed resistance to cucumber vein yellowing virus, zucchini yellow mosaic virus (ZYMV) and papaya ring spot mosaic virus-W (Chandrasekaran et al., 2016) in cucumber, cassava brown streak virus (Gomez et al., 2018) in cassava, Rice tungro spherical virus (Macovei et al., 2018) in rice. Induced resistance to viral diseases like cotton leaf curl Kokhran virus; Merremia mosaic virus and tomato yellow leaf curl virus (Ali et al., 2016), tomato yellow leaf curl virus; beet curly top virus and Merremia mosaic virus (Ali et al., 2015), turnip mosaic virus (Aman et al., 2018), cucumber mosaic virus (CMV) and tobacco mosaic virus (TMV) (Zhang et al., 2018) was achieved in *Nicotiana benthamiana*. Extensive review on the use of CRISPR-Cas9 for disease resistance has been made by Ahmad et al. (2020). A total of 50 reports on the development of disease-resistant crops have been mentioned. Among them, 15 are on bacterial diseases and fungal diseases each, and 20 on viral diseases. Crop-wise reports on disease resistance are 10 in tobacco, 14 in rice, 10 in tomato, 5 in Arabidopsis, 2 each in citrus, grapes, and cassava, and one each in cocoa, papaya, apple, banana, barley, and cucumber. In bacterial disease resistance, 8 reports in rice, 4 in tomato, 2 in citrus, and 1 in apple have been mentioned. In fungal disease resistance, 4 in tomato, 2 in wheat, 2 in grapes, and 3 in rice have been reported. In viral disease resistance, 10 in tobacco, 4 in Arabidopsis, 2 in tomato, and 1 each in banana, barley, cucumber, and cassava have been reported.

KEYWORDS

- **cereals**
- **CRISPR**
- **disease resistant**
- **mutagenesis**
- **mutation breeding**
- **oilseeds**
- **pulse crops**
- **pulses**
- **reverse genetics**
- **TILLING**

REFERENCES

Acevedo-Garcia, J., Spencer, D., Thieron, H., Reinstädler, A., Hammond-Kosack, K., Phillips, A. L., & Panstruga, R., (2017). mlo-based powdery mildew resistance in hexaploid bread wheat generated by a non-transgenic TILLING approach. *Plant Biotechnol. J., 15*(3), 367–378.

Agnihotri, A., Gupta, K., Prem, D., Sarkar, G., Mehra, V. S., & Zargar, S. M., (2009). Genetic enhancement in rapeseed- mustard for quality and disease resistance through *in-vitro* techniques. In: *Proc. of the 16th Australian Research Assembly on Brassicas* (p. 28). Ballarat, Australia.

Ahloowalia, B. S., Maluszynski, M., & Nichterlein, K., (2004). Global impact of mutation-derived varieties. *Euphytica, 135*(2), 187–204.

Ahmad, S., Wei, X., Sheng, Z., Hu, P., & Tang, S., (2020). CRISPR/Cas9 for development of disease resistance in plants: Recent progress, limitations, and future prospects. *Brief. Funct. Genomics*, 26–39. https://doi.org/10.1093/bfgp/elz041.

Ali, Z., Abulfaraj, A., Idris, A., Ali, S., Tashkandi, M., & Mahfouz, M. M., (2015). CRISPR/Cas9-mediated viral interference in plants. *Genome Biol., 16*, 238. doi: 10.1186/s13059-015-0799-6).

Ali, Z., Ali, S., Tashkandi, M., Zaidi, S. S., & Mahfouz, M. M., (2016). CRISPR/Cas9-mediated immunity to Gemini viruses: Differential interference and evasion. *Sci. Rep., 6*, 26912. doi: 10.1038/srep26912.

Aman, R., Ali, Z., Butt, H., Mahas, A., Aljedaani, F., Khan, M. Z., Ding, S., & Mahfouz, M., (2018). RNA virus interference via CRISPR/Cas13a system in plants. *Genome Biol., 19*(1), 1. doi: 10.1186/s13059-017-1381-1.

Bakshi, S., Singh, J., & Jambhulkar, S. J., (2021). Isolation and characterization of yellow rust resistant mutants in wheat. In: Sivasankar, S., et al., (eds.), *Genetic Diversity and Crop Adaptation to Climate Change*. doi: 10.1079/9781789249095.0010.

Bhatia, C. R., & Swaminathan, M. S., (1963). Frequency and spectrum of mutation induced by radiation in some varieties of bread wheat. *Euphytica, 12*, 97–112.
Bhatia, C. R., (1991). Economic impact of mutant varieties in India. In: *Plant Mutation Breeding for Crop Improvement* (Vol. 1, pp. 33–47). IAEA ISBN 92-0-010091-0, ISSN 0074-1884.
Bhatia, C. R., Maluszynski, M., Nichterlein, K., & Van, Z. L., (2001). Grain legume cultivars derived from induced mutations, and mutations affecting nodulation. *Mutation Breeding Review, 13*, 1–44.
Bhullar, N. K., Zhang, Z., Wicker, T., & Keller, B., (2010). Wheat gene bank accessions as a source of new alleles of the powdery mildew resistance gene Pm3: A large-scale allele mining project. *B.M.C. Plant Biol., 10*, 88.
Busungu, C., Taura, S., Sakagami, J. I., & Ichitani, K., (2016). Identification and linkage analysis of a new rice bacterial blight resistance gene from XM14, a mutant line from IR24. *Breed. Sci., 66*(4), 636–645. doi: 10.1270/jsbbs.16062.
Campbell, J., Zhang, H., Giroux, M. J., Feiz, L., Jin, Y., Wang, M., Chen, X., & Huang, L., (2012). A mutagenesis-derived broad-spectrum disease resistance locus in wheat. *Theor. Appl. Genet., 125*(2), 391–404. doi: 10.1007/s00122-012-1841-7.
Chandrasekaran, J., Brumin, M., Wolf, D., Leibman, D., Klap, C., Pearlsman, M., Sherman, A., Arazi, T., & Gal-On, A., (2016). Development of broad virus resistance in non-transgenic cucumber using CRISPR/Cas9 technology. *Mol. Plant Pathol., 17*(7), 1140–1153. doi: 10.1111/mpp.12375.
Chandrashekhar, N., Guptha, O., Yelshetty, S., Sharma, O. P., Bhagat, S., Chattopadhay, C., Sehgal, M., et al., (2014). *Integrated Pest Management for Chickpea* (p. 43). National Centre for Integrated Pest Management, New Delhi, India.
Chattopadhyay, C., Kolte, S. J., & Waliyar, F., (2016). *Diseases of Edible Oilseed Crops* (pp. 3–9). CRC Press.
Chopra, V. L., (2005). Mutagenesis: Investigating the process and processing the outcome for crop improvement. *Cur. Sci., 89*, 353–359.
Coutinho, M. P., (1977). Use of radiation for improving vines regarding their resistance to mildew. In: Murray, M. J., (ed.), *Induced Mutation for Improving Disease Resistance in Crop Plants IAEA* (pp. 233–240). Viena.
Das, M. L., & Rahman, A., (1988). Induced mutagenesis for the development of high yielding varieties in mustard. *J. of Nucl. Agri. and Biol., 17*(1), 1–4.
De Toledo, T. D. P., Brail, Q., Dahlbeck, D., & Staskawicz, B. J., (2016). *CRISPR-Cas9 Mediated Mutagenesis of a DMR6 Ortholog in Tomato Confers Broad-Spectrum Disease Resistance*. bioRxiv. doi: 10.1101/064824.
Duggal, V., Jellis, G. J., Hollins, T. W., & Stratford, R., (2000). Resistance to powdery mildew in mutant lines of the susceptible wheat cultivar hobbit 'Sib.' *Plant Pathol., 49*(4), 468–476.
Fang, Y. F., & Ramasamy, R. P., (2015). Current and prospective methods for plant disease detection. *Biosensors, 4*, 537–561. doi: 10.3390/bios5030537.
Forster, B. P., (2014). The development of 'Eldo Ngano I': The world's first Ug99 resistant mutant wheat variety. *I.A.E.A. Bull., 55*, 18, 19.
Fu, D., Uauy, C., Distelfeld, A., Blechl, A., Epstein, L., Chen, X., Sela, H., et al., (2009). A Kinase-START gene confers temperature-dependent resistance to wheat stripe rust. *Science, 323*(5919), 1357–1360.

Gomez, M. A., et al., (2018). *Simultaneous CRISPR/Cas9-Mediated Editing of Cassava eIF4E Isoforms nCBP-1 and nCBP-2 Reduces Cassava Brown Streak Disease Symptom Severity and Incidence*. bioRxiv. doi: 10.1101/209874. 209874. (all authors).

Hancock, C. N., Zhang, F., & Floyd, K., (2011). The rice miniature inverted repeat transposable element mPing is an effective insertional mutagen in soybean. *Plant Physiol, 157*, 552–562.

Hidalgo, O. B., Pires De, M. A., Cabral, R. S., Tussel, R. T., Arzola, M., Santos, R., & Pérez, M. C., (1998). Phytotoxic effect of culture filtrate from *Fusarium subglutinans* the causal agent of fusariose of pineapple (*Anana scomosus* [L.] Merr.). *Euphytica, 104*(2), 73–77.

Hoffman, T. J., Schmidt, S., Zheng, X., & Bent, A. F., (1999). Isolation of ethylene-insensitive soybean mutants that are altered in pathogen susceptibility and gene-for-gene disease resistance. *Plant Physiology, 119*(3), 935–950.

Humphry, M., Reinstadler, A., Ivanov, S., Bisseling, T., & Panstruga, R., (2011). Durable broad-spectrum powdery mildew resistance in pea er1 plants is conferred by natural loss-of-function mutations in PsMLO1. *Molecular Plant Pathology, 12*(9), 866–878.

Hurni, S., Scheuermann, D., Krattinger, S. G., Kessel, B., Wicker, T., Herren, G., Fitze, M. N., et al., (2015). The maize disease resistance gene *Htn1* against Northern corn leaf blight encodes a wall-associated receptor-like kinase. *Proc. Natl. Acad. Sci. U. S. A., 112*(28), 8780–8785.

IAEA, (1975). *Induced Mutations for Disease Resistance in Crop Plants.* IAEA-181.

IAEA/MVD, (2021). https://mvd.iaea.org (accessed on 12 December 2023).

Jambhulkar, S. J., (2007). Mutagenesis: Generation and evaluation of induced mutations. In: Gupta, P. C., (ed.), *Advances in Botanical Research (Rapeseed Breeding)* (Vol. 45, pp. 418–428). Academic Press: Elsevier Publication Ltd., USA.

Jia, H., Zhang, Y., Orbović, V., Xu, J., White, F. F., Jones, J. B., & Wang, N., (2017). Genome editing of the disease susceptibility gene CsLOB1 in citrus confers resistance to citrus canker. *Plant Biotechnol. J., 15*(7), 817–823. doi: 10.1111/pbi.12677.

Jung, Y. H., Lee, J. H., Agrawal, G. K., Rakwal, R., Kim, J. A., Shim, J. K., et al., (2005). The rice (*Oryza sativa*) blast lesion mimic mutant, *blm*, may confer resistance to blast pathogens by triggering multiple defense-associated signaling pathways. *Plant Physiol. Biochem., 43*, 397–406. doi: 10.1016/j.plaphy.2005.03.002.

Kara, A., & Arici, S. E., (2019). *Determination of Gamma Rays Efficiency Against Rhizoctonia Solani in Potatoes, 17*(1), 254–259.

Kara, A., & Arici, S. E., (2019). Determination of gamma rays efficiency against *Rhizoctonia solani* in potatoes. *Open Chem, 17*, 254–225.

Khush, G. S., (2005). What it will take to feed 5.0 billion rice consumers in 2030. *Plant Mol. Biol., 59*(1), 1–6.

Kim, Y. A., Moon, H., & Park, C. J., (2019). CRISPR/Cas9-targeted mutagenesis of *Os8N3* in rice to confer resistance to *Xanthomonas oryzae pv. Oryzae. Rice, 12*, 67. https://doi.org/10.11 86/s12284-019-0325-7.

Konzak, C. F., (1984). Role of induced mutations. In: Vose, P. B., & Blixt, S. G., (eds.), *Crop Breeding* (pp. 216–292). Pergamon Press: Oxford.

Krattinger, S. G., Lagudah, E. S., Spielmeyer, W., Singh, R. P., Huerta-Espino, J., McFadden, H., Bossolini, E., et al., (2009). A putative ABC transporter confers durable resistance to multiple fungal pathogens in wheat. *Science, 323*(5919), 1360–1363.

Lapins, K. O., (1973). Induced mutation in fruit tree. In: *Induced Mutation in Vegetatively Propagated Plants* (pp. 1–19). IAEA, Vienna.

Lellis, A. D., Kasschau, K. D., Whitham, S. A., & Carrington, J. C., (2002). Loss-of-susceptibility mutants of *Arabidopsis thaliana* reveal an essential role for eIF (iso) 4E during potyvirus infection. *Curr. Biol., 12*(12), 1046–1051. doi: 10.1016/S0960-9822(02)00898-9.

Liu, D., Chen, X., Liu, J., Ye, J., & Guo, Z., (2012). The rice ERF transcription factor OsERF922 negatively regulates resistance to *Magnaporthe oryzae* and salt tolerance. *J. Exp. Bot., 63*(10), 3899–3911. doi: 10.1093/jxb/ers079.

Mabrouk, Y., Charaabi, K., Mahiout, D., Rickauer, M., & Belhadj, O., (2018). Evaluation of chickpea (*Cicer arietinum* L.) irradiation-induced mutants for resistance to *Ascochyta* blight in controlled environment. *Brazilian Journal of Botany, 41*, 311–318.. https://doi.org/10.1007/s40415-018-0458-8.

Macovei, A., Sevilla, N. R., Cantos, C., Jonson, G. B., Slamet-Loedin, I., Čermák, T., Voytas, D. F., et al., (2018). Novel alleles of rice eIF4G generated by CRISPR/Cas9-targeted mutagenesis confer resistance to rice tungro spherical virus. *Plant Biotechnol. J., 16*(11), 1918–1927. doi: 10.1111/pbi.12927.

Maluszynski, M., (1990). Induced mutations – an integrating tool in genetics and plant breeding. In gene manipulation in plant improvement. Gustafson, J. P., (ed.), *Proceedings of the 19th Stadler Genetics Symposium* (Vol. II, pp. 127–162). Plenum Press: New York.

Maluszynski, M., Ahloowalia, B. S., & Sigurbjörnsson, B., (1995). Application of *in-vivo* and *in-vitro* mutation techniques for crop improvement. *Euphytica, 85,* 303–315. doi: 10.1007/BF000.

Masuda, T., et al., (1997). Selection of mutants resistant to black spot disease by chronic irradiation of gamma –rays in Japanese pear •osanijissciki. *J. Jpn. Soc. Hortic. Sci., 66*, 88–92.

Meena, P. D., Sujith, K. M. S., Meena, H. S., Jambhulkar, S., Gohartaj, D. P., Snigdha, S., Riteka, G., et al., (2020). Confirmation of induced tolerance to *Alternaria* blight disease in Indian mustard (*Brassica juncea* L.). *Applied Biochemistry and Biotechnology*. https://doi.org/10.1007/s12010-020-03362-2.

Mejlhede, N., Kyjovska, Z., Backes, G., Burhenne, K., Rasmussen, S. K., & Jahoor, A., (2006). EcoTILLING for the identification of allelic variation in the powdery mildew resistance genes Mlo and Mla of barley. *Plant Breed., 125*(5), 461–467.

Mew, T. W., Alvarez, A. M., Leach, J. E., & Swings, J., (1993). Focus on bacterial blight of rice. *Plant Dis., 77*(1), 5–12.

Micke, A., (1991). Induced mutations for crop improvement. *Gamma Field Symp., 30*, 1–21.

Mohapatra, T., Robin, S., Sarla, N., Sheshasayee, M., Singh, A. K., Singh, K., Singh, N. K., et al., (2014). EMS induced mutants of upland rice variety Nagina22: Generation and Characterization. *Proc. Indian Natl. Sci. Acad., 80*(1), 163–172. doi: 10.16943/ptinsa/2014/v80i1/55094.

Mothilal, A., Karunanithi, K., & Babu, C., (2010). Gamma ray induced Spanish bunch mutant with foliar disease resistance in groundnut (*Arachis hypogaea* L.) *Electronic Journal of Plant Breeding, 1*(3), 342–345.

Murray, M. J., (1977). Successful use of irradiation breeding to obtain verticillium resistant strains of peppermint, *Mentha piperita*; In: *Induced Mutation for Improving Disease Resistant in Crop Plants* (pp. 233–240). IAEA, Vienna.

Nakajima, K., (1973). Induction of useful mutation in mulberry by gamma rays. In: *Induced Mutation in Vegetatively Propagated Plants* (Vol. 34, No. 17–21, pp. 105–117). Vienna.

Nekrasov, V., Wang, C., Win, J., Lanz, C., Weigel, D., & Kamoun, S., (2017). Rapid generation of a transgene free powdery mildew resistant tomato by genome deletion. *Sci. Rep., 7*(1), 482. doi: 10.1038/s41598-017-00578-x.

Oerke, E. C., (2006). Crop losses to pests. *J. Agric. Sci., 144*(1), 31–43.

Oladosu, Y., Rafii, M. Y., Abdullah, N., Hussin, G., Ramli, A., Rahim, H. A., Miah, G., & Usman, M., (2016). Principle and application of plant mutagenesis in crop improvement: A review. *Biotechnol. Biotechnol. Equip., 30*(1), 1–16.

Oliva, R., Ji, C., Atienza-Grande, G., Huguet-Tapia, J. C., Perez-Quintero, A., Li, T., Eom, J. S., et al., (2019). Broad-spectrum resistance to bacterial blight in rice using genome editing. *Nat. Biotechnol., 37*(11), 1344–1350. doi: 10.1038/s41587-019-0267-z.

Peiris, R., Wickramasinghe, T. K., & Indrasena, S. P., (2009). M 127—A promising tomato variety developed through induced mutation technique. In: Shu, Q. Y., (ed.), *Induced Plant Mutation in the Genomics Era* (pp. 379, 380). Proceeding of an International joint FAOIAEA Symposium, Rome; Food and Agriculture Organization of the United Nation.

Peng, A., Chen, S., Lei, T., Xu, L., He, Y., Wu, L., Yao, L., & Zou, X., (2017). Engineering canker-resistant plants through CRISPR/Cas9-targeted editing of the susceptibility gene CsLOB1 promoter in citrus. *Plant Biotechnol. J., 15*(12), 1509–1519. doi: 10.1111/pbi.12733.

Piron, F., Nicolaï, M., Minoïa, S., Piednoir, E., Moretti, A., Salgues, A., Zamir, D., et al., (2010). An induced mutation in tomato eIF4E leads to immunity to two potyviruses. *PLOS One, 5*(6), e11313.

Prasad, P., Bhardwaj, S. C., Gangwar, O. P., & Kumar, S., (2021). Wheat rust research: Impact, thrusts, and road map to sustained wheat production. In: Sindhu, S. P. S., Singh, C., Poonam, J. G. P. S., & Sarial, A. K., (eds.) *Improving Cereal Productivity Through Climate Smart Practices* (pp. 177–203). Wood head Publishing. doi: 10.1016/B978-0-12-821316-2.00011-X.

Pulse India, (2019). *An India Pulses and Grains Association Publication* (Vol. IV, No. 2).

Pulses: Status and Contribution to Food Security. Govt. of India, Directorate of pulses development Min. of Agri. And FW, (DAC&FW), Bhopal (MP).

Pyott, D. E., Sheehan, E., & Molnar, A., (2016). Engineering of CRISPR/Cas9-mediated potyvirus resistance in transgene-free *Arabidopsis* plants. *Mol. Plant Pathol., 17*(8), 1276–1288. doi: 10.1111/mpp.12417.

Raihan, A., Hasan, M. M., Islam, S., & Rahman, L., (2018). Development of yellow mosaic virus resistance in mung bean through EMS mutagenesis. *Asian Journal of Advances in Agricultural Research, 7*(1), 1–6.

Raza, A., Razzaq, A., Mehmood, S. S., Zou, X., Zhang, X., Lv, Y., & Xu, J., (2019). Impact of climate change on crops adaptation and strategies to tackle its outcome: A review. *Plants (Basel), 8*(2), 34. doi: 10.3390/plants8020034.

Reyes-Borja, W. O., Sotomayor, I., Garzón, I., Vera, D., Cedeño, M., Castillo, B., Tanaka, A., et al., (2007). Alteration of resistance to black sigatoka (*Mycosphaerell afijiensis Morelet*) in banana by *in-vitro* irradiation using carbon-ion beam. *Plant Biotechnol, 24*(3), 349–353.

Ribeiro, F. X., Parlevliet, J. E., & Zambolim, L., (2001). Concepts in plant disease resistance. *Fitopatol. Bras., 26*, 577–589.

Rózewicz, M., Wyzinska, M., & Grabinski, J., (2021). The most important fungal diseases of cereals—Problems and possible solutions. *Agronomy, 11*, 714. https://doi.org/ 10.3390/agronomy11040714.

Saito, A., Nakazawa, N., & Suzuki, M., (2001). Selection of mutants resistant to Alternaria blotch from *in vitro*-cultured apple shoots irradiated with X- and γ-rays. *J. Plant Physiol., 158*(3), 391–400.

Saito, A., Nakazawa, N., & Suzuki, M., (2001). Selection of mutants resistant to Alternaria blotch from *in vitro*-cultured apple shoots irradiated with X- and γ-rays. *J. Plant Physiol., 158*, 391–400.

Seeholzer, S., Tsuchimatsu, T., Jordan, T., Bieri, S., Pajonk, S., Yang, W., Jahoor, A., et al., (2010). Diversity at the Mla powdery mildew resistance locus from cultivated barley reveals sites of positive selection. *Mol. Plant Microbe Interact., 23*(4), 497–509.

Sharma, B., (2016). Multiple allelism at Er locus for powdery mildew resistance in pea (*Pisum sativum* L.). *Indian J. Genet., 76*(4), 512–514.

Sharma, M., Gupta, S. K., & Pratap, A., (2012). Induced mutagenesis and *in vitro* cell selection for resistance against *Alternaria blight* in *Brassica napus* L. *Ind. J. of Agri.l Res., 46*(4), 369–373.

Simmons, C., Hantke, S., Grant, S., Johal, G. S., & Briggs, S. P., (1998). The maize lethal leaf spot 1 mutant has elevated resistance to fungal infection at the leaf epidermis. *Molecular Plant-Microbe Interaction, 11*, 1110–1118.

Singh, D. P., Sharma, B. K., & Banerjee, C., (1973). X-ray induced mutation in jute (*C. capsularis* and *C. olitorius*). *Genetica Agaria, 27*, 115–147.

Smith, P. H., Howie, J. A., Worland, A. J., Stratford, R., & Boyd, L. A., (2004). Mutations in wheat exhibiting growth-stage-specific resistance to biotrophic fungal pathogens. *Mol. Plant Microbe Interact., 17*(11), 1242–1249. doi: 10.1094/MPMI.2004.17.11.1242.

Swarup, V., & Raghava, S. P. S., (1974). Induced mutation for resistance to leaf curl virus and its inheritance in garden zinnia. *Indian Genet, J. and Plant Breed., 34*, 17–21.

Takahashi, A., Kawasaki, T., Henmi, K., ShiI, K., Kodama, O., Satoh, H., & Shimamoto, K., (1999). Lesion mimic mutants of rice with alterations in early signaling events of defense. *Plant J., 17*(5), 535–545. doi: 10.1046/j.1365-313x.1999.00405.x.

Talamè, V., Bovina, R., Sanguineti, M. C., Tuberosa, R., Lundqvist, U., & Salvi, S., (2008). TILLMore, a resource for the discovery of chemically induced mutants in barley. *Plant Biotechnol. J., 6*(5), 477–485. https://doi.org/10.1111/j.1467-7652.2008.00341.x.

Taura, S., Ogawa, T., Yoshimura, A., & Omura, T., (1991a). Induction of mutants resistant to bacterial blight in rice. *Jpn. J. Breed., 41*(2), 279–288.

Taura, S., Ogawa, T., Yoshimura, A., Ikeda, R., & Lwata, N., (1992). Identification of a recessive resistance gene to rice bacterial blight of mutant line XM6, *Oryza sativa* L. *Jpn. J. Breed., 42*(1), 7–13.

Taura, S., Ogawa, T., Yoshimura, A., Ikeda, R., & Omura, T., (1991b). Identification of a recessive resistance gene in induced mutant line XM5 of rice-to-rice bacterial blight. *Jpn. J. Breed., 41*(3), 427–432.

Todd, A. J., (1990). Performance of disease resistant peppermint mutants in the USA. *Mutat. Breed. Newsletter, 36*, 14, 15.

Vanniarajan, C., Ganeshram, S., Souframanien, J., Veni, K., Anandhi, L. S., & Kuralasaran, V., (2019). Gamma rays induced urdbbean [*Vigna mungo* (L.) hepper] mutants with YMV resistance, good batter quality and bold seeded type. *Legume Research, 42*(1), 25–31.

Wang, F., Wang, C., Liu, P., Lei, C., Hao, W., Gao, Y., Liu, Y. G., & Zhao, K., (2016). Enhanced rice blast resistance by CRISPR/Cas9-targeted mutagenesis of the ERF transcription factor gene OsERF922. *PLOS One, 11*(4), e0154027. doi: 10.1371/journal.pone.0154027.

Wang, Y., Cheng, X., Shan, Q., Zhang, Y., Liu, J., Gao, C., & Qiu, J. L., (2014). Simultaneous editing of three homoeoalleles in hexaploid bread wheat confers heritable resistance to powdery mildew. *Nat. Biotechnol., 32*(9), 947–951. doi: 10.1038/nbt.2969.

Wu, J. L., Wu, C., Lei, C., Baraoidan, M., Bordeos, A., Madamba, M. R. S., Ramos-Pamplona, M., et al., (2005). Chemical and irradiation-induced mutants of indica rice IR64 for forward and reverse genetics. *Plant Mol. Biol., 59*(1), 85–97. doi: 10.1007/s11103-004-5112-0.

Xiao, N., Wu, Y., Pan, C., Yu, L., Chen, Y., Liu, G., Li, Y., et al., (2017). Improving of rice blast resistances in japonica by pyramiding major R genes. *Frontiers in Plant Science, 7*, 1918. https://doi.org/10.3389/fpls.2016.01918.

Zeng, L., Yin, Z., Chen, J., Leung, H., & Wang, G. L., (2002). Fine genetic mapping and physical delimitation of the lesion mimic gene Spl11 to a 160-kb DNA segment of the rice genome. *Mol. Genet. Genomics, 268*, 253–261.

Zhang, M. X., Xu, J. L., Luo, R. T., Shi, D., & Li, Z. K., (2003). Genetic analysis and breeding use of blast resistance in a japonica rice mutant R917. *Euphytica, 130*(1), 71–76.

Zhang, T., Zheng, Q., Yi, X., An, H., Zhao, Y., Ma, S., & Zhou, G., (2018). Establishing RNA virus resistance in plants by harnessing CRISPR immune system. *Plant Biotechnol. J., 16*(8), 1415–1423. doi: 10.1111/pbi.12881.

Zhang, Y., Bai, Y., Wu, G., Zou, S., Chen, Y., Gao, C., & Tang, D., (2017). Simultaneous modification of three homoeologs of TaEDR1 by genome editing enhances powdery mildew resistance in wheat. *Plant J., 91*(4), 714–724. doi: 10.1111/tpj.13599.

Zhong-Hua, W., Xin-Chen, Z., & Yu-Lin, J., (2014). Development and characterization of rice mutants for functional genomics studies and breeding. *Rice Sci., 21*(4). doi: 10.1016/S1672-6308(13)60188-2.

CHAPTER 10

Recent Advances in Developing Disease Resistance in Pulse Crops

VINOD JANARDAN DHOLE, J. SOUFRAMANIEN, and P. DHANASEKAR

Nuclear Agriculture and Biotechnology Division, Bhabha Atomic Research Center, Mumbai, Maharashtra, India

ABSTRACT

Diseases are one of the major biotic stresses limiting the pulse crops production all over the world. Increase in the production of pulses with the pace of increasing population is a big challenge to agriculture scientists in the era of climate change. Development of varieties with genetic resistance is the only feasible solution which is achieved to some extent with the help of conventional methods of plant breeding. The advances in genomics, genetic engineering and other throughput molecular techniques have paved the way to faster, efficient, and effective development of resistant varieties irrespective of availability of resistance gene sources. The whole genome sequencing of most of the pulse crops has resulted in the availability of abundant genomic resources like SNPs, SSR, and INDELs which will be very useful for fine mapping of resistance genes. The closely linked DNA markers are very useful in marker-assisted selection as demonstrated in chickpea through the development of disease-resistant varieties like Pusa Chickpea-10216 and Super Annigiri-1. Mutation breeding and genetic engineering approaches are going to play an important role when resistance gene sources are not available for some of the serious diseases. Genetically engineered common bean transgenic line EMBRAPA 5.1 is the first transgenic pulse crop variety released for commercial cultivation. Genomics, genetic engineering, and modern mutation breeding will change the scenario of disease resistance breeding of pulses in the near future.

New Advances in Crop Disease Management. Sangita Sahni, Ganesh Thapa, and Bishun Deo Prasad (Eds)

10.1 INTRODUCTION

Pulses belong to the family Leguminosae, which includes major pulse crops like chickpea, pigeon pea, dry pea, lentil, cowpea, mung bean, blackgram, and common bean, etc. The majority of pulse crops are consumed as split grain (dahl), dehulled split grain, whole grain, flour, and sprouts. Their plant parts are used as dry fodder for animals, which are rich in nutrients. Pulses are a very important component in the diet of the vegetarian population owing to their rich and good quality protein, along with slow-releasing carbohydrates, vitamins, minerals, high fiber, and low-fat content. Pulses ensure the nutritional security of the low-income group as they are a cheap source of protein, especially for vegetarian, rural, and poor people all over the world. Pulses play a very important role in sustainable agriculture as they are less demanding, fix atmospheric nitrogen, and their crop residue adds valuable organic matter to the soil. Due to these qualities, they can fit into many cropping systems and crop rotations, thus helping to improve soil health. Pulses are grown in all regions of the world, with Asia and Africa contributing around half of the world production. India, Canada, Myanmar, China, Brazil, and Australia are major countries sharing the maximum production of pulses around the globe. Compared to cereals, the productivity of pulses is very low. Low productivity in pulses is mainly due to biotic and abiotic stresses, which were aggravated due to climate change and their cultivation in marginal land with less input. Among the biotic stresses, diseases caused by fungi, bacteria, viruses, and nematodes are a major threat to pulse production and productivity throughout the world. To ensure nutritional security in a changing environment, it is very important to reduce the losses that occur due to diseases in various pulse crops. The use of chemical, cultural, and bio-control methods of disease control not only increases the cost of cultivation but also pollutes the environment and is still not one hundred percent effective. Pulses are grown mainly in rain-fed conditions with low input by small and marginal farmers in underdeveloped and developing countries. They cannot afford an increase in the cost of cultivation. Hence, the development of disease-resistant pulse crop varieties is the only effective and efficient way to reduce losses and the cost of cultivation. In this chapter, we will discuss present methods and recent advances that will enhance the efficiency of different breeding approaches to develop disease resistance in pulse crops.

10.1.1 MAJOR DISEASES OF PULSE CROPS AND THEIR ECONOMIC IMPORTANCE

Demand for pulses increasing with increasing population and awareness about the importance of protein in vegetarian diets. But due to low productivity, the production of pulses is not able to cope up with increasing demand. Biotic stresses are reducing the productivity of pulses crops considerably. Among the biotic stresses, diseases caused by fungi, bacteria, and viruses are of most economic importance as they are causing the huge yield losses from 5% to 100% depending on the stage at which the crop get infected by the disease (Table 10.1). *Fusarium wilt*, Ascochyta blight and botrytis gray mold are major diseases of concern in chickpea causing 10% to 100% yield losses (Singh & Reddy, 1991; Pande et al., 2006); while fungal diseases like *fusarium wilt*, phytophthora blight, Alternaria blight and viral disease viz., sterility mosaic disease (SMD) are economically most important in pigeon pea causing up to 100% yield losses depending on disease severity and stage of infection (Sharma et al., 2013; Elbeaino et al., 2014; Sharma & Ghosh, 2016; Sharma & Ghosh, 2018). In South-East Asian countries, mung bean, and blackgram are mainly infected by viral diseases like Yellow Mosaic Disease (YMD) (Figure 10.1) and Leaf Crinkle Virus (LCV) along with fungal diseases like powdery mildew and Cercospora Leaf Spot (CLS) reducing yield from 5% to 100% (Nene, 1972; Chenulu et al., 1979; Fernandez & Shamugasundaram, 1987; Chand et al., 2012). Cowpea, another crop belonging to genus *Vigna* is mostly prone to viral diseases. Among them, Bean Common Mosaic Disease (BCMD) and Cowpea Yellow Mosaic Diseases (CYMD) are most important while cowpea bacterial blight caused by *Xanthomonas axonopodis* pv. *vignicola* causing huge yield losses (Williams, 1977; Sikirou, 1999; Jordan & Hammond, 2020). Pea and lentil are important winter season crops in which fungal diseases viz., powdery mildew, rust, and *fusarium wilt* are most devastating (Sepulveda, 1985; Kumar et al., 2010; Upadhyay et al., 2015). Worldwide huge yield losses due to diseases were reported every year and we can imagine the situation with the example of Australia alone. The loss from diseases in the Australian pulse industry alone averages $74 million per year, or 14.8% of the gross value of pulse production and losses would be far higher without the current range of controls, which include the use of resistant varieties, rotation, paddock management and the use of pesticides (Murray & Brennan, 2012). The losses are much more in the underdeveloped and developing countries where farmers could not afford the cost of plant protection against these diseases.

TABLE 10.1 Losses in Important Pulse Crops Due to Major Diseases

Crop	Disease	Causal Organism	Transmission/ Vector	Yield Losses (%)	References
Chickpea (*Cicer arietinum*)	Wilt/root rot	*Fusarium oxysporum* f. sp. *ciceris*	Soil borne	10 to 90	Singh & Reddy (1991)
	Ascochyta blight	*Ascochyta rabiei*	Seed borne	10 to 80	Reddy & Singh (1990)
	Botrytis grey mold (BGM)	*Botrytis cinerea* Pers. Ex. Fr.	Seed borne	10 to 100	Pande et al. (2006)
Pigeon pea (*Cajanus cajan* L. Druce)	*Fusarium wilt*	*Fusarium udum* Butler	Soil borne	50 to 70	Sharma & Ghosh (2016)
	Phytophthora blight	*Phytophthora drechsleri* f. sp. *Cajani*	Soil borne	Up to 100	Sharma & Ghosh (2018)
	Sterility mosaic disease (SMD)	*Pigeon pea sterility mosaic virus* (PSMV)	Eriophyid mites	95 to 100	Elbeaino et al. (2014)
	Alternaria blight	*Alternaria tenuissima*	Seedborne	20 to 80	Sharma et al. (2013)
Mungbean (*Vigna radiata* L. Wilczek) & Blackgram (*Vigna mungo* L. Hepper)	Yellow mosaic disease (YMD)	Mungbean yellow mosaic virus (MYMV)	White flies	5 to 100	Nene (1972); Chenulu et al. (1979)
	Powdery mildew	*Erysiphe polygoni*	Air borne	20 to 40	Fernandez & Shamugasundaram (1987)
	Cercospora leaf spot	*Cercospora canescens* Ellis	Soil borne	50 to 70	Chand et al. (2012)
	Leaf crinkle virus (LCV)	Urd bean leaf crinkle virus (ULCV)	Seed/sap/aphid	62 to 100	Nene (1972)
Cowpea (*Vigna unguiculata* L. Walp.)	Cowpea bacterial blight	*Xanthomonas axonopodis* pv. *vignicola*	Seed	64 to 100	Sikirou (1999)
	Bean common mosaic.	Bean common mosaic virus (BCMV)	Seed/aphids	6 to 98	Jordan & Hammond (2020)
	Cowpea yellow mosaic virus	Cowpea yellow mosaic virus	Aphids	80 to 100	Williams (1977)

TABLE 10.1 *(Continued)*

Crop	Disease	Causal Organism	Transmission/ Vector	Yield Losses (%)	References
Pea (*Pisum sativum* L.)	Powdery mildew	*Erysiphe polygoni*	Air borne	25 to 50	Warkentin et al. (1996)
	Rust	*Uromyces viciae fabae* (Pers.) de Bary	Air borne	20 to 100	Upadhyay et al. (2015)
Lentil (*Lens culinaris* Medic.)	*Fusarium Wilt*	*Fusarium oxysporum* f. sp. *lentis*	Soil borne	Up to 100	Kumar et al. (2010)
	Rust	*Uromuces febae* (Pers) de Bary	Air borne	70 to 100	Sepulveda (1985)
Common bean/Rajmash (*Phaseolus vulgaris* L.)	Bean common mosaic	Bean common mosaic virus	Seed/aphids	Up to 100	Worrall et al. (2015)

FIGURE 10.1 Comparison of resistant and susceptible varieties of mung bean to yellow mosaic disease in field conditions.

10.1.2 WHY TO BREED RESISTANT VARIETIES?

Agrochemicals are mainly used to control plant diseases and their vectors for a long time. Now a day, people are more concerned about their effects on environment and health due to awareness and information available from modern research. It leads to tighter rules and regulation on their use, while the list of banned fungicides is also becoming long, e.g., European Union banned one of the world's most commonly used fungicides mancozeb while recently captan and carbendazim are proposed to be banned in India. Therefore, it is crucial to find out such a solution which can take care of plant health without increasing cost of cultivation, without any hazards to health of consumers and free from pollution. The solution lies in breeding cultivars with multiple disease resistance against major diseases.

10.2 BREEDING STRATEGIES FOR THE DEVELOPMENT OF DISEASE RESISTANT VARIETIES IN PULSES

Breeding strategies for the development of disease resistance is mainly based on availability of sources of resistance genes in the germplasm. When resistant sources are easily crossable with recipient parent then genes can be easily transferred to develop resistant varieties. Wild species are very good reservoir of disease resistance genes, but gene transfer is quite difficult due to crossability barriers and linkage drags. The recent biotechnology tools like embryo rescue, chromosome doubling, and use of bridging species can be used to overcome crossability barriers. While the new approaches like genetic engineering, transgenic, and mutation breeding can be applied for the development of disease resistance varieties when the sources of resistance genes are not available in the germplasm (Figure 10.2).

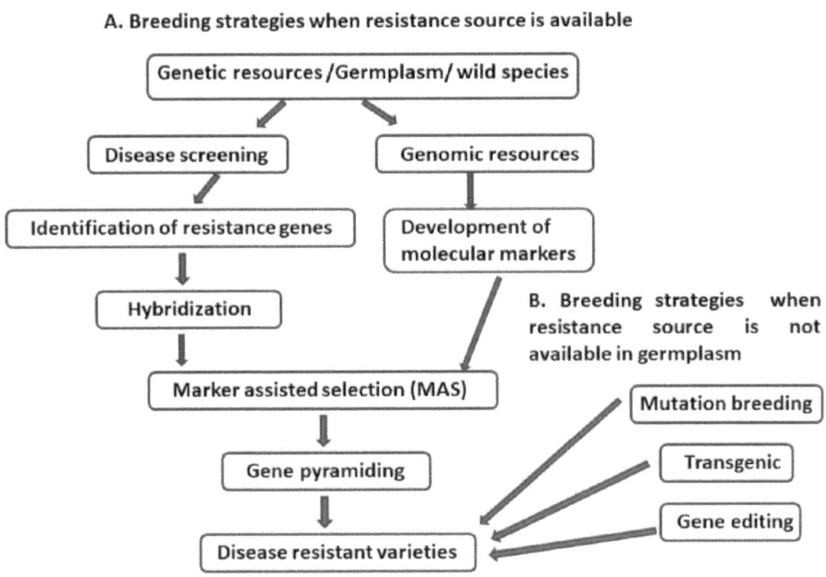

FIGURE 10.2 Different breeding strategies for the development of disease resistance in pulse crops.

10.2.1 BREEDING STRATEGIES WHEN RESISTANCE SOURCE IS AVAILABLE

10.2.1.1 CONVENTIONAL BREEDING

Conventional breeding involves hybridization between resistant and susceptible parents and subsequently resistant varieties developed by following methods like pedigree or back cross methods. Conventional methods are mostly depending on available sources of resistance genes which may be present in land races, germplasm, or wild relatives. This approach has two main disadvantages, i.e., take a lot of time to deliver a final product and linkage drag while introgression of resistance gene from wild relatives. Breeding resistance against the pathogens needed prior information on sources of resistance gene in the germplasm, efficient screening techniques and inheritance of resistance genes.

10.2.1.2 SOURCES OF RESISTANCE GENES IN GERMPLASM

Disease resistance genes present in the cultivated germplasm are easy to transfer without any crossability barriers; hence it is the most efficient way to use these sources for the development of resistant varieties. Disease resistance sources are available in various pulse crops and some of these sources were used successfully as a donor for the development of resistant varieties (Table 10.2). In chickpea, IC275447, EC 267301, and IC248147 are resistance sources against Ascochyta blight (Gayacharan et al., 2020), while Sanford, WR-315, ICCV-07105, ICCV 07111 and IPC2010-21 (INGR20077) possessed resistance genes to *fusarium wilt* (Saabale et al., 2019; Sharma et al., 2019). Against botrytis gray mold, ICC1069, P349-2 and NEC2451 are the resistance sources (Haware et al., 1993), while inter-specific derivative 'line 04067-81-2-1-1' is a reliable source for phytophthora root-rot to develop resistant varieties in chickpea (Amalraj et al., 2019). In pigeon pea, resistance sources were identified and used in breeding against SMD (BSMR 736, ICP 7869, ICP9045 and ICP 14076) (Sharma et al., 2012), phytophthora blight (METH12, Pant A3, ICP 580, and UC2515/2) (Mohar et al., 2014) and *fusarium wilt* (KPL 43, DPPA 85-12, ICP 7903, and ICP 12775) (Sharma & Pande, 2011). The YMD resistance sources were identified in mung bean viz., Pusa-0672, MH-521, WGG42, ML-1451, ML-1299 and blackgram viz., KU-9921, DKU-4, Mash-114, TU-17-4, Vamban-6, TU-94-2 (Anonyms, 2021). In mung bean, NM-98, 98-cmg-003, C2/94-4-42, NM-1, NM-2, 98 cmg-018, BRM-188 and CO-3 are the resistance sources for CLS (Anonyms, 2021) and TARM-1, TARM-2 and TARM-18, TM-96-2 and Mulmarada were reported to be resistant to different races of powdery mildew (Reddy, 2009). The germplasm of pea viz., P660-4, IFPI3260, IFPI3232, IFPI3261, and IFPI3253 had resistance against powdery mildew (Fondevilla et al., 2011) while Kinnauri is possessing resistance against Ascochyta blight (Rastogi & Saini, 1984). Rust resistance was found in P 9-77, P 2432, P2572m, and P 2930 germplasm lines which can be utilize for the development of rust resistance in pea (Mishra et al., 2009). In lentil, resistance genes against major diseases like Ascochyta blight (ILL 5588, Indianhead and ILL7537), *fusarium wilt* (Pant L-4, Pant L-6, Pant L-8 and Noori) and rust (Morocco, FLIP2004-7L, Pant L-639, Pant L-406) are identified (Garkoti et al., 2013). In rajmash, germplasm lines Ring Bean, Arkasuvida, and Black Turtle-1 are important resistance sources against Bean Common Mosaic Virus (BCMV) (Patil et al., 2017b; Meziadi et al., 2017) and the germplasm lines like EC565673A, LD-2-1, IC-1435, AMBER were the resistance sources for powdery mildew (Mishra et al., 2019).

10.2.1.3 SOURCES OF RESISTANCE GENES IN WILD RELATIVES

When resistance sources for important diseases are not available in cultivated species, wild relatives are second choice for the plant breeders. Wild species of pulse crops are very good reservoirs for resistance genes against various important diseases (Table 10.3). In chickpea, wild species *Cicer judaicum, Cicer innatifidum, Cicer montbretii, Cicer bijugum, Cicer echinospermum* and *Cicer reticulatum* are sources of resistance genes to diseases like Ascochyta blight, *fusarium wilt*, Botrytis gray mold and Phytophthora root rot (van der Maesen & Pundir, 1984: Singh & Reddy, 1993; Singh et al., 1994; Ahmad et al., 2005; Knights et al., 2008). Pigeon pea is very rich in wild germplasm in which *Cajanus sericeus, Cajanus albicans, Cajanus acutifolius* and *Cajanus platycarpus* are the sources for resistance genes against sterility mosaic disease, phytophthora blight and *fusarium wilt* (Akinola et al., 1975; Singh et al., 1993; Mallikarjuna & Saxena, 2002; Mallikarjuna et al., 2011). In pigeon pea, phytophthora blight resistance genes were successfully introgressed from *Cajanus platycarpus* into short duration variety (Mallikarjuna et al., 2005). Two cultivated crop species viz., rice bean (*Vigna umbellata*) and blackgram (*Vigna mungo*), while two wild species *Vigna umbellate* and *Vigna trilobata* are the sources of resistance gene for YMD and CLS, respectively for mung bean (Singh & Dikshit, 2002; Pandiyan et al., 2008; Sehrawat & Yadav, 2014; Bhanu et al., 2017). In interspecific cross of *Vigna radiata* x *Vigna umbellate*, QTLs associated with mung bean yellow mosaic virus (MYMV) resistance were identified (Mathivathana et al., 2019). The wild species *Pisum fulvum* possesses resistance genes for powdery mildew and Ascochyta blight diseases in pea (Clulow et al., 1991; Fondevilla et al., 2011). In lentil, resistance genes for anthracnose, Ascochyta blight, *fusarium wilt*, powdery mildew and rust disease can be transferred from *Lens ervoides, Lens lamottei, Lens nigricans, Lens odemensis, Lens montbretti*, and *Lens culinaris* ssp. *orientalis* (Bayaa et al., 1994, 1995; Gupta & Sharma, 2006;

TABLE 10.2 Sources of Resistance Genes in Germplasm and Their Inheritance

Crop	Name of Disease	Sources of Resistance Genes	Inheritance	Prevalent Races	References
Chickpea	Ascochyta blight	IC275447, IC117744, EC267301, IC248147	A single dominant gene, combination of a recessive and a dominant gene and QTLs.	Pathotypes I, II, III, and IV.	Singh & Reddy (1983); Cho et al. (2004); Peever et al. (2004); Taleei et al. (2009); Atik (2013); Gayacharan et al. (2020)
–	*Fusarium wilt*	SANFORD, WR-315, IPC2010-21 (INGR20077) ICCV 98505, ICCV 07105, ICCV 07111, ICCV 07305, ICCV 08113, and ICCV 93706.	Single recessive, two recessives.	8 races: 0, 1A, 1B/C, 2, 3, 4, 5, and 6.	Sharma & Muehlbauer (2007); Saabale et al. (2019); Sharma et al. (2019)
–	Botrytis gray mold.	ICC1069, P349-2 and NEC2451	Single dominant gene, duplicate dominant epistasis.	8 pathotypes	Haware et al. (1993); Chaturvedi et al. (1995); Kishore (2005); Bhardwaj et al. (2018)
–	Phytophthora root rot. *P. medicaginis*	*C. echinospermum* interspecific breeding line 04067-81-2-1-1.	Major QTL QRBprrsi01, QYBprrsi01, QRBprrsi03 and QYBprrsi02.	No confirmed information regarding pathogen variability is available.	Amalraj et al. (2019)
Pigeon pea	Sterility mosaic disease.	BSMR 736, ICPs 7869, 9045, 11015, 11230, 11281, 11910, and 14976.	Two genes, SV1 and SV2, with inhibitory gene interaction, QTLs: four QTLs for Patancheru PSMD isolate and two QTLs for Bangalore PSMD isolate.	Five different isolates.	Reddy et al. (1993); Gnanesh et al. (2011); Sharma et al. (2012); Daspute et al. (2014)

TABLE 10.2 *(Continued)*

Crop	Name of Disease	Sources of Resistance Genes	Inheritance	Prevalent Races	References
–	Phytophthora blight	METH12, UC2515/2, UC796/1, UC2113/1, UC2568/1, Pant A83-14, Pant A3, ICPL161, Comp 1ESR6, ICP4726-8, ICP113 (IC306508), ICP580 (IC306495)	Monogenic dominant	No confirmed information regarding pathogen variability is available.	Mohar et al. (2014)
–	*Fusarium wilt*	KPL 43, DPPA 85-12, DPPA 85-13 and DPPA 85-8. ICP 7903, ICP 12031, ICP 12059, ICP 12771, ICP 12775.	Two or more recessive genes (1:15) with duplicate dominant epistasis.	14 isolate	Sharma & Pande (2011); Datta & Lal (2013); Patil et al. (2017a)
Mungbean	Yellow mosaic disease.	Pusa-0672, MH-521, WGG42, ML-1451, ML-1299.	A single dominant gene, recessive gene inheritance, two recessive genes, digenic dominant interaction with some modifier genes, five QTLs.	Strain-A (Indonesia) strain-B (Vietnam isolates) MYMV and MYMIV (from India).	Khattak et al. (2000); Dhole & Reddy (2012, 2013); Mahalingam et al. (2018); Markam et al. (2018); Dikshit et al. (2020); Anonyms (2021)
–	Cercospora leaf spot	NM-98, 98-cmg-003, C2/94-4-42, NM-1, NM-2, 98-cmg-018, BRM-188, CO-3, Basanti, PDM-11, BARIMung-2, VC3960-88, V4718, LGG 460.	A single dominant gene, Duplicate epistatic interaction, two genes.	–	Chankaew et al. (2011); Choudhary et al (2021); Anonyms (2021)
–	Powdery mildew	TARM-1, TARM-2 and TARM-18, TM-96-2 and Mulmarada.	Two QTLs on two LGs, *qPMR-1*, and *qPMR-2*, three dominant gene Pm1 and dominant gene Pm2, Pm3.	Three races race-1, race 2 and race 3.	Reddy et al. (1994); Reddy (2009); Kasettranan et al (2010)

TABLE 10.2 *(Continued)*

Crop	Name of Disease	Sources of Resistance Genes	Inheritance	Prevalent Races	References
Blackgram	Yellow mosaic disease	KU-9921, DKU-4, Mash-114, TU-17-4, Vamban-6, TU-94-2.	Monogenic dominant, digenic recessive, monogenic recessive.	MYMIV-VSKN	Singh (1980); Gupta et al. (2005); Naimuddin et al. (2011); Rambabu et al. (2018); Anonyms (2021)
–	Powdery mildew	LBG-17, K-5-572, LBG-645, IC-281977, BPMR-145, TU-17-4.	A single recessive gene, three duplicate recessive genes.	Three races	Kaushal & Singh (1989); Kute et al. (2003); Khandappagol & Rangaiah (2019); Kanwade et al. (2019)
Cowpea	Cowpea bacterial blight.	Danila, IT81D-1228-14.	Three QTLs, CoBB-1, CoBB-2, and CoBB-3.	Two strains Xav18 and Xav19.	Agbicodo et al. (2010)
–	Bean common mosaic.	V-5 and IC 8996	Partial or incomplete dominant gene.	BCMV-NL1, -NL3 and -NY15.	Hobbs et al. (1987); Lana et al. (1988)
–	Cowpea yellow mosaic virus.	V-57817	Single recessive gene.	–	Dinesh et al. (2018)
Pea	Powdery mildew	P660-4, IFPI3260, IFPI3232, IFPI3261 and IFPI3253.	Two recessive genes er1, er2 and single dominant gene Er3.	–	Fondevilla et al. (2011); Bobkov & Selikhova (2021)
–	Ascochyta blight	Kinnauri	3 to 14 QTLs	–	Rastogi & Saint (1984); Tar'an et al. (2003); Timmerman-Vaughan et al. (2004)
–	Rust	P 9-77, P 2432; P2572 and P 2930.	A single dominant gene, a single oligo-gene showing partial dominance, involvement of one to two major genes; two QTLs viz., *Qruf* and *Qruf1*, *Qruf2* and *Qruf3*.	Two species *U. pisi* (Persoon) de Bary (European Countries) and *U. fabae* (Pers.) de Bary (India).	Tyagi & Srivastava (1999); Vijayalakshmi et al. (2005); Singh & Ram (2001); Mishra et al. (2009); Singh et al. (2015); Rai et al. (2016)

TABLE 10.2 *(Continued)*

Crop	Name of Disease	Sources of Resistance Genes	Inheritance	Prevalent Races	References
Lentil	Ascochyta blight	ILL5588, Indianhead, ILL7537	Ral2 (dominant) and ral2 (recessive) dominant gene for foliar resistance (AbR1), Two dominant complementary genes.	Five/six pathotypes	Tay & Slinkard (1989); Andrahennadi (1994); Nasir & Bretag (1998); Chowdhury et al. (2001); Nguyen et al. (2001); Sambasivam et al. (2017)
–	*Fusarium wilt*	Pant L-4, Pant L-6, Pant L-8 and Noori.	Single dominant gene.	Eight races	Eujayl et al. (2006); Garkoti et al. (2013); Hiremani & Dubey (2018)
–	Rust	Morocco, FLIP2004-7L, s Pant L-639, Pant L-406, Pant L-6, pant L-7 and Pant L-8.	Duplicate dominant, monogenic dominant, two duplicate non-allelic and non-linked dominant genes, single recessive gene.	–	Lal et al. (1996); Chahota et al. (2002); Mishra et al. (2008); Garkoti et al. (2013); Kumar & Gill (2020)
Common bean/ Rajmash	Bean common mosaic virus.	Ring bean, Arka-Suvidha, Black Turtle-1.	Two dominant genes, four recessive genes bc-1, bc-2, bc-3, and bc-u.	Four strains (NL-1, NL-1n, NL-7 and NL-7n).	Kyle et al. (1986); Milkas et al. (2000); Sharma et al. (2011); Trabanco et al. (2012); Patil et al. (2017b); Meziadi et al. (2017); Feng et al. (2018)
–	Powdery mildew	EC565673A, LD-2-1, IC-1435, AMBER.	Two dominant genes, two genes with a double recessive epistasis, single dominant gene.	–	Rezende et al. (1999); Murube et al. (2017); Mishra et al. (2019)

Tullu et al., 2006). In rajmash, *Phaseolous acutifolius* and *Phaseolous coccineus* are the sources of resistance genes to diseases like common blight, Bean Golden Yellow Mosaic Disease (BGYMD), root rot, anthracnose, and angular leaf spot (Singh & Munoz, 1999; Osorno et al., 2003; Silbernagel & Hannan, 1992; Mahuku et al., 2003). These resistance genes to various diseases were accumulated in wild species due to natural mutation followed by natural selection which took a very long period. Although the resistance sources are available in wild species, these sources are not used very

TABLE 10.3 Sources of Resistance Genes in Wild Species of Pulse Crops

Sl. No.	Crop	Wild Species	Resistance Against	References
1.	Chickpea	*Cicer judaicum, Cicer pinnatifidum, Cicer montbretii, Cicer bijugum, Cicer echinospermum, Cicer reticulatum*	*Ascochyta* blight, *Fusarium wilt, Botrytis* gray mold, *Phytophthora* root-rot.	van der Maesen & Pundir (1984): Singh & Reddy Pundir (1993); Singh et al. (1994); Ahmad et al. (2005); Knights et al. (2008)
2.	Pigeon pea	*Cajanus sericeus, Cajanus albicans, Cajanus sericeus, Cajanus acutifolius, Cajanus platycarpus*	Sterility mosaic disease, *Phytophthora* blight, *Fusarium wilt.*	Akinola et al. (1975); Singh et al. (1993); Mallikarjuna & Saxena (2002); Mallikarjuna et al. (2011)
3.	Mungbean and Blackgram	*Vigna umbellata, Vigna trilobata, Vigna mungo var. silvestris*	Yellow mosaic disease, *Cercospora* leaf spot, bacterial leaf spot.	Singh & Dikshit (2002); Pandiyan et al. (2008); Sehrawat & Yadav (2014); Bhanu et al. (2017)
4.	Pea	*Pisum fulvum* (Two genes conferring resistance to powdery mildew, *er1* and *er2, Er3*).	Powdery mildew, *Ascochyta* blight.	Clulow et al. (1991); Fondevilla et al. (2011)
5.	Lentil	*Lens ervoides, Lens lamottei, Lens nigri-cans, Lens odemensis, Lens montbretti, Lens culinaris* ssp. *orientalis*	Anthracnose, *Ascochyta* blight, *Fusarium wilt,* powdery mildew, rust.	Bayaa et al. (1994, 1995); Gupta & Sharma (2006); Tullu et al. (2006)
6.	Common bean/Rajmash	*Phaseolus acutifolius, Phaseolus coccineus*	Common blight, BGYMV root rot, anthracnose, angular leaf spot.	Singh & Munoz (1999); Osorno et al. (2003); Silbernagel & Hannan (1992); Mahuku et al. (2003)

frequently in breeding programs. The transfer of resistance genes from wild species is very tedious and time consuming, which is also associated with linkage drag, crossability barriers, lack of homology between chromosomes of cultivated and wild species. The crossability barriers have been overcome by the modern methods like embryo rescue technique, chromosome doubling and use of bridging species. The recent molecular techniques like molecular markers are very useful for eliminating linkage drag and also reduce time required to transfer resistance genes by foreground and background selections. These techniques will encourage the alien gene transfer from wild to cultivated species which can help to reduce the yield losses due to diseases in pulse crops.

10.2.1.4 INHERITANCE OF RESISTANCE GENES

Information on inheritance of disease resistance genes is pre-requisite for development of resistant varieties. Inheritance of gene will decide the breeding method to be followed for efficient and effective development as well as deployment of disease resistance varieties. Gene pyramiding for the development of durable as well as multiple disease resistance is the present need for stabilizing the food production. Gene pyramiding is not possible without knowledge of inheritance of genes to be pyramided. In pulse crops, information on inheritance of resistance genes to different diseases are available but with contradictory reports for some diseases (Table 10.2). In chickpea, a single dominant gene, combination of a recessive and a dominant gene and QTLs were reported for Ascochyta blight (Singh & Reddy, 1983; Cho et al., 2004; Taleei et al., 2009), while resistance to YMD in mung bean is reported to be controlled by a single dominant gene, recessive gene, two recessive genes, digenic dominant interaction with some modifier genes and five QTLs (Khattak et al., 2000; Dhole & Reddy, 2012, 2013; Mahalingam et al., 2018; Markam et al., 2018; Dikshit et al., 2020; Win et al., 2021). Such type of results were also noticed for number of genes governing resistance to diseases like *fusarium wilt* of chickpea (single recessive gene and two recessive genes) (Sharma & Muehlbauer, 2007; Saabale et al., 2019; Sharma et al., 2019), sterility mosaic of pigeon pea (two and four QTLs) (Gnanesh et al., 2011; Daspute et al., 2014) and YMD of blackgram (single recessive gene and two recessive genes) (Singh, 1980; Gupta et al., 2005; Rambabu et al., 2018). Gene interactions were also involved in governing resistance to *fusarium wilt* of chickpea (duplicate dominant epistasis) (Sharma & Muehlbauer, 2007; Saabale et al., 2019), cercospora leaf spot of mung bean (duplicate epistatic interaction) (Chankaew et al., 2011; Choudhary et al., 2021), Ascochyta blight (Two dominant complementary genes) (Andrahennadi, 1994; Chowdhury et al., 2001) and rust (duplicate dominant) of lentil (Lal et al., 1996; Mishra et al., 2008; Kumar & Gill, 2020). Transfer of simply inherited resistance gene is quite easy as compared to multiple genes and polygenes with gene interactions. There are various reasons behind the different inferences drawn for the same disease in a single crop. Genetics of disease resistance is depending on the background of genetic material and the source of resistance whether it is from cultivated or wild species. Disease scoring also decides the number of genes involved, e.g., sometimes the plants are scored simply as resistant and susceptible irrespective of its intensity which results into identification of single or few genes governing disease resistance. When disease rating scale is used, it will take care of disease intensity and disease progress from which we can estimate type of inheritance, i.e., qualitative or quantitative with different types of gene interactions.

10.2.1.5 SCREENING TECHNIQUES

Screening techniques are the most important for phenotyping the breeding population which decide the accuracy of identification resistance genotypes, genes, and their genetics. It is playing crucial role in tagging the disease resistance genes, identification of QTLs with the help of molecular markers. Thus, authenticity of genotyping is based on accurate phenotyping. Screening techniques are developed according to type of disease (root diseases or foliar diseases), pathogen (obligate or facultative) and transmission (sap/seed/insect vectors). Disease screening can be done in natural or field conditions and artificially in green houses and laboratories. Sick plot method is commonly used for soil born and root disease like *fusarium wilt* in pigeon pea and phytophthora blight in chickpea. In case of air borne or foliar diseases, spore suspension is sprayed to develop disease while for seed borne diseases, dry spores are dusted on seeds or seeds may be soaked in a pathogen spore suspension. The field screening of large population at hot spot is followed for foliar diseases like rust, Ascochyta blight and YMD in pulse crops. Infector/spreader rows are also employed in field screening of diseases like YMD, powdery mildew and rust diseases to increase the disease spread. The disease scoring is more accurate when carried out at the stage of 90% infection in infector rows. Some diseases like YMD which is transmitted only by insect vector whiteflies and hence cannot be created artificially. Large scale screening of such diseases is only possible in the field conditions at hot spot. For small scale confirmation of disease reaction, the genotypes can be screened in net houses using viruliferous whiteflies (Shivaprasad et al., 2006). Agroinoculation, another technique was evolved to identify MYMV resistant mung bean genotypes in which tandem viral construct of MYMV mobilized in *Agrobacterium tumefaciens* and then agroinoculation are performed on sprouted seeds (Sudha et al., 2015; Sai et al., 2017). It will reduce the chances of disease escape due to uniform disease conditions. In laboratory screening, excised leaf technique was successfully used for powdery mildew disease screening in mung bean (Reddy, 2009). In field conditions, there is a possibility of mixed infection of two or more races of pathogen. In excised leaf technique, purified single spore is used to infect the leaves in controlled conditions hence race specific screening is possible. The new screening technique of 'pollen bioassay' developed to screen chickpea genotypes for *fusarium wilt* by using 'Fusaric acid' the toxins produced by the *fusarium* (Ratna Babu & Ravikumar, 2010). It is a simple and effective screening technique which is based on inhibition of pollen tube growth with different concentration of Fusaric acid.

10.2.1.6 GENETIC VARIABILITY IN PATHOGENS

Development of durable resistance is most important in disease resistance breeding for which information on prevailing races or strains of pathogen in that locality is necessary. The identification of different races of pathogen is mainly based on availability of differential varieties. Multiple races of pathogens are identified for diseases like *fusarium wilt* of chickpea (eight races) and powdery mildew of mung bean (three races), while different strains of viruses were identified in MYMV of mung bean (four strains) and BCMV of rajmash (four strains) (Table 10.2). In some pathogens, information on number of races and genetic variability is not available, e.g., diseases like phytophthora blight of pigeon pea and chickpea. Variability in pathogen often analyzed on the basis of morphology of pathogen viz., conidial/spore size, culture type, growth rate, pigmentation, radial growth, and colony characters. Pathogenic variability is also assessed on the basis of bio-chemicals produced by pathogens, e.g., phytotoxins like fusaric acid and pectinolytic enzymes produced by *fusarium* (Iftikhar et al., 2004). Aggressive and more virulent isolates of *Fusarium udum* reported to be produced more amounts of pectinolytic enzymes and rich in total sugar (Kumar et al., 2007). Isolates of *Botrytis cinerea* with higher concentration of oxalic acid and greater activity of lytic enzymes were found to be more pathogenic to chickpea crop (Kumari et al., 2014). The pathological, morphological, and biochemical basis of pathogen variability is not providing the clear picture about the existence of races or strains of pathogen which is needed for disease resistance breeding program. Recent advances in molecular biology made it possible to study the pathogen variability at genetic/DNA level. Initially, molecular markers like random amplified polymorphic DNA (RAPD) and restriction fragment length polymorphism (RFLP) were applied to study variability in pathogens infecting pulse crops. Isolates of *Cercospora canescens* causing cercospora leaf spot in pulses were differentiated at DNA level by using RAPD and Internal transcribed spacer region (ITS) markers of ribosomal DNA (rDNA) markers (Joshi et al., 2006). Further, the new race of *C. canescens* (NFCCI-2370) was differentiated from other isolate (MTCC-10835) by comparing ITS sequence data (Chand et al., 2012). The Botrytis cinerea isolates were grouped on the basis of geographical region and host by using RAPD markers (Kumari et al., 2014). Simple sequence repeats (SSR) markers were used to study genetic diversity in different isolates of *Ascochyta rabiei* causing Ascochyta blight in chickpea (Bayraktar et al., 2007). Race specific markers like Sequence Characterized Amplified Region (SCAR) markers were developed to identify the race *Foc* of *Fusarium oxysporum* f. sp. *ciceris* (Jimenez-Gasco & Jimenez-Diaz, 2003). The SSR and ISSR markers differentiated the isolates of *Fusarium oxysporum* f. sp. *ciceris* which corresponds to geographic locations and racial distribution (Dubey et al., 2012). The race specific DNA markers should be developed for each prevailing races/strains of pathogens. It is very much needed for quick and reliable screening of diseases there by helping in the identification of differential varieties scientifically. Apart from DNA markers, a diagnostic multiplex-RT-PCR method was developed to detect and differentiate the isolates of PPSMV-1 and PPSMV-2 causing sterility mosaic virus in pigeon pea (Patil et al., 2017a). The recent studies on powdery mildew of mung bean and blackgram based on internal transcribed spacer (ITS) and large subunit sequences of the ribosomal DNA revealed that this disease in both the crops is caused by two species viz., *Podosphaera xanthii* and a newly recognized taxon *Erysiphe vignae* sp. nov. (Kelly et al., 2021).

10.2.2 MOLECULAR BREEDING APPROACHES FOR THE DEVELOPMENT OF DISEASE RESISTANCE

10.2.2.1 GENOMICS RESEARCH IN PULSE CROPS

Most of the pulse crops are grown in developing and underdeveloped countries of Asia and Africa, where the funds, infrastructure, and expertise are the limitations for the genomic research. Hence genomic research was lagging behind in the pulse crops as compared to cereals and oilseed crops. Until second decade of 21st century, these crops had limited genomic resources and regarded as 'orphans' of the genomic revolution (Varshney et al., 2009). The new generation sequencing and high throughput approaches in genomic research are cost effective and very quick which encouraged the genomic research in the pulse crops of poor countries. The first draft genome sequence came in the pigeon pea, and it was the beginning of new era of genomic studies in pulse crops. The draft genome sequence of pigeon pea variety Asha (ICPL 87119) was carried out and published by two independent groups in 2012 (Sing et al., 2012; Varshney et al., 2012). Very next year, this was followed by the whole genome sequencing of genotype CDC Frontier, a Canadian Kabuli chickpea variety which is resistant to Ascochyta blight (Varshney et al., 2013) and desi type variety ICC 4958 of chickpea (Jain et al., 2013). Subsequently, the draft of whole genome sequence was published in mung bean (Kang et al., 2014), common bean (Schmutz et al., 2014), pea (Kreplak et al., 2019), cowpea (Lonardi et al., 2019) and blackgram (Souframanien et al., 2021).

10.2.2.2 GENOMIC RESOURCES AFTER WHOLE GENOME SEQUENCING OF PULSES

After whole genome sequencing, pulse crops became very rich in genomic resources which can be applied to develop tightly linked markers with disease resistance genes, identification of QTLs, foreground, and background selection for rapid transfer of resistance gene. The drafts of pigeon pea whole genome sequencing resulted into annotation of 48,680 genes along with 3,09,052 SSRs and 12,164 to 16,651 SNPs from which 23,410 primer pairs were designed (Varshney et al., 2012). In the same year 2012, another draft genome sequencing of pigeon pea was published which revealed the presence of 47,004 protein coding genes and 1,213 disease resistant (R like) genes and defense response (DR) genes (Sing et al., 2012). These disease resistant genes were classified into five classes, out of which 98 DR genes belongs to chitinases (31 genes), glucanases (56 genes) and thaumatin-like proteins (11 genes), while 219 R like genes belongs to viral resistance, verticillium wilt resistance, bacterial blight resistance and genes containing LRR motif but without NBS, CC or LZ motifs (Sing et al., 2012). The large number of disease resistance genes and defense response genes are playing important role to make the pigeon pea a hardy crop with fever diseases (Sing et al., 2012). Whole genome sequencing of chickpea was depicted 27,571 protein coding genes, 729 putative resistance related genes including receptor like kinase and 30,000 monomeric SSRs (Jain et al., 2013). In another draft of chickpea whole genome sequencing; 28,269 genes, 81,845 SSRs, 76,084 SNPs in 15,526 genes were reported. Out of the SSRs, 48,298 SSRs are suitable for PCR primer design which can be used as markers. The chickpea genome assembly contained 187 disease resistance genes homologs (Varshney et al., 2013). Mungbean draft genome sequence information helped in the identification of 22,427 protein coding genes, 2,922,833 SNPs, 2,00,808 SSRs and 464 LLR genes. From the LRR genes, 73 LRR genes with NB-ARC domain and 19 genes without NB-ARC domain were identified as genes for disease resistance along with 30 SNP markers flanking resistance genes (Kang et al., 2014). In common bean, out of 27,197 protein coding genes, 376 genes are NB-LLR (NL) genes related to disease resistance and 1,397,405 SNPs were also found (Schmutz et al., 2014). In pea, 44,756 protein coding genes and 17,212,424 SNPs were reported (Kreplak et al., 2019), while in cowpea 29,773 protein coding loci, 3,36,285 SNPs and 402 genes annotated belonging to NBS-LLR super family (Lonardi et al., 2019). Blackgram genome analysis identified 42,115 genes, 34,816 SSRs, 1,659 genes related to R-genes which can be used to enhance the genetic improvement (Souframanien et al., 2021). This genomic information can be applied more efficiently with the help of bioinformatics-based comparative and predictive approaches to design various markers and identification of candidate genes imparting disease resistance in pulse crops.

10.2.2.3 MARKERS LINKED TO DISEASE RESISTANCE AND MARKER ASSISTED SELECTION

Marker assisted selection (MAS) is one of the promising molecular tools which will be the choice of future for plant breeders to develop disease resistant varieties. Success of MAS is mainly depending on type of marker, its reproducibility, linkage with resistance gene, ease of screening, cost, and time required. Identification of tightly linked markers and their use can be categorized in to two phases. One is before the era of whole genome sequencing and other is after whole genome sequencing of pulse crops. Before the whole genome sequencing, due to scarce genomic resources only few markers like RFLP, AFLP, RAPD, and ISSR markers were mostly used for the development of linkage maps, while cross transferability of SSR from other related crops were studied (Table 10.4). The PCR based DNA markers like RAPD and ISSR were most commonly used to tag resistance gene viz., *fusarium wilt* resistance gene in pigeon pea (Kotresh et al., 2006; Prasanthi et al., 2009), YMD resistance gene in mung bean (Dhole & Reddy, 2013; Kalaria et al., 2014) and blackgram (Souframanien & Gopalakrishna, 2006) and rajmash (Ravikumar et al., 2009), while in pea, RAPD markers were reported be linked with powdery mildew resistance (Frew et al., 2002) and rust resistance (Vijayalaksmi et al., 2005). The reproducibility of RAPD and ISSR markers is always questionable and hence some of these markers were converted to SCAR markers for better reproducibility, i.e., for *fusarium wilt* in pigeon pea (Prasanthi et al., 2009), YMD resistance in mung bean (Dhole & Reddy, 2013) and blackgram (Souframanien & Gopalakrishna, 2006). Another type of markers like RFLP linked to SMV resistance in pigeon pea (Ganapathy et al., 2009) and AFLP markers were applied to detect the QTLs imparting resistance to YMD in mung bean (Chen et al., 2013) and Cowpea Golden Mosaic Virus in cowpea (Rodrigues et al., 2012). The application of these markers in MAS was limited due to lack of very tight linkage with resistance genes. The Resistance Gene Analog (RGA) markers were very useful for tagging the resistance genes and reported to be tightly linked to CLS resistance in mung bean (Tantasawat et al., 2020), YMD resistance gene in blackgram (Basak et al., 2004; Maiti et al., 2011). Further these markers were validated in blackgram (Sowmini & Jayamani, 2014).

After the whole genome sequencing of various pulse crops, plenty of genomic resources like SSR and SNP markers were available which can be applied to increase the efficiency of selection. Particularly the SSR markers flanking the resistance gene or QTLs which were reported for *fusarium wilt* resistance in chickpea (Gowda et al., 2009), CLS resistance in mung bean (Chankaew et al., 2011) and QTLs conferring rust resistance in pea (Rai et al., 2011, 2016). Some of the SSR markers were found to be very useful in marker assisted selection and backcrossing, e.g., *fusarium wilt* resistance (Varshney et al., 2014; Pratap et al., 2017) and Ascochyta resistance in chickpea (Varshney et

TABLE 10.4 Molecular Markers Linked with Disease Resistant Genes in Pule Crops

Crop	Disease	Markers Linked with Resistant Genes	Remark	References
Chickpea	*Fusarium wilt*	SSR	SSR markers H3A12 and TA110, flanked the Foc1 locus at 3.9 and 2.1 cM, respectively, while Foc2 was mapped 0.2 cM from TA96 and 2.7 cM from H3A12. The H1B06y and TA194 markers flanked the Foc3 locus at 0.2 and 0.7 cM, respectively.	Gowda et al. (2009)
–	–	SSR	UBC-864_{400} and UBC-811_{1250} linked with *Fusarium wilt*.	Haji-Allahverdipoor et al. (2011)
–	–	Six SSR markers.	Useful for marker assisted backcrossing.	Varshney et al. (2014)
–	–	SSR	SSR marker TA194 tightly linked to resistance gene.	Ahmad et al. (2014)
–	–	STMS	TA200 and TR2 flanking markers linked to *Foc* 1 at 1.0 and 3.0 cM.	Barman et al. (2014)
–	–	SSR markers.	A stable QTL (GSSR 18-TC14801) for wilt resistance.	Jingade & Ravikumar (2015)
–	–	Two SSR markers.	Useful for MABC for resistance gene *Foc* 2.	Pratap et al. (2017)
–	–	Three SSR markers.	The SSR markers CESSR433, NCPGR21 and ICCM0284 linked to loci controlling FW (Foc2) resistance.	Jha et al. (2021)
–	Ascochyta blight.	STMS markers.	Useful for MAS.	Castro et al. (2013)
–	–	Eight SSR marker.	Used for marker assisted backcrossing.	Varshney et al. (2014)
–	–	SNP marker.	SNP (Ca4: 15,920,939), located in the conserved catalytic domain of the LRR receptor-like kinase (Ca_05515), led to an amino acid substitution (Gly/Ala).	Li et al. (2017)
Pigeon pea	*Fusarium wilt*	SCAR	RAPD marker converted to $SCAR_{937}$.	Prasanthi et al. (2009)
–	–	SSR	Five SSR markers associated with resistance.	Khalekar et al. (2014)
–	–	SSR	ASSR-1, ASSR-23, ASSR-148, ASSR-229, ASSR-363, and ASSR-366 associated with FW resistance.	Sing et al. (2016)

TABLE 10.4 *(Continued)*

Crop	Disease	Markers Linked with Resistant Genes	Remark	References
–	–	SSR	Six SSR marker associated with FW.	Patil et al. (2017a)
–	–	10 SNP markers.	Diagnostic marker kit based on Kompetitive allele-specific PCR genotyping assay.	Saxena et al. (2021)
–	Sterility mosaic	AFLP	Four markers (E-CAA/M-GTG150, E-CAA/M-GTG60, E-CAG/M-GCC120 and E-CAG/M-GCC150) are linked to SMD.	Ganapathy et al. (2009)
–	–	10 SNP markers.	Diagnostic marker kit based on Kompetitive allele-specific PCR genotyping assay.	Saxena et al. (2021)
–	–	SNP markers.	One SNP at 19,958,148 bp position on CcLG11 present within the QTL (*qSMD11.1*) by GBS.	Saxena et al. (2017)
Mungbean	Powdery mildew	SSR	SSR marker-based QTLs Viz., *qPMR-1* and *qPMR-2* associ-ated with resistance.	Kasettranan et al. (2010); Chankaew et al. (2013)
–	–	ISSR-RGA	I85420 and I42PL222 markers linked to PM resistant allele.	Chathiranrat et al. (2018)
–	Yellow mosaic disease	SCAR	SCAR marker MYMVR-583 is associated with MYMV resistance.	Dhole & Reddy (2013); Alam et al. (2014); Binyamin et al. (2015)
–	–	AFLP and SSR	Four major QTLs on three different linkage groups for MYMIV resistance using AFLP and SSR markers.	Chen et al. (2013)

TABLE 10.4 *(Continued)*

Crop	Disease	Markers Linked with Resistant Genes	Remark	References
–	–	SSR	Composite interval mapping identified five QTLs for MYMIV resistance: three QTLs for India (qYMIV1, qYMIV2 and qYMIV3) and two QTLs for Pakistan (qYMIV4 and qYMIV5).	Kitsanachandee et al. (2013)
–	–	RAPD and ISSR	Linked to YMD resistance.	Kalaria et al. (2014)
–	–	SNPs	A QTL on chromosome 4, designated as qMYMV4-1 by using GBS.	Mathivathana et al. (2019)
–	Cercospora leaf spot	SSR	Markers CEDG117 and VR393 flanking qCLS	Chankaew et al. (2011)
–	–	SSR and ISSR-RGA.	Two ISSR markers, 1 ISSR-RGA marker and 3 SSR markers closely linked with resistance gene.	Tantasawat et al. (2020)
Blackgram	Yellow mosaic disease	RGA markers	VMYR1 marker tightly linked to resistant gene.	Basak et al. (2004(
–	–	SCAR	Linked to resistance gene at 6.8 cM.	Souframanien & Gopalakrishna (2006)
–	–	RGA	RGA markers, YR4 and CYR1 associated with resistance to MYMIV.	Maiti et al. (2011)
–	–	SSR	CEDG180 linked to MYMIV resistance at a distance of 12.9 cM.	Gupta et al. (2013)
–	–	RGA and SCAR	Validated in blackgram genotypes.	Sowmini & Jayamani (2014)
Cowpea	Bacterial blight	SNP markers	Markers associated with three QTLs.	Agbicodo et al. (2010)
–	Cowpea yellow mosaic virus.	SSR markers	Linked to QTL.	Gioi et al. (2012)

TABLE 10.4 *(Continued)*

Crop	Disease	Markers Linked with Resistant Genes	Remark	References
–	Cowpea golden mosaic virus.	AFLP	Associated with three QTLs.	Rodrigues et al. (2012)
–	*Fusarium wilt* resistance	SNPs	Associated with *Fot* race 3 and 4.	Pottorff et al. (2012, 2014)
Common bean/ Rajmash	Yellow mosaic disease	RAPD marker	OPP-07730 linked to MYMV resistance gene.	Ravishankar et al. (2009)
Lentil	Rust	SSR	SSR marker Gllc 527 was estimated to be linked to rust resistant locus at a distance of 5.9 cM.	Dikshit et al. (2016)
Pea	Powdery mildew	RAPD	Sc-OPO-181200, Sc-OPE-161600 linked to resistance gene er3.	Frew et al. (2002)
–	–	SNPs	SNP marker linked with *er1-6* allele.	Sun et al. (2017)
–	–	InDel	Linked to *er1-7* allele.	Sun et al. (2016)
–	–	PsMlo1	A near-diagnostic marker (*PsMlo1*) for PM.	Sudheesh et al. (2014)
–	–	KASPar markers	Eight markers for eight resistance allele.	Ma et al. (2017)
–	*Mycosphaerella pinodes*	SSR markers	Linked to resistance gene *mp*.	Dita et al. (2006)
–	Rust (*Uromyces fabae*)	2 RAPD markers	Two markers flanking *Ruf* gene.	Vijayalakshmi et al. (2005)
–	–	SSR	Flanking markers AA505, AA446 for 2 QTLs.	Rai et al. (2016)
–	–	SSR	Flanking markers AA505, AA446, AA121, AD147 for 4 QTLs.	Rai et al. (2011)
–	Rust (*Uromyces pisi*)	SNPs	SNP markers 3567800, AD2803563695, 3569323 for 3 QTLs.	Barilli et al. (2018)

al., 2014). The diagnostic SNP markers kit based on Kompetitive allele-specific PCR genotyping assay is available for *fusarium wilt* and SMD resistance in pigeon pea (Saxena et al., 2021). Application of Genome Wide Association (GWAS) paved the way for the identification SNP located in the exons and the QTLs of the resistance genes, e.g., Ascochyta resistance in chickpea (Li et al., 2014) and SMD resistance QTL in pigeon pea (Saxena et al., 2017). Day by day, the new markers are going to be available in pulse crops and application of marker assisted selection in the development of disease resistant variety will became very common. The Translational Pigeon pea Genomic Consortium (TPGC) involving nine research institutions/agricultural universities of six different states of India was established in 2017 to improve 11 mega varieties through MAS including resistance for *fusarium wilt* and SMD (Saxena et al., 2020). In blackgram, MABC was successfully applied to develop nine lines possessing resistance to MYMV (Subramaniyan et al., 2021). Recently, two chickpea varieties viz., Pusa Chickpea-10216 (multiple disease resistant) and MBC-WR-SA-1 (Super Annigeri 1) highly resistant to wilt disease were developed through MAS (Bharadwaj et al., 2021).

10.3 BREEDING STRATEGIES WHEN SOURCES OF RESISTANCE GENES ARE NOT AVAILABLE

Conventional breeding methods are very successful in the development of disease resistant varieties for last several decades when the resistant gene source is available in the germplasm. There are several diseases for which such resistant sources are not available, e.g., for MYMIV in cowpea, resistant gene sources are not available and direct chemical control of disease is not possible causing huge yield losses. In such situations, mutation breeding and genetic engineering approaches are the alternative methods for the development of disease resistance in pulse crops. Whole genome sequencing is going to make the task easier for the genetic engineering approaches like gene editing.

10.3.1 TRANSGENIC APPROACHES AND GENETIC ENGINEERING AND FOR DISEASE RESISTANCE

Transgenics is one of the most successful approaches of biotechnology applied for the crop improvement. Pulse crops are lagging behind in transgenic development just like genomic studies due to lack of regeneration protocols in pulse crops. In general, pulse crop belonging to grain legumes are mostly recalcitrant and difficult to regenerate *in-vitro* due to slow morphogenesis, development of albinos, vitreous tissues, and no response in dedifferentiated Calli (Veltcheva et al., 2005; Pratap et al., 2010). More recently, regeneration protocols are developed in various pulse crops like mung bean (Mahalakshmi et al., 2006; Himabindu et al., 2014; Mojumder et al., 2015), blackgram (Saini & Jaiwal, 2005; Prasad et al., 2014), chickpea (Shukla et al., 2015), lentil (Akcay et al., 2009), pea (Das et al., 2014), pigeon pea (Asande et al., 2016; Pawar et al., 2014), cowpea (Raveendar et al., 2009) and common bean (Thao et al., 2013). The recent research and development in tissue culture research in the pulse crops will pave the way for the transgenic development for disease resistance. In last few years number of transgenic were developed in pulse crops and mainly concentrated on insect resistance. Common bean became the first pulse crop having commercial transgenic variety (EMBRAPA 5.1) possessing resistance to bean golden mosaic virus (BGMV) and it was de-regularized and released for commercial cultivation in Brazil (Tollefson, 2011). It was developed through RNA interference (RNAi) construct which induced post transcriptional gene silencing of the *AC1* viral gene (Bonfim et al., 2007). Now a day, disease resistance is becoming one of the prime objectives in the transgenic research and several attempts of transgenics developments in pulse crops for disease resistance were reported. In cowpea, transgenic cowpea plants expressing AC2-hp and AC2+AC4-hp RNA and showing 100% resistance against MYMIV infection were developed through RNAi approach (Kumar et al., 2017). The mung bean transgenic plants with the non-expressor of pathogenesis related gene-1 of *Brassica juncea* (BjNPR1) showed resistance against the seedling rot disease caused by *Rhizoctonia solani* (Vijayan & Kirti, 2011).

Genetic engineering is fast developing science and it can be applied to develop multi-layer defense against pathogens causing serious diseases of crops. Several approaches like RNAi, resistance (R) gene stacking, R gene enrichment sequencing (RenSeq) and modifying susceptibility genes to attenuate pathogenicity are going to change the future of disease resistance research for crop breeding (Dong & Ronald, 2019). Conventionally, the transgenics development is based on transfer of desired gene from unrelated organism into the plants via agrobacterium mediated and other methods. The emergence of gene editing tools has reduced the dependency of transgenic development on other organism for desirable genes, which is always questionable by environmental and health point of view. In gene editing, precise and targeted changes to the genome of an organism is brought through the use of engineered nucleases and cellular DNA repair pathways (Bowmik et al., 2020). The gene editing involved changes to the endogenous plant

DNA, such as deletions, insertions, and replacements of DNA of various lengths at designated targets and may or may not contain foreign DNA depending on the type of gene edits introduced (Barrangou & Doudna, 2016). The advances in genome editing systems like zinc finger nucleases (ZFNs), transcription activator-like effector nucleases (TALENs) and the clustered regularly interspaced short palindromic repeat (CRISPR)-associated protein-9 (Cas9)-based genome editing tool (CRISPR/Cas9) made the precise gene editing applicable to the crop plants. These genome editing tools can be used for generating qualitative as well as quantitative genetic variation. The quantitative variation can be generated through Cas 9 activity in heterozygous loss of function mutant backgrounds which can be applied to the screening of QTLs imparting resistant to diseases (Rodriguez-Leal et al., 2017). The plant defense pathways can be identified more efficiently by using the genome sequencing and bioinformatics tools for detailed study of plant-pathogen interactions. The application of gene editing tools is very limited in pulse crops but in future it will be used more frequently in the development of genetically engineered disease resistant pulse crop varieties. The future of genetically modified (GM) crops will be depending on the regularity norms and awareness in consumers about the environmental and heath safety. Uncertainty of regulations on GM crops developed through gene editing tools has been clarified up to certain extent as some leading GM crop producing nations like USA, Australia, Argentina, and Brazil put the GM crops developed through gene editing in conventionally bred crops if no foreign genes was present in final crop variety (Smyth, 2019). Such moves in national policies of developed nations in favor of GM crops will encourage the investment in research and development of genetic engineering.

10.3.2 MUTATION BREEDING AND ADVANCES IN MUTANT SCREENING FOR DISEASE RESISTANCE

Induced mutation is one of the most effective and efficient tool of plant breeding when the sources for disease resistance genes are not available in germplasm. In pulse crops, induced mutation has contributed immensely to the development disease resistance varieties. Some of the varieties possessing resistance to important diseases of pulse crop are enlisted in Table 10.5. In chickpea, wilt resistant variety CM-2008 was developed through ethyl methane sulfonate (EMS), while BGM-547 (Tolerant to wilt, root rot and stunt diseases) and TAEK-SAGEL (resistant to *Ascochyta* blight) were the outcome of gamma rays induced mutation. Against most serious viral disease like YMD, resistant varieties like Pant Moong-2, Binamoong-6, TAP-7 in mung bean and TU-94-2 in blackgram were developed through gamma rays induced mutagenesis. Anthracnose resistant varieties viz., FT-Paulistinha and Zornitsa were the gamma rays induced mutants of rajmash and lentil, respectively. Multiple disease resistant varieties in cowpea were developed through gamma rays viz., Lukusuzi (tolerant to Ascochyta, cercospora, and Cowpea Aphid Born Mosaic Virus) and TC-901 (resistant to Cowpea Mosaic and root-rot diseases). Apart from varieties, mutants resistant to various diseases in pulse crop were also developed which add the valuable genetic resources in the germplasm. Several YMD resistant mutants were isolated in mung bean (Sing & Chaturvedi, 1981; Reddy, 2009; Vairam et al., 2016; Anuragi et al., 2022) and blackgram (Kundagrami et al., 2009; Vanniarajan et al., 2017).

TABLE 10.5 Disease Resistant Pulse Crop Varieties Developed Through Mutation Breeding

Crop	Name of Variety	Disease Reaction	Mutagen Used	Country
Chickpea	CM-2008	Resistance to wilt.	EMS	Pakistan
	BGM-547	Tolerant to wilt, root rot and stunt diseases.	Gamma rays	India
	TAEK-SAGEL	Ascochyta blight resistance.	Gamma rays	Turkey
Mungbean	Pant Moong-2	Resistant to YMD	Gamma rays	India
	Binamoong-6	Resistant to YMD	Gamma rays	Bangladesh
	Binamoong-8	Resistant to YMD	Gamma rays	Bangladesh
	TAP-7	Resistant to YMD	Gamma rays	India
Blackgram	Vamban-2	Resistant to YMD	Spontaneous mutant	India
	TU-94-2	Resistant to YMD	Gamma rays	India
	CO-4	Resistant to YMD	MMS	India
Common bean/Rajmash	FT-Paulistinha	Anthracnose resistant	Gamma rays	Brazil
Lentil	Zornitsa	Anthracnose resistant	EMS	Bulgaria
	NIAB MASOOR-2006	Resistance to blight and rust.	Gamma rays	Pakistan
Cowpea	Lukusuzi	Tolerant to Ascochyta, Cercospora and cowpea aphid borne mosaic virus.	Gamma rays	Zambia
	TC-901	Resistance to cowpea mosaic and root-rot diseases.	Gamma rays	India

Source: www.mvd.iaea.org.

In conventional mutation breeding, the mutant screening is mainly depending on phenotypic expression of mutated gene, while the mutations with very small phenotypic changes are very difficult to identify. The recent advances in genomics paved the ways for the identification of even mutations in single base pair, i.e., point mutations. The reverse genetic approach like TILLING (Targeting Induced Local Lesion in Genomes) is the PCR based screening which make it possible to identify the point mutations induced by chemical mutagenesis (McCallum et al., 2000; Till et al., 2004). The EcoTILLING (Ecotype TILLING) is the variation of TILLING approach which can be applied to identify the natural genetic variation, i.e., SNPs and IN/DELs (Greene et al., 2003). The high throughput approaches like HRM (High Resolution Melting) genotyping, KeyPoint™ Technology and EMAIL (Endonucleolytic Mutation Analysis by Internal Libeling) will enhance the speed of TILLING to identify the mutants for different traits including disease resistance (Cross et al., 2008; Rigola et al., 2009; Botticella et al., 2011).

10.4 DISEASE RESISTANT VARIETIES DEVELOPED DURING LAST 10 YEARS IN MAJOR PULSE CROPS

Conventional plant breeding approaches are quite successful in the development of disease-resistant varieties in various pulse crops. Disease resistance has become one of the integral breeding objectives of national and international breeding programs. Recently, it is becoming mandatory for the release of a new variety that the newly developed genotype must possess resistance against the serious disease along with the high. The latest varieties developed in Mungbean (IPM-512-1, MH112, IPM312-20, and IPM409-4) and blackgram (KPU12-1735, VBN 10, OBG 33 AND KPU 524-65) are possessing resistance against the most serious diseases like YMD, cercospora leaf spot, and powdery mildew (Table 10.6). The varieties of pigeon pea like TDRG 59, Bheema, ICPL 332 WR, and one hybrid variety ICPH 2740 are resistant to *fusarium wilt* and sterility mosaic disease. The powdery mildew-resistant and rust-tolerant varieties viz., IPF 5-19, IPF 4-9, and IPF 99-25 were released in pea while BCMV-resistant variety IIPR 96-4 was developed in rajmash. In lentil, the multiple disease-resistant varieties like IPL 220 and IPL526 were released for commercial cultivation. In the most important pulse crop chickpea, GJG 0809 (multiple disease resistant), Indira Chana 1 (resistant to wilt), and Pant Kabuli Gram 2 (resistant to BGM) were

TABLE 10.6 Disease Resistant Varieties Recently Released in India

Crop	Varieties	Disease Reaction	Breeding Methods	References
Mungbean	IPM 512-1 (Soorya)	Resistant to MYMV, Cercospora leaf spot and Anthracnose.	Pedigree method.	Project Coordinator's Report, AICRP on MULLaRP, ICAR, IIPR, Kanpur (2021)
	MH 1142	Moderately resistance to Anthracnose and powdery mildew and resistant to MYMV.	Selection	
	IPM 312-20 (Vashudha)	Moderately resistance to Anthracnose and powdery mildew and resistance to MYMV.	Interspecific hybridization (*Vigna radiata x Vigna mungo*).	
	IPM 409-4 (Heera)	Resistant to MYMV	Pedigree method	
Blackgram	KPU 12-1735 (Kota Urd 4)	Moderately resistant to MYMV and Cercospora leaf spot.	Pedigree method	
	VBN 10 (VBG 12-034)	Resistance to MYMV, ULCV, and LC disease.	Pedigree method	
	OBG 33 (Shashi)	Moderately resistant to MYMV.	Pedigree method	
	KPU 524-65 (Kota Urd 3)	Resistant to MYMV, leaf spot and anthracnose.	Pedigree method	
Chickpea	GJG 0809	Resistant to wilt and stunt, root rot and tolerant to Ascochyta blight.	–	ICAR-All India Coordinated Research Project on Chickpea, IIPR, Kanpur (2020)
	Indira Chana 1	Resistant to wilt	–	
	Pant Kabuli gram-2 (PG 071)	Tolerant to wilt and BGM	–	
	Pusa Chickpea-10216	Moderately resistant to *Fusarium wilt*, dry root rot and stunt diseases.	Marker assisted selection.	Bharadwaj et al. (2021)
	MABC-WR-SA-1 (Super Annigeri-1)	Highly resistant to *Fusarium wilt* disease.	Marker assisted introgression resistance segment for *Fusarium wilt* from WR 315 in the genetic background of Annigeri-1.	

TABLE 10.6 *(Continued)*

Crop	Varieties	Disease Reaction	Breeding Methods	References
Pigeon pea	TDRG 59	Resistant to *Fusarium wilt* and sterility mosaic diseases.	–	www.icrisat.org
	Bheema	Resistant to *Fusarium wilt* and sterility mosaic diseases.	–	
	ICPH 2740	Resistant to *Fusarium wilt* and sterility mosaic diseases.	Hybrid	–
	ICPL 332 WR (TDRG 4)	Resistant to *Fusarium wilt.*	–	–
Pea	IPF 5-19 (Aman)	Powdery mildew resistance, tolerant to rust.	–	www.iipr.icar.gov.in
	IPF 4-9	Powdery mildew resistance, tolerant to rust.	–	–
	IPF 99-25 (Adarsh)	Powdery mildew resistance.	–	–
Common bean/Rajmash	IIPR 96-4 (Amber)	Resistant to BCMV.	–	www.iipr.icar.gov.in
Lentil	IPL 220	Resistance major diseases including rust, *Fusarium wilt* and *Stemphylium* blight.	–	www.iipr.icar.gov.in
	IPL 526	Resistant to rust and tolerant to wilt.	–	–

released recently in India. Most of the varieties developed in the pulse crop are the output of conventional breeding methods like hybridization followed by the pedigree method while very few varieties were developed through interspecific hybridization like mung bean variety IPM 312-20.

Until recently, the marker-assisted selection was not applied successfully in the development of disease-resistant varieties in pulse crops. The recent development in genomics and availability of new genomic resources are encouraging the use of MAS. In chickpea, Pusa Chickpea-10216 (multiple disease resistant) and MBC-WR-SA-1 (Super Annigeri 1) highly resistant to wilt disease were the outcome of MAS (Bharadwaj et al., 2021).

KEYWORDS

- **disease resistance**
- **disease screening**
- **gene editing**
- **genetic engineering**
- **marker assisted selection**
- **mutant screening**
- **pulses**
- **transgenic development**

REFERENCES

Agbicodo, E. M., Fatokun, C. A., Bandyopadhyay, R., Wydra, K., Diop, N. N., Muchero, W., Ehlers, J. D., et al., (2010). Identification of markers associated with bacterial blight resistance loci in cowpea (*Vigna unguiculata* L. Walp.). *Euphytica, 175*, 215–226.

Ahmad, F., Gaur, P. M., & Croser, J. S., (2005). Chickpea (*Cicer arietinum* L.). In: Singh, R. J., & Jauhar, P. P., (eds.), *Genetic Resources, Chromosome Engineering and Crop Improvement* (Vol. 1, pp. 187–217). Grain Legumes. CRC Press, Boca Raton, Florida.

Ahmad, Z., Mumtaz, A. S., Ghafoor, A., Ali, A., & Nisar, M., (2014). Marker assisted selection (MAS) for chickpea *Fusarium oxysporum* wilt resistant genotypes using PCR based molecular markers. *Mol. Biol. Rep., 41*, 6755–6762.

Akcay, U. C., Mahmoudian, M., Kamci, H., Yucel, M., & Oktem, H., (2009). Agrobacterium tumefaciens-mediated genetic transformation of a recalcitrant grain legume, lentil (*Lens culinaris* Medik). *Plant Cell Rep., 28*, 407–417.

Akinola, J. O., Whiteman, P. C., & Wallis, E. S., (1975). *The Agronomy of Pigeon Pea (Cajanus Cajan)* (p. 57). Review Series no.1/1975. CAB International, Wallingford, UK.

Alam, A. K., Somta, M. M., & Srinives, P., (2014). Identification and confirmation of quantitative trait loci controlling resistance to mungbean yellow mosaic disease in mungbean (*Vigna radiata* L. Wilczek). *Mol. Breed., 34*, 1497–1506.

Amalraj, A., Taylor, J., Bithell, S., Li, Y., Moore, K., Hobson, K., & Sutton, T., (2019). Mapping resistance to phytophthora root rot identifies independent loci from cultivated (*Cicer arietinum* L.) and wild (*Cicer echinospermum* P.H. Davis) chickpea. *Theor. Appl. Genet., 132*, 1017–1033.

Andrahennadi, C., (1994). *Genetics and Linkage of Isozyme Markers and Resistance to Seedborne Ascochyta Infection in Lentil*. M. Sc. Thesis, Department of Crop Science and Plant Physiology; University of Saskatchewan.

Anonyms, (2020). *Project Coordinator's Report*, ICAR-All India Coordinated Research Project on Chickpea, IIPR, Kanpur, India.

Anonyms, (2021). *Project Coordinator's Report*, AICRP on MULLaRP, ICAR, IIPR, Kanpur, India.

Anuragi, H., Yadav, R., & Sheoran, R., (2022). Gamma-rays and EMS induced resistance to mungbean yellow mosaic India virus in mungbean (*Vigna radiata* L. Wilczek) and its validation using linked molecular markers. *Int. J. Rad. Biol., 98*(1), 69–81.

Asande, L. K., Indieka, A. S., Adero, M. O., Kiboi, S., & Amugune, N. O., (2016). *In vitro* regeneration of pigeon pea using leaf explants. *Afr. Crop Sci. J., 24*(2), 191–201.

Atik, O., Ahmed, S., Abang, M. M., Imtiaz, M., Hamwieh, A., Baum, M., El-Ahmed, A., et al., (2013). Pathogenic and genetic diversity of *Didymella rabiei* affecting chickpea in Syria. *Crop Prot., 46*, 70–79.

Barilli, E., Cobos, M. J., Carrillo, E., Kilian, A., Carling, J., & Rubiales, D. A., (2018). High-density integrated DArTseq SNP-based genetic map of *Pisum fulvum* and identification of QTLs controlling rust resistance. *Front. Plant Sci., 9*, 167.

Barman, P., Handique, A. K., & Tanti, B., (2014). Tagging STMS markers to *Fusarium* wilt race 1 resistance in chickpea (*Cicer arietinum* L.). *Mol. Breed.* doi: 10.1007/s11032-009-9282-y.

Barrangou, R., & Doudna, J. A., (2016). Applications of CRISPR technologies in research and beyond. *Nat. Biotechnol., 34*, 933–941.

Basak, J., Kundargrami, S., Ghose, T. K., & Pal, A., (2004). Development of yellow mosaic virus (YMV) resistance linked DNA marker in *Vigna mungo* from populations segregating for YMV- reaction. *Mol. Breed., 14*, 375–383.

Bayaa, B., Erskine, W., & Hamdi, A., (1994). Response of wild lentil to *Ascochyta fabae* f. sp. *lentis* from Syria. *Genet. Resour. Crop Evol., 41*, 61–65.

Bayaa, B., Erskine, W., & Hamdi, A., (1995). Evaluation of a wild lentil collection for resistance to vascular wilt. *Genet. Resour. Crop Evol., 42*, 231–235.

Bayraktar, H., Dolar, F. S., & Tor, M., (2007). Determination of genetic diversity within *Ascochyta rabiei* (pass.) Labr., the cause of Ascochyta blight of chickpea in Turkey. *J. Plant Pathol. 89*(3), 341–347.

Bhanu, A. N., Kumar, P., Singh, M. N., Srivastava, K., & Hemantaranjan, A., (2017). Assessment of genetic purity of inter-specific F_1 hybrids involving *Vigna radiata* and *Vigna umbellate. J. Exp. Biol. Agric. Sci., 5*(5), 636–643.

Bharadwaj, C., Tripathi, S., Soren, K. R., et al., (2021). Introgression of "QTL-hotspot" region enhances drought tolerance and grain yield in three elite chickpea cultivars. *Plant Genome*, e20076.

Bhardwaj, R., Panwar, R. K., Nehra, M., Gaur, A. K., Verma, S. K., & Arora, A., (2018). Inheritance studies on resistance to botrytis grey mould in chickpea (*Cicer arietinum* L.). *Int. J. Chem. Stud., 6*(6), 2255–2257.

Bhowmik, P., Konkin, D., & Polowick, P., (2021). CRISPR/Cas9 gene editing in legume crops: Opportunities and challenges. *Legume Sci., 3*, e96.

Binyamin, R., Khan, M. A., Khan, N. A., & Khan, A. I., (2015). Application of SCAR markers linked with mungbean yellow mosaic virus disease-resistance gene in Pakistan mungbean germplasm. *Gen. Mol. Res., 14*, 2825–2830.

Bobkov, S. V., & Selikhova, T. N., (2021). Introgession of powdery mildew resistance into cultural pea from wild accession of *P. fulvum. IOP Conf. Ser. Earth Environ. Sci., 650*, 012091.

Bonfim, K., Faria, J. C., Nogueira, O. P. L. E., Mendes, E. A., & Aragao, J. L. F., (2007). RNAi-mediated resistance to bean golden mosaic virus in genetically engineered common bean (*Phaseolus vulgaris*). *Mol. Plant-Microbe Interact., 20*(6), 717–726.

Botticella, E., Sestili, F., Hernandez-Lopez, A., Phillips, A., & Lafiandra, D., (2011). High resolution melting analysis for the detection of EMS induced mutations in wheat SbeIIa genes. *BMC Plant Biol., 11*, 156.

Castro, P., Rubio, J., Madrid, E., Fernández-Romero, M. D., Millán, T., & Gil, J., (2013). Efficiency of marker-assisted selection for Ascochyta blight in chickpea. *J. Agric. Sci., 153*, 56–67.

Chand, R., Singh, V., Pal, C., Kumar, P., & Kumar, M., (2012). First report of a new pathogenic variant of *Cercospora canescens* on mungbean (*Vigna radiata*) from India. *New Dis. Rep., 26*, 6.

Chankaew, S., Somta, P., Isemura, T., Tomooka, N., Kaga, A., Vaughan, D. A., et al., (2013). Quantitative trait locus mapping reveals conservation of major and minor loci for powdery mildew resistance in four sources of resistance in mungbean (*Vigna radiata* L. Wilczek). *Mol. Breed., 32*, 121–130.

Chankaew, S., Somta, P., Sorajjapinun, W., & Srinives, P., (2011). Quantitative trait loci mapping of *Cercospora* leaf resistance in mungbean, *Vigna radiata* L. Wilczek. *Mol. Breed., 28*, 255–264.

Chathiranrat, N., Suphattra, N., Chadaporn, C., Napatsawan, W., Pitchakon, P., & Tantasawat, P. A., (2018). Selection of mungbean resistant to powdery mildew in BC_1F_1 progenies based on ISSR and ISSR-RGA Markers. *Int. J. Adv. Sci. Eng. Technol., 6*(2), 73–77.

Chaturvedi, R., Singh, I. S., & Gupta, A. K., (1995). Inheritance of resistance to botrytis grey mold in chickpea (*Cicer arietinum*). *Legume Res., 18*, 1–4.

Chen, H. M., Ku, H. S., Schafleitner, R., Bains, T. S., Kuo, G. C., Liu, C. A., & Nair, R. M., (2013). The major quantitative trait locus for mungbean yellow mosaic Indian virus resistance is tightly linked in repulsion phase to the major bruchid resistance locus in a cross between mungbean (*Vigna radiata* L. Wilczek) and its wild relative *Vigna radiata* ssp. *sublobata*. *Euphytica, 192*, 205–216.

Chenulu, V., Venkateshwarlu, V., & Rangaraju, R., (1979). Studies on yellow mosaic disease of mungbean. *Indian Phytopath., 32*, 230–235.

Cho, S., Chen, W., & Muehlbauer, F. J., (2004). Pathotype-specific genetic factors in chickpea (*Cicer arietinum* L.) for quantitative resistance to Ascochyta blight. *Theor. Appl. Genet., 109*, 733–739.

Choudhary, P., Chand, R., & Singh, A. K., (2021). Genetics of *Cercospora* leaf spot resistance in mungbean (*Vigna radiata* L. Wilczek) through generation mean analysis. *Legume Res.* doi: 10.18805/LR-4456.

Chowdhury, M. A., Andrahennadi, C. P., Slinkard, A. E., & Vandenberg, A., (2001). RAPD and SCAR markers for resistance to Ascochyta blight in lentil. *Euphytica, 118*, 331–337.

Clulow, S. A., Matthews, P., & Lewis, B. G., (1991). Genetical analysis of resistance to *Mycosphaerella pinodes* in pea seedlings. *Euphytica, 58*, 183–189.

Cross, M. J., Waters, D. L., Lee, L. S., & Henry, R. J., (2008). Endonucleolytic mutation analysis by internal labeling (EMAIL). *Electrophoresis, 29*(6), 1291–1301.

Das, A., Kumar, S., Nandeesha, P., Yadav, I. S., Saini, J., Chaturvedi, S. K., & Datta, S., (2014). An efficient *in vitro* regeneration system of fieldpea (*Pisum sativum* L.) via. shoot organogenesis. *J. Plant Biochem. Biotechnol., 23*(2), 184–189.

Daspute, A., Fakrudin, B., Bhairappanavar, S. B., Kavil, S. P., Narayana, M. Y. D., Kaumar, A., Krishnaraj, P. U., et al., (2014). Inheritance of pigeonpea sterility mosaic disease resistance in pigeonpea. *Plant Pathol. J., 30*(2), 188–194.

Datta, J., & Lal, N., (2013). Genetic diversity of *Fusarium* wilt races of pigeonpea in major regions of India. *Afr. Crop Sci. J., 21*(3), 201–211.

Dhole, V. J., & Reddy, K. S., (2012). Genetic analysis of resistance to mungbean yellow mosaic virus in mungbean (*Vigna radiata*). *Plant Breed., 131*, 414–417.

Dhole, V. J., & Reddy, K. S., (2013). Development of a SCAR marker linked with a MYMV resistance gene in mungbean (*Vigna radiata* L. Wilcek). *Plant Breed., 132*, 127–132.

Dikshit, H. K., Mishra, G. P., Somta, P., Shwe, T., A. Alam, K. M. M., & Bains, T. S., et al., (2020). Classical genetics and traditional breeding in mungbean. In: Nair, R. M., (ed.), *The Mungbean Genome, Compendium of Plant Genomes* (pp. 43–54). Switzerland AG: © Springer Nature.

Dikshit, H. K., Singh, A., Singh, D., Aski, M., Jain, N., Hegde, V. S., Basandrai, A. K., et al., (2016). Tagging and mapping of SSR marker for rust resistance gene in lentil (*Lens culinaris* Medikus sub. sp. *culinaris*). *Indian J. Exp. Biol., 54*(6), 394–399.

Dinesh, H. B., Lohithaswa, H. C., Viswanatha, K. P., Poonam, S., Manjunatha, L., Ambika, D. S., & Prasanna, K. M. K., (2018). Genetic analysis and marker assisted backcrossing for transfer of mosaic virus resistance in cowpea (*Vigna unguiculata* L. Walp.). *Legume Res., 41*, 663–668.

Dita, M. A., Rispail, N., Prats, E., Rubiales, D., & Singh, K. B., (2006). Biotechnology approaches to overcome biotic and abiotic stress constraints in legumes. *Euphytica, 147,* 1–24.

Dong, O. X., & Ronald, P. C., (2019). Genetic engineering for disease resistance in plants: Recent progress and future perspectives. *Plant Physiol., 180*(1), 26–38.

Dubey, S. C., Priyanka, K., Singh, V., & Singh, B., (2012). Race profiling and molecular diversity analysis of *Fusarium oxysporum* f.sp. *ciceris* causing wilt in chickpea. *J. Phytopathol., 160*, 576–587.

Elbeaino, T., Digiaro, M., Uppala, M., & Sudini, H., (2014). Deep sequencing of pigeonpea sterility mosaic virus discloses five RNA segments related to *Emaraviruses*. *Virus Res., 188*, 27–31.

Eujayl, I., Erskine, W., Bayaa, B., Baum, M., & Pehu, E., (2006). *Fusarium* vascular wilt in lentil: Inheritance and identification of DNA markers for resistance. *Plant Breed., 17*(5), 497–499.

Feng, X., Orellana, G. E., Myers, J. R., & Karasev, A. V., (2018). Recessive resistance to bean common mosaic virus conferred by the bc-1 and bc-2 genes in common bean (*Phaseolus vulgaris* L.) affects long distance movement of the virus. *Phytopathology, 108*(8), 1011–1018.

Fernandez, G. C., & Shamugasundaram, S., (1987). The AVRDC mungbean improvement program: The past, present and future. *Proc. Second Int. Mungbean Symposium* (pp. 558–595). Bangkok, Thailand.

Fondevilla, S., Cubero, J. I., & Rubiales, D., (2011). Confirmation that the *Er3* gene, conferring resistance to *Erysiphepisi* in pea, is a different gene from *er1* and *er2*genes. *Plant Breed., 130*, 281, 282.

Frew, T. J., Russell, A. C., & Timmerman-Vaughan, G. M., (2002). Sequence tagged microsatellite markers linked to the *sbm1* gene for resistance to pea seed borne mosaic virus in pea. *Plant Breed., 121*, 512–516.

Ganapathy, K. N., Byregowda, M., Venkatesha, S. C., Ramachandra, R., Gnanesh, B. N., & Girish, G., (2009). Identification of AFLP markers linked to sterility mosaic disease in pigeonpea *Cajanus cajan* (L.) Millsp. *Int. J. Integr. Biol., 7*(3), 145–149.

Garkoti, A., Kumar, S., Lal, M., & Singh, V., (2013). Major diseases of lentil: Epidemiology and disease management: A review. *Agriways, 1*(1), 62–64.

Gayacharan, Rani, U., Singh, S., Basandrai, A. K., Rathee, V. K., Tripathi, K., et al., (2020). Identification of novel resistant sources for Ascochyta blight (*Ascochyta rabiei*) in chickpea. *PLoS One, 15*(10), e0240589.

Gioi, T. D., Boora, K. S., & Chaudhary, K., (2012). Identification and characterization of SSR markers linked to yellow mosaic virus resistance gene(s) in cowpea (*Vigna unguiculata). Int. J. Plant Rese., 2*(1), 1–8.

Gnanesh, B. N., Bohra, A., Sharma, M., Byregowda, M., Pandey, S., Wesley, V., Saxena, R. K., Saxena, K. B., et al., (2011). Genetic mapping and quantitative trait locus analysis of resistance to sterility mosaic disease in pigeonpea (*Cajanus cajan* L. Millsp.). *Field Crop Res. 123*, 53–61.

Gowda, S. J. M., Radhika, P., Kadoo, N. Y., Mhase, L. B., & Gupta, V. S., (2009). Molecular mapping of wilt resistance genes in chickpea. *Mol. Breed.* doi 10.1007/s11032-009-9282-y.

Greene, E. A., Codomo, C. A., Taylor, N. E., Henikoff, J. G., Till, B. J., Reynolds, S. H., Enns, L. C., et al., (2003). Spectrum of chemically induced mutations from a large-scale reverse-genetic screen in *Arabidopsis*. *Genetics, 164*, 731–740.

Gupta, D., & Sharma, S. K., (2006). Evaluation of wild *Lens* taxa for agro-morphological traits, fungal diseases, and moisture stress in northwestern Indian hills. *Genet. Resour. Crop Evol., 53*, 1233–1241.

Gupta, S. K., Singh, R. A., & Chandra, S., (2005). Identification of a single dominant gene for resistance to mungbean yellow mosaic virus in blackgram (*Vigna mungo* (L.) Hepper). *Sabrao J. Breed. Genet., 37*(2), 85–89.

Gupta, S., Gupta, D. S., Anjum, T. K., Pratap, A., & Kumar, J., (2013). Inheritance and molecular tagging of MYMIV resistance gene in blackgram (*Vigna mungo* L. Hepper). *Euphytica, 193*, 27–37.

Haji-Allahverdipoor, K., Bahramnejad, B., & Amini, J., (2011). Selection of molecular markers associated with resistance to *Fusarium* wilt disease in chickpea (*Cicer arietinum* L.) using multivariate statistical techniques. *Aust. J. Crop Sci., 5*(13), 1801–1809.

Haware, Y. P., & Mcdonald, D., (1993). Botrytis gray mould of chickpea. In: Haware, M. P., Faris, D. G., & Gowda, C. L. L., (eds.) *Recent Advances in Research on Botrytis Gray*

Mould of Chickpea (pp. 3–6). International Crops Research Institute for Semi – Arid Tropics, Patancheru, Andhra Pradesh, India.

Himabindu, Y., Reddy, M. C., & Chandrasekhar, T., (2014). *In vitro* regeneration of green gram (*Vigna radiata* L. Wilczek) cultivar vamban-2 using cotyledonary nodes. *CIBTech J. Biotechnol., 3*, 11–15.

Hiremani, N. S., & Dubey, S. C., (2018). Race profiling of *Fusarium oxysporum* f. sp. *lentis* causing wilt in lentil. *Crop Prot., 108*, 23–30.

Hobbs, H. A., Kuhn, C. W., Papa, K. E., & Brantley, B. B., (1987). Inheritance of non-necrotic resistance to Southern Bean mosaic Virus in cowpea. *Phytopathology, 77*(12), 1624–1629.

IAEA, (2021). *Mutant Variety Database*. https://mvd.iaea.org (accessed on 12 December 2023).

ICRISAT, (2021). www.icrisat.org (accessed on 12 December 2023).

IIPR, Kanpur, India: https://iipr.icar.gov.in (accessed on 12 December 2023).

Jain, M., Misra, G., Patel, R. K., et al., (2013). A draft genome sequence of the pulse crop chickpea (*Cicer arietinum* L.). *The Plant J., 74*, 715–729.

Jha, U. C., Jha, R., Bohra, A., Manjunatha, L., Saabale, P. R., Parida, S. K., Chaturvedi, S. K., et al., (2021). Association mapping of genomic loci linked with *Fusarium* wilt resistance (*Foc*2) in chickpea. *Plant Genet. Resour., 19*(3), 195–202.

Jimenez-Gasco, M. M., & Jimenez-Díaz, R. M., (2003). Development of a specific polymerase chain reaction-based assay for the identification of *Fusarium oxysporum* f. sp. *ciceris* and its pathogenic races 0, 1A, 5, and 6. *Phytopathology, 93*, 200–209.

Jingade, P., & Ravikumar, R. L., (2015). Development of molecular map and identification of QTLs linked to *Fusarium* wilt resistance in chickpea. *J. Genet., 94*(4), 723–729.

Jordan, R. L., & Hammond, J., (2020). Bean common mosaic virus and bean common mosaic necrosis virus (*Potyviridae*). *Encyclopedia of Virology*. doi.org/10.1016/B978-0-12-809633-8.21293-4.

Joshi, A., Souframanien, J., Chand, R., & Pawar, S. E., (2006). Genetic diversity study of *Cercospora canescens* (Ellis & Martin) isolates, the pathogen of *Cercospora* leaf spot in legumes. *Curr. Sci., 90*(4), 564–568.

Kalaria, R. K., Chauhan, D., Mahatma, M., & Mahatma, L., (2014). Identification of RAPD and ISSR makers for resistance against mungbean yellow mosaic virus in mungbean (*Vigna radiata*) under south Gujarat agro-climatic condition of India. *The Bioscan, 9*(3), 1177–1182.

Kang, Y. J., Sue, K. K., Kim, M. Y., et al., (2014). Genome sequence of mungbean and insights into evolution within *Vigna species*. *Nat. Commun., 5*, 5443. doi: 10.1038/ncomms6443.

Kanwade, D. G., Kute, N. S., Bedis, M. R., Pawar, V. Y., Deshmukh, G. P., Jaybhaye, C. P., & Giri, M. D., (2019). Inheritance of powdery mildew resistance in mung bean (*Vigna radiata* L. Wilczek). *J. Pharmacog. Phytochem., 8*(3), 3642–3644.

Kasettranan, W., Somta, P., & Srinives, P., (2010). Mapping of quantitative trait loci controlling powdery mildew resistance in mungbean (*Vigna radiata* L. Wilczek). *J. Crop Sci. Biotechol. 13*(3), 155–161.

Kaushal, R. P., & Singh, B. M., (1989). Genetics of resistance in black gram to powdery mildew caused by *Erysiphie polygoni*. *Indian J. Agric. Sci.*, 648–685.

Kelly, L. A., Vaghefi, N., Bransgrove, K., Fechner, N. A., Stuart, K., Pandey, A. K., Sharma, M., et al., (2021). *One Crop Disease, How Many Pathogens? Podosphaera xanthii and Erysiphe Vignae sp. Nov. Identified as the Two Species That Causes Powdery Mildew of*

Mungbean (Vigna radiata) and Black Gram (V. mungo) in Australia. doi.org/10.1094/PHYTO-12-20-0554-R.

Khalekar, G. D., Akhare, A. A., Gahukar, S. J., Sing, N. K., & Kumar, M., (2014). Identification of simple sequence repeat marker associated with wilt resistance in pigeonpea. *J. Environ. Biol. 35*, 955–960.

Khan, I. A., Alam, S. S., & Jabbar, A., (2004). Purification of phytotoxin from culture filtrates of *Fusarium oxysporum* f. sp. *ciceri*s and its biological effects on chickpea. *Pak. J. Bot., 36*(4), 871–880.

Khandappagol, M., & Rangaiah, S., (2019). Evaluation of blackgram (*Vigna mungo* L. Hepper) Genotypes for resistance against powdery mildew (*Erysiphe polygoni* DC) under natural and artificially inoculated conditions. *Int. J. Curr. Microbiol. App. Sci., 9*, 171–182.

Khattak, G. S. S., Haq, M. A., Ashraf, M., & Elahi, T., (2000). Genetics of mungbean yellow mosaic virus (MYMV) in mungbean (*Vigna radiata* L. Wilczek). *J. Genet. Breed., 54*(4), 237–243.

Kishore, G. K., (2005). *Cultural, Morphological, Pathogenic and Genetic Variation in Botrytis Cinerea, Causal Agent of Gray Mold in Chickpea.* Ph. D. thesis, Jawaharlal Nehru Technological University, Hyderabad, Andhra Pradesh, India.

Kitsanachandee, R., Somta, P., Chatchawankanphanich, O., Akhtar, K. P., Shah, T. M., Nair, R. M., Bains, T. S., et al., (2013). Detection of quantitative trait loci for mungbean yellow mosaic India virus (MYMIV) resistance in mungbean (*Vigna radiata* L. Wilczek) in India and Pakistan. *Breed. Sci., 63*, 367–373.

Knights, E. J., Southwell, R. J., Schwinghamer, M. W., & Harden, S., (2008). Resistance to *Phytophthora medicaginis* Hansen and Maxwell in wild *Cicer* species and its use in breeding root rot resistant chickpea (*Cicer arietinum* L.) *Aust. J. Agric. Res., 59*, 383–387.

Kotresh, H., Fakrudin, B., Punnuri, S. M., Rajkumar, B. K., Thudi, M., Paramesh, H., Lohithaswa, H., & Kuruvinashetti, M. S., (2006). Identification of two RAPD markers genetically linked to a recessive allele of a *Fusarium* wilt resistance gene in pigeonpea (*Cajanus cajan* L. Millsp.). *Euphytica, 149*, 113–120.

Kreplak, J., Madoui, M. A., Petr, C., et al., (2019). A reference genome for pea provides insight into legume genome evolution. *Nat. Genet., 51*, 1411–1422.

Kumar, A., & Gill, R. K., (2020). Inheritance of rust resistance, yield performance in irrigated condition and identification of high yielding rust resistant recombinant inbred lines in lentil. *Int. J. Curr. Microbiol. App. Sci., 9*(2), 2393–2399.

Kumar, S., Kumar, J., Singh, S., Ahmed, S., Chaudhary, R., & Sarker, A., (2010). Vascular wilt disease of lentil: A review. *Journal of Lentil Research, 4*, 1–14.

Kumar, S., Tanti, B., Patil, B. L., Mukherjee, S. K., & Sahoo, L., (2017). RNAi-derived transgenic resistance to Mungbean yellow mosaic India virus in cowpea. *PLoS One, 12*(10), e0186786.

Kumar, V., Chauhan, V. B., & Srivastava, J. P., (2007). Pathogenic and biochemical variability in *Fusarium udum* causing pigeon pea wilt. *Indian Phytopath., 60*, 281–288.

Kumari, S., Tayal, P., Sharma, E., & Kapoor, R., (2014). Analyses of genetic and pathogenic variability among *Botrytis cinerea* isolates. *Microbiol. Res., 169*, 862–872.

Kundagrami, S., Basak, J., Maiti, S., Kundu, A., Das, B., Ghose, T. K., & A. Pal, A., (2009). Agronomic, genetic, and molecular characterization of MYMV tolerant mutant lines of *Vigna mungo. Int. J. Plant Breed. Genet., 3*(1), 1–10.

Kute, N. S., Deshmukh, R. B., & Patil, J. V. (2003). Inheritance resistance to powdery mildew (*Erysiphe polygoni* D.C.) in mungbean (*Vigna radiata* L. Wilczeck). *Legume Res., 26*(1), 24–27.

Kyle, M. M., Dickson, H. M., & Provvidenti, R., (1986). Linkage analysis of hypersensitive resistance to four viruses in *Phaseolus vulgaris*. *Bean Improv. Coop., 29*, 80, 81.

Lal, C., Sharma, S. K., Chahota, R. K., & Lal, C., (1996). Inheritance of rust resistance in lentil. *Indian J. Genet., 56*, 350, 351.

Lana, A. F., Lohuis, H., Bos, L., & Dijkstra, J., (1988). Relationships among strains of bean common mosaic virus and blackeye cowpea mosaic virus- members of the potyvirus group. *Ann. Appl. Biol., 113*(3), 493–505.

Li, Y., Ruperao, P., Batley, J., Edwards, D., Davidson, J., Hobson, K., & Sutton, T., (2017). Genome analysis identified novel candidate genes for Ascochyta blight resistance in chickpea using whole genome re-sequencing data. *Front. Plant Sci., 8*, 359.

Lonardi, S., Munoz-Amatria, M., Liang, Q., et al., (2019). The genome of cowpea (*Vigna unguiculata* L. Walp.). *The Plant J., 98*, 767–782.

Ma, Y., Coyne, C. J., Main, D., Pavan, S., Sun, S., Zhu, Z., Zong, X., et al., (2017). Development and validation of breeder friendly KASPar markers for er1, a powdery mildew resistance gene in pea (*Pisum sativum* L.). *Mol. Breed., 37*, 151.

Mahalakshmi, L. S., Leela, T., Kumar, S. M., Kumar, B. K., Naresh, B., & Devi, P., (2006). Enhanced genetic transformation efficiency of mungbean by use of primary leaf explants. *Curr. Sci., 91*, 93–99.

Mahalingam, A., Satya, V. K., Manivannan, N., Narayanan, S. L., & Sathya, P., (2018). Inheritance of mungbean yellow mosaic virus disease resistance in greengram (*Vigna radiata* L. Wilczek). *Int. J. Curr. Microbiol. Appl. Sci., 7*(1), 880–885.

Mahuku, G., Jara, C., Cajiao, C., & Beebe, S., (2003). Sources of resistance to angular leaf spot (*Phaeoisariopsis griseola*) in common bean core collection, wild *Phaseolus vulgaris* and secondary gene pool. *Euphytica, 130*, 303–313.

Maiti, S., Basak, J., Kundagrami, S., Kundu, A., & Pal, A., (2011). Molecular marker-assisted genotyping of mungbean yellow mosaic India virus resistant germplasm of mungbean and urdbean. *Mol. Biotechnol., 47*, 95–104.

Mallikarjuna, N., & Saxena, K. B., (2002). Production of hybrids between *Cajanus acutifolius* and *C. cajan*. *Euphytica, 124*, 107–110.

Mallikarjuna, N., Jadhav, D. R., Reddy, M. V., & Datta, U., (2005). Introgrssion of phytophthora blight disease resistance from *Cajanus platy-carpus* in to short duration pigeonpea. *Indian J. Genet., 65*, 261–264.

Mallikarjuna, N., Senthilvel, S., Jadhav, D. R., Saxena, K. B., Sharma, H. C., Upadhyaya, H. D., et al., (2011). Progress in utilization of *Cajanus platy-carpus* (Benth.) Maesen in pigeonpea improvement. *Plant Breed., 130*, 507–514.

Markam, N. K., Nair, S. K., Nanda, H. C., & Lakpale, N., (2018). Studies on allelic relationship for resistance to mungbean yellow mosaic virus disease in mungbean genotypes. *Int. J. Chem. Stud., 6*(2), 2401–2403.

Mathivathana, M. K., Murukarthick, J., Karthikeyan, A., Jang, W., Dhasarathan, M., Jagadeeshselvam, N., Sudha, M., et al., (2019). Detection of QTLs associated with mungbean yellow mosaic virus (MYMV) resistance using the interspecific cross of *Vigna radiata* x *Vigna umbellata*. *J. Appl. Genet., 60*(3, 4), 255–268.

McCallum, C. M., Comai, L., Greene, E. A., & Henikoff, S., (2000). Targeted screening for induced mutations. *Nat. Biotechnol., 18*, 455–457.

Meziadi, C., Blanchet, S., Geffroy, V., & Pflieger, S., (2017). Genetic resistance against viruses in *Phaseolus vulgaris* L.: State of the art and future prospects. *Plant Sci., 265*, 39–50.

Miklas, P. N., Larsen, R. C., Riley, R., & Kelly, J. D., (2000). Potential marker-assisted selection for *bc-12* resistance to bean common mosaic potyvirus in common bean. *Euphytica, 116*, 211–219.

Mishra, G. P., Mishra, S. K., Tiwari, S. K., & Kumar, A., (2008). Inheritance of lentil (*Lens culinaris*) resistance to rust. *Indian J. Agric. Sci., 78*, 994–996.

Mishra, R. K., Pandey, K. K., & Pandey, P. K., (2009). Screening of pea (*Pisum sativum*) genotypes against rust caused by *Uromyces fabae*. *Indian J. Agr. Sci., 79*, 402–403.

Mishra, R. K., Parihar, A. K., Basvaraj, T., & Krishna, K., (2019). Identification of new sources of resistance in Rajmash (*Phaseolus vulgaris* L.) against powdery mildew (*Erysiphe polygoni*) and stem rot (*Sclerotinia sclerotiorum*) diseases. *Legume Res., 42*(3), 430–433.

Mohar, S., Narinder, K. G., Mukesh, K. R., Om, P. D., Dutta, M., et al., (2014). Pigeon pea genetic resources and its utilization in India: Current status and future prospects. *J. Plant Sci. Res. 1*(2), 107.

Mojumder, S., Hossain, M. D., Haque, M. S., & Nasiruddin, K. M., (2015). *In vitro* regeneration of BINA mungbean varieties. *Journal of Environmental Science and Natural Resources, 7*(2), 47–52.

Murray, G. M., & Brennan, J. P., (2012). *The Current and Potential Costs from Diseases of Pulse Crops in Australia*. ACT, Australia, Grains Research and Development Corporation.

Murube, E., Ana, C., & Juan, J. F., (2017). Identification of new resistance sources to powdery mildew, and the genetic characterization of resistance in three common bean genotypes. *Crop Pasture Sci., 68*(11), 1006–1012.

Naimuddin, K., Akram, M., & Sanjeev, G., (2011). Identification of mungbean yellow mosaic India virus infecting *Vigna mungo* var. *silvestris* L. *Phytopathol. Mediterr., 50*, 94–100.

Nasir, M., & Bretag, T. W., (1998). Reactions of lentil accessions from 25 different countries to Australian isolates of *Ascochyta lentis*. *Genet. Resour. Crop Evol., 45*, 297–299.

Nene, Y. L., (1972). A survey of the viral diseases of pulse crops in Uttar Pradesh, G.B. Pant University of Agriculture and Technology, Pantnagar. *Res. Bull., 4*, 191.

Nguyen, T., Taylor, P., Brouwer, J., Pang, E., & Ford, R., (2001). A novel source of resistance in lentil (*Lens culinaris* ssp. *culinaris*) to Ascochyta blight caused by *Ascochyta lentis*. *Australas. Plant Pathol., 30*, 211–215.

Osorno, J. M., Beaver, J. S., Ferwerda, F. H., & Miklas, P. N., (2003). Two genes from *Phaseolus coccineus* L. Confer resistance to bean golden yellow mosaic virus. *Ann. Rep. Bean Improv. Coop., 46*, 147–148.

Pande, S., Galloway, J., Gaur, P. M., Siddique, K. H. M., Tripathi, H. S., Taylor, P., MacLeod, M. W. J., et al., (2006). Botrytis grey mould of chickpea: A review of biology, epidemiology and disease management. *Aust. J. Agric. Res., 57*, 1137–1150.

Pandiyan, M., Ramamoorthi, N., Ganesh, S. K., Jebaraj, S., Pagarajan, P., & Balasubramanian, P., (2008). Broadening the genetic base and introgression of MYMV resistance and yield improvement through unexplored genes from wild relatives in mungbean. *Plant Mutation Reports, 2*, 33–38.

Patil, P. G., Bohra, A., Dubey, J., Saabale, P. R., Mishra, R. K., Pange, G., Das, A., et al., (2017a). Genetic analysis and molecular resistance to race 2 of *Fusarium* wilt in pigeonpea (*Cajanus cajan* L. Millsp.). *Crop Prot., 100*, 117–123.

Patil, R. K., Satish, D., & Jagadeesha, R. C., (2017b). Screening of recombinant inbred lines of French bean (*Phaseolus vulgaris* L.) for bean common mosaic virus (BCMV) resistance. *Plant Arch., 17*(1), 479–482.

Pawar, B. D., Jadhav, A. S., Pawar, S. V., Chimote, V. P., & Kale, A. A., (2014). An efficient regeneration system for pigeonpea (*Cajanus cajan* L.) *J. Crop Improv., 28*(6), 825–833.

Peever, T. L., Salimath, S. S., Su, G., Kaiser, W. J., & Muehlbauer, J., (2004). Historical and contemporary multi-locus population structure of *Ascochyta rabiei* (teleomorph: *Didymella rabiei*) in the Pacific Northwest of the United States. *Mol. Ecol., 13*, 291–309.

Pottorff, M., Roberts, A. P., Close, T. J., Lonardi, S., Wanamaker, S., & Ehlers, J. D., (2014). Identification of candidate genes and molecular markers for heat induced brown discoloration of seed coat in cowpea (*Vigna ungiculata* L. Walp.). *BMC Genom., 15*, 328. doi: 10.1186/1471-2164-15-328.

Pottorff, M., Wanamaker, S., Ma, Y. Q., Ehlers, J. D., Roberts, A. P., & Close, T. J., (2012). Genetic and physical mapping of candidate genes for resistance to *Fusarium* oxysporium f. sp. *tracheiphilum* race 3 in cowpea (*Vigna ungiculata* L. Walp.). *PLos One,* 7, e41600.

Prasad, M. G., Sridevi, V., & Satish, K. M., (2014). Efficient plant regeneration from cotyledonary node of blackgram (*Vigna mungo* L. Hepper). *Int. J. Adv. Biotechnol. Res., 5*, 20–24.

Prasanthi, L. R., Reddy, B. V. B., Rani, K. R., Prasad, Y. S., Rajeshwari, T., & Reddy, K. R., (2009). Development of sequence characterized amplified region (SCAR) marker for *Fusarium* wilt resistance gene in pigeon pea (*Cajanus cajan* L. Millsp.). *Int. J. Plant Breed., 3*(2), 134–138.

Pratap, A., Chaturvedi, S. K., Tomar, R., Rajan, N., Malviya, N., Thudi, M., Saabale, P. R., et al., (2017). Marker-assisted introgression of resistance to *Fusarium* wilt *race 2* in Pusa 256, an elite cultivar of Desi chickpea. *Mol. Genet. Genomics*. doi 10.1007/s00438-017-1343-z.

Pratap, A., Choudhary, A. K., & Kumar, J., (2010). *In vitro* techniques towards genetic enhancement of food legumes: A review. *Journal of Food Legumes, 23*, 169–185.

Rai, R., Singh, A. K., Chand, R., Srivastava, C. P., Joshi, A. K., & Singh, B. D., (2016). Genomic regions controlling components of resistance for pea rust caused by *Uromyces fabae* (Pers.) de-Bary. *J. Plant Biochem. Biotechnol., 25*, 133–141.

Rai, R., Singh, A. K., Singh, B. D., Joshi, A. K., Chand, R., & Srivastava, C. P., (2011). Molecular mapping for resistance to pea rust caused by *Uromyces fabae* (Pers) de-Bary. *Appl. Genet. 123*, 803–813.

Rambabu, E., Ch. Anuradha, Sridhar, V., & Sokka, R., (2018). Genetics of mungbean yellow mosaic virus (MYMV) resistance in blackgram (*vigna mungo* (l.) hepper). *Int. J. Curr. Microbiol. App. Sci., 7*(4), 3869–3872.

Rastogi, K. B., & Saint, S. S., (1984). Inheritance of resistance to pea blight (*Ascochyta pinodella*) and induction of resistance in pea (*Pisum sativum*). *Euphytica, 33*, 9–11.

Ratna, B. D., & Ravikumar, R. L., (2010). Parallel response between gametophyte and sporophyte for *Fusarium* wilt resistance in the recombinant inbred lines of chickpea (*Cicer arietinum* L.). *Curr. Sci., 99*(4), 513–518.

Raveendar, S., Premkumar, A., Sasikumar, S., Ignacimuthu, S., & Agastian, P., (2009). Development of a rapid, highly efficient system of organogenesis in cowpea *Vigna unguiculata* (L.) Walp. *S. Afr. J. Bot., 75*(1), 17–21.

Ravishankar, K. V., Aghora, T. S., Mohan, N., Naveen, A. H., & Krishnareddy, M., (2009). Identification of RAPD marker linked to mungbean yellow mosaic virus resistance in French bean (*Phaseolus vulgaris* L.). *J. Hortl. Sci., 4*(2), 167–169.

Reddy, K. S., (2009). Identification and inheritance of a new gene for powdery mildew resistance in mungbean (*Vigna radiata* L. Wilczek). *Plant Breed., 128*, 521–523.

Reddy, K. S., Pawar, S. E., & Bhatia, C. R., (1994). Inheritance of powdery mildew (*Erysiphe polygoni* D.C.) resistance in mungbean (*Vigna radiata* L. Wilczek). *Theor. Appl. Genet., 88*, 945–948.

Reddy, M. V., & Singh, K. B., (1990). Relationship between Ascochyta blight severity and yield loss in Chickpea and identification of resistant lines. *Phytopath. Medit., 29*(1), 32–38.

Reddy, M. V., Raju, T. N., Nene, Y. L., Ghanekar, A. M., Amin, K. S., Arjunan, G., Astaputre, J. V., et al., (1993). Variability in sterility mosaic pathogen in pigeonpea in India. *Indian Phytopathol., 46*, 206–212.

Rezende, V. F., Antonio, M., Ramalho, P., & Hercules, R. C., (1999). Genetic control of common bean (*Phaseolus vulgaris*) resistance to powdery mildew (*Erysiphe polygoni*). *Genet. Mol. Biol., 22*(2), 233–236.

Rigola, D., Van, O. J., Janssen, A., Bonne, A., Schneiders, H., Van, D. P. H. J. A., Van, O. N. J., et al., (2009). High-throughput detection of induced mutations and natural variation using KeyPoint™ technology. *PLoS One, 4*(3), e4761.

Rodrigues, M. A., Santos, C. A. F., & Santana, R. F. J., (2012). Mapping of AFLP loci linked to tolerance to cowpea golden mosaic virus. *Genet. Mol. Res., 11*, 3789–3797.

Rodriguez-Leal, D., Lemmon, Z. H., Man, J., Bartlett, M. E., & Lippman, Z. B., (2017). Engineering quantitative trait variation for crop improvement by genome editing. *Cell, 171*, 470–480.

Saabale, P. R., Manju, N. J., Bhat, S., Biradar, R. S., Mishra, R. K., Naimuddin, Chaudhary, R. G., et al., (2019). Elucidation of host resistance for chickpea wilt (*Fusarium oxysporium* f. sp. *ciceris*). *Journal of Food Legumes, 32*(3), 174–177.

Sai, C. B., Nagarajan, P., Raveendran, M., Rabindran, R., Kannan, B. J. R., & Senthil, N., (2017). Understanding the inheritance of mungbean yellow mosaic virus (MYMV) resistance in mungbean (*Vigna radiata* L. Wilczek). *Mol. Breed., 37*, 63.

Saini, R., & Jaiwal, P. K., (2005). Transformation of a recalcitrant grain legume, *Vigna mungo* L. Hepper, using *Agrobacterium tumefaciens*-mediated gene transfer to shoot apical meristem cultures. *Plant Cell Rep., 24*, 164–171.

Sambasivam, P., Taylor, P. W. J., & Ford, R., (2017). Pathogenic variation and virulence related responses of *Ascochyta lentis* on lentil. *Eur. J. Plant Pathol., 147*, 265–277.

Saxena, R. K., Hake, A., Bohra, A., Khan, A. W., Hingane, A., Sultana, R., Singh, I. P., et al., (2021). A diagnostic marker kit for *Fusarium* wilt and sterility mosaic diseases resistance in pigeonpea. *Theor. App. Genet., 134*, 367–379.

Saxena, R. K., Hake, A., Hingane, A. J., et al., (2020). Translational pigeonpea genomics consortium for accelerating genetic gains in pigeonpea (*Cajanus cajan* L.). *Agron., 10*, 1289.

Saxena, R. K., Kale, S. M., Kumar, V., Parupali, S., Joshi, S., Singh, V., Garg, V., Das, R. R., Sharma, M., Yamini, K. N., et al., (2017). Genotyping-by-sequencing of three mapping populations for identification of candidate genomic regions for resistance to sterility mosaic disease in pigeonpea. *Sci. Rep., 7*, 1–10.

Schmutz, J., McClean, P. E., Mamidi, S., et al., (2014). A reference genome for common bean and genome-wide analysis of dual domestications. *Nat. Genet.* doi: 10.1038/ng.3008.

Sehrawat, N., & Yadav, M., (2014). Screening and cross-compatibility of various Vigna species for yellow mosaic virus resistance. *J. Innov. Biol., 1*, 31–34.

Sepulveda, R. P., (1985). Effect of rust caused by *Uromuces febae* (Pers) de Bary on the yield of lentils. *Agricultural Technical, 45,* 335–339.

Sharma, K. D., & Muehlbauer, F. J., (2007). *Fusarium* wilt of chickpea: Physiological specialization, genetics of resistance and resistance gene tagging. *Euphytica, 157*, 1–14.

Sharma, M., & Ghosh, R., (2016). An update on the host plant resistance to pigeonpea diseases. *Legume Perspective, 11*, 21–23.

Sharma, M., & Ghosh, R., (2018). Isolation, identification, and pathogenicity of *Phytophthora* blight of pegionpea. *Plant Health Prog., 19*, 233–236.

Sharma, M., & Pande, S., (2011). New sources of resistance to *Fusarium* wilt, sterility mosaic disease and phytophthora blight in vegetable pigeon pea germplasm. *Ind. J. Plant Prot., 39*, 288–293.

Sharma, M., Ghosh, R., & Pande, S., (2013). Occurrence of *Alternaria alternata* causing Alternaria blight in pegionpea in India. *Adv. Biosci. Biotechnol., 4*, 702–705.

Sharma, M., Ghosh, R., Tarafdar, A., Rathore, A., Chobe, D. R., Kumar, A. V., Gaur, P. M., et al., (2019). Exploring the genetic cipher of chickpea (*Cicer arietinum* L.) through identification and multi-environment validation of resistant sources against *Fusarium* wilt (*Fusarium oxysporum* f. sp. *ciceris*). *Front. Sustain. Food Syst., 3*, 78.

Sharma, M., Rathore, A., Mangala, U. N., Ghosh, R., Sharma, S., Upadhyay, H. D., & Pande, S., (2012). New sources of resistance to *Fusarium* wilt and sterility mosaic disease in a mini-core collection of pigeonpea germplasm. *Eur. J. Plant Pathol.* DOI 10.1007/s10658-012-9949-9.

Sharma, P., Sharma, P. N., Kapil, R., Sharma, S. K., & Sharma, O. P., (2011). Analysis of 30-terminal region of bean common mosaic virus strains infecting common bean in India. *Indian J. Virol. 22*(1), 37–43.

Shivaprasad, P. V., Thillaichidambaram, P., Balaji, V., & Veluthambi, K., (2006). Expression of full-length and truncated rep genes from mungbean yellow mosaic virus-Vigna inhibits viral replication in transgenic tobacco. *Virus Genes, 33*, 365–374.

Shukla, A., Das, A., Ansari, J., & Datta, S., (2015). *In vitro* regeneration of chickpea (*Cicer arietinum* L.) via somatic embryogenesis. *Journal of Food Legumes, 28*(3), 199–202.

Sikirou, R., (1999). *Epidemiological Investigations and Development of Integrated Control Methods of Bacterial Blight of Cowpea Caused by Xanthomonas Campestris pv. Vignicola* (p. 218). Ph. D. Thesis. University of Go¨ttingen, Germany.

Silbernagel, M. J., & Hannan, R. M., (1992). Use of plant introductions to develop U.S. bean cultivars. In: Shands, H. L., & Wiesner, L. E., (eds.), *Use of Plant Introductions in Cultivar Development, Part-2* (pp. 1–8). CSSA Special Publication No. 20, CSSA, Madison, Wisconsin.

Singh, A. K., Rai, R., Singh, B. D., Chand, R., & Srivastava, C. P., (2015). Validation of SSR markers associated with rust (*Uromyces fabae*) resistance in pea (*Pisum sativum* L.). *Physiol. Mol. Biol. Plant, 21*, 243–247.

Singh, B. B., & Dikshit, H. K., (2002). Possibilities and limitations of interspecific hybridization involving green gram (*Phaseolus radiatus*) and blackgram (*Phaseolus mungo*). *Indian J. Agric. Sci., 72*, 676–678.

Singh, D. P., (1980). Inheritance of resistance to yellow mosaic virus in blackgram (*Vigna mungo* L. Hepper). *Theor. Appl. Genet., 57*(5), 233–235.

Singh, D., Sinha, B., Rai, V. P., Singh, M. N., Singh, D. K., Kumar, R., & Singh, A. K., (2016). Genetics of *Fusarium* wilt resistance in pigeonpea (*Cajanus cajan*) and efficacy of associated SSR markers. *Plant Pathol. J., 32*(2), 95–101.

Singh, J., Sidhu, P. S., Verma, M. M., Gosal, S. S., & Singh, J., (1993). Wide cross hybridization in *Cajanus*. *Crop Imp., 20*, 27–30.
Singh, K. B., & Reddy, M. V., (1983). Inheritance of resistance to Ascochyta blight in chickpea. *Crop Sci., 23*, 9, 10.
Singh, K. B., & Reddy, M. V., (1991). Advances in disease-resistance breeding in chickpea. *Adv. Agron., 45*, 191–222.
Singh, K. B., & Reddy, M. V., (1993). Sources of resistance to *Ascochyta* blight in wild *Cicer* species. *Neth. J. Plant Pathol., 99*, 163–167.
Singh, K. B., Malhotra, R. S., Haldia, H., Knights, E. J., & Verma, M. M., (1994). Current status and future strategy in breeding chickpea for resistance to biotic and abiotic stresses. *Euphytica, 73*, 137–149.
Singh, N. K., Gupta, D. K., Jayaswal, P. K., et al., (2012). The first draft of the pigeonpea genome sequence. *J. Plant Biochem. Biotechnol., 21*(1), 98–112.
Singh, R., & Ram, H., (2001). Inheritance of powdery mildew and rust resistance in pea. *Ann. Biol. 15*, 13–16.
Singh, S. P., & Munoz, C. G., (1999). Resistance to common bacterial blight among *Phaseolus* species and common bean improvement. *Crop Sci., 39*, 80–89.
Singh, V. P., & Chaturvedi, S. N., (1981). Gamma-ray induced male sterile mutants in mungbean. *Pulse Crops Newsletter, 1*, 41, 42.
Smyth, S. J., (2019). Global status of the regulation of genome editing technologies. *CAB Reviews, 14*, 1–6.
Souframanien, J., & Gopalakrishna, T., (2006). ISSR and SCAR markers linked to the mungbean yellow mosaic virus (MYMV) resistance gene in blackgram (*Vigna mungo* L. Hepper). *Plant Breed., 125*, 619–622.
Souframanien, J., Raizada, A., Dhanasekar, P., & Suprasanna, P., (2021). Draft genome sequence of the pulse crop blackgram (*Vigna mungo* L. Hepper) reveals potential R-genes. *Sci. Rep., 11*, 11247.
Sowmini, K., & Jayamani, P., (2014). Validation of molecular markers linked with yellow mosaic disease resistance in blackgram *Vigna mungo* (L.). *Legume Genomics and Genetics, 5*(4), 25–30.
Subramaniyan, R., Kulanthaivel, V., Manivannan, N., Mahalingam, A., & Satya, V. K., (2021). Marker-assisted backcross breeding for enhancing Mungbean Yellow Mosaic Virus (MYMV) disease resistance in blackgram (*Vigna mungo* L. Hepper) cv. MDU 1. *Physiol. Mol. Plant Pathol., 116*, 101732.
Sudha, M., Karthikeyan, A., Shobhana, V. G., Nagarajan, P., Raveendran, M., Senthil, N., et al., (2015). Search for *Vigna species* conferring resistance to mungbean yellow mosaic virus in mungbean. *Plant Genet. Resour., 13*(2), 162–167.
Sudheesh, S., Lombardi, M., Leonforte, A., Cogan, N. I., Materne, M., Forster, J., et.al., (2014). Consensus genetic map construction for field pea (*Pisum sativum* L.), trait dissection of biotic and abiotic stress tolerance and development of a diagnostic marker for the *er1* powdery mildew resistance gene. *Plant Mol. Biol. Rep., 33*, 1–13.
Sun, S., Deng, D., Wang, Z., Duan, C., Wu, X., Wang, X., Zong, X., & Zhu, Z., (2017). A novel *er1* allele and the development and validation breeder friendly KASPar markers for *er1*, a powdery mildew resistance gene in pea (*Pisum sativum* L.). *Mol. Breed., 37*, 151.
Sun, S., Fu, H., Wang, Z., Duan, C., Zong, X., & Zhu, Z., (2016). Discovery of a novel *er1* allele conferring powdery mildew resistance in Chinese pea (*Pisum sativum* L.) landraces. *PLoS One, 11*, e0147624.

Taleei, A., Kanouni, H., & Baum, M., (2009). Genetical analysis of Ascochyta blight resistance in chickpea. In: Slezak, D., & Arsalan, T., et al., (eds.). *Bioscience and Biotechnology, Communications in Computer, and Information Science*. Springer-Verlag: Berlin, Germany.
Tantasawat, P. A., Poolsawat, O., Arsakit, K., & Papan, P., (2020). Identification of ISSR, ISSR-RGA and SSR markers associated with *Cercospora* leaf spot resistance gene in mungbean. *Intl. J. Agric. Biol., 23*, 447–453.
Tar'an, B., Warkentin, T., Somers, D. J., Miranda, D., Vandenberg, A., Blade, S., Woods, S., Bing, D., Xue, A., DeKoeyer, D., et al., (2003). Quantitative trait loci for lodging resistance, plant height and partial resistance to Ascochyta blight in field pea (*Pisum sativum* L.). *Appl. Genet., 107*, 1482–1491.
Tay, J., & Slinkard, A., (1989). Transgressive segregation for Ascochyta resistance in lentil. *Can. J. Plant Sci., 69*, 547.
Thao, N. T., Thao, N. T. P., Hassan, F., & Jacobsen, H. J., (2013). *In vitro* propagation of common bean (*Phaseolus vulgaris* L.). *Journal of Science Development, 11*, 868–876.
Till, B. J., Reynolds, S. H., Weil, C., et al., (2004). Discovery of induced point mutations in maize genes by TILLING. *BMC Plant Biol., 4*, 1–8.
Timmerman-Vaughan, G. M., Frew, T. J., Butler, R., Murray, S., Gilpin, M., Falloon, K., Johnston, P., et al., (2004). Validation of quantitative trait loci for Ascochyta blight resistance in pea (*Pisum sativum* L), using populations from two crosses. *Appl. Genet., 109*, 1620–1631.
Tollefson, J., (2011). Brazil cooks up transgenic bean. *Nature, 478*, 168.
Trabanco, N., Pe´rez-Vega, E., Campa, A., Rubiales, D., & Ferreira, J. J., (2012). Genetic resistance to powdery mildew in common bean. *Euphytica, 186*, 875–882.
Tullu, A., Buchwaldt, L., Lulsdorf, M., Banniza, S., Barlow, B., Slinkard, A. E., et al., (2006). Sources of resistance to anthracnose (*Colletotrichum truncatum*) in wild *Lens* species. *Genet. Resour. Crop Evol., 53*, 111–119.
Tyagi, M. K., & Srivastava, C. P., (1999). Inheritance of powdery mildew and rust resistance in pea. *Ann. Bio., 15*, 13–16.
Upadhyay, V., Kushwaha, K. P. S., & Pandey, P., (2015). Inspection of different localities in Uttarakhand and Uttar Pradesh for the prevalence of rust disease of pea (*Pisum sativum*). *Trends in Biosciences, 8*(16), 4336–4340.
Vairam, N., Lavanya, S. A., Muthamilan, M., & Vanniarajan, C., (2016). Screening of M_3 mutants for yellow vein mosaic virus resistance in greengram (*Vigna radiata* L. Wilczek). *Int. J. Plant Sci., 11*(2), 265–269.
Van, D. M. L. L. G., & Pundir, R. P. S., (1984). Availability and use of wild *Cicer* germplasm. *Plant Genet. Resour. Newsl., 57*, 19–24.
Van, D. M. L. L. G., Maxted, N., Javad, F., Coles, S., & A. Davies, M. R., (2007). Taxonomy of the genus *Cicer* revisited. In: Yadav, S. S., Redden, R. J., Chen, W., & Sharma, B., (eds.), *Chickpea Breeding and Management* (pp. 14–46). CAB International, Wallingford, UK.
Vanniarajan, C., Ganeshram, S., Souframanien, J., Veni, K., Anandhi, L. S., & Kuralarasan, V., (2017). Gamma rays induced urdbbean (*Vigna mungo* L. Hepper) mutants with YMV resistance, good batter quality and bold seeded type. *Legume Res.* doi: 10.18805/LR-3824.
Varshney, R. K., Chen, W., Li, Y., Bharti, A. K., Saxena, R. K., Schlueter, J. A., et al., (2012). Draft genome sequence of pigeonpea (*Cajanus cajan*), an orphan legume crop of resource-poor farmers. *Nat. Biotechnol., 30*, 83–89.

Varshney, R. K., Chi, S., Saxena, R. K., et al., (2013). Draft genome sequence of chickpea (*Cicer arietinum*) provides a resource for trait improvement. *Nat. Biotechnol.* doi: 10.1038/nbt.2491.

Varshney, R. K., Close, J. T., Singh, N. K., Hoisington, D. A., & Cook, D. R., (2009). Orphan legume crops enter the genomics era! *Curr. Opin. Plant Biol., 12*, 202–210.

Varshney, R. K., Mohan, S. M., Gaur, P. M., et al., (2014). Marker-assisted backcrossing to introgress resistance to *Fusarium* wilt race 1 and Ascochyta blight in C 214, an elite cultivar of chickpea. *Plant Genome, 7*, 1–11.

Veltcheva, M., Svetleva, D., Petkova, S. P., & Perl, A., (2005). *In-vitro* regeneration and genetic transformation of common bean (*Phaseolus vulgaris* L.) –problems and progress. *Sci. Hortic., 107*(1), 2–10.

Vijayalakshmi, S., Yadav, K., Kushwaha, C., Sarode, S. B., Srivastava, C. P., Chand, R., & Singh, B. D., (2005). Identification of RAPD markers linked to rust (*Uromyces fabae*) resistance gene in pea (*Pisum sativum*). *Euphytica, 144*, 265–274.

Vijayan, S., & Kirti, P. B., (2011). Mungbean plants expressing BjNPR1 exhibit enhanced resistance against the seedling rot pathogen, *Rhizoctonia solani. Transgenic Res.* DOI 10.1007/s11248-011-9521-y.

Warkentin, T. D., Rashid, K. Y., & Xue, A. G., (1996). Fungicidal control of powdery mildew in field pea. *Can. J. Plant Sci., 76*, 933–935.

Williams, R. J., (1977). Identification of multiple disease resistance in cowpea. *Trop. Agric., 54*(1), 53–59.

Win, K. S., Win, S., Htun, T. M., Win, N. K. K., & Oo, K. S., (2021). Inheritance and gene effect of resistance to mungbean yellow mosaic virus (MYMV) in mungbean (*Vigna radiata* L. Wilczek). *Legume Res., 44*, 486–490.

Worrall, E. A., Wamonje, F. O., Mukeshimana, G., Harvey, J. J., Carr, J. P., & Mitter, N., (2015). Bean common mosaic virus and bean common mosaic necrosis virus: Relationships, biology, and prospects for control. *Adv. Virus Res., 93*, 1–46.

CHAPTER 11

Mining Resistance Genes for Disease Management

MANKESH KUMAR,[1] SATYENDRA,[1] RABIYA PARVEEN,[1] GAUTAM KUNAL,[1] KSHITIZ,[2] SANGITA SAHNI,[3] and BISHUN DEO PRASAD[4]

[1]*Department of Plant Breeding and Genetics, Bihar Agricultural University, Sabour, Bhagalpur, Bihar, India*

[2]*Division of Nematology, Indian Agricultural Research Institute, New Delhi, India*

[3]*Department of Plant Pathology, TCA, Dholi, Dr. Rajendra Prasad Central Agricultural University, Pusa, Samastipur, Bihar, India*

[4]*Department of Agricultural Biology and Molecular Biology, College of Basic Sciences and Humanities, Dr. Rajendra Prasad Central Agricultural University, Pusa, Samastipur, Bihar, India*

ABSTRACT

Plants have a natural system for defense against different pests and diseases. When a pathogen attacks a host plant, it tries to override the physical barriers to infection imparted by various defense signaling mechanisms triggered by plant receptors and defense response signaling genes. Recent advances in sequencing technologies, high throughput genotyping, and precise phenotyping for disease resistance in plants have paved the way for gene mining to discover new alleles and their role in plant immunity. In this chapter, we review the various components of the immune system in plants and different aspects of allele mining in the exploitation of the existing genetic variation for disease resistance in crops. We conclude that the rapid advances in

New Advances in Crop Disease Management. Sangita Sahni, Ganesh Thapa, and Bishun Deo Prasad (Eds)

molecular marker techniques have revolutionized cereal genomics and the understanding of the structure and behavior of different resistance genes present in cereal genomes. The discovery and characterization of novel genes for plant resistance will assist in the development of superior genotypes that can meet the ever-increasing demands of food grains.

11.1 INTRODUCTION

Occurrence of pests and diseases in crops globally reduces the production by about 20–40% every year. It is projected that the world population will rise from 7.4 billion to 8.4 billion in 2030 and 9.5 billion in 2050. Hence, to feed the ever-growing population, the greatest challenge would be to increase crop productivity and quality, and to reduce the loss of crops both at pre- and post-harvest stages. Since plant survival is mostly dependent on productivity, earlier farmers would have chosen species based on a variety of qualities, including disease resistance. The most essential defense mechanisms used by host plants, including our crops, against diseases are resistance mechanisms. Pesticides provide efficient protection, but unfavorable environmental impacts and the emergence of resistant pathogen strains can limit their use. Chemical pesticides are frequently out of reach for farmers in impoverished countries. For these reasons, substantial effort has been expended into figuring out how plants develop natural resistance mechanisms. Plants have a powerful arsenal of inducible defense mechanisms, including genetically programmed infected cell death (the hypersensitive response, HR), tissue strengthening, and antibiotic synthesis at the infection site (Hammond-Kosack & Jones, 1996). These local reactions can then induce a long-lasting systemic response (systemic acquired resistance (SAR)) that prepares the plant to fight a wide range of diseases (Flor, 1942; Dong, 2001). The early 20th century witnessed the discovery of resistance genes (R genes) in plants, which was made possible by knowing about their heritability under different environments as well as overall genetics. Heinrich Anton de Bary identified pathogenic microorganisms as the cause of plant diseases in the 19th century (Kutschera & Hossfeld, 2012).

Plants have a multilayered defense mechanism against pathogens. Once a pathogen overcomes the mechanical barrier, the plant's immune system defends against the pathogens by recognizing the molecules of the enemy and conducting signal transduction and responding accordingly. Like in animals, the immune response in plants gets alerted upon pathogen infection through specialized pattern recognition receptors (PRRs) localized at the cell peripheral area. These receptors recognize two types of molecular patterns: one associated with the conserved motifs in the pathogen itself (known as pathogen-associated molecular patterns (PAMPs)) or the patterns aroused after the damage caused by the pathogen to the host cell (known as damage-associated molecular patterns, DAMPs). DAMs are related to the components of host cells that are released in the course of cell damage by the pathogen, and because these are non-functional parts of the cell and can be harmful, the autoimmune system of the host also tries to eliminate these DAMs from the cells. Effectors that facilitate the infection of pathogens are detected by plant receptors which have nucleotide-binding domains and also leucine-rich repeats (NLRs). Expression of pathogenic-related genes in the host is regulated by mitogen-activated protein kinases (MAPKs), ubiquitin, hormones, calcium, G-proteins, transcription factors (TFs), and epigenetic modifications, which lead to various responses, such as the production of reactive oxygen species (ROS), alteration in the cell wall, promoting stomatal closure, hypersensitive response (HR), or the production of different anti-pest proteins and compounds) to prevent further infection. They also affect various other signaling systems, such as growth and abiotic stress response.

Growing varieties with resistance to a set of prevailing pathogens has been the most effective and economical way to control disease. However, this resistance is often broken or lost shortly after the release of cultivars due to the high variability of different pathogen races. The emergence of new virulent races of pathogens, caused by high mutation rates in changing environmental conditions, has also been influenced by erratic climatic conditions in recent times. This has led to the occurrence of new diseases in areas where they were not previously prevalent. To develop varieties with more durable resistance, it is necessary to incorporate multiple and novel genes for resistance into target genotypes. This requires exploring new alleles available in the diverse gene pool of crop species and implementing efficient selection and breeding strategies to introduce these genes into high-yielding backgrounds. Conventional breeding, which relies on phenotypic screening, is affected by environmental variation and therefore, field screening in natural conditions is not reliable. However, molecular breeding techniques utilize genomic information of disease resistance genes and their flanking regions to precisely transfer them into high-yielding crop plants in a shorter time and with less need to handle large mapping populations.

This chapter discusses various aspects of pathogen recognition and effectors for virulence and resistance factors in disease interactions. It also explores different breeding strategies and aspects to discover novel resistance genes in plants. Specifically, the chapter focuses on how different approaches of allele mining can be utilized to explore disease resistance genes, which can then be introduced into genotypes through molecular plant breeding approaches.

11.2 VIRULENCE AND AVIRULENCE GENES

Plants have 'self' from 'nonself' mechanism for imparting resistance to pathogens which is monitored by the extracellular and intracellular environment of the host. However, pathogens can overcome the host defense by triggering the production of certain proteins and molecules known as effectors which hinders the host immunity and change the plant cell environment to facilitate colonization (Toruño, Stergiopoulos, & Coaker, 2016). Pathogens secrete effectors into the extracellular environment or deliver it in an orchestrated way into the host. Type III bacterial secretion, formation of fungal haustoria and oomycetes and stylets in nematodes are different modes to suppress host defense. In response to pathogen attack, plants have two main surveillance systems which are discussed in subsections.

11.2.1 HOST RESPONSE IN THE EXTRACELLULAR ENVIRONMENT

Before the invasion of pathogen into host cell, the pattern-recognition receptors (PRRs) found on the host cell surface gets triggered for recognition of conserved molecular sequences of the microbes. These receptors are basically act as sensor machines to detect molecules typical for the pathogen and can differentiate between self and non-self. These PPRs sense specific parts of virus, bacteria, and parasites. Bacterial flagellin and fungal cell wall chitin are the two best-studied PPRs. The microbes triggered immunity (MTI) at the infection site results in several responses like deposition of callose, alteration in cell membrane ion flux and phytohormone concentrations, downstream induction or repression of resistant genes, enhancement in the production of ROS and nitric oxide (NO). The microbe's infestation triggers the innate immunity and is universal for entire plant species against all known races of a specific plant pathogen (Samira, Kimball, Samayoa et al., 2020).

11.2.2 HOST RESPONSE IN THE INTRACELLULAR ENVIRONMENT

R proteins constitute a large family of receptors found in the intracellular environment of the host which are recognized by specific pathogen effectors. The R proteins are characterized by presence of a nucleotide binding site (NBS) and leucine-rich repeats (LRR). These are known as NOD-like receptors (NLRs) proteins as well as variable amino- and carboxy-terminal domains. This large family is divided into two major groups depending upon the features at their N terminus: one set has a Toll/interleukin-1 receptor-like (TIR) domain and the other a coiled coil (CC) domain (Bayless & Nishimura, 2020), producing different signaling response (Figure 11.1). The NBS domain seems to play role in pathogen selection, whereas the LRR region plays role in pathogen variation due to its highly variable nature (McHale, Tan, Koehl et al., 2006). LRR domain is integral part of almost all immune receptors in plants and animals which is characterized by the presence of hydrophobic leucine residues.

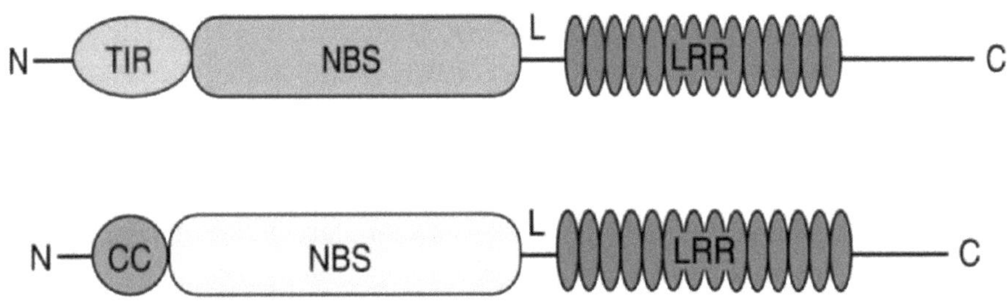

FIGURE 11.1 Main features of NBS-LRR domain. [N: amino terminus; TIR: toll/interleukin-1 receptor-like domain; CC: coiled-coil domain; NBS: nucleotide binding site; L: linker; LRR: leucine-rich repeat domain; and C: carboxyl terminus].

Source: Modified from McHale et al., 2006

The bulk of known R proteins are divided into a few distinct classes based on its structural motifs. Most R proteins are race-specific because they only control pathogen races that express the matching effector protein (s). Effective resistance can be given against numerous races and even different pathogen species on rare occasions which is termed as race-non-specific R proteins. In most of the R genes leucine zipper (LZ) at N-terminal putative or other CC sequence is commonly found. Effectors are the virulence factors secreted by specialized plant pathogens that can evade or suppress the host resistance. However, the avirulence factors (AVRs) which is a type of effectors, are matched with the resistance proteins in host plants which triggers a second layer of host defense and are known as effector triggered immunity (ETI) or *R*-gene mediated defense. *Hypersensitive response* (HR) of the host plants characterized by a rapid, localized cell death at the infection site is mediated by AVR mediated pathogenicity. The HR is a type of programmed cell death (PCD), the signal of which generally activates other distal part of the host plant and provide protection against subsequent infection by starving the pathogen to feed on the nutrients at the early stage of infection. Thus, for causing the infection the pathogen should have the ability to suppress the HR through the development of genetic variants of AVRs which wouldn't be recognized by the receptors and activate the downstream HR of the host plant.

11.3 THE GENE-FOR-GENE CONCEPT

According to Flor's pioneering work on gene-for-gene concept, R genes of host are associated with avirulence genes present in the pathogen that cause an incompatible response. Phytopathogenic bacteria have specialized type of secretion system known as Type III for directly entering into plant cell, which was discovered in the late 1990s while studying plant–bacterial interactions. These effectors either change the plant physiological stage to promote pathogen colonization or suppress the activation of the host plant's defenses. In this response, plants develop particular antidotes to protect itself by the emergence of R genes, whose products led to the direct or indirect detection of a specific subset of pathogen effector gene products and the activation of defense responses in the attacked tissues (LeBlanc, Fink, Perkins, & Sousa, 2021).

Flor officially examined the genetics of flax (*Linum usitatissimum*) and the flax rust (causal fungus: *Melampsora lini*) pathogen in the early 1940s and suggested the "gene-for-gene concept." Both host and pathogen co-exist in nature in which pathogens are present in two forms, i.e., virulent and avirulent whereas host exist in two forms, i.e., resistant and susceptible. Any changes in the virulence and avirulence nature of the pathogen is balanced by the resistant and susceptible nature of the host and pathogen and vice versa. According to this concept, the changes that converse to the virulence nature of the pathogen is corresponding to the genes that are present in the host that confers resistant nature of the host plant. Plant resistance is predicted to arise only when a plant has a dominant "resistance gene" (R) and the pathogen has a comparable dominant "avirulence gene" (*Avr*). Complementary pairings of dominant genes so determine incompatibility. Disease is caused by an alteration in plant resistance gene (dominant R changing to r) or the pathogen avirulence gene (dominant *Avr* changing to *avr*) (compatibility). Although the gene-for-gene concept does not apply to all host–pathogen interactions, it does when the invading pathogen makes its establishment in living plant tissue and subsequently, the host immune response triggers resulting in a highly localized cell death response called the 'hypersensitive response' at the site of infection (HR). A 'boom and bust cycle' was more frequently observed, in which the R gene ceased to protect the crop a few years after it was brought into commercial production, and catastrophic disease followed by epidemics.

Plant Genotype	**Pathogen Genotype**	
	Avr	**avr**
R	Incompatible	Compatible
r	Compatible	Compatible

Resistance (R) proteins are frequently used to activate plant defenses in order to prevent pathogen invasion in which, the leucine-rich repeats (LRRs) play role of central NBS during host pathogen interaction. The variable amino terminal domain in LRRs is the most common features of R proteins. Extra-cellular LRR domains, transmembrane domains, and intracellular serine/threonine kinase domains are found in other classes. Pathogen sensing and/or the activation of conserved defense signaling networks are both functions of R proteins. During infection, pathogens release specific effectors that are thought to promote their growth in host tissue. These effectors are either directly identified by R proteins or are recognized by a targeted plant protein that is itself guarded by R proteins. Various defense signaling networks are then activated as a result of phosphorylation, oligomerization, degradation, conformational changes of R protein and shuttling between the cytoplasm and nucleus which results in profound cellular changes and the initiation of several defense responses both locally at the infection site and throughout the plant.

When a pathogen (Bacteria, fungi, oomycetes, and viruses) attacks a plant, the pattern/pathogen recognition receptors (PRRs) lying on the cell membrane of host, recognizes *elicitors* that are encoded by a conserved molecular sequence of pathogen known as pathogen/microbe-associated molecular patterns (PAMPs/MAMPs) (Figure 11.2). Similarly, wall-associated kinases (WAKs) in the host detect damage-associated molecular patterns (DAMPs) that results from cellular damage by the pathogen at the time of infection. Upon the recognition of PPRs with PAMPs, triggered immunity in plants induces production of ROS and reactive nitrogen species. At the time of infection, PAMPs also changes the host cell wall, induces antimicrobial compounds, and begin the production of pathogenesis-related

(PR) proteins (Jagodzik et al., 2018). In the arm race between host and pathogen, the resistance is often broken down by overcoming or suppressing this MAMP triggered immunity (MTI) through secretion of virulence factors called effectors by the pathogens. A second layer of defense is then activated in which the resistance proteins in host induces an ETI or R-gene mediated defense against virulence effectors, also referred to as AVRs. It is characterized by a hypersensitive response (HR) which rapid and results in localized cell death at the infection site thus deprive the pathogen from nutrition and to survive on the infected patch of plant.

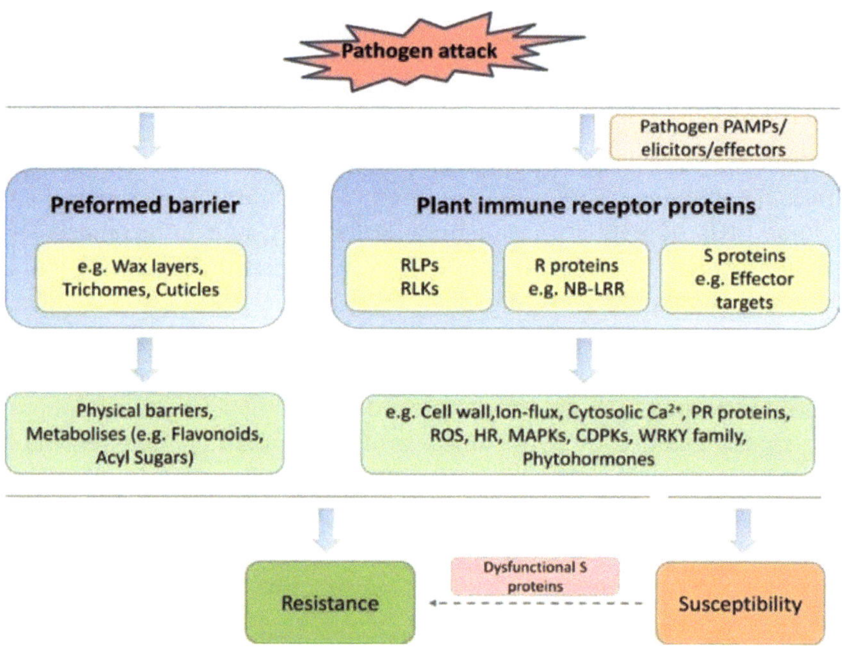

FIGURE 11.2 Various components of plant disease resistance mechanism involved in pathogen recognition, signaling, and defense response.

11.4 QUANTITATIVE AND QUALITATIVE GENES FOR DISEASE RESISTANCE IN PLANTS

Generally, disease resistance in plants is controlled by Mendelian genes and they follow the gene-for-gene relationship (Flor's theory). The host plants imply resistance through hypersensitive reaction, which is qualitative in nature, targeted to specialized races of pathogen and aroused due to interaction of avirulence genes in pathogens and resistant genes in hosts.

The quantitative resistance in other hand is controlled by polygenes, non-hypersensitive, and are race non-specific (Newman, Sundelin, Nielsen, & Erbs, 2013). Quantitative resistance confers an incomplete or partial level of resistance giving a continuous distribution between resistance and susceptible phenotypes among progenies while either a complete or higher level of resistance giving two distinct: resistant and susceptible classes of progenies is achieved through qualitative resistance.

The genes controlling quantitative resistance are governed by more than one gene and are associated with QTL (quantitative trait loci). Each loci contribute variable and minor effect to the phenotype to counter more than one races of pathogen. However, such type of resistance may lead to total resistance specially in case of strong effect QTLs (Niks, Qi, & Marcel, 2015). For example, in Tomato, a strong level of resistance to *Xanthomonas campestris* is governed by three QTLs namely *rx1, rx2*, and *rx3* (Stall, Jones, & Minsavage, 2009). Quantitative resistance lasts long as compared to qualitative resistance (Parlevliet, 2002) because of three factors: (i) the quantitative resistance exerts a low selection pressure on the pathogen population thus preventing the emergence of new pathogen races (ii) there is very low chances of simultaneous mutations in all pathogens which can completely overcome multiple resistant QTLs in the host, and (iii) a combination of different resistance-associated mechanisms, which can act successively at different phases of pathogen life cycle or throughout plant development together are more difficult to overcome (Palloix, Ayme, & Moury, 2009). However, the vertical resistance conferred by a major gene like *Mlo* in case of spring barley which has been widely used to control powdery mildew in Europe since 1979 (Acevedo-Garcia, Kusch, & Panstruga, 2014) has been reported to be overcome by virulent isolates of the pathogen.

11.5 MOLECULAR MARKERS AND ALLELE MINING FOR DISEASE RESISTANCE

The DNA markers which are the basis of molecular breeding approaches, was developed in late 1980s and since then, it has made a drastic change in the conventional breeding strategy by enhancing precision in QTL mapping, marker assisted selection and gene pyramiding. The markers act as 'gene tag' which lie in very close proximity to the resistant gene (tightly linked) or may be located inside the cassette of the gene itself (functional or perfect marker). The genetic transformation through transgenics strategies have also been used to develop new resistant cultivars, particularly through the

Bt-technology in different crops such as cotton, soyabean, maize, papaya, etc. Molecular breeding is a low-cost, high throughput as compared to conventional breeding methods which has been used to develop resistant cultivars in several crops. The limitation in conventional breeding is overcome by molecular breeding by allowing precise introgression of single or multiple disease resistance genes into individual lines to develop a more durable resistance in less time. Through the use of molecular markers, the genetics of quantitative resistance can also be determined, and the QTL involved in resistance involving complex mechanism can also be deciphered. These molecular markers are efficiently used in marker assisted selection in reduced time to stake multiple resistance in individual genotype without need to wait for the phenotype (resistance) development at the time of maturity (Ashkani et al., 2015).

Next generation sequencing (NGS) technologies are powerful platform that has enabled the sequencing of thousands to millions of DNA molecules simultaneously. Through high throughput sequencing option with capability to sequence multiple individuals simultaneously, the sequencing of accession panels and multi-parental populations in several crops in very less time is now possible. The discovery of greater number of QTL with small/minor effects at higher sequencing depth in most of the crops like, rice, *Brassica,* Chickpea, Pigeon pea, Soybean, etc., is possible and are now available in public database for utilization of breeding programs. This has also helped to discover the genes controlling important agronomical traits along with the genomic regions conferring resistance to major disease and pests. Sequencing data of large and diverse panels in germplasm collections has helped in the discovery of allelic variants as well as novel alleles that can be targeted for trait improvement, like disease resistance in high yielding background in crop plants.

Alleles are alternative forms of a gene and mining is nothing but searching the new alleles or genes in the entire gene pool of crop species which includes wide range of cultivars, related species and even across species. The main objective of allele or gene mining is to identify allelic variation that affects plant phenotype in order to utilize the genetic resource (Sarika & Amruta, 2014). Gene mining in crops helps in identification of new haplotypes and development of functional markers for use in Marker Assisted Breeding programs. It enables the breeders to directly assess the key alleles conferring resistance to biotic stress, tolerance to abiotic stress, greater nutrient use efficiency, enhanced yield, and improved quality. It also provides an insight into molecular basis of novel trait variations and identifies the nucleotide sequence changes associated with superior alleles.

As the name implies, allele mining is the process of detecting the new superior alleles out of several million base pair sequence of the genomes for traits of agronomic importance, disease resistance, quality, etc. It includes the analysis of coding and regulatory regions of the candidate genes and study the potential genetic diversity lying in collection of large number of germplasm lines across the species for a particular candidate gene to know the functional variation of relevance. The new alleles identified can be further used in development of new high varieties through marker assisted breeding programs. There are several reports for the use of new mined alleles for disease resistance to develop crop varieties in crops like rice, wheat maize, etc. The bacterial blight resistance gene *Xa21* from *O. longistaminata* (Ronald, Albano, & Tabien, 1992) and different blast resistance genes including *Pi9* from *O. minuta* (Qu et al., 2006), *Pid3-A4* from *O. rufipogon* (Lv et al., 2013), and *Pi-hk1* from a landrace Heikezijing (Wu et al., 2013) are among the resistant genes that have been introgressed from wild relatives and landraces to improve the rice cultivars. Hence, the molecular breeding for disease resistance assisted by modern NGS techniques can be applied to develop resistant cultivars in crop plants in more efficient manner compared to conventional breeding strategies.

Allele mining mainly aims at identifying a small size of genomes and accessions controlling resistance from a large collection of gene pool based on either sequence information or phenotypic information. The known sequence can be used as reference to search the genome database and identify accessions with unique sequences or haplotypes. Main steps of allele miming to decipher new genes for resistance in plants have been shown in Figure 11.3.

11.6 STRATEGIES AND APPROACHES FOR ALLELE MINING

Scientists all over the world have used different breeding strategies like linkage mapping, genome-wide association study (GWAS), mutation breeding (TILLING, EcoTILLING) to discover genetic loci associated with disease resistance in plants. The association between phenotype and genotype at specific locations across the genome can be done by QTL mapping and statistical association analysis can be performed either on a population aroused from biparental mapping or a set of diverse collection of germplasm in association analysis. Genetic variation across the genome can be described well by using molecular markers that can distinguish easily between the two or more closely related strains of pathogen. Modern genomics tools have

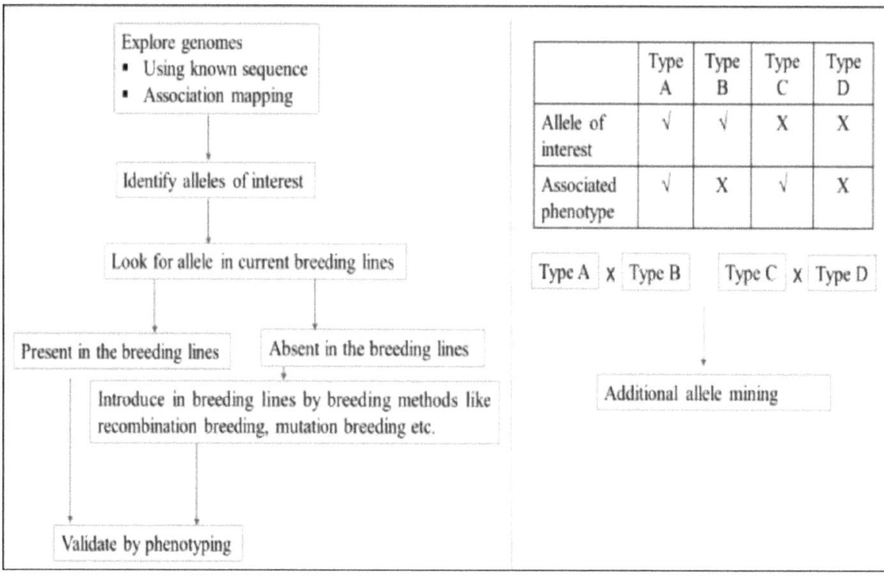

FIGURE 11.3 Allele mining workflow for resistant genes in crops. Explore the sequenced genomes to identify similar alleles in current breeding lines. This is followed by validation through precise phenotyping, i.e., appropriate disease screening techniques. Additional step of allele mining can be applied by crossing breeding lines for confirming the association of alleles with target phenotype.

provided breeders with a set of new techniques that allow the study of the whole genome in very less time with more accuracy than the conventional pre-genomics scientific breeding methodologies. One of the most important factors for the success of modern genomics is the development of high-throughput DNA sequencing technologies which is also known as NGS methods. However, functional validation of identified genic and non-genic regions of the genomes is still great challenges which have been done through reverse and forward genetics approach in most of the crops.

11.6.1 FORWARD AND REVERSE GENETICS APPROACH FOR GENE IDENTIFICATION

Forward genetics is used where a large number of individuals are screened for morphological differences for a particular trait and a causative gene is assigned. This approach is used to identify genes responsible for disease resistance in plants and seeks to find its genetic basis. The process of forward genetics starts with getting mutants either through natural variation or induced mutation followed by selection of mutant phenotype and later gene identification and its functional characterization. On the other hand, in reverse genetics approach, we try to find out the generation of disease resistance phenotype aroused as a result of changes at the molecular level in particular gene sequences. The forward genetics approach is very cumbersome, time-consuming, and labor-intensive, and functionally defective alleles can't be identified as this will be unable to express (Aklilu, 2021). Forward genetics approaches are not practical for genome-wide analysis primarily due to the effort and time involved in identifying each gene coding for a particular phenotype, and also the mutant phenotype may or may not be distinguished on the basis of field-level screening of the mutants.

Both the forward and reverse genetics approaches start with the creation of a mutagenized population for which appropriate plant material, nature and strength of mutagen, and dosage treatment are important factors for getting a large number of mutants in the population. In contrast to reverse genetics, which starts with a protein or DNA for which there is no genetic information and then works backward to create a mutant gene, forward genetics proceeds to show the existence of the resistant gene by progeny ratio analysis before concluding with cloning and sequencing to determine its DNA and protein.

TILLING (targeting induced local lesions in genomes) is one of the non-transgenic reverse genetics tools which uses artificial mutagenesis and sequencing techniques for allele mining. In TILLING, chemical mutagen, EMS is widely used to create point mutations resulting in non-sense, missense or silent mutations. Ethyl methane sulfonate (EMS; $CH_3SO_2OC_2H_5$), a monofunctional alkylating is able to induce changes in nucleotides which results in transition or transversion, deletion, and mispairing (Kim, Schumacher, & Zhu, 2006). When multiple alleles are amplified by PCR, heated, and then gradually cooled, DNA heteroduplexes are created, which are the basis of the TILLING method. The mismatch between the two DNA strands creates a "bubble," which single stranded nucleases then cleave. The products are then arranged on several platforms according to size. When mutant and wild-type PCR products anneal together, heteroduplex molecules are created, and these molecules become the targets for a single-strand-specific nuclease present in celery juice extract (CJE). CEL I, isolated from celery, is the first known endonuclease which has the ability to recognize mismatches in the heteroduplex and also cleaves DNA on the 3' side of the mismatch (Barkley & Wang,

2008). EcoTILLING is a variant of TILLING and detect naturally occurring SNPs in genes in accessions, variety, or cultivar. It is the best method of allele mining without causing any damage to the nature and is useful for the association mapping study and linkage disequilibrium (LD) analysis. EcoTILLING is useful not only to determine the extent of variation, but also to assay the level of heterozygosity within a gene.

Seeds are mutagenized in the initial step of the TILLING and EcoTILLING processes, preferably by EMS, utilizing chemical mutagens (Figure 11.4). It is permitted for M1 plants to self in order to produce M2 individuals. DNA is extracted from either a naturally occurring population (EcoTILLING) or a mutagenized population (TILLING). Then, using gene-specific primers labeled with a dye, PCR is conducted to amplify a target region of up to 1.5 kb and detected under infrared light. This is followed by pooling (usually eightfold) and screening for mutations. DNAs are pooled for TILLING, although sample pooling is not used in a conventional EcoTILLING test. The pooled samples are then arrayed onto a 96-well configuration. Gene-specific primers with fluorescent tags are used to amplify the pooled DNA. IRDye 700, a fluorescent dye that is detectable at 700 nm, is used to label the 5' end of the forward primer, whereas IRDye 800, a fluorescent dye, is used to label the reverse primer. Samples are denatured and annealed after PCR amplification in order to create heteroduplexes between mutant and wild-type DNA strands. After that, samples are treated with Cel I, a single-strand nuclease designed specifically to digest mismatched base pairs. After the reaction has ended, the kit's buffer components are used to purify DNA, and sample volumes are decreased. Then, a piece of each sample is placed onto a denaturing polyacrylamide gel. Fluorescently marked DNA is seen with the LI-COR DNA analyzer instrument, and the LI-COR analyzer's gel data is analyzed with the use of specialist software, and the precise nucleotide alteration is then identified using conventional DNA sequencing techniques. A single LI-COR analyzer can screen 1,600 samples per day for mutations in a single 1.5 kb gene target utilizing eightfold pooling.

By using the TILLING technique, researchers are able to identify several genes of agronomic importance and also validated them functionally. By this approach, the role of *PFT1* gene in resistance against fungal pathogens in wheat was demonstrated (Fitzgerald, Kazan, Li, Morell, & Manners, 2010). Similarly, this technique was used in tomato to demonstrate role of translation initiation factors *eIF4E* and *eIF4G* which in virus resistance (Piron et al., 2010).

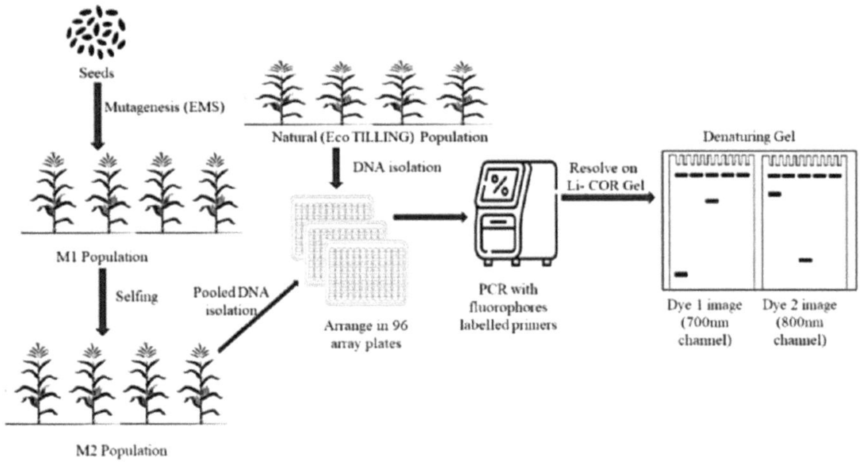

FIGURE 11.4 Basic steps of conducting a typical TILLING and EcoTILLING assays.
Source: Modified from Simsek and Kacar, 2010

However, a successful TILLING project also depends on the suitable quality DNA pooled from set of individual plants is needed. In theory, mutation discovery through TILLING is limited to the target sequence of up to 1,500 bp length while most of the eukaryotic genes are more than 2,000 bp or much higher in length. Additionally, the point mutations occurred at non-coding regions such as introns, untranslated, and promoter regions have no clear-cut effect at phenotypic level, care should take to select the coding regions of the target sequence to see the mutation effect at phenotypic level.

Loss-of-function mutations caused by transferred DNA (T-DNA) insertion or Transposon Elements Mediated insertion have advantage over traditional mutation approach of gene discovery due to easier detection of random mutagenized sequence. The T-DNA plasmid of a gram-negative bacterium, *Agrobacterium tumefaciens,* has been widely used in reverse genetic research in plant systems. The Ti plasmid of this bacterium contains genes for tumor of crown gall formation upon infection on wounded plants. The bacterium genome gets integrated into plant's genome by transferring its T-DNA located on (Tissier & Bourgeois, 2001). By using the recombinant DNA (rDNA) technology, *Agrobacterium tumefaciens* have been successfully engineered to transfer resistant genes from distantly related species to create transgenic rice plants. Through T-DNA insertion, a large mutant population can be created an to decipher genes related to disease resistance in crop plants. The mutagenized sites can be explored by specialized PCR based methods such as TAIL-PCR, plasmid rescue, inverse PCR or adapter PCR using T-DNA border and flanking

DNA specific primers. The agro-infiltration technique to test transient expression of potato leaf blight pathogen in sub-cellular location in order to decipher resistance pathway mediated by the resistance gene *RB* (Bhaskar, Venkateshwaran, Wu, Ané, & Jiang, 2009). However, inefficient transgene expression and genotype specific transformation efficiency are the disadvantage with this system. Activation T-DNA tagging is another approach in plants to overcome the limitation of functional redundancy of mutated genes by using tandem copies of cauliflower mosaic virus (CaMV) 35S enhancer sequence.

Transposons, also known as jumping genes, are one of the important reverse genetics tools which were first discovered by Barbara McClintock in the 1940's when she was studying the mosaic pigmentation patterns of maize kernels. It can cause mutation in the adjacent sequence upon mobilization and act as gene tag in plants and animals when associated with a marker gene (s). The Mutator (*Mu*) transposon has been widely used by Pioneer Hi-Bred in 1995 on a large scale to screen for *Mu* insertions in Maize (Brutnell, 2002). The transposons known as P-elements have been vastly used to examine gene function in *Drosophilia.* A series of insertional mutants in a gene of interest can be ordered and stocks can be prepared to know the function. The disrupted genes through TE insertion can easily be mapped and the affected genome can be cloned. However, this approach requires the creation of transgenic material which is often not feasible in certain species.

Another strategy that has been extensively used and researched in plant and animal reverse genetics systems is RNA interference technology (RNAi). The discovery of numerous important biological activities and roles for genes involved in resistance to significant biotic stressors has been transformed by RNAi. It is a gene silencing mechanism, which is mediated by the cell's detection and destruction of double stranded RNA, prevents a gene from functioning (dsRNA) (Moffat & Sabatini, 2006). One of the first instances of post-transcriptional gene silencing (PTGS) in plants was discovered in the petunia plant (Napoli, Lemieux, & Jorgensen, 1990), where genetic constructs were added in an effort to boost the expression of chalcone synthase (CHS), a crucial enzyme in the flavonoid biosynthesis process that results in the pigmentation of plants and flowers. Unexpectedly, the experiment caused both endogenous and exogenous CHS gene expression to be suppressed. Recently, RNAi has developed into a potent tool for silencing the expression of genes, examining their loss-of-function phenotype, and enabling research of gene function in the absence of mutant alleles. The RISC complex subsequently degrades specific target mRNAs based on sequence complementarily with the siRNA, which can then go on to recognize homologous transcripts for degradation or target genes for DNA methylation (Aklilu, 2021). RNAi-based silencing uses the abbreviated siRNAs, from double-stranded RNA (dsRNA). The cyst nematodes *Heterodera glycines* and *Globodera pallida* were the first to exhibit RNAi of plant-parasitic nematode (PPN) genes (Urwin, Lilley, & Atkinson, 2002). Then, in the egg and larval stages of several worms, gene knockdown has been accomplished using dsRNA (Iqbal, Fosu-Nyarko, & Jones, 2020). It was shown that silencing GhNDR1 and GhMKK2 genes reduced cotton's resistance to Verticillium wilt using a VIGS-mediated RNAi method (Gao et al., 2013).

Genome-editing technologies have expanded quickly and widely, which has been crucial in modifying the sequence of nucleotides. Examples include exact alterations to ORFs (open reading frames) or the highly precise introduction of a premature codon in the 5' region. Although there are many different kinds of genome-editing techniques utilized with crop plants, clustered regularly interspaced short palindromic repeats (CRISPR)-CRISPR associated (Cas) is a popular platform for genome editing. The primary benefit of CRISPR/Cas technology is the ability to specifically alter the DNA of genes in diploid or polyploid genomes. Both functioning alleles and mutant alleles can be produced as a result. There is a positive association between the level of Cas9 expression and the frequency of mutations in crop species, and CRISPR/Cas has been used to create mutations in rice and wheat. With an efficiency of 13.1%, this method was utilized to create specific mutations in the protoplast of maize. Rice's ERF transcription factor has been interfered with using CRISPR/Cas9 to improve blast resistance. Both CRISPR/Cas and TILLING are genome editing techniques that primarily alter or change particular genes. They have both proven helpful in studying point mutations to detect single locus heterosis. The main difficulty, however, lies in enhancing the influence on the phenotypic trait at desired levels and simultaneously reducing the off-target phenotypes, for which the information of genome sequence is necessary to predict their effects. Since TILLING causes mutations across the entire genome, while CRISPR/Cas employs just a few sites, these technologies can be used to introduce allelic variants into the same gene.

In Genome-Wide Association Studies (GWAS), the whole genome of individual or set of individual plants are scanned using large number of markers to find out the candidate gene or loci or even tightly linked markers based on the association between phenotypic and genotypic polymorphism present in the individual or population. In contrast to the traditional biparental mapping, which is based on linkage, GWAS is based on LD between marker and gene responsible for a particular trait. It exploits high-density marker for single-nucleotide polymorphism (SNP) polymorphisms across all chromosomes for QTL mapping and novel gene discovery in plants (Varshney, Terauchi, & McCouch, 2014). This approach overcomes several limitations of conventional gene mapping by providing higher resolution and also decipher the minor effect QTLs allowing to identify often small haplotype blocks which are significantly correlated with particular QTL/gene (Brachi, Morris, & Borevitz, 2011). Information about thousands of functional single nucleotide polymorphisms (SNPs) in crops is now available for utilization in breeding programs due to advancements in NGS technologies as well as automation in high-throughput genotyping systems. By the use of GWAS, greater proportion of phenotypic variation can be explained at molecular level which has enabled us to know the genetic architecture of traits. This has led to dissection of complex traits and identification of novel and superior alleles for disease resistance to be utilized for breeding (Sun et al., 2017).

The power of analysis in GWAS depends largely on the large and representative sample size consisting of hundreds to thousands of individuals of with the maximum variability for the traits under study in addition to a high-density saturated panel of markers (preferentially thousands of SNPs). The reliability of markers trait association results through GWAS is based mainly on three factors, viz. (i) markers with high-density single nucleotide polymorphisms (SNPs); (ii) understanding of population dynamics and the allele frequency spectrum (Q matrix); and (iii) phenotyping accuracy for every individual in the population. Manhattan plots displaying genome-wide p-values of SNP(s)-trait relationships are used to showcase the reliability of GWAS results. The association analysis is generally carried out using general linear model (GLM) or a mixed linear model (MLM) and population sampling is very crucial to cover the untapped sources of variation in the population (D'Agostino, Tripodi, & Pasquale, 2017).

A sorghum association mapping population's MAMP response to variation was assessed using GWAS, which consistently found variance for the reaction to chitin and flg22, an epitope of flagellin (Samira, Kimball, Samayoa et al., 2020). Two SNP loci were found to be connected to variations in the flg22 response, while one was connected to variations in the chitin response. Additionally, they evaluated resistance to the Bipolaris cookei-caused necrotrophic fungus's target leaf spot (TLS) disease, and they discovered one strong correlation on chromosome 5 close to a previously reported disease resistance gene. GWAS has been applied to wheat for a variety of characteristics, including disease resistance (Ledesma-Ramírez et al., 2019). In recent years, GWAS techniques for mine rice blast resistance have been used. Six of the 21 blast resistance relationships were found to be co-localized with cloned blast R genes in a GWAS analysis using 151 accessions and a poor coverage of SSR markers (Wang, Jia, Ghai, Lee, & Jia, 2015). To better understand the complexity of the resistance mechanisms and links between the potential genes providing resistance, additional fine mapping investigations using larger biparental populations and/or allele mining based on the sequencing of accessions in a diversity panel would be necessary.

11.7 CONCLUSION

Breeding for a resistant and productive variety is an effective, economical, environmentally friendly, and sustainable approach to controlling the yield losses due to various diseases in crops. Allele mining for disease resistant genes is an important component to develop varieties with durable resistance through the marker-assisted introgression of new genes into a variety. For mining new genes, various forward and reverse breeding tools are available that have facilitated the discovery of many genes and QTL of interest, and identification of their sources with ease. The new genes identified through allele mining can be effectively employed in gene pyramiding, genome editing, and other novel molecular breeding approaches which present possibilities to attain durable resistance against the pathogens. NGS has enabled the reduction of the cost and time required for sequencing the whole genome and now sequence data is available for the whole genome in most of the cultivated species of crops. With the vast and immense potential application of allele mining for disease resistance genes in plants, efforts are underway to develop a dedicated effort for allele mining by various research institutes.

KEYWORDS

- **allele mining**
- **disease resistance**
- **extracellular environment**
- **gene identification**
- **next generation sequencing**
- **nucleotide-binding leucine-rich repeat**
- **reverse genetics**
- **TILLING**

REFERENCES

Acevedo-Garcia, J., Kusch, S., & Panstruga, R., (2014). Magical mystery tour: MLO proteins in plant immunity and beyond. *New Phytol.*, *204*, 273–281. doi: 10.1111/nph.12889.

Aklilu, E., (2021). Review on forward and reverse genetics in plant breeding. *All Life*, *14*, 127–135.

Ashkani, S., Rafii, M. Y., Shabanimofrad, M., Miah, G., Sahebi, M., Azizi, P., Tanweer, F. A., et al., (2015). Molecular breeding strategy and challenges towards improvement of blast disease resistance in rice crop. *Front. Plant Sci.*, *6*, 886. doi: 10.3389/fpls.2015.00886.

Barkley, N. A., & Wang, M. L., (2008). Application of TILLING and EcoTILLING as reverse genetic approaches to elucidate the function of genes in plants and animals. *Current Genomics*, *9*, 212–226. https://doi.org/10.2174/138920208784533656.

Bayless, A. M., & Nishimura, M. T., (2020). Enzymatic functions for toll/interleukin-1 receptor domain proteins in the plant immune system. *Front. Genet.*, *11*, 539. doi: 10.3389/fgene.2020.00539.

Bhaskar, P. B., Venkateshwaran, M., Wu, L., Ané, J. M., & Jiang, J., (2009). *Agrobacterium*-mediated transient gene expression and silencing: A rapid tool for functional gene assay in potato. *Plos One*, *4*, 6, e5812. https://doi.org/10.1371/journal.pone.0005812.

Brachi, B., Morris, G. P., & Borevitz, J. O., (2011). Genome-wide association studies in plants: The missing heritability is in the field. *Genome Biol.*, *12*, 232. https://doi.org/10.1186/gb-2011-12-10-232.

Brutnell, T. P., (2002). Transposon tagging in maize. *Funct. Integr. Genomics*, *2*, 4–12.

D'Agostino, N., & Tripodi, P., (2017). NGS-based genotyping, high-throughput phenotyping, and genome-wide Association Studies Laid the Foundations for next-generation breeding in horticultural crops. *Diversity*, *9*, 38. doi: 10.3390/d9030038.

Dong, X., (2001). Genetic dissection of systemic acquired resistance. *Curr. Opin. Plant Biol.*, *4*, 309–314.

Fitzgerald, T. L., Kazan, K., Li, Z., Morell, M. K., & Manners, J. M., (2010). A high-throughput method for the detection of homologous gene deletions in hexaploid wheat. *BMC Plant Biol.*, *10*, 264. doi: 10.1186/1471-2229-10-264.

Flor, H. H., (1942). Inheritance of pathogenicity in *Melampsora lini*. *Phytopathology*, *32*, 653–669.

Gao, X., Li, F., Li, M., Kianinejad, A. S., Dever, J. K., Wheeler, T. A., et al., (2013). Cotton *GhBAK1* mediates *Verticillium* wilt resistance and cell death. *J. Integr. Plant Biol.*, *55*, 586–596. doi: 10.1111/jipb.12064.

Hammond-Kosack, K. E., & Jones, J. D. G., (1996). Inducible plant defense mechanisms and resistance gene function. *Plant Cell*, *8*, 1773–1791.

Iqbal, S., Fosu-Nyarko, J., & Jones, M. G. K., (2020). Attempt to silence genes of the RNAi pathways of the root-knot nematode, *Meloidogyne incognita* results in diverse responses including increase and no change in expression of some genes. *Front. Plant Sci.*, *11*, 328. doi: 10.3389/fpls.2020.00328.

Jagodzik, P., Tajdel-Zielinska, M., Ciesla, A., Marczak, M., & Ludwikow, A., (2018). Mitogen-activated protein kinase cascades in plant hormone signaling. *Front. Plant Sci.*, *9*, 1387.

Kim, Y., Schumacher, K. S., & Zhu, J. K., (2006). EMS mutagenesis of *Arabidopsis*. In: Salinas, J., & Sanchez-Serrano, J. J., (eds.), *Methods in Molecular Biology, Arabidopsis Protocols* (2nd edn., Vol. 323). Humana Press Inc., Totowa, NJ.

Kutschera, U., & Hossfeld, U., (2012). Physiological phytopathology: Origin and evolution of a scientific discipline. *J. Appl. Bot. Food Qual.*, *85*, 1–5.

LeBlanc, M. A., Fink, M. R., Perkins, T. T., & Sousa, M. C., (2021). Type III secretion system effector proteins are mechanically labile. *PNAS*, *118*(12), e2019566118. https://doi.org/10.1073/pnas.2019566118.

Ledesma-Ramírez, L., Solís-Moya, E., Iturriaga, G., Sehgal, D., Reyes-Valdes, M. H., Montero-Tavera, V., Sansaloni, C. P., et al., (2019). GWAS to identify genetic loci for resistance to yellow rust in wheat pre-breeding lines derived from diverse exotic crosses. *Front. Plant Sci.*, *10*, 1390. doi: 10.3389/fpls.2019.01390.

Lv, Q., et al., (2013). Functional analysis of *Pid3-A4*, an ortholog of rice blast resistance gene *Pid3* revealed by allele mining in common wild rice. *Phytopathology*, *103*, 594–599.

McHale, L., Tan, X., Koehl, P., et al., (2006). Plant NBS-LRR proteins: Adaptable guards. *Genome Biol.*, *7*, 212, https://doi.org/10.1186/gb-2006-7-4-212.

Moffat, J., & Sabatini, D. M., (2006). Building mammalian signaling pathways with RNAi screens. *Nat. Rev. Mol. Cell Biol.*, *7*, 177–187.

Napoli, C., Lemieux, C., & Jorgensen, R., (1990). Introduction of a chimeric chalcone synthase gene into petunia results in reversible co-suppression of homologous genes in trans. *Plant Cell*, *2*, 279–289.

Newman, M. A., Sundelin, T., Nielsen, J. T., & Erbs, G., (2013). MAMP (microbe-associated molecular pattern) triggered immunity in plants. *Front. Plant Sci.*, *4*, 139. doi: 10.3389/fpls.2013.00139.

Niks, R. E., Qi, X. Q., & Marcel, T. C., (2015). Quantitative resistance to biotrophic filamentous plant pathogens: Concepts, misconceptions, and mechanisms. In: Van, A. N. K., (ed.), *Annual Review of Phytopathology* (pp. 445–470). Palo Alto, CA: Annual Reviews.

Palloix, A., Ayme, V., & Moury, B., (2009). Durability of plant major resistance genes to pathogens depends on the genetic background, experimental evidence, and consequences for breeding strategies. *New Phytol.*, *183*, 190–199. doi: 10.1111/j.1469-8137.2009.02827.x.

Parlevliet, E. J., (2002). Durability of resistance against fungal, bacterial, and viral pathogens; present situation. *Euphytica*, *124*, 147–156. doi: 10.1023/a:1015601731446.

Piron, F., Nicolaï, M., Minoïa, S., Piednoir, E., Moretti, A., Salgues, A., Zamir, D., Caranta, C., & Bendahmane, A., (2010). An induced mutation in tomato *eIF4E* leads to immunity to two potyviruses. *PLoS One*, *5*, 6, e11313. doi: 10.1371/journal.pone.0011313.

Qu, S., et al., (2006). The broad-spectrum blast resistance gene *Pi9* encodes a nucleotide-binding site-leucine-rich repeat protein and is a member of a multigene family in rice. *Genetics*, *172*, 1901–1914.

Ronald, P., Albano, B., & Tabien, R., (1992). Genetic and physical analysis of the rice bacterial blight disease resistance locus, *Xa21*. *Mol. Gen. Genet.*, *236*, 113–120.

Samira, R., Kimball, J. A., Samayoa, L. F., et al., (2020). Genome-wide association analysis of the strength of the MAMP-elicited defense response and resistance to target leaf spot in sorghum. *Sci. Rep*, *10*, 20817. https://doi.org/10.1038/s41598-020-77684-w.

Sarika, G. U., & Amruta, N., (2014). Allele mining and its importance in seed quality. *International Journal and Developmental Research*, *4*, 819–824.

Stall, R. E., Jones, J. B., & Minsavage, G. V., (2009). Durability of resistance in tomato and pepper to *Xanthomonas* causing bacterial spot. *Annu. Rev. Phytopathol.*, *47*, 265–284. doi: 10.1146/annurev-phyto-080508-081752.

Sun, C., Zhang, F., Yan, X., Zhang, X., Dong, Z., Cui, D., et al., (2017). Genome-wide association study for 13 agronomic traits reveals distribution of superior alleles in bread wheat from the yellow and Huai Valley of China. *Plant Biotechnol. J., 15*, 953–969. doi: 10.1111/pbi.12690.

Tissier, A., & Bourgeois, P., (2001). Reverse genetics in plants. *Curr. Genomics*, *2*, 269–284.

Toruño, T. Y., Stergiopoulos, I., & Coaker, G., (2016). Plant-pathogen effectors: Cellular probes interfering with plant defenses in spatial and temporal manners. *Annual Review of Phytopathology*, *54*, 419–441.

Urwin, P. E., Lilley, C. J., & Atkinson, H. J., (2002). Ingestion of double-stranded RNA by preparasitic juvenile cyst nematodes leads to RNA interference. *Mol. Plant-Microbe Int.*, *15*, 747–752. doi: 10.1094/MPMI.2002.15.8.747.

Varshney, R. K., Terauchi, R., & McCouch, S. R., (2014). Harvesting the promising fruits of genomics: Applying genome sequencing technologies to crop breeding. *PLoS Biol.*, *12*, e1001883.

Wang, X., Jia, M. H., Ghai, P., Lee, F. N., & Jia, Y., (2015). Genome wide association of rice blast disease resistance and yield-related components of rice. *Mol. Plant Microbe Interact*, *28*, 1383–1392. https://doi.org/10.1094/MPMI-06-15-0131-R.

Wu, Y., et al., (2013). Fine mapping and identification of blast resistance gene *Pi-hk1* in a broad-spectrum resistant japonica rice landrace. *Phytopathology*, *103*, 1162–1168.

CHAPTER 12

Molecular Mapping of Resistant Genes

SUROJIT SEN,[1] SUNAYANA RATHI,[2] SANGITA SAHNI,[3] and
BISHUN DEO PRASAD[4]

[1]*Department of Zoology, Mariani College, Mariani, Jorhat, Assam, India*

[2]*Department of Biochemistry and Agricultural Chemistry, Assam Agricultural University, Jorhat, Assam, India*

[3]*Department of Plant Pathology, TCA, Dholi, Samastipur, Bihar, India*

[4]*Department of AB&MB, College of Basic Sciences and Humanities, Dr. Rajendra Prasad Central Agricultural University, Pusa, Samastipur, Bihar, India*

ABSTRACT

Knowing the location or position of the gene is the most important step in isolation and cloning of resistance genes. Since the classical approach of mapping genes by linkage mapping, many milestones have been crossed and this has been possible only after the discovery of molecular markers and more specifically DNA markers. These molecular markers are now used for the construction of a molecular map so that the location of the resistant genes is identified and can be used in upcoming breeding endeavors. The need for new resistant genotypes for improvement in the yield of agricultural crops has been addressed by breeders so far by the classical approach, but further success in developing resistant varieties needs fast breeding techniques with the implementation of modern biotechnological tools. It is now possible to identify, tag, map, and clone the resistance genes for the creation of better varieties that will address future food security concerns. This is made possible by new molecular technology and created molecular tools. This chapter deals with some of the procedures involved in molecular mapping of resistance genes.

New Advances in Crop Disease Management. Sangita Sahni, Ganesh Thapa, and Bishun Deo Prasad (Eds)

12.1 INTRODUCTION

Molecular mapping involves identification of allelic variations linked to a particular trait, tagging, and tracking them through molecular markers and then mapping them to a particular locus in the chromosome helps in further studies of inheritance and breeding activities. To cope with the tremendous biotic and abiotic stresses plants have naturally evolved resistance (R) genes in them. Natural selection has played a significant role for selection of these resistant genes' generation after generations since presence of these naturally occurring resistant genes provided added advantage for survival. These dominant and quantitative sub-dominant R genes occurring naturally can be exploited to enhance plants' immune response against pathogens and has been practiced in traditional agricultural breeding practices to pile up these resistant genes in a single breed so that durable resistance is developed, and yield is increased.

Knowledge about the positions of the genes becomes a very important step in identification, isolation, selection, cloning or pyramiding these R genes. This became feasible only after the advancement of modern molecular biology techniques and discovery of molecular markers. Since the isolation of first molecular markers, many milestones have been passed from the isolation and development of genetic linkage maps to physical map and finally the sequencing of the genome. The classical or morphological markers, which are just the phenotypic trait itself, are the earliest of the numerous types of markers that are currently being employed. Biochemical markers are the second type of markers which are allelic variants of protein or enzymes (isoenzymes) detected electrophoretically due to differences in their net electrical charges. Polymorphism shown by isoenzymes can be easily seen as distinct bands and their use has led to development of the principles applicable to molecular markers. However, due to limited number of isoenzyme loci in most of the species, need for the development of DNA based molecular markers arose. Attempts were made to find polymorphism in nucleic acid (NA) and protein associated with the trait and finally molecular markers based on DNA were discovered. The development of DNA sequence-based markers started after Sanger invented the sequencing technique. Consequently, many dominant as well as co-dominant markers were developed. All these markers are located at a particular location in the chromosome (loci). They can be categorized as random, gene-targeted, or functional markers depending on the location (Andersen & Lübberstedt, 2003). Most of the markers belong to the category of random markers and they have no effect on target trait expression and mostly they are located in the intergenic region. Gene targeted markers are developed from information on gene sequence. They are markers identified within the gene or located in close proximity to the gene. Functional markers are those which cause phenotypic variations. Based on the genotyping tools used to find them, molecular markers can be categorized. The first group consists of markers that need to be genotyped using PCR (SSR, AFLP, RAPD), whereas the second group consists of markers based on hybridization (such as arrays) or sequencing (e.g., genotyping by sequencing, mini sequencing). The majority of the SNPs in the latter group. The most widespread application of DNA markers is for creating genetic (or linkage) maps, which is in addition to diversity and pedigree research.

The genomics, genetics, and evolutionary background of the genes influencing a trait must be thoroughly understood in order to use genetic resources for breeding. In order to generate better varieties to address future food security concerns, resistance genes are being identified, tagged, mapped, and cloned using new molecular technologies and molecular tools. The need for new genotypes is enormous today for improvement in agricultural crop yield. Breeders are continuously working for the development of resistant cultivars and most of the achievement has come from classical approach, further success in developing resistant varieties needs implementation of modern biotechnological interventions. In this chapter, we made an effort to describe the genetic advancement and molecular mapping investigations related to the emergence of resistant plant kinds.

A genetic map is the illustration of various genetic markers as located in a specific order in the chromosome as well as the relative distance between them in cM (centimorgan). Genetic markers are used for map construction which can further be used for marker-assisted selection.

12.2 STEPS INVOLVED IN MOLECULAR MAP CONSTRUCTION

In plants, molecular maps are prepared on the basis of basic genetic principles by use of some steps which are discussed in subsections.

12.2.1 MAPPING POPULATIONS DEVELOPMENT

First and foremost, requirement for construction of molecular map is a mapping population. Mapping population is generated from a population of different genotypes having contrasted genetic diversity for the traits of interest. The genotypes should have sufficient polymorphism so that they can be distinguished and should not be distantly related. In case of species which are self-pollinating, homozygous inbred lines such as F_2, backcross (BC), recombinant inbred lines (RILs), double haploid (DH), and near-isogenic lines (NILs), and are used as mapping population (Figure 12.1). Selfing of F_1 hybrids give rise to F_2 population while cross between F_1 and any one of the recurrent parents produces the BC population. Production of RILs is time consuming and it needs more than six to eight generations of single-seed selections of individual specific plants of F_2 populations. Repeated back cross selection for more than six generations, recovers the genome of recurrent parent. Then the selected individual from BC_7F_1 upon selfing, produces BC_7F_2 lines. These BC_7F_2 have homozygous alleles of the target gene and are known as nearly isogenic lines (NILs). In self-pollinating species, DH population is usually produced by doubling the gametes of either F_1 or F_2 population. Generally, pollen grains or haploid embryos are regenerated by tissue culture technique after the chromosome numbers are doubled. In cross pollinated species-controlled crosses are carried out so that half or full sib progenies are derived for DH production.

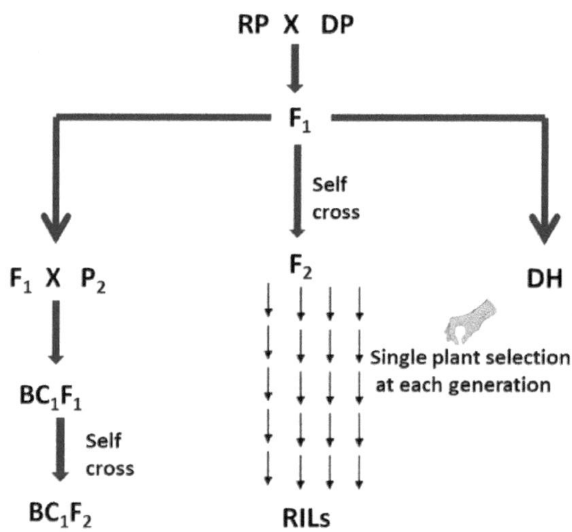

FIGURE 12.1 Development of mapping populations [RP: recipient parent; and DP: donor parent].

Selection of type of mapping population is a critical step for linkage mapping success. Depending upon the objectives of the breeder and the nature of the molecular marker used mapping populations are constructed. Although F_2 and BC populations can be generated within a very short period of time, but they sometimes provide false linkage as they are derived simply after only one or two rounds of recombination events. RILs and NILs populations on the other hand, are derived from several rounds of recombination events so the chance to get the spurious linkage is less hence they are more preferred for genetic mapping. All these mapping populations like RIL, NILs, and DH are immortal, with high reproducibility rate. Although they require several years to develop, but once developed they can be used successfully.

12.2.2 POPULATION SIZE FOR MAPPING

A sufficient number of individuals should be selected in the mapping population to increase the accuracy of the genetic map. It has been found that the larger the population size, the more possibility there is of getting recombinants and the more accurate the genetic map becomes. The mapping population size (F2, BC or DH) usually ranges from 50 to 1,000 individuals, and the resolution of the map depends on the population size. Normally, genetic maps are constructed using a population size of 50 to 100 individuals in crop plants. A population of 200 individuals is considered appropriate for any type of mapping population because it has been discovered that as the mapping population size increases, so does the accuracy of the map. In cases where the molecular markers are tightly linked, a population size of more than 1,000 is considered for screening.

12.2.3 SELECTION OF MOLECULAR MARKERS SYSTEMS

Among the various marker systems available the most commonly used markers for molecular mapping are SNPs (single nucleotide polymorphisms), RFLPs (restriction fragment length polymorphism), SSRs (simple sequence repeats), AFLPs (amplified fragment length polymorphisms), RAPDs (random amplified polymorphic DNAs), and DArT (diversity arrays technology).

All these markers available are classified into different groups based on:

1. **Transmission Mode:** Nuclear inheritance or organellar inheritance of both the parents or either of male or female parent.
2. **Inheritance Mode:**
   i. **Dominant:** ISSR, AFLP, RAPD.
   ii. **Co-Dominant:** SNP, STR, SSR, RFLP.
3. **Analysis Mode:**
   i. **Hybridization-based:** RFLP.
   ii. **PCR-based:**

a. Without prior sequence information (e.g., ISSR, RAPD, AP-PCR, DAF, AFLP).
b. Site targeted, from known sequences (e.g., SSR, EST, CAPS, SCAR, STS).

4. **DArT (Diversity Arrays Technology):** Microarray-based: First molecular marker used for molecular mapping was RFLP and after its successful application, it has been extensively used for construction of molecular maps in many plant species. Although, RFLP markers being co-dominant in nature with good transferability rate and locus-specificity, the major limitations of RFLP markers are high-cost involvement and low polymorphism. The process's high time requirement prevents automation, which is another downside. As opposed to other marker systems that do not entail radioactivity, radioactive probes were initially employed to identify RFLP markers. However, both radioactive and non-radioactive probes are now available for RFLP markers. For high density molecular map construction AFLP, PCR-based markers, is popularly used for genetic mapping. These markers are polymorphic, easily reproducible, can be used without sequence information and primer development. RAPD markers on the other hand, are PCR dependent and used for constructing linkage maps without the prior genomic sequence information using short primers (10 bp). However, they are not widely accepted because of poor reproducibility. Co-dominant markers are more accurate in map construction than the dominant ones. Hence, among the PCR-based molecular markers, SSR markers which are co-dominant and highly reproducible gradually became the marker of choice among the breeders for map construction. The main advantage of SSR markers (microsatellites) is that they are highly polymorphic, locus specific, and easy to use.

SNPs (Single Nucleotide Polymorphism), efficient genetic markers, can be defined as difference in single base in the DNA of two individuals. Because there are four nucleotides in a single DNA molecule, each SNP has the potential to have four alleles, but the majority are bi-allelic, meaning they only come in two forms. Any two individuals with a single base difference at a particular locus can have either of the two bases C/G or A/T. The benefits of SNPs include their great abundance, the fact that they may be typed without the need of gel electrophoresis, and the ease with which they can be automated. SNPs might be dominant or co-dominant depending on the type of methodology utilized.

Diversity arrays technology (DArT) markers are sequence markers used for genotyping, which is based on microarray hybridization. This marker system is used in genetic mapping as well as fingerprinting. The significant advantage of DArT is its ability to investigate several marker loci at a time and that is why high throughput screening can be done with it.

Properties that are desired to be present ideally in molecular markers are high rate of polymorphism, co-dominant inheritance, high frequency of occurrence throughout the genome, high reproducibility and transferability and the assay technique should be of low cost and fast. But none of the markers that are currently on the market meet all the requirements. All markers have their own advantages and disadvantages. As a result, marker systems are chosen in accordance with the study's goals, nature, and the genetic variety of the species under examination. To choose the best marker system, it is necessary to critically evaluate each one while taking into account the cost and amount of analysis time needed.

12.2.4 MAPPING POPULATION GENOTYPING

Parental genotypes in the mapping population are then screened with a selected molecular marker system so that sufficient polymorphic markers are identified for molecular map construction. Then genotyping of all individuals of mapping population is carried out with the help of these polymorphic markers. The construction of the molecular maps depends upon genotyping data of each individual as scored on the basis of the type of markers and mapping populations used. In case of F_2 population as heterozygotes is present, genotyping data is scored as 1 for parent with one allele, 2 for F_1 with both allele and 0 for parent with two allele. In marker system having a high level of polymorphism, the genotyping data and methods adopted for scoring, determines the quality of the genetic map. Hence, it is very essential to check all the possible errors, including typographic, wrong scoring and error due to missing data.

12.2.5 LINKAGE ANALYSIS AND MAP CONSTRUCTION

The software packages in Table 12.1 are used to prepare the molecular map between the makers on the basis of linkage from the genotyping data recorded in a mapping population.

TABLE 12.1 List of Molecular Mapping Software Programs

Name	URL	Description
GENEHUNTER	ftp://ftp.broadinstitute.org/distribution/software/genehunter/gh2.1/.	Linkage and QTL analysis with models and without models. Able to manage several markers.
GMENDEL	http://www.genetics.ucla.edu/software/mendel5	Calculates G statistics, using the maximum likelihood method, determines recombination frequencies.
LINKAGE	https://lab.rockefeller.edu/ott/programs	Multipoint linkage analysis using a model. Capable of handling a small number of markers.
MAPMAKER/EXP	https://www.softpedia.com/get/Science-CAD/MapMaker.shtml ftp://ftp.broadinstitute.org/distribution/software/mapmaker3/	Carries out multipoint linkage analysis for co-dominant, dominant, and recessive markers.
MAP manager QTX	http://mapmgr.roswellpark.org/mmQTX.html.	Mendelian and quantitative trait loci are mapped.
JoinMap	https://www.kyazma.nl/index.php/JoinMap/	Analyzes all different kinds of mapping populations and combines the maps of those populations with those that share common markers.
LINKAGE. FASTLINK	https://www.ncbi.nlm.nih.gov/CBBresearch/Schaffer/fastlink.html	A modified and enhanced version of LINKAGE's primary programs that runs sequentially considerably faster and is used to map genes and determine their general location.
MERLIN	http://csg.sph.umich.edu/abecasis/Merlin/	Linkage/QTL study with models and without models.
SOLAR	http://solar-eclipse-genetics.org/ https://www.nitrc.org/frs/?group_id=558	Software for quantitative genetic analysis, SNP association analysis, and genetic variance and linkage analysis.
QTL IciMapping	https://www.isbreeding.net/	Integrated software for quantitative trait locus mapping and building genetic linkage maps in biparental populations.
R/ASMap	https://cran.r-project.org/web/packages/ASMap/index.html	Using the MSTmap algorithm, linkage maps of doubled haploid, backcross, and recombinant inbred plants are constructed, modified, and diagnosed.
MapDisto	http://mapdisto.free.fr/	Mapping genetic markers in populations that exhibit experimental segregation, such as F2, backcross, single-seed descent, doubled haploids, and highly recombinant lines.

Two steps are crucial for creating a genetic map based on the segregation of DNA markers. Based on the frequencies of two-point recombination, markers must first be organized into linkage groups. The next step is to rank the markers inside each linkage group and estimate the distances between them using multipoint recombination rates. The basic principles adopted to construct the molecular map as described in subsections are almost similar and are used in these programs as listed in Table 12.1.

12.2.5.1 SEGREGATION DISTORTION TEST

Molecular markers normally segregate following the Mendel's law in 3:1 (dominant); 1:2:1 (co-dominant), in case of F_2 or 1:1 in case of DH, RIL or BC population. But sometimes deviations from these ratios are observed. This deviation from the expected ration is termed as segregation distortion which may be due to statistical bias or scoring errors during genotyping or there may be biological reasons also. Segregation distortion is an error which can influence the mapping process and hence they have to be taken care of separately. There are various methods by which recombination frequencies are calculated, tested separately for genes/markers showing segregation distortion. Independent χ^2 test and LOD (logarithm of odds) values are used to estimate the effects of segregation distortion on linkage. Significantly distorted segregation loci may cause the length of the map to be cut down. Hence, during calculation, these loci showing distortion should be excluded. However, it has been found that dominant markers are more affected by segregation distortion than co-dominant markers. Software like 'Mapdisto' can take care of segregation data with distorted markers and correct the results of linkage groups.

12.2.5.2 LINKAGE GROUPS ESTABLISHMENT

Molecular markers are grouped into linkage groups based on two-point recombination rates estimating statistically by odds ratio. This ratio is termed as the LOD which is the ratio of the probability of two loci that are linked over those not linked with a given value of recombination. A LOD score of 3 between two markers means odds are 1,000 to 1, i.e., linkage is 1,000 times more likely than no linkage. In order to establish these linkage groups researcher specified critical LOD score (linklod), and marker pairs with are combination LOD score above this critical LOD are considered to be linked.

Usually a 'linklod' value of three is used as the minimum threshold value to determine linkage between markers. Fragmented linkage groups will be produced by higher LOD values. The significance level of LOD score is assigned depending upon the sample size of the linkage groups.

12.2.5.3 DETERMINATION OF MAP DISTANCE AND LOCUS ORDER

The researcher must specify parameters such the minimum "maplod," the recombination threshold value, the jump threshold value, and the mapping function in order to determine the map distance. A LOD score above the 'maplod' is used for calculating map distance. Value of 'maplod' can range from 0.1 to 3.0 and lower the 'maplod' value weaker is the linkage information. Hence, for fine mapping more stringent approach is applied.

A genetic mapping function describes the relationship between map distance and recombination frequency. The number of recombination events determines the map distance between two loci. The map distance equals the recombination frequency when map distance is less than 10 cM, but this is not true when map distance is more than 10 cM. In these circumstances, the mapping functions (mf) that are most frequently used to translate recombination frequency (r) to map distance (cM) are:

i. Haldane mapping function = $(-\ln(1 - 2r) \times 50$ (assumes absence of interference)
ii. Kosambi mapping function= $(1/4\ln[(1 + 2r)/(1 - 2r)])$ (assumes positive interference)

Interference is the effect of crossover occurring in a certain region which reduces the probability of occurrence of crossover in adjacent region. It is assessed as a coefficient of coincidence and is the ratio of observed double crossovers to expected crossovers that are predicted by a map (CC). The maps are essentially constructed in two steps: first, markers are assigned to various linkage groups, and then, second, they are arranged in order of positions within the group.

During map building process most, informative loci are added first and consequently others are added up one by one on the basis of linkage information with the earlier markers in the map. The procedure becomes easier if the positions of few markers are known beforehand. After including a marker on the map, the goodness of fit for the position is calculated and it is retained or discarded depending upon the measure of goodness of fit. This process continues for several rounds until mapping of all the markers are complete.

Determination of marker order within linkage groups is very similar to the classical traveling salesman problem. Essentially, the problem is to choose the best order among m!/2 possible orders, where m = number of markers. With an increase in the number of markers, the number of orders becomes so large that it is not feasible to determine with the current computing power available. Therefore, approximate solutions are adopted for large-scale genetic mapping. Locus ordering is done traditionally from testcross data by two- or three-point way. In situations where multiple crossovers occur at a much higher frequency than random (localized negative interference), gene order for sites which are closely linked is determined using three-factor reciprocal crosses. Various software uses any of the statistical algorithms like MLE (maximum likelihood estimator), WLSE (weighted least squares estimate), UG (unidirectional growth), RECORD (recombination counting and ordering) as well as SARF (sum of adjacent recombination fractions), PARF (product of adjacent recombination fractions), LHMC (likelihood through hidden Markov chains), and SALOD (sum of adjacent LOD scores) are used along with the RIPPLE algorithm for verification of the error.

12.3 QTL MAPPING

Quantitative trait loci (QTL) are genomic regions associated with a quantitative trait of choice. The variance in quantitative features, which are continuous characters, is caused by a number of loci, each of which has a small but varied effect.

Using the following criteria, quantitative genetics tries to define quantitative traits:

- How many loci are responsible for a character?
- The gene action on average.
- The degree of interaction between polygenes.
- The degree of interaction with the environment.

These characteristics are frequently influenced by multiple genes, as well as by the environment. Thus, mapping QTL is not as simple as mapping a single gene. Analysis of such genomic regions affecting phenotypic expression can be done with or without molecular maps. Basic steps in QTL mapping involves phenotypic evaluation of a relatively high number of sample plants from a segregating population for polymorphic genetic markers followed by genotyping the population and then statistical analyzes to identify the loci that are affecting the trait(s) of interest (Figure 12.2). Such studies are done to detect association between genetic marker and a phenotype.

QTL mapping requirements:

- Traits which can be measured.
- Molecular markers RFLP, AFLP, CAPS, SSR, etc.
- Segregating population F_2, F_3, RILs, DHL, NILs, etc.
- QTL mapping software: QTL Cartographer, MQTL, MapMaker/QTL, etc.

The fundamental approach used in all marker-based QTL mapping investigations is the same.

- Choose parents who are different from one another in a particular trait.
- Check for polymorphic marker loci in the two parents.
- Create inbred recombinant lines (F_2-derived lines can be used).
- Phenotype is screened in field.
- Compare the MM and mm lines' means at each marker locus.
- Declare the QTL with the highest MM-mm value.

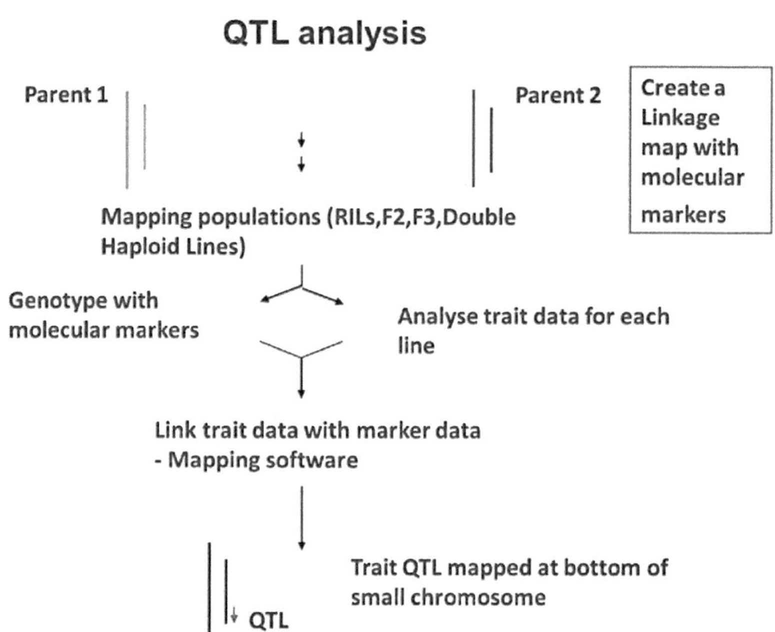

FIGURE 12.2 QTL mapping approach.

12.3.1 QTL ANALYSIS WITHOUT USING MOLECULAR MAP (SINGLE MARKER ANALYSIS)

Marker's association with the trait of interest can also be determined without map construction by statistical methods like AMOVA, t-test, single locus analysis, and regression analysis of the F_2, BC, RILs, and DHLs. Here phenotype of progeny of each marker class is compared and mean difference is calculated. Then from the difference in mean phenotypic effect of substituting an allele at the QTL is estimated. Simple statistical tests like t-test or F-test are used to test the significance of phenotypic effect. The close proximity of the QTL to the marker is shown by a significant value. Single marker analysis is used when orders of markers are not known, or the mapping is incomplete to find the best possible QTLs. However, the major limitations of this analysis are that it results in a number of false positives, underestimates the number and effects of the QTL as well as the positions of the QTL cannot be determined with precision.

12.3.2 QTL INTERVAL MAPPING BY USING MOLECULAR MAPS

A marker interval is the segment in between two markers. In order to estimate the site of the QTL, interval mapping techniques require data on the values of two flanking markers. Between two flanking markers, the likelihood that the data will be obtained at various QTL loci is computed, and the locus where the likelihood of acquiring the observed data is highest is designated as a QTL. Interval mapping can be systematically used for scanning the whole genome to mark evidence of QTL. Like linkage mapping, a map function (Haldane or Kosambi) is used to convert recombination frequency to distance or vice versa. The LOD score is calculated at each increment (walking step) in the interval until the LOD score of the whole genome is profiled and locations wherever the peak exceeds the threshold value are declared to be QTL.

Interval mapping may be simple, composite, simplified composite or multiple type depending on the marker intervals used to detect the QTL.

12.3.2.1 SIMPLE INTERVAL MAPPING (SIM)

In SIM, to detect the presence of QTL, screening of individual interval is done at a time. Hence it becomes difficult to resolve two QTLs in the given marker interval. Further, the cumulative effect of additional QTLs present can contribute to sampling variance and if there is linkage between two QTL_S it may lead to biases.

12.3.2.2 COMPOSITE INTERVAL MAPPING (CIM)

Here variance from other QTL as well as biasness due to QTL linked to a particular position being tested is taken into account. Genetic markers are incorporated as co-factors which partially absorb genetic variance associated with unlinked and linked markers and greatly increase the power for detecting QTL. One of the major drawbacks of CIM is that it is highly dependent on background markers.

12.3.2.3 SIMPLIFIED COMPOSITE INTERVAL MAPPING (CIM)

This approach is much faster than CIM and can address larger issue. It uses interval mapping by regression and constant coefficient for background markers. Simplified composite interval mapping (CIM) can be done by software, like MQTL.

12.3.2.4 MULTIPLE INTERVAL MAPPING (MIM)

Specialty of MIM is that it can use multiple marker intervals simultaneously and can be accessed by the model to find potential QTLs for mapping. Estimation of parameters is based on maximum likelihood and can identify a number of QTLs, their interaction and contribution to genetic variance.

12.3.2.5 INTERVAL MAPPING APPROACHES

Interval mapping is done by following either of the two approaches:

1. **Maximum Likelihood Approach:** In order to systematically scan the entire genome for QTL, Lander & Botstein (1989) pioneered the interval mapping method via maximum likelihood approach. By maximizing a likelihood function, it is possible to statistically estimate the QTL parameters of a probability distribution, making the observed data the most likely under the presumptive statistical model. While LOD scores are used to forecast the map position of QTL, models fitted with or without the QTL impact can be used to assess the likelihood of the observed distributions. LOD score is the

ratio of the likelihood that the effect occurs by linkage to the likelihood that it occurs by chance. The maximum likelihood approach of searching is one-dimensional, considering a single QTL at a time and interactions between multiple QTLs are not considered. The quantities of mixture components and likelihood factors substantially increase as the number of QTLs taken into account in the model rises. Hence maximization of likelihood becomes difficult to obtain by the algorithm.

2. **Interval Mapping by Regression:** This method is very similar but simplified form of maximum likelihood method. QTL mapping positions are estimated similarly, except in the positions of gaps or missing data. Unlike the basic method of regression on coded marker genotypes in QTL analysis, here regression of the phenotypes is done on QTL genotypes.

The difference between maximum likelihood method and regression becomes significant when QTLs are closely linked or there is interaction between QTLs. However, maximum likelihood interval mapping seems to be more effective and predicts more precisely with a lower error rate than REG interval mapping, while being a computationally expensive method (Table 12.2).

TABLE 12.2 QTL Detection Methods

Method	Program	Significance	References
Interval mapping	MAPMAKER/ QTL	Places QTL in intervals of markers.	Lander & Botstein (1989)
Composite interval mapping	CIM in QTL cartographer.	Use nearby markers and/or additional QTL as cofactors.	Zeng (1994); Basten et al. (2001)
Multiple interval mapping	MIM in QTL cartographer.	Uses a single chromosomal test to simultaneously look for several QTL.	Kao et al. (1999)
Bayesian interval mapping	BMAPQTL, also BIM in QTL cartographer.	One can estimate QTL effect and position separately.	Satagopan et al. (1996)
Outbred QTL	QTL express	Least squares regression in several sibling and pedigree designs involve outbred crossover.	Seaton et al. (2002)

12.3.3 PHYSICAL MAPS AND ITS RELATION TO GENETIC MAPS

Physical maps are actual positions of the genes on the DNA molecules and genetic maps sometimes show discrepancies with physical maps. The number of crossovers that are scored influences the resolution of a genetic map and there are regions in the chromosome which have more chances of recombination than others. Therefore, even though in an ideal world both markers are equal, genetic map distance does not always match the actual distance between two markers.

In physical mapping, entire genome is cloned into large insert vector and then all the cloned fragments are arranged in terms of their chromosomal positions by using genetic markers as a probe then the minimum path of the large insert clone is constructed.

The basic requirements for the construction of physical map are:

- Availability of highly saturated genetic and molecular maps;
- Large insert cloning vectors are required;
- High throughput methods of clone packing, culturing, and selection;
- Powerful software to construct minimum tiling path (MTPs).

Since genetic mapping is not accurate and there are limitations, alternative mapping procedures are used to check the accuracy of genetic mapping. The most popular methods for physical mapping are:

1. **Restriction Mapping:** It locates the relative positions of recognition sequences on a DNA molecule with the help of different restriction endonucleases. Restriction maps are constructed by comparing the fragment sizes produced after restriction digestion.

2. **Fluorescent In Situ Hybridization (FISH):** Performed by hybridizing marker probes with intact chromosome to detect the positions of markers located.

3. **Sequence Tagged Site (STS) Mapping:** Using the methods of PCR and/or hybridization, STS analyzes genomic fragments to map the locations of short sequences.

A well-prepared physical map when compared with genetic map helps to create a genome map which can be used as a guide for further DNA sequencing experiments. Preparation of a single contig for the whole genome, mapping each nucleotide being the ultimate goal.

12.4 CONCLUSION

The process of resistance gene identification to finally molecular breeding for development of durable resistant varieties involves searching for new resistance sources which can be done by collecting landraces as well as wild relatives and then screening and finding resistance genes in them. After new sources of resistance are identified, they can be transferred from alien species to common cultivar via hybridization or translocation. These resistance genes' chromosomal position can be determined via C-banding or FISH. Then these newly identified R genes are transferred to some elite varieties where the trait is desired. Identification of the transferred resistance genes in the elite varieties is then carried out by molecular markers like RFLPs, RAPDs, AFLP, SSR or STS. From these selected elite cultivar resistant varieties are developed by backcrossing and selection with the help of molecular markers or by gene pyramiding by marker-assisted selection or by map-based cloning and by transformation.

KEYWORDS

- **genetic map**
- **MapMaker**
- **mapping population**
- **molecular mapping**
- **molecular marker**
- **physical map**
- **QTL mapping**

REFERENCES

Andersen, J. R., & Lübberstedt, T., (2003). Functional markers in plants. *Trends Plant Sci.*, *8*, 554–560.

Basten, C. J., Weir, B. S., & Zeng, Z. B., (2001). *QTL CARTOGRAPHER. Version 1.15*. Raleigh, NC: Department of Statistics, North Carolina State University.

Desta, Z. A., Koning, D., & Ortiz, R., (2020). Molecular mapping and identification of quantitative trait loci for domestication traits in the field cress (*Lepidium campestre* L.) genome. *Heredity*, *124*, 579–591.

Garrido-Cardenas, A. J., Mesa-Valle, C., & Manzano-Agugliaro, F., (2018). Trends in plant research using molecular markers. *Planta, 247*, 543–557.

Goutam, U., Kukreja, S., Yadav, R., Salaria, N., Thakur, K., & Goyal, A. K., (2015). Recent trends and perspectives of molecular markers against fungal diseases in wheat. *Front. Microbiol., 6*, 861.

Huang, X. Q., & R¨oder, M. S., (2004). Molecular mapping of powdery mildew resistance genes in wheat: A review. *Euphytica., 137*, 203–223.

Kao, C. H., Zeng, Z. B., & Teasdale, R. D., (1999). Multiple interval mapping for quantitative trait loci. *Genetics, 152*(3), 1203–1216.

Lander, E. S., & Botstein, D., (1989). Mapping mendelian factors underlying quantitative traits using RFLP linkage maps. *Genetics, 121*, 185–199.

Satagopan, J. M., & Yandell, B. S., (1996). *Estimating the Number of Quantitative Trait Loci Via Bayesian Model Determination.* Special Contributed Paper Session on Genetic Analysis of Quantitative Traits and Complex Disease, Biometric Section, Joint Statistical Meetings; Chicago.

Seaton, G., Haley, C. S., Knott, S. A., Kearsey, M., & Visscher, P. M., (2002). QTL express: Mapping quantitative trait loci in simple and complex pedigrees. *Bioinformatics Applications Note, 18*(2), 339, 340.

Semagn, K., Bjørnstad, A., & Ndjiondjop, M. N., (2006). Principles, requirements, and prospects of genetic mapping in plants. *African Journal of Biotechnology, 5*(25), 2569–2587.

Staub, J. E., Felix, C., Serquen, F. C., & Gupta, M., (1996). Genetic markers, map construction, and their application in plant breeding. *Hort. Science, 31*(5), 729–741.

Steiner, B., Lemmens, M., Griesser, M., Scholz, U., Schondelmaier, J., & Buerstmayr, H., (2004). Molecular mapping of resistance to *Fusarium* head blight in the spring wheat cultivar Frontana. *Theor. Appl. Genet., 109*, 215–224.

Wang, Y., Ning, Z., Hu, Y., Chen, J., Zhao, R., Chen, H., Ai, N., et al., (2015). Molecular mapping of restriction-site associated DNA markers in allotetraploid upland cotton. *PLoS One, 10*(4), e0124781.

Yu, X., Ren, S., Zhao, L., Guo, J., Bao, Y., Ma, Y., Hongwei, W. H., et al., (2018). Molecular mapping of a novel wheat powdery mildew resistance gene Ml92145E8-9 and its application in wheat breeding by marker-assisted selection. *The Crop Journal, 6*, 621–627.

Zeng, Z. B., (1994). Precision mapping of quantitative trait loci. *Genetics, 136*, 1457–1468.

CHAPTER 13

Plant Disease Management Through Gene Pyramiding

SUROJIT SEN,[1] SUNAYANA RATHI,[2] SANGITA SAHNI,[3] and BISHUN DEO PRASAD[4]

[1]*Department of Zoology, Mariani College, Mariani, Jorhat, Assam, India*

[2]*Department of Biochemistry and Agricultural Chemistry, Assam Agricultural University, Jorhat, Assam, India*

[3]*Department of Plant Pathology, TCA, Dholi, Samastipur, Bihar, India*

[4]*Department of AB&MB, College of Basic Sciences and Humanities, Dr. Rajendra Prasad Central Agricultural University, Pusa, Samastipur, Bihar, India*

ABSTRACT

The primary goal of all agricultural activities is to increase crop productivity. Traditional practices for combating the various pathogenic plagues suffered by our crops are based on the principles of prevention or therapy. The best approach for prevention is the utilization of cultivars that are resistant or tolerant to infection. As resistance to pathogens naturally occurs in plants, breeding for resistance seems to be a more sustainable approach for mitigating the damage caused by pathogens and optimizing the plant production system for better yield in limited fertile areas. Plant diseases are a significant barrier to agricultural output and continue to result in significant losses. Hence, for faster development of resistant varieties, gene pyramiding through trans- and cis-genic biotechnologies is now the preferred approach for incorporating durable disease resistance in plants. This chapter discusses the various gene pyramiding strategies for the management of plant diseases.

New Advances in Crop Disease Management. Sangita Sahni, Ganesh Thapa, and Bishun Deo Prasad (Eds)

13.1 INTRODUCTION

Disease management may be considered as a proactive process initiated in order to forecast or model disease occurrence for anticipation and use of effective measures to control susceptible stages of the disease cycle. It is rather actions taken ahead for disease control than reactive control on a prescription basis. However, the distinction between the proactive and reactive approach is blurred, especially when it comes to application of specific measures. Therefore, in order to effectively manage any disease, a thorough understanding of the pathogen, its identification, the mode of interaction between host and pathogen, the disease cycle, environmental factors influencing the disease cycle, host plant's cultural requirements are necessary (Servin & White, 2016).

Among the various practices for combating the various pathogenic plagues suffered by our crops, use of chemical pesticides, physical barriers, using predators and parasitoids for biological pest management, in cultural practices some crucial pathogen management strategies include the use of plant material devoid of viruses, seasons without crop, weed control (by using different hosts), and rouging of infected plants.

The two guiding principles of traditional pathogen management systems are prevention and treatment. Infection can be avoided by choosing a season or location that isn't conducive to infection, or it can be prevented by excluding people or by other strategies. But if somehow this is escaped then eradication is done by means of destroying or inactivating the inoculums. Further, protection is provided by means of toxicant or other barriers to infection. However, best approach for prevention is utilization of cultivars that are resistant or tolerant to infection. Resistance to pathogen naturally occurs in plants among the varieties and line as well as near relatives in the wild. A sustainable strategy for addressing diseases brought on by infections must include breeding cultivars with resistance to these pathogens. However, if the prevention tactics fails and the plant gets infected, therapy or treatment is adopted on a need basis.

Improving disease resistance in plants is crucial prevention strategy of disease management. Specific or qualitative resistance is a term used to describe the race-specific disease resistance provided by a single major gene. This sort of resistance, which is normally temporary, can be fully eliminated when a pathogenic race capable of attacking that genotype appears. Quantitative resistance on the other hand derives resistance from many different genes with cumulative effects that promote more long-lasting resistance to pathogens (Singh et al., 2013).

Development of plants that are resistant to disease has been relatively successful by stacking of these qualitative as well as quantitative resistance genes into a single genotype by traditional breeding techniques. This stacking technique, also called gene pyramiding, strategy is to create a larger base of long-lasting or durable disease resistance by combining multiple genes in a single variety.

The traditional method of gene pyramiding requires a lot of effort and is frequently unsuccessful in consistently combining and detecting many resistance genes in a single genotype. However, breeding programs have gotten quicker at selecting and screening particular alleles as a result of the development of DNA-based genetic technology and molecular biology techniques. Breeders can now easily detect the trait of interest in the offspring at an early stage and eliminate unwanted plant material. Use of MAS has revolutionized the breeding process as well as pyramiding of desired resistant genes.

Due to the continuous challenge of global food security and the growing global population, the need to significantly boost agricultural productivity is urgent. Plant diseases are still a major cause of losses and a major barrier to crop output.

13.2 RESISTANCE THEORY AND PYRAMIDING

There are several models proposed to describe host resistance to pathogen infection. Harold Henry Flor proposed the first theory while working on rust disease of flax. According to this theory, inheritance of host's resistance and parasite's pathogenicity is controlled by matching pair of genes. The resistance gene (R) that the host possesses forms the basis of the gene-for-gene theory, which produces a protein that recognizes a protein from the pathogen's *Avr* (avirulence) gene (Flor, 1971). Gene pyramiding is a natural extension of the gene-for-gene theory.

Another theory called the guard/decoy or indirect recognition theory is an alternative to the theory of gene-for-gene. The guard theory, first put out by van der Biezen & Jones in 1998, contends that proteins for plant resistance function by keeping an eye on (guarding) the target of the associated pathogen effector. The interaction between the pathogen and intermediary components presents in the host results into modification of intermediary components, which are then recognized by plant nucleotide-binding site leucine-rich repeat (NBS-LRR) proteins (Agrios, 2005). The guarded effector target, also known as the guardee, is crucial for pathogenicity by the effector protein in the absence of the cognate R protein, claims the guard model.

Decoy molecules resemble a component at the host-pathogen interface that changes as an infection progresses. Decoys can be manipulated in the same way as the component they mimic, but instead of serving the intended purpose, they serve the opposite one by either preventing the manipulation of the mimicked component or by generating molecular recognition event.

A third theory, the bait model, works similarly to the guard/decoy model but the so-called bait directly interacts with the NLR (NOD-like receptor family of proteins) (Jo, 2013). The sensing decoy "trickles" the effector in the Bait mechanism, resulting in a recognition event. These lures don't stop the effector from coming into contact with its operative target; rather, they just allow recognition to outweigh the benefits of the effector manipulating its operative target.

All of the above theories could be combined into a single 'helper model.' The helper protein can then be depicted within its own context as a guard, decoy, bait, or anything else which helps in regulation of recognition in host pathogen interaction.

13.3 PYRAMIDING METHODS

13.3.1 GENE PYRAMIDING THROUGH TRADITIONAL BACKCROSSING

Generation of an ideal resistant genotype by gene pyramiding technique, bringing all desirable genes from different sources into an elite variety is done traditionally by recurrent back crossing. The goal is to introduce the desired gene into the recurrent parent's DNA from the donor parent. However, along with the transfer of desired resistant gene, some unwanted genes also get transferred due to linkage drag. So, to recover the parent genome and eliminate the unwanted linkage drag six to eight recurrent backcrossing is done. Then, the resulting filial populations are selected phenotypically and self-crossed to produce two resistant genotype (homozygous and heterozygous) and one susceptible genotype. Homozygous progenies for the resistant gene are identified after one generation of field screening and maintained as improved line for resistance.

13.3.2 MUTATION BREEDING

Mutation breeding or variation breeding is also used to complement conventional breeding. In this technique seeds are exposed to radiation in order to create improved agronomic trait and resistance to pest and diseases. Although mutants with desired traits are created by chance, they are then bred with normal cultivar to create new lines. Induced mutations are employed in traditional mutation breeding to create new varieties. Mutated genes can be easily tagged and pyramided into single elite breeding line. The lines can be maintained by subsequent breeding program, but durability of these traits developed is low because of segregation.

13.3.3 MARKER-ASSISTED SELECTION (MAS)

Since the beginning of agricultural practice, crop plants have been domesticated and grown in order to take use of the genetic diversity in plants. However, this diversity's potential richness is still not being fully tapped. Numerous economically significant allelic variants of characteristics are still underutilized and can be investigated to address current and future threats to plant productivity. Screening of these important traits of interest can be aided by marker-assisted selection (MAS) for utilization of remains of the unutilized genetic diversity of plants of agro-economic importance.

The most useful, economical, and environmentally friendly approach to disease prevention is to exploit the host's natural resistance. Breeding work is traditionally focused on utilization of the wide spectrum of resistant genes present naturally and piling these genes/QTLs into desired cultivar. Pyramiding disease resistance through conventional breeding takes a long time and is entirely influenced by the environment. Particularly, MAS in molecular breeding is faster, well-structured and precise. By using standard techniques, it is challenging to combine several genes and detect their existence due to their epistatic effects. Marker assisted selection provides potential solution (better selection strategies) in a short span of time, to many problems not resolved by conventional breeding.

There are many different kinds of marker systems, including morphological, molecular, and biochemical markers. The most popular type of marker for linkage mapping is the molecular marker, commonly known as a DNA-based marker, because morphological and biochemical indicators have numerous limitations. The molecular markers used for marker analysis are: PCR based molecular markers like AFLP (amplified fragment length polymorphism), RAPD (random amplified polymorphic DNAs), AP-PCR (arbitrarily primed-PCR), ALP (amplicon length polymorphism), SAP (specific amplicon polymorphism), DAF (DNA amplification fingerprinting), AS-PCR (allele specific PCR), SCAR (sequence-characterized amplified region), SSCP (single strand conformation polymorphism), SSR (simple sequence repeats), SSLP (microsatellite simple sequence length polymorphism), EST (end sequence tagged), STS (sequence tagged sites), VNTR (variable length nucleotide tandem repeats), and SNPs (single nucleotide polymorphisms). Any of these DNA markers connected to the resistance loci may be used to pick plants that are then backcrossed to the recurrent parent to recover the disease. Genomic DNA is generally extracted and isolated from progenies were sampled from each succeeding generation. DNA amplification is performed by PCR and results are analyzed using agarose gel electrophoresis. The presence or absence of an interest-specific resistance gene is then determined by comparing the sizes of bands on the gel to a molecular marker.

Strategies used for pyramiding by MAS are almost common in plant with slight variations in the basic workflow (Figure 13.1). Two commonly used strategies are step wise transfer and simultaneous transfer.

To create the F1 hybrid in a step-by-step transfer, the parent of the donor (DP1) and the repeating parent (RP1) are crossed. The F_1 hybrids are repeatedly backcrossed till the third generation of backcross (BC_3) and improved repeating parent (IRP_1) are produced. Further, to pyramid several genes, repeating parent is crossed with other parent of the donor (DP_2).

The repeating parent (RP1) is crossed with several different parents of the donor like DP1, DP2, etc., in order to create first filial hybrids by simultaneous transfer. Intercrossing of F_1 hybrids produces improved F_1 (IF_1). After backcrossing with the recurrent parent, this improved F_1 becomes the enhanced recurrent parent (IRP). A third strategy commonly used is a combination of the two known as simultaneous and stepwise transfer. Here, multiple donor parents are crossed simultaneously with a recurrent parent (RP_1) before being backcrossed up to the BC_3 generation. The pyramided line is created by intercrossing the backcross populations with the specific gene of interest.

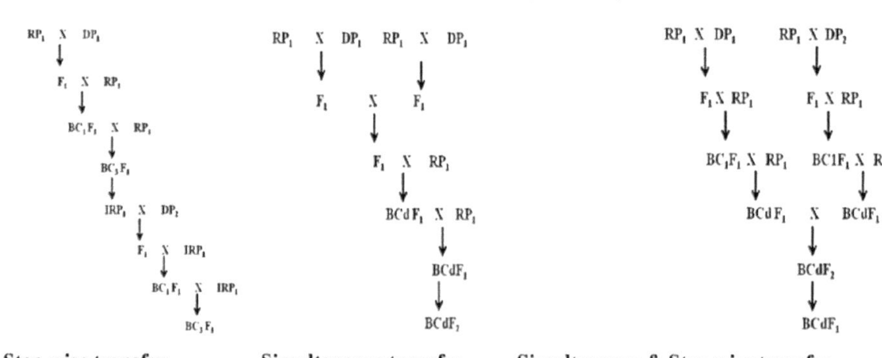

FIGURE 13.1 Strategies for pyramiding by MAS.

Source: Adapted from: Sen & Rathi (2020).

Different independent crossing streams can be used in the marker-assisted backcrossing approach to follow R/RQTL loci that have been molecularly marked (Figure 13.2):

- First, under screening house conditions, the donor parents with the R gene and resistant QTL loci are independently bred with the recurrent parent cultivar.
- Then, to produce the BC1F1 or BC1, the F1 and RP are crossed.
- Then, each crossing stream of BC1F1 populations is used to verify the targeted genes using markers that are closely connected to the resistance loci.
- Then, chosen BC1F1 plants from each crossing stream that have a specified resistance trait (R gene and RQTL loci) are crossed with the RP to create the BC2F1 populations.
- Selected BC plants are repeatedly crossed with the RP to produce subsequent backcross populations.
- Selection of the backcross progeny having target traits is done during each round of backcrossing based on genotype as well as phenotype.
- Further screening of targeted homozygous plants from each backcross stream is done by selfing of (BC_4) selected plants.
- Then, randomly chosen targeted R genes homozygous genotypes are taken from each segregating population.
- Next step is harvesting of BC_4F_3 seeds individually from each selected BC_4F_2 plants.
- These selected lines are monogenic or near isogenic lines.
- Further integration of the resistance genes into a single background is done by intercrossing and pyramided lines with different gene combinations are created.
- Finally, further integration all of the R genes by crossing, polygenic lines with broad spectrum resistance are created.

Marker-aided selection benefits:

- Very reliable;
- At the seedling stage, plants can be chosen;
- Saves the breeder's time;
- Plants lacking the target genes can be eliminated early by the breeder;
- Cost effective in long run.

Marker-assisted selection's limitations:

- Initial cost of equipment and consumables is high;

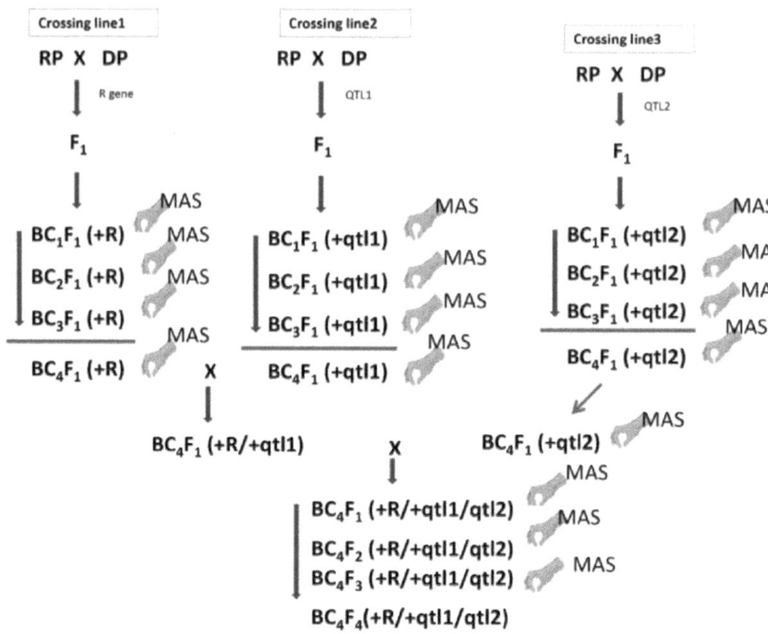

FIGURE 13.2 Breeding strategy for pyramiding major R gene and QTLs for different disease using marker assisted selection [RP: recurrent parent; DP: donor parent; QTL: quantitative trait loci; R: resistant gene; and MAS: marker assisted selection].

- Sampling bias may affect detection;
- Recombination or crossing-over may occur, creating problem in making decision.

13.3.4 SOMATIC FUSION AND TISSUE CULTURE TECHNIQUES

Problems that cannot be solved by conventional breeding methods can be dealt with various tissue culture techniques which significantly shorten the breeding process. Among the different alternative tools, tissue culture is also used for development of resistance to diseases in susceptible cultivars. The genetic variation produced *in vitro* during clonal regeneration is known as somaclonal variation. These variations can be exploited for selection of disease resistant clones and further developed into resistant lines. During the past few decades extensive work has been done for selection of disease resistant plant by tissue culture technique. Tissue culture filtrate free of cells, toxins derived pathogens, inoculating the culture cells directly with pathogen or all of them can be used for selection purpose. However, information of resistance due to somaclonal variation is limited for some diseases and discrepancies are found in variations for resistance among the somaclones. These differences can be ascribed to the test conditions, such as the growth medium's composition, the cultivation period, or the presence of illness, etc. However, somaclonal variation may contribute to planning and achievement of breeding objectives for rapid development of resistant lines from the prevailing cultivars.

Somatic fusion or protoplast fusion is another technique of genetic manipulation where distinct species of plant are fused to from a hybrid with characteristics of both the species. These hybrids can further be cultured and selected as per need. Protoplast fusion can be exploited for introgression of resistant genes from wild and distant sources and pyramiding them in a single plant. Hybridization can be both intra or inter specific hence by this process genes can be transferred even from unrelated species.

13.3.5 GENE PYRAMIDING THROUGH GENETIC ENGINEERING

In order to develop broad-spectrum lasting resistance, genetic engineering (transgenic method) enables the simultaneous insertion of many genes into the genome. This is another crucial strategy in the control of plant diseases. Genetic modification or engineering has the flexibility of transferring the gene of interest from eukaryotic, prokaryotic, viruses or even from synthetic sequences into crop plants. Additionally, it enables the introduction and overexpression of genes that code for proteins that directly limit the development of infections.

An effective and useful alternative to traditional methods for genetic pathogen enhancement is the transgenic approach. Transgenes stacking in a single locus helps them to co-inherit, leading to more effective tolerance in transgenic plant in successive generations.

Major steps in plant genetic engineering:

- Identification of target gene and isolation of DNA from the desired organism;
- The extraction of DNA fragments from the donor organism;
- Inserting a selected vector with an isolated DNA segment;
- Screening of vectors;
- Gene cloning;
- Screening of the transformed host cells;
- Regeneration of the transgenic plant;
- Backcross breeding, if necessary.

There are two methods for molecular gene stacking in plants: Simultaneous introduction method or Sequential introduction method:

1. **Simultaneous Introduction Method:** This process, known as co-transformation, is effective at introducing many genes into plants. This is done either by employing numerous plasmids (for unlinked genes or distinct DNA fragments) or by a solo plasmid (for linked transgenes) and using techniques for transformation as the biolistic approach or Agrobacterium-mediated, they are concurrently changed into the plant (Segolela et al., 2019).

   i. **Single Plasmid Co-Transformation:** A single construct containing all the genes, having their own promoter and terminator, is inserted into the same plasmid to carry out this operation. All genes frequently co-localize in the same regions of the plant genome, which can be exploited for genetic transformation by *Agrobacterium tumefaciens* or by bombardment.

   ii. **Co-Transformation Using Multiple Plasmids:** During this procedure, a number of constructs are made, each of which contains one or more genes. Following this, each construct is put into a distinct plasmid. To change genetic material, plasmids can be blasted or delivered using *A. tumefaciens*.

➢ **Advantages:**

- Highly efficient and faster process.
- Time consuming multiple events of transformation can be avoided.
- Transgenes integrate in same chromosomal position hence they co-inherit.

➢ **Limitations:**

- Complex plasmids with many gene cassettes using the same promoters might be difficult to build.
- Integrating a large number of copies.
- Unwanted T-DNA inclusion.

2. **Sequential Introduction Method:**

   i. **Re-Transformation:** The addition of a new trait to a transgenic plant is known as a multi-trait or combined trait event. This happens when numerous genes are repeatedly inserted into a plant that already harbors a transgenic.

- **Advantages:**
    - Usually used in vegetatively propagating plants.
    - Stacking of transgenes is independent of sexual crossing.
    - Can transfer characters to plants which are unable to produce seeds by sexual reproduction.
- **Limitations:**
    - With every transformation, a new marker gene is added, which causes an accumulation of markers or many copies of the same marker gene, which impairs the regulatory mechanism.
    - After genetic segregation in sexually reproducing crops, transgenes are difficult to keep in the offspring because genes get introduced in numerous loci.

ii. **Sexual Crossing:** The sexual fusion of two or more events is another method of gene stacking. In sexually reproducing transgenic plants, individuals are crossed for stacking of genes.

- **Advantages:**
    - The method is flexible;
    - Usually previously tested and approved events are used;
    - Events can be accumulated as per market need.
- **Limitations:**
    - Cannot be used in vegetative propagated plants;
    - Chances of segregation in the next generations are there.

With the advancement of molecular biology, efficient tool like gene assembly in *Agrobacterium* by nucleic acid transfer using recombinase technology (GAANTRY) has been developed to overcome the limitations of gene pyramiding. Here multiple genes are stacked in the T-DNA of *Agrobacterium* plasmids in a controlled orientation. Genes are stacked sequentially to produce high-quality transgenic occurrences with few copies and no backbone contaminations. The process is straightforward, effective, and affordable. It can also construct and steadily maintain enormously large T-DNAs (Collier et al., 2018).

13.4 CONCLUSION

Resistance durability is a key requirement for sustainable management of plant diseases through pyramiding of genes. The main objective of a gene pyramiding method is to integrate all the desirable genes into one genotype, where at every locus there is homozygosity for all the desired alleles. Gene pyramiding is a successful method for creating long-lasting, broad-spectrum resistance and reducing the production loss brought on by plant disease. Gene pyramiding through traditional breeding is time-consuming, inefficient, and expensive in the long run as it takes years to develop a new cultivar. The use of modern molecular biology techniques like MAS and genetic engineering has boosted the stacking technique for the development of durable resistance and management of crop diseases. Each strategy has advantages and drawbacks of its own. However, selecting the right target genes and a cost-effective strategy is, of course, at the breeder's discretion.

KEYWORDS

- **disease management**
- **gene pyramiding**
- **genetic engineering**
- **marker-assisted selection**
- **molecular marker**
- **quantitative trait loci**
- **resistance theory**
- **traditional backcrossing**

REFERENCES

Adugna, A., (2004). Alternate approaches in deploying genes for disease resistance in crop plants. *Asian Journal of Plant Sciences, 3*(5), 618–623.

Agrios, G. N., (2005). *Plant Pathology* (5th edn.). Elsevier Academic Press.

Collier, R., Thomson, J. G., & Thilmony, R., (2018). A versatile and robust agrobacterium-based gene stacking system generates high-quality transgenic *Arabidopsis* plants. *Plant J.*, *95*(4), 573–583.

Flor, H. H., (1971). Current status of the gene-for-gene concept. *Annual Review of Phytopathology*, *9*(1), 275–296.

Jo, K. R., (2013). *Unveiling and Deploying Durability of Late Blight Resistance in Potato: From Natural Stacking to Cisgenic Stacking*. Wageningen University. https://library.wur.nl/WebQuery/wurpubs/442388 (accessed on 12 December 2023).

Prasanna, H. C., Sinha, D. P., Rai, G. K., Krishna, R., Kashyap, S. P., Singh, N. K., Singh, M., & Malathi, V. G., (2014). Pyramiding Ty-2 and Ty-3 genes for resistance to monopartite and bipartite tomato leaf curl viruses of India. *Plant Pathology*. doi: 10.1111/ppa.12267.

Ramalingam, J., Raveendra, C., Savitha, P., Vidya, V., Chaithra, T. L., Velprabakaran, S., Saraswathi, R., et al., (2020). Gene pyramiding for achieving enhanced resistance to bacterial blight, blast, and sheath blight diseases in rice. *Front. Plant Sci.*, *11*, 591457.

Renier, A. L., Van, D. H., & Kamoun, S., (2008). From guard to decoy: A new model for perception of plant pathogen effectors. *The Plant Cell.*, *20*, 2009–2017.

Segolela, J. C., Wokadala, O. C., & Feto, N. A., (2019). Bacillus thuringiensis-based gene pyramiding: A way forward for a combined horizontal and vertical resistance in plant. In: *Bacilli and Agrobiotechnology: Phytostimulation and Biocontrol*. (pp. 291–304). Springer, Cham.

Sen, S., & Rathi, S., (2020). Pyramiding and cloning of R genes. In: Sangita, S., & Bishun, D. P., (eds.). *Host-Pathogen Interaction: A Molecular Approach* (pp. 127–136). Published by: Today and tomorrow's Printers and publishers, New Delhi. ISBN: 10: 81-7019-695-X, 13: 9788170196952.

Servin, A. D., & White, J. C., (2016). Nanotechnology in agriculture: Next steps for understanding engineered nanoparticle exposure and risk. *NanoImpact*, *1*, 9–12.

Singh, S., Singh, R. P., Bhavani, S., Huerta-Espino, J., & Eugenio, L. V. E., (2013). QTL mapping of slow-rusting, adult plant resistance to race Ug99 of stem rust fungus in PBW343/Muu RIL population *Theoretical and Applied Genetics*, *126*(5), 1367–1375.

Van, D. B. E. A., & Jones, J. D. G., (1998). Plant disease resistance proteins and the gene-for-gene concept. *Trends Plant Sci.*, *23*, 454–456.

Werner, K., Friedt, W., & Ordon, F., (2005). Strategies for pyramiding resistance genes against the barley yellow mosaic virus complex (BaMMV, BaYMV, BaYMV-2). *Molecular Breeding*, *16*, 45–55.

CHAPTER 14

Lactic Acid Bacteria as Biocontrol of Fusarium Head Blight

GANESH THAPA[1] and MICHAEL B. BYRNE[2]

[1]*Biohazards and Biosafety, Trinity College of Dublin, The University of Dublin, College Green, Dublin, Ireland*

[2]*School of Agriculture and Food Science, University College, Dublin, Belfield Dublin [4], Ireland*

ABSTRACT

Fusarium head blight (FHB) caused mainly by *Fusarium graminearum* is a devastating economic disease, which affects kernel development, and human/animal health with mycotoxins contaminated grains. To tackle the ineffective strategies of FHB management and the rising concerns of food safety and mycotoxins, new sustainable control approaches must be developed, validated, and implemented. The use of biocontrol agents to control FHB has been of utmost interest lately. Lactic acid bacteria (LAB) present the greatest potential for both fungal growth inhibition and detoxification of mycotoxins. The ability to inhibit fungal development and remove several mycotoxins, along with its recognized status as safe with probiotic potential, make LAB a promising biological control candidate for control of *F. graminearum* in crops. Considering the potential of LAB, as well as the growing consumer demand for foods with reduced levels of chemical additives and contamination, this chapter offers insight into the LAB biocontrol strategies, effectiveness, and environmental innocuousness against *F. graminearum*.

New Advances in Crop Disease Management. Sangita Sahni, Ganesh Thapa, and Bishun Deo Prasad (Eds)

14.1 INTRODUCTION

FHB is incessant hurdle of economic sustainability of most cereal crops' growers viz., barley and wheat worldwide (Maiorano, Blandino, Reyneri, & Vanara, 2008). It is widely caused by *Fusarium graminearum* (Summerell, Laurence, Liew, & Leslie, 2010; Starkey et al., 2007). The FHB usually results in reduced germination and poor plant establishment leading to yield and quality losses, sterile florets, and discoloration, withered, and light-weight kernels (Scherm et al., 2013). FHB has been accountable for extensive losses of barley and wheat yield in North America and Canada in past and has been a nagging problem in Europe also resulting in severe economic and sociological difficulties (Karlsson, Persson, & Friberg, 2021). The yield loss is further aggravated by the mycotoxins contamination viz., trichothecenes deoxynivalenol (DON), nivalenol (NIV), and T-2 toxin (T2) and HT-2 (HT2) toxin, and the estrogenic mycotoxin zearalenone (ZEA) (Bottalico & Perrone, 2002), which are known instigators of chronic and acute mycotoxicosis in humans and animals (Haidukowski et al., 2005; Rai, Das, & Tripathi, 2020).

Though, chemicals have been used to reduce FHB disease levels and mycotoxin contamination. The resistant cultivars are the most effective, economic, and sustainable way to control FHB in field (Buerstmayr et al., 2002; Gilbert & Fernando, 2004). However, the complexity of the FHB disease cycle, resistance mechanisms and the combined deadlock of both agronomic and resistant traits prompted to use combinatorial approach of fungicides, crop rotation and crop residue management by tillage practices (Maiorano, Blandino, Reyneri, & Vanara, 2008; Wegulo, Bockus, Nopsa, Peiris, & Dowell, 2013). This is what we called integrated pest management (IPM), and its implementation leads to economically and ecologically justified means to reduce human and environmental risk. The efficacy of the IPM approach remain highly variable (Perez, Dill-Macky, & Kinkel, 2008) and there is need to find alternatives with reduced detrimental effects.

Biological control is now considered as an emerging "environmentally friendly and sustainable" alternative to reduce or control a targeted pathogen by organisms/biologics. The narrow efficacy range of biocontrol agent (BCA) ensures safety to other non-target beneficial organisms maintaining ecobalance (Alabouvette, Olivain, & Steinberg, 2006). Though there are tremendous scientific endeavors from scientists to develop proficient biological control agents (BCAs) applicable for FHB control with daily progression (Wachowska et al., 2014; Gimeno, Leimgruber, Kägi, Jenny, & Vogelgsang, 2021). But there are various stages such as, regulatory issues, scientific and technological challenges, screening, formulation which needs to be successfully completed prior to commercialization of effective BCA for FHB control. The application of formulations, its design and monitoring approach of the BCA friendly to agroecosystem also needs to be develop simultaneously. The validation of the efficacy and its stability in diverse environmental and agronomic conditions must be done to replace the usage of chemicals. Keeping all these in mind, hereon the FHB disease cycle, symptoms, LAB as FHB biological control, mechanism, limitations, and future directions will be discussed.

14.2 FHB DISEASE PROGRESSION, COMPLEXITY, INFLUENTIAL FACTORS

Though the *Fusarium spp.,* infection symptoms appear same in most of cereal crops, but infection by *Fusarium* spp. with reduced pathogenicity may not give typical ear blight symptoms particularly in barley and oats. This variation in symptoms depends on the species of *Fusarium* causing infection in infected plants (Parry, Jenkinson, & McLeod, 1995). Though most of the head are susceptible to *Fusarium* infection once flowering has commenced, but various symptoms may pervade depending on the constituents of the *Fusarium* spp. complex. Initial infection by most *Fusarium* spp. is characterized by dark purple to brown water-soaked lesions starting from the exterior surface of the florets and glume, coinciding with the presence of perithecium initials (Parry, Jenkinson, & McLeod, 1995; Trail, 2009). The spread of fungal hyphae is characterized by brown/purplish coloration in severe cases with remaining kernels also becoming blighted or bleached due to loss of chlorophyll (Logrieco, Bottalico, Mulé, Moretti, & Perrone, 2003). The impaired nutrients supply through glumes and pericarp due to excess fungal proliferation also lead to orange, salmon pink and red fungal growth ascertained by FHB epidemics (Scherm et al., 2013; Logrieco, Bottalico, Mulé, Moretti, & Perrone, 2003). Extensive fungal proliferation leads to premature death of spikelets. The FHB infected bleached spikelets differ from bleaching due to heat and drought stress where the entire cereal spikelet perishes, with the green color fading to a bleached white shade akin to FHB infection (Parry, Jenkinson, & McLeod, 1995; Langevin, Eudes, & Comeau, 2004).

Anthesis is a critical stage in fungal infection, known to stimulate initial hyphal spread through the presence of pollen grains (Kang & Buchenauer,

1999; Strange & Smith, 1971). Histological investigations of infected wheat spikelets suggest that only extruded or retained anthers are sites for initial infection with spread throughout the grains, glumes, lemma, and rachis (Hallmann, Quadt-Hallmann, Mahaffee, & Kloepper, 1997; To, 1930). Much like wheat, barley has differing symptoms depending on the antagonistic *Fusarium* spp., grayish to tan brown lesions were reported on heads infected with *F. graminearum*, *F. avenaceum*, and *F. sporotrichoides* with difficulty in distinguishing between the preceding species (Salas et al., 1999). Though it is difficult to reliably differentiate disease caused by the multiple *Fusarium* spp. with similar pathogenicity and the visually identification remains questionable (Salas et al., 1999). But ultimately, Grains with severe *Fusarium* infection leads to soft empty grains.

Even though this disease can progress with asymptomatic spikelets but the content of mycotoxins within it was found very high (Osborne & Stein, 2007). Thus, FHB poses a significant threat to global food security from a myco-toxicity perspective. Therefore, a novel technology viz., Fluorescence polarization immunoassays, hyperspectral imaging, Infrared spectroscopy, and biosensors, which can quantify mycotoxin resistance perchance may aids in the identification of the infectious fungal agent with scope for their use in field situations (Gilbert & Haber, 2013; Liu, Elen, Sundheim, Langseth, & Skinnes, 1997).

The FHB disease complexity in cereals can vary within and between years as different species may occupy greater segments of the infected plant during different phases of plant growth (Pettitt & Parry, 1996; Parry, 1990) (Figure 14.1). *Fusarium* can survive individually or coexists in the same plant with changing temperature, precipitation, and cropping systems (Tunali et al., 2008; Fernandez & Jefferson, 2004). This existence ability has led to shift in species migration and thus unearthing a new problem for farmers (Balmas et al., 2015). There may be more than one significant competitive interaction between antagonistic species in *F. culmorum*, *F. avenaceum*, *M. nivale var. majus*, and *var. nivale* and *F. pseudograminearum*, but each *Fusarium* sp. have different requirements of the plant with *F. culmorum* able to exclude subsequent pathogenic colonizers (Pettitt, Xu, & Parry, 2003). It is not yet fully understood the role played by so-called weaker antagonistic species viz., *F. subglutinans*, *F. oxysporum*, *F. acuminatum*, *F. solani*, and *F. verticilloides* and how they may influence disease incidence and severity (Tunali et al., 2008). Perhaps they may facilitate individual pathogenic species rather than causing direct damage to the plant, however, this requires further study to assigned concrete role of these species.

FIGURE 14.1 FHB disease progression in wheat. FHB symptoms quantified as the number of diseased spikelets per head to assess the disease progression in wheat both CM82036 (resistant) and Remus (susceptible) cultivars.

14.3 FACTORS INFLUENCING PRODUCTION, SURVIVAL, AND SPREAD OF *FUSARIUM*

The ability of *Fusarium* inoculum to persist through extended time under hostile conditions has led to its dominance as significant pathogen of small grain cereals with enhanced production and survivability of fungal inoculum (Leplat, Friberg, Abid, & Steinberg, 2013). Hence, while attempting to constrain fungal diseases, eliminating the source of inoculum is crucial and the critical factors of inoculum survival is important (Bateman, 2005). *Fusarium* inoculum overwinters as chlamydospores or mycelium on crop debris and soil until germination stimuli presents and the conditions required for inoculum persistence depends on the reproductive strategies of individual fungal species (Keller, Bergstrom, & Shields, 2014; Doohan, Brennan, & Cooke, 2003). As along with temperature, light, carbon dioxide (CO_2), water availability and other physio-chemical factors do affect the infectivity and virulence of *Fusarium* inoculum (Leplat, Friberg, Abid, & Steinberg, 2013; Doohan, Brennan, & Cooke, 2003; Xu, 2003).

The inoculum production is rapid under warm humid conditions, with ascospore production of *F. graminearum* favored at temperatures of 25°C–28°C under high water activity (Leplat, Friberg, Abid, & Steinberg, 2013). It was noted that water activities between 0.950 and 0.995 allows DON production (McMullen et al., 2012). Higher CO_2 levels may result in condensed reproductive cycles, increasing the likelihood for development of new virulent species as they adapt to changing environs (Melloy et al., 2010). Light is a critical signal for *Fusarium* growth as it influences many physiological responses including sexual development and asexual conidiation (Doohan, Brennan, & Cooke, 2003; Fanelli et al., 2012). Anemophilous distribution of fungal spores is the primary conduit to ensure reasonable dissemination (Abbott, 2002). Ascospores and conidia are also reliant on rain-splash dispersal, insect vectors and perithecia mechanisms (Parry, Jenkinson, & McLeod, 1995; Keller, Bergstrom, & Shields, 2014; Doohan, Brennan, & Cooke, 2003). Ascospores have been established as a main source of inoculum dispersed mainly by wind (Osborne & Stein, 2007). Splash-dispersal has been identified as a key component for macroconidia dispersal (Parry, Jenkinson, & McLeod, 1995). Insect vectors are equally involved in the dispersal of fungal inoculum (Parry, Jenkinson, & McLeod, 1995). The adaption of fungal reproductive structures to produce inoculum contained within sticky droplets facilitates attachment to the bodies of the insect (Abbott, 2002). Certain vertebrates (e.g., Hummingbirds (*Lampornisam ethystinus*) does play role in the spread of *Fusarium* inoculum (Lara & Ornelas, 2003). Mechanisms employed by perithecia initials are said to aid dispersal of ascospores which are known to bring about FHB epidemics.

14.4 THE PERSISTENT PROBLEM OF *FUSARIUM* MYCOTOXINS CONTAMINATION

Mycotoxins are naturally formed secondary compounds of *Fusarium* spp. (Steyn, 1995), produced depending on various growing conditions (Kokkonen, Ojala, Parikka, & Jestoi, 2010). This includes, trichothecenes (DON, NIV, T2-toxin, and acetyl deoxynivalenols (ADONs), ZEA, and fumonisins (FUM) (Shala-Mayrhofer et al., 2013; Malachova et al., 2010). Other *Fusarium* mycotoxins include enniatins, moniliformin, beauvericin, and fusaproliferin (Ferrigo, Raiola, & Causin, 2016). Contamination in the field or during processing of human and animal grain-based food is concerning due to its inherent toxicity caused by these metabolites (Bottalico & Perrone, 2002; Zachariasova et al., 2014). Mycotoxin poisoning through consumption of spoiled food has been described as the comparable to pathologies caused by exposure to pesticides or heavy metal residues (Bennett & Klich, 2003). Consequently, over 100 countries and in EU have permissible tolerable limits of mycotoxin in human and animal food (Van & Jonker, 2004; Cheli, Battaglia, Gallo, & Dell'Orto, 2014).

14.5 ECONOMIC IMPACT OF *FUSARIUM* HEAD BLIGHT (FHB)

FHB results in reduced crop yield and quality and thereby leads to economic consequences at all socio-economic levels (Stenglein, 2009). There have been several cases of economic losses across globe and this disease needs serious attention. In extreme situations estimated losses of up to 80% of the crop have been recorded (Windels, 2000).

The pecuniary extent to which farmers were penalized with grain merchants offering ca. \$5.36/bu for quality grain while mycotoxin presence reduced the price received by \$0.96 cents/bu (McMullen, Jones, & Gallenberg, 1997). Many hidden costs are incurred post epidemics including screening/cleaning of grain, recuperation costs from planting infected seeds to establish the subsequent crops and the human and animal health costs associated with consumption of contaminated grain.

Obtaining an accurate evaluation of the financial costs associated with FHB outbreaks in Northwest Europe remains problematic as most recent estimates fail to combine facets in relation to the disease. However, relative studies between 1998 through 2000 in the United States estimated that through combined primary and secondary losses, a total of \$2.7 billion cost was incurred for wheat and barley due to FHB and combined with Canadian losses, it is estimated that costs upwards of \$5 billion were suffered (Matny, 2015; Nganje, Bangsund, Leistritz, Wilson, & Tiapo, 2002). Estimates suggest FHB epidemics occur two to three times every 10 years resulting in austere financial complications. Combined with doubt over the financial benefits of applying fungicides under certain weather conditions, climate change and with increasing resistance to fungicides, the sector is left with no option but accelerate the pursuit of economic driven solutions (Wegulo, Bockus, Nopsa, Peiris, & Dowell, 2013).

14.6 MANAGEMENT STRATEGIES OF FHB IN CEREALS

Fungal diseases of barley pose serious yield, quality, and economic threats to farmers. Implementation of up-to-date plant protection techniques from sowing until spikelet formation is crucial to limit *Fusarium* borne diseases. To date, much work has focused on the development of management strategies of FHB on wheat. However, plant protection practices require further exploration with respect to the control of FHB in different season grown barley (Hückelhoven, Hofer, Coleman, & Heß, 2018; Wegulo et al., 2015). The control of FHB in cereals were primarily based on the disruption of pathogenesis and reduction of spikelets exposure to pathogenic inoculum. A portfolio of approaches including biological/cultural/chemical and breeding/growing resistant cultivars will be effective in FHB disease management than one individual approach (McMullen et al., 2012; Wegulo et al., 2015). Development of innovative practices at farm level reducing FHB disease and mycotoxin level will lead to greater control of our food chain (Hückelhoven, Hofer, Coleman, & Heß, 2018). Different strategies such as crop rotation, tillage practices, fungicides application, resistant varieties were used that partially reduced the impact of FHB (Vogelgsang et al., 2019; Pan, Mionetto, Tiscornia, & Bettucci, 2015). Managing FHB with biological control using antagonistic microorganisms could be a promising strategy to reduce FHB and mycotoxin production (Degrassi & Carpentieri-Pipolo, 2020).

14.6.1 LACTIC ACID BACTERIA (LAB)

Lactic acid bacteria (LAB) are non-spore forming, rod or coccus-shaped organisms that ferment alcohols and carbohydrates to produce lactic acid (Stiles & Holzapfel, 1997; Orla-Jensen, Orla-Jensen, & Spur, 1926). An aerobic respiration mechanism has since been revealed among LAB (Pedersen, Gaudu, Lechardeur, Petit, & Gruss, 2012), which can be found in the enteric system of humans, animals, on the surfaces of fresh fruit, cereals, vegetables, and in processed dairy goods (Oliveira, Zannini, & Arendt, 2014; Trias, Bañeras, Montesinos, & Badosa, 2008). LAB was first identified using traditional bacterial taxonomy until the late 20th century when molecular techniques became more frequently used (Stiles & Holzapfel, 1997). The first identification and classification of new LAB using molecular weight standards 16S and 23SrRNA concluded two ancestral lines for all Gram-positive bacteria (Schleifer et al., 1995). With advancement of molecular methods and the growing importance of LAB to the food and agricultural sectors stemmed the first complete genome sequencing of *Lactococcus lactis* ssp. *Lactis* (Bolotin et al., 2001). However, many strains used in industrial applications have not yet been sequenced or their sequences not yet released (Douillard & De Vos, 2014), while a phylogenetic update based on 232 genes from 28 LAB genome sequences has been carried out (Zhang, Ye, Yu, & Shi, 2011).

14.6.2 THE ROLE OF LAB IN PLANT AND HUMAN HEALTH

LAB is recognized for their role in bio-preservation of food and feed because of their ability to produce many antimicrobial compounds (Magnusson & Schnürer, 2001). This largely involves the fermentation of a carbohydrate source to produce antibacterial proteins like bacteriocin and reuterin among others (Schnürer, & Magnusson, 2005; Cleveland, Montville, Nes, & Chikindas, 2001). In agriculture, LAB have long been established as key players in the ensiling process of forage crops through their production of antifungal metabolites (Broberg, Jacobsson, Ström, & Schnürer, 2007). Plant surfaces are naturally colonized by lactic acid producing bacteria leaving the micro-ecosystem of plant surfaces preponderant to LAB (Oliveira, Zannini, & Arendt, 2014). Such is the interest in their application as bio-preservative agents that various works has been carried out on their efficacy to detoxify mycotoxins from human food (Fuchs et al., 2008). A noteworthy justification for LAB implementation as BCA's is that they are generally recognized as safe (GRAS; USA) and have QPS (EU) status, albeit that *Enterococcus* and *Streptococcus* show pathogenic potentials (Oliveira, Zannini, & Arendt, 2014; Collins, Cotter, Hill, & Ross, 2010).

The potential health benefits of living organisms added to the diet consumed by both humans and animals is well established. One such example of a probiotic is Yakult® comprised of *Lactobacillus casei* strain Shirota (Ou et al., 2019). Today, bacteria considered to have probiotic properties have established their own niche sector worth over $100 billion annually (Bull et al., 2014). This success of LAB has been attributed in their ability to produce various organic acids (butyric, lactic, phenyl lactic, formic, and propionic) antimicrobial compounds (hydrogen peroxide (H_2O_2), propionate, ethanol), bacteriocins (reuterin, nisin, lacticin), or bacteriocin-like substances following the fermentation process (Oliveira, Zannini, & Arendt, 2014; Muhialdin, Hassan, & Saari, 2013). The complex antagonistic mechanisms by which many LAB work *in vitro* have been reviewed (Dalié, Deschamps, & Richard-Forget, 2010; De Vuyst, & Vandamme, 1994). It remains challenging to distinguish the complex and often synergistic role played by the individual compounds as most fungi are naturally sensitive (particularly during conidial germination) to the acidic conditions these bacteria are cultured under (Schnürer, & Magnusson, 2005; Gerez, Font De, & Rollan, 2008). Extensive work has been done to understand the complex interactions between the conditions for growth (e.g., temperature, incubation period, growth media, pH, nutritional factors) and how they might influence the propensity and effectiveness of compounds. Organic acids can cause drop in pH which increases the plasma membrane permeability and thus neutralizing the electrochemical proton gradient which facilitates fungal death (Oliveira, Zannini, & Arendt, 2014; Schnürer, & Magnusson, 2005). However, only a minimum inhibitory concentration of these compounds alone is not enough to account for their antimicrobial activity (Ndagano, Lamoureux, Dortu, Vandermoten, & Thonart, 2011).

14.7 LAB AS FHB BIOCONTROL AGENT

The use of BCAs has emerged as alternative to noxious and costly chemical controls (Bach et al., 2016; Jackson & Schisler, 2002) to control pathogens. Their importance must not be underestimated as progressive resistance can be seen among fungal isolates towards widely used groups of fungicides (Becher, Miedaner, & Wirsel, 2013). BCA comprised of bacteria, fungi, soil microorganisms, endophytes or combination thereof offer plants amelioration to harmful pathogens and healthy plant (Rabiey, Ullah, Shaw, & Shaw, 2017; Henkes et al., 2011). So far, several potential BCA have been collected and rigorously tested for their proficiency to limit *Fusarium* diseases in economically important crops. Their role as a BCA against *Fusarium.* has been inspected, however, auxiliary research is required to evaluate if LAB are equal to or better than commercially available fungicides (Lowe & Arendt, 2004). LAB have been attributed to inhibit growth of *F. oxysporum in vitro* and *in vivo* experiments involving tomato (*Solanum lycopersicum*). LAB soil drench and seed treatments have demonstrated increased secondary root growth of up to 358 and 181% over the control (Hamed, Moustafa, & Abdel-Aziz, 2011). *In vitro* studies of over 496 LAB strains showed them to have a moderate to high antifungal activity against the growth of *Botrytis cinerea* strain CECT 2100 (Trias, Bañeras, Montesinos, & Badosa, 2008).

It is well established that *Lactobacillus* strains to inhibit *F. graminearum* conidial germination by ≥ 70% (Gerez, Font De, & Rollan, 2008). In microtiter plate well assays, *L. reuteri* strain R29 and *L. amylovorus* strain DSM19280 displayed strong inhibition of *F. culmorum* for up to 7 days (Oliveira, Zannini, & Arendt, 2014). High antifungal activity was reported for *L. amylovorus* strain DSM19280 against *F. culmorum* growth in sourdough (Ryan et al., 2011). Through the production of various acids, LAB have proven to be useful additive for the control of *Alternaria alternata* in mango and persimmon fruit while several other LAB spp. has been shown to remove mycotoxins from *in vitro* studies (Oliveira, Zannini, & Arendt, 2014; Franco, Garcia, Hirooka, Ono, & Dos, 2011; Prusky & Lichter, 2007).

Not only LAB, but a strain of *Pseudomonas fluorescens* has been identified as a possible BCA of FHB in small grain cereals in both field and glasshouse trials (Schisler, Khan, Boehm, Lipps, Slininger, & Zhang, 2006) with no deleterious effects on developing spikelets. *In vitro* testing of the *Bacillus subtilis* strain SG6 resulted in 87.9%, 95.6% and 100% inhibition of mycelium growth, sporulation, and DON production of *F. graminearum* (Zhao et al., 2014). It is well reported of LAB strains with ability to inhibit *F. graminearum* growth and DON production (Franco, Garcia, Hirooka, Ono, & Dos, 2011). Despite their potential, there has been several reported obstacles facing BCA use, including delivery mechanisms, viability issues and compatibility in formulation mixes (Yuen & Schoneweis, 2007). Although promising, further work is required to unearth future BCA which provide control at levels like or exceeding that of commercially available fungicides. The different LAB as biocontrol agents have been studied and some of its relevant uses as BCA is summarized in Table 14.1.

14.8 POSSIBLE LAB BIOCONTROL MECHANISMS OF FHB

The LAB as bacterial endophytes can promote plant growth and FHB control (Hallmann, Quadt-Hallmann, Mahaffee, & Kloepper, 1997; Chen et al., 2018). The mode of action of LAB in control of diseases may be immediate or delayed response. The LAB can directly control *F. graminearum* by (i) the production of inhibitory secondary metabolites such as lipopeptide antibiotics, phenazine derivatives; (ii) competitive blockage of pathogen invasion sites; (iii) biofilm formation and colonization; (iv) inhibition of spores germination and growth; and (v) detoxification of the virulence factors of the phyto-pathogen (Chen et al., 2018; Compant, Duffy, Nowak, Clément, & Barka, 2005; Baffoni et al.,

TABLE 14.1 Lactic Acid Bacteria (LAB); Compound(s) Produced and Their Antimicrobial Activity Spectrum

LAB Isolate	Compound(s)	Antimicrobial Spectrum	Reference(s)
L. brevis P68	*Para* hydroxyl Benzoic acid.	Broad spectrum.	Arasu et al. (2015)
L. brevis B696	Acetic acid	*Penicillium disco, Aspergillus versicolor.*	Cabo et al. (2002)
L. brevis AM7	Five antifungal peptides	*Penicillium roqueforti* DPPMAF1.	Coda et al. (2008)
L. reuteri CRL 1098	Lactic acid, acetic acid, phenyllactic acid.	*Aspergillus niger* CH 101, *Penicillium* sp. CH 102, *F. graminearum* CH 103, *Geotrichum citriaurantii* INTA1, *Penicillium digitatum* INTA2.	Gerez et al. (2013)
L. reuteri 1100	Acetic acid, phenyllactic acid.	*F. graminearum*	Gerez et al. (2009)
L. brevis R2Δ	Cyclic dipeptides cyclo(leu-pro), cyclo(pro-pro), cyclo(met-pro) and cyclo(phe-pro).	ND	Axel et al. (2014)
L. amylovorus DSM19280	Carboxylic acids, Sodium decanoate, *Nucleosides, Cyclic-dipetides.*	*Aspergillus fumigatus* J9, *Aspergillus niger* FST 4.21, *F. culmorum* FST 4.05.	Ryan et al. (2011)
L. brevis (B23)	Bacteriocin like substance.	*Staphylococcus aureus, Escherichia coli*, and *salmonella typhi*	Chaudhary et al. (2021); Rushdy & Gomaa (2013)
L. amylovorus DCE 471	Bacteriocinamylovorin L471	*L. delbrueckii* subsp. *Bulgaricus* LMG690lT.	DeVuyst et al. (1996)
L. amylovorus FST2.11	15 Organic acids	ND	Brosnan et al. (2012)

2015). Else, LAB can indirectly induce plant systemic resistance responses (induction of systemic resistance (ISR)), to react faster and more efficiently upon subsequent pathogen invasion (Griffin, 2014). Another form of induced resistance is the systemic acquired resistance (SAR). In both SAR and ISR, the defenses of the plants are primed by previous infections that provoke resistance (or tolerance) against future pathogens attack.

LAB can be potential biocontrol agents since they produce bioactive compounds with antimicrobial properties against *Fusarium*. In addition,

LAB Has been reported of mycotoxin detoxification (Baffoni et al., 2015) ability. *L. plantarum* SLG17 and *B. amyloliquefaciens* FLN13 were used as BCA during heading period until anthesis to reduce FHB index in durum wheat (Baffoni et al., 2015). Therein, *Fusarium* growth and trichothecene accumulation was significantly reduced in grain by antifungal compounds from LAB (Zhao, 2013). The LAB inhibited *F. graminearum* spread probably due to the nutrient competition and/or production of antagonistic compounds and DON detoxification (Franco, Garcia, Hirooka, Ono, & Dos, 2011). Though the antifungal activity of LAB is a complex process, but the production of antifungal compounds is well documented. The LAB is reported to produced inhibitory compounds originating from the proteins (Magnusson & Schnürer, 2001), including low molecular weight compounds (phenyl lactic acid, reuterin, cyclic dipeptides, benzoic acid, hydroxylated fatty acids, methyl hydantoin, and mevalonolactone) (Schnürer, & Magnusson, 2005; Niku-Paavola, Laitila, Mattila-Sandholm, & Haikara, 1999; Okkers, Dicks, Silvester, Joubert, & Odendaal, 1999). New preservation systems of antifungal LAB strains with stabilizer/enhancer agents could be the future use as bio-strategies for controlling FHB growth (Garrote et al., 2010). LAB isolated from wine fermentations, particularly *L. plantarum*, *L. hilgardii*, *L. paracasei*, and *L. lactis* species were effective to protect Lycopersicon esculentum plants against the fungus *F. oxysporum* and promote its growth.

Recently there is report of application of *C. rosea* as BCA against *F. graminearum* on maize mulch to degrade inaccessible FHB (Gimeno et al., 2020) inoculum. The inoculum control may be due to inhibited perithecia development, ascospore discharge, reduction in DON and ZEN content (López-Seijas, García-Fraga, Da Silva, & Sieiro, 2020). Not only protection from fungal diseases, the LAB may have role to play in germination as LAB were isolated from germinated Norwegian barley (Østlie, Porcellato, Kvam, & Wicklund, 2021).

14.9 FACTORS AND CHALLENGES AFFECTING BIOCONTROL EFFICIENCY IN THE FIELD

Despite numerous studies of microorganisms as potential BCA for control of FHB (Table 14.1), there are many problems to overcome for field applications. One of the greatest challenges being sustainable maintenance of cell viability, which has been key factor of very less availability of commercial products in market.

The progression of disease and its biocontrol in field is determined by isolate virulence, formulation efficacy and prevailing environmental conditions. The efficacy and stability of BCA across crop seasons and location depends on formulation (granular, powder or liquid), delivery and viability of BCAs. Hence, the most effective BCAs will be stable across seasons/ environment maintaining efficacy, adhesiveness, viability, and proper distribution.

Additionally, the success of biocontrol also depends on the biology of the pathogen, viability across season, application time, storage, and distribution condition. BCA application prior to infection effectively control the phytopathogen, irrespective of magnitude of disease control. However, the limits and percentage of crop loss before and after using BCA should be considered to determine the sustainability FHB control.

14.10 CONCLUSION AND FUTURE PERSPECTIVES

To cope with the increasing world population, the major concern is to secure food availability with enhanced yield per area and decrease in yield losses. Therefore, reducing crop yield in field and storage is another way to maintain food sustainability.

The implementation of stricter legislation regarding chemicals, environmental safety, and human health has reduced chemical use. Hence, control of FHB with BCAs could be an alternative environmentally sustainable strategy to be added to an IPM.

The use of LAB as BCA to control FHB infection and DON reduction in wheat is reported. However, there are still issues regarding formulation, time of application, and selectivity of BCA. Though the improvement in efficacy of biological fungicides is a continual process, which involves repetitive screening, validation, optimization, upscaling, dose, and application time recommendation.

Future BCAs should be selective, stable with high efficacy to control pathogens. Hence, one way to obtain such BCA is to use LABs of broad activity with the potential to reduce pathogenicity of *Fusarium*.

The sustainable use and stability of BCAs must overcome the issues of natural climate change in field conditions. Therefore, the search for lead BCAs, its validation and optimization will continue until sustainable increase of crop yield and food availability.

KEYWORDS

- **biocontrol Fusarium head blight**
- **biocontrol mechanisms**
- **cereals**
- **food safety**
- **Fusarium graminearum**
- **lactic acid bacteria**
- **mycotoxins**

REFERENCES

Abbott, S. P., (2002). Mycotoxins and indoor molds. *Indoor Environment Connections*, *3*(4), 14–24.

Alabouvette, C., Olivain, C., & Steinberg, C., (2006). Biological control of plant diseases: The European situation. *European Journal of Plant Pathology*, *114*(3), 329–341.

Bach, E., Dos Santos, S. G. D., De Carvalho, F. G., Lisboa, B. B., & Passaglia, L. M. P., (2016). Evaluation of biological control and rhizosphere competence of plant growth promoting bacteria. *Applied Soil Ecology*, *99*, 141–149.

Baffoni, L., Gaggia, F., Dalanaj, N., Prodi, A., Nipoti, P., Pisi, A., Biavati, B., & Di Gioia, D., (2015). Microbial inoculants for the biocontrol of *Fusarium* spp. in durum wheat. *BMC Microbiology*, *15*(1), 1–10.

Balmas, V., Scherm, B., Marcello, A., Beyer, M., Hoffmann, L., Migheli, Q., & Pasquali, M., (2015). *Fusarium* species and chemotypes associated with *fusarium* head blight and *fusarium* root rot on wheat in Sardinia. *Plant Pathology*, *64*(4), 972–979.

Bateman, G., (2005). The contribution of ground-level inoculum of *Fusarium culmorum* to ear blight of winter wheat. *Plant Pathology*, *54*(3), 299–307.

Becher, R., Miedaner, T., & Wirsel, S. G., (2013). 8 Biology, diversity, and management of FHB-causing *Fusarium* species in small-grain cereals. In: *Agricultural Applications* (pp. 199–241). Springer.

Bennett, J., & Klich, M., (2003). Chotoxins. *Clin. Microbiol. Rev*, *16*, 497–516.

Bolotin, A., Wincker, P., Mauger, S., Jaillon, O., Malarme, K., Weissenbach, J., Ehrlich, S. D., & Sorokin, A., (2001). The complete genome sequence of the lactic acid bacterium *Lactococcus lactis* ssp. lactis IL1403. *Genome Research*, *11*(5), 731–753.

Bottalico, A., & Perrone, G., (2002). Toxigenic *Fusarium* species and mycotoxins associated with head blight in small-grain cereals in Europe. In: *Mycotoxins in Plant Disease* (pp. 611–624). Springer.

Broberg, A., Jacobsson, K., Ström, K., & Schnürer, J., (2007). Metabolite profiles of lactic acid bacteria in grass silage. *Applied and Environmental Microbiology*, *73*(17), 5547–5552.

Buerstmayr, H., Lemmens, M., Hartl, L., Doldi, L., Steiner, B., Stierschneider, M., & Ruckenbauer, P., (2002). Molecular mapping of QTLs for *Fusarium* head blight resistance in spring wheat. I. Resistance to fungal spread (type II resistance). *Theoretical and Applied Genetics*, *104*(1), 84–91.

Bull, M. J., Jolley, K. A., Bray, J. E., Aerts, M., Vandamme, P., Maiden, M. C., Marchesi, J. R., & Mahenthiralingam, E., (2014). The domestication of the probiotic bacterium *Lactobacillus acidophilus*. *Scientific Reports*, *4*(1), 1–8.

Cheli, F., Battaglia, D., Gallo, R., & Dell'Orto, V., (2014). EU legislation on cereal safety: An update with a focus on mycotoxins. *Food Control*, *37*, 315–325.

Chen, Y., Wang, J., Yang, N., Wen, Z., Sun, X., Chai, Y., & Ma, Z., (2018). Wheat microbiome bacteria can reduce virulence of a plant pathogenic fungus by altering histone acetylation. *Nature Communications*, *9*(1), 1–14.

Cleveland, J., Montville, T. J., Nes, I. F., & Chikindas, M. L., (2001). Bacteriocins: Safe, natural antimicrobials for food preservation. *International Journal of Food Microbiology*, *71*(1), 1–20.

Collins, B., Cotter, P. D., Hill, C., & Ross, R. P., (2010). Applications of lactic acid bacteria-produced bacteriocins. *Biotechnology of Lactic Acid Bacteria*, 89–109.

Compant, S., Duffy, B., Nowak, J., Clément, C., & Barka, E. A., (2005). Use of plant growth-promoting bacteria for biocontrol of plant diseases: Principles, mechanisms of action, and future prospects. *Applied and Environmental Microbiology*, *71*(9), 4951–4959.

Dalié, D., Deschamps, A., & Richard-Forget, F., (2010). Lactic acid bacteria–potential for control of mould growth and mycotoxins: A review. *Food Control*, *21*(4), 370–380.

De Vuyst, L., & Vandamme, E. J., (1994). Antimicrobial potential of lactic acid bacteria. In: *Bacteriocins of Lactic Acid Bacteria* (pp. 91–142). Springer.

Degrassi, G., & Carpentieri-Pipolo, V., (2020). *Biological Control of Fusarium Head Blight by Bacterial Endophytes and Reduction of Deoxynivalenol in Wheat.* Embrapa Trigo-Artigo em Periódico Indexado (ALICE).

Doohan, F., Brennan, J., & Cooke, B., (2003). Influence of climatic factors on *Fusarium* species pathogenic to cereals. In: *Epidemiology of Mycotoxin Producing Fungi* (pp. 755–768). Springer.

Douillard, F. P., & De Vos, W. M., (2014). Functional genomics of lactic acid bacteria: From food to health. *Microbial Cell Factories*, *13*(1), 1–21.

Fanelli, F., Schmidt-Heydt, M., Haidukowski, M., Geisen, R., Logrieco, A., & Mulè, G., (2012). Influence of light on growth, fumonisin biosynthesis and FUM1 gene expression by *Fusarium proliferatum*. *International Journal of Food Microbiology*, *153*(1, 2), 148–153.

Fernandez, M., & Jefferson, P., (2004). Fungal populations in roots and crowns of common and durum wheat in Saskatchewan. *Canadian Journal of Plant Pathology*, *26*(3), 325–334.

Ferrigo, D., Raiola, A., & Causin, R., (2016). *Fusarium* toxins in cereals: Occurrence, legislation, factors promoting the appearance and their management. *Molecules*, *21*(5), 627.

Franco, T., Garcia, S., Hirooka, E., Ono, Y., & Dos, S. J., (2011). Lactic acid bacteria in the inhibition of *Fusarium graminearum* and deoxynivalenol detoxification. *Journal of Applied Microbiology*, *111*(3), 739–748.

Fuchs, S., Sontag, G., Stidl, R., Ehrlich, V., Kundi, M., & Knasmüller, S., (2008). Detoxification of patulin and ochratoxin A, two abundant mycotoxins, by lactic acid bacteria. *Food and Chemical Toxicology*, *46*(4), 1398–1407.

Garrote, G., Abraham, A., De Antoni, G., Mozzi, F., Raya, R., & Vignolo, G., (2010). *Biotechnology of Lactic Acid Bacteria: Novel Applications.* Hoboken New Jersey.

Gerez, C. L., Font De, V. G., & Rollan, G. C., (2008). Functionality of lactic acid bacteria peptidase activities in the hydrolysis of gliadin-like fragments. *Letters in Applied Microbiology*, *47*(5), 427–432.
Gilbert, J., & Fernando, W., (2004). Epidemiology and biological control *of Gibberella zeae/ Fusarium graminearum*. *Canadian Journal of Plant Pathology*, *26*(4), 464–472.
Gilbert, J., & Haber, S., (2013). Overview of some recent research developments in *Fusarium* head blight of wheat. *Canadian Journal of Plant Pathology*, *35*(2), 149–174.
Gimeno, A., Kägi, A., Drakopoulos, D., Bänziger, I., Lehmann, E., Forrer, H. R., Keller, B., & Vogelgsang, S., (2020). From laboratory to the field: Biological control of *Fusarium graminearum* on infected maize crop residues. *Journal of Applied Microbiology*, *129*(3), 680–694.
Gimeno, A., Leimgruber, M., Kägi, A., Jenny, E., & Vogelgsang, S., (2021). UV protection and shelf life of the biological control agent *Clonostachys rosea* against *Fusarium graminearum*. *Biological Control*, *158*, 104600.
Griffin, M. R., (2014). Biocontrol and bioremediation: Two areas of endophytic research which hold great promise. *Advances in Endophytic Research*, 257–282.
Haidukowski, M., Pascale, M., Perrone, G., Pancaldi, D., Campagna, C., & Visconti, A., (2005). Effect of fungicides on the development of *Fusarium* head blight, yield and deoxynivalenol accumulation in wheat inoculated under field conditions with *Fusarium graminearum* and *Fusarium culmorum*. *Journal of the Science of Food and Agriculture*, *85*(2), 191–198.
Hallmann, J., Quadt-Hallmann, A., Mahaffee, W., & Kloepper, J., (1997). Bacterial endophytes in agricultural crops. *Canadian Journal of Microbiology*, *43*(10), 895–914.
Hamed, H. A., Moustafa, Y. A., & Abdel-Aziz, S. M., (2011). *In vivo* efficacy of lactic acid bacteria in biological control against *Fusarium oxysporum* for protection of tomato plant. *Life Science Journal*, *8*(4), 462–468.
Henkes, G. J., Jousset, A., Bonkowski, M., Thorpe, M. R., Scheu, S., Lanoue, A., Schurr, U., & Röse, U. S., (2011). Pseudomonas fluorescens CHA0 maintains carbon delivery to *Fusarium graminearum*-infected roots and prevents reduction in biomass of barley shoots through systemic interactions. *Journal of Experimental Botany*, *62*(12), 4337–4344.
Hückelhoven, R., Hofer, K., Coleman, A., & Heß, M., (2018). *Fusarium* infection of malting barley has to be managed over the entire value chain. *Journal of Plant Diseases and Protection*, *125*(1), 1–4.
Jackson, M., & Schisler, D., (2002). In selecting fungal biocontrol agents amenable to production by liquid culture fermentation. In: Elad, Y., Kohl, J., & Shtienberg, D., (eds.), *Proceedings of the 7th Biocontrol Working Group Meeting, Influence of Abiotic and Biotic Factors on Biocontrol Agents, IOBC WPRS Bulletin* (pp. 387Á391).
Kang, Z., & Buchenauer, H., (1999). Immunocytochemical localization of *Fusarium* toxins in infected wheat spikes by *Fusarium culmorum*. *Physiological and Molecular Plant Pathology*, *55*(5), 275–288.
Karlsson, I., Persson, P., & Friberg, H., (2021). *Fusarium* head blight from a microbiome perspective. *Frontiers in Microbiology*, *12*, 371.
Keller, M. D., Bergstrom, G. C., & Shields, E. J., (2014). The aerobiology of *Fusarium graminearum*. *Aerobiologia*, *30*(2), 123–136.
Kokkonen, M., Ojala, L., Parikka, P., & Jestoi, M., (2010). Mycotoxin production of selected *Fusarium* species at different culture conditions. *International Journal of Food Microbiology*, *143*(1, 2), 17–25.

Langevin, F., Eudes, F., & Comeau, A., (2004). Effect of trichothecenes produced by *Fusarium graminearum* during *Fusarium* head blight development in six cereal species. *European Journal of Plant Pathology*, *110*(7), 735–746.

Lara, C., & Ornelas, J. F., (2003). Hummingbirds as vectors of fungal spores in *Moussonia deppeana* (*Gesneriaceae*): Taking advantage of a mutualism? *American Journal of Botany*, *90*(2), 262–269.

Leplat, J., Friberg, H., Abid, M., & Steinberg, C., (2013). Survival of *Fusarium graminearum*, the causal agent of *Fusarium* head blight: A review. *Agronomy for Sustainable Development*, *33*(1), 97–111.

Liu, W., Elen, O. N., Sundheim, L., Langseth, W., & Skinnes, H., (1997). Comparison of visual head blight ratings, seed infection levels, and deoxynivalenol production for assessment of resistance in cereals inoculated with *Fusarium culmorum*. *European Journal of Plant Pathology*, *103*(7), 589–595.

Logrieco, A., Bottalico, A., Mulé, G., Moretti, A., & Perrone, G., (2003). Epidemiology of toxigenic fungi and their associated mycotoxins for some Mediterranean crops. In: *Epidemiology of Mycotoxin Producing Fungi* (pp. 645–667). Springer.

López-Seijas, J., García-Fraga, B., Da Silva, A. F., & Sieiro, C., (2020). Wine lactic acid bacteria with antimicrobial activity as potential biocontrol agents against *Fusarium Oxysporum* f. sp. *lycopersici*. *Agronomy*, *10*(1), 31.

Lowe, D. P., & Arendt, E. K., (2004). The use and effects of lactic acid bacteria in malting and brewing with their relationships to antifungal activity, mycotoxins, and gushing: A review. *Journal of the Institute of Brewing*, *110*(3), 163–180.

Magnusson, J., & Schnürer, J., (2001). *Lactobacillus coryniformis* subsp. *coryniformis* strain Si3 produces a broad-spectrum proteinaceous antifungal compound. *Applied and Environmental Microbiology*, *67*(1), 1–5.

Maiorano, A., Blandino, M., Reyneri, A., & Vanara, F., (2008). Effects of maize residues on the *Fusarium* spp. infection and deoxynivalenol (DON) contamination of wheat grain. *Crop Protection*, *27*(2), 182–188.

Malachova, A., Hajslova, J., Ehrenbergerova, J., Kostelanska, M., Zachariasova, M., Urbanova, J., Cerkal, R., et al., (2010). *Fusarium* mycotoxins in spring barley and their transfer into malt. *Kvasny Prumysl (Czech Republic)*.

Matny, O., (2015). *Fusarium* head blight and crown rot on wheat & barley: Losses and health risks. *Adv. Plants Agric. Res.*, *2*(1), 00039.

McMullen, M., Bergstrom, G., De Wolf, E., Dill-Macky, R., Hershman, D., Shaner, G., & Van, S. D., (2012). A unified effort to fight an enemy of wheat and barley: *Fusarium* head blight. *Plant Disease*, *96*(12), 1712–1728.

McMullen, M., Jones, R., & Gallenberg, D., (1997). Scab of wheat and barley: A re-emerging disease of devastating impact. *Plant Disease*, *81*(12), 1340–1348.

Melloy, P., Hollaway, G., Luck, J., Norton, R., Aitken, E., & Chakraborty, S., (2010). Production and fitness of *Fusarium pseudograminearum* inoculum at elevated carbon dioxide in FACE. *Global Change Biology*, *16*(12), 3363–3373.

Muhialdin, B. J., Hassan, Z., & Saari, N., (2013). Lactic acid bacteria in biopreservation and the enhancement of the functional quality of bread. *Lactic Acid Bacteria R & D for Food, Health and Livestock Purposes*, 155–172.

Ndagano, D., Lamoureux, T., Dortu, C., Vandermoten, S., & Thonart, P., (2011). Antifungal activity of 2 lactic acid bacteria of the *Weissella* genus isolated from food. *Journal of Food Science*, *76*(6), M305–M311.

Nganje, W. E., Bangsund, D. A., Leistritz, F. L., Wilson, W. W., & Tiapo, N. M., (2002). In estimating the economic impact of a crop disease: The case of *Fusarium* head blight in US wheat and barley. *National Fusarium Head Blight Forum Proceedings* (pp. 275–281). Citeseer.

Niku-Paavola, M. L., Laitila, A., Mattila-Sandholm, T., & Haikara, A., (1999). New types of antimicrobial compounds produced by *Lactobacillus plantarum*. *Journal of Applied Microbiology*, *86*(1), 29–35.

Okkers, D., Dicks, L., Silvester, M., Joubert, J., & Odendaal, H., (1999). Characterization of pentocin TV35b, a bacteriocin-like peptide isolated from *Lactobacillus pentosus* with a fungistatic effect on *Candida albicans*. *Journal of Applied Microbiology*, *87*(5), 726–734.

Oliveira, P. M., Zannini, E., & Arendt, E. K., (2014). Cereal fungal infection, mycotoxins, and lactic acid bacteria mediated bioprotection: From crop farming to cereal products. *Food Microbiology*, *37*, 78–95.

Orla-Jensen, S., Orla-Jensen, A., & Spur, B., (1926). The butter aroma bacteria. *Journal of Bacteriology*, *12*(5), 333–341.

Osborne, L. E., & Stein, J. M., (2007). Epidemiology of *Fusarium* head blight on small-grain cereals. *International Journal of Food Microbiology*, *119*(1, 2), 103–108.

Østlie, H. M., Porcellato, D., Kvam, G., & Wicklund, T., (2021). Investigation of the microbiota associated with ungerminated and germinated Norwegian barley cultivars with focus on lactic acid bacteria. *International Journal of Food Microbiology*, *341*, 109059.

Ou, Y., Chen, S., Ren, F., Zhang, M., Ge, S., Guo, H., Zhang, H., & Zhao, L., (2019). Lactobacillus casei strain Shirota alleviates constipation in adults by increasing the pipecolinic acid level in the gut. *Frontiers in Microbiology*, *10*, 324.

Pan, D., Mionetto, A., Tiscornia, S., & Bettucci, L., (2015). Endophytic bacteria from wheat grain as biocontrol agents of *Fusarium graminearum* and deoxynivalenol production in wheat. *Mycotoxin Research*, *31*(3), 137–143.

Parry, D., (1990). The incidence of *Fusarium* spp. in stem bases of selected crops of winter wheat in the Midlands, UK. *Plant Pathology*, *39*(4), 619–622.

Parry, D., Jenkinson, P., & McLeod, L., (1995). *Fusarium* ear blight (scab) in small grain cereals—A review. *Plant Pathology*, *44*(2), 207–238.

Pedersen, M. B., Gaudu, P., Lechardeur, D., Petit, M. A., & Gruss, A., (2012). Aerobic respiration metabolism in lactic acid bacteria and uses in biotechnology. *Annual Review of Food Science and Technology*, *3*, 37–58.

Perez, C., Dill-Macky, R., & Kinkel, L. L., (2008). Management of soil microbial communities to enhance populations of *Fusarium graminearum*-antagonists in soil. *Plant and Soil*, *302*(1), 53–69.

Pettitt, T. R., & Parry, D. W., (1996). Effects of climate change on Fusarium foot rot of winter wheat in the United Kingdom. In: Frankland, J. C., Magan, N., Gadd, G. M., (eds.). *Fungi and Environmental Change.* British Mycological Society Symposia. Cambridge: Cambridge University Press, 20-31. doi:10.1017/CBO9780511753190.003.

Pettitt, T., Xu, X., & Parry, D., (2003). Association of *Fusarium* species in the wheat stem rot complex. *European Journal of Plant Pathology*, *109*(7), 769–774.

Prusky, D., & Lichter, A., (2007). Mechanisms modulating fungal attack in post-harvest pathogen interactions and their control. In: *Sustainable Disease Management in a European Context* (pp. 281–289). Springer.

Rabiey, M., Ullah, I., Shaw, L. J., & Shaw, M. W., (2017). Potential ecological effects of *Piriformospora indica*, a possible biocontrol agent, in UK agricultural systems. *Biological Control, 104*, 1–9.

Rai, A., Das, M., & Tripathi, A., (2020). Occurrence and toxicity of a *fusarium* mycotoxin, zearalenone. *Critical Reviews in Food Science and Nutrition, 60*(16), 2710–2729.

Ryan, L. A., Zannini, E., Dal Bello, F., Pawlowska, A., Koehler, P., & Arendt, E. K., (2011). *Lactobacillus amylovorus* DSM 19280 as a novel food-grade antifungal agent for bakery products. *International Journal of Food Microbiology, 146*(3), 276–283.

Salas, B., Steffenson, B., Casper, H., Tacke, B., Prom, L., Fetch, Jr. T., & Schwarz, P., (1999). *Fusarium* species pathogenic to barley and their associated mycotoxins. *Plant Disease, 83*(7), 667–674.

Scherm, B., Balmas, V., Spanu, F., Pani, G., Delogu, G., Pasquali, M., & Migheli, Q., (2013). *Fusarium culmorum*: Causal agent of foot and root rot and head blight on wheat. *Molecular Plant Pathology, 14*(4), 323–341.

Schisler, D., Khan, N., Boehm, M., Lipps, P., Slininger, P., & Zhang, S., (2006). Selection and evaluation of the potential of choline-metabolizing microbial strains to reduce *Fusarium* head blight. *Biological Control, 39*(3), 497–506.

Schleifer, K. H., Ehrmann, M., Beimfohr, C., Brockmann, E., Ludwig, W., & Amann, R., (1995). Application of molecular methods for the classification and identification of lactic acid bacteria. *International Dairy Journal, 5*(8), 1081–1094.

Schnürer, J., & Magnusson, J., (2005). Antifungal lactic acid bacteria as biopreservatives. *Trends in Food Science & Technology, 16*(1–3), 70–78.

Shala-Mayrhofer, V., Varga, E., Marjakaj, R., Berthiller, F., Musolli, A., Berisha, D., Kelmendi, B., & Lemmens, M., (2013). Investigations on *Fusarium* spp. and their mycotoxins causing *Fusarium* ear rot of maize in Kosovo. *Food Additives & Contaminants: Part B, 6*(4), 237–243.

Starkey, D. E., Ward, T. J., Aoki, T., Gale, L. R., Kistler, H. C., Geiser, D. M., Suga, H., et al., (2007). Global molecular surveillance reveals novel *Fusarium* head blight species and trichothecene toxin diversity. *Fungal Genetics and Biology, 44*(11), 1191–1204.

Stenglein, S., (2009). *Fusarium poae*: A pathogen that needs more attention. *Journal of Plant Pathology*, 25–36.

Steyn, P. S., (1995). Mycotoxins, general view, chemistry, and structure. *Toxicology Letters, 82*, 843–851.

Stiles, M. E., & Holzapfel, W. H., (1997). Lactic acid bacteria of foods and their current taxonomy. *International Journal of Food Microbiology, 36*(1), 1–29.

Strange, R., & Smith, H., (1971). A fungal growth stimulant in anthers which predisposes wheat to attack by *Fusarium graminearum*. *Physiological Plant Pathology, 1*(2), 141–150.

Summerell, B. A., Laurence, M. H., Liew, E. C., & Leslie, J. F., (2010). Biogeography and phylogeography of *Fusarium*: A review. *Fungal Diversity, 44*(1), 3–13.

To, C., (1930). Physiologic specialization in *Fusarium* spp. causing headblight of small grains. University of Minnesota. Minnesota Agricultural Experiment Station. University Farm. Retrieved from the University of Minnesota Digital Conservancy. https://hdl.handle.net/11299/204213.

Trail, F., (2009). For blighted waves of grain: *Fusarium graminearum* in the postgenomics era. *Plant Physiology, 149*(1), 103–110.

Trias, M. R., Bañeras, V. L., Montesinos, S. E., & Badosa, R. E., (2008). Lactic acid bacteria from fresh fruit and vegetables as biocontrol agents of phytopathogenic bacteria and fungi. *International Microbiology, 2008*(11), 231–236.

Tunali, B., Nicol, J. M., Hodson, D., Uçkun, Z., Büyük, O., Erdurmuş, D., Hekimhan, H., et al., (2008). Root and crown rot fungi associated with spring, facultative, and winter wheat in Turkey. *Plant Disease, 92*(9), 1299–1306.

Van, E. H., & Jonker, M., (2004). Current regulations governing mycotoxin limits in food. *Mycotoxins in Food: Detection and Control*, 49–68.

Vogelgsang, S., Beyer, M., Pasquali, M., Jenny, E., Musa, T., Bucheli, T. D., Wettstein, F. E., & Forrer, H. R., (2019). An eight-year survey of wheat shows distinctive effects of cropping factors on different *Fusarium* species and associated mycotoxins. *European Journal of Agronomy, 105*, 62–77.

Wachowska, M., Gabrysiak, M., Muchowicz, A., Bednarek, W., Barankiewicz, J., Rygiel, T., Boon, L., et al., (2014). 5-Aza-2′-deoxycytidine potentiates anti-tumor immune response induced by photodynamic therapy. *European Journal of Cancer, 50*(7), 1370–1381.

Wegulo, S. N., Baenziger, P. S., Nopsa, J. H., Bockus, W. W., & Hallen-Adams, H., (2015). Management of *Fusarium* head blight of wheat and barley. *Crop Protection, 73*, 100–107.

Wegulo, S. N., Bockus, W. W., Nopsa, J. F. H., Peiris, K. H., & Dowell, F. E., (2013). Integration of fungicide application and cultivar resistance to manage *Fusarium* head blight in wheat. *Fungicides-Showcases of Integrated Plant Disease Management from Around the World*, 35–54.

Windels, C. E., (2000). Economic and social impacts of *Fusarium* head blight: Changing farms and rural communities in the Northern great plains. *Phytopathology, 90*(1), 17–21.

Xu, X., (2003). Effects of environmental conditions on the development of *Fusarium* ear blight. In: *Epidemiology of Mycotoxin Producing Fungi* (pp. 683–689). Springer.

Yuen, G. Y., & Schoneweis, S. D., (2007). Strategies for managing *Fusarium* head blight and deoxynivalenol accumulation in wheat. *International Journal of Food Microbiology, 119*(1, 2), 126–130.

Zachariasova, M., Dzuman, Z., Veprikova, Z., Hajkova, K., Jiru, M., Vaclavikova, M., Zachariasova, A., et al., (2014). Occurrence of multiple mycotoxins in European feeding stuffs, assessment of dietary intake by farm animals. *Animal Feed Science and Technology, 193*, 124–140.

Zhang, Z. G., Ye, Z. Q., Yu, L., & Shi, P., (2011). Phylogenomic reconstruction of lactic acid bacteria: An update. *BMC Evolutionary Biology, 11*(1), 1–12.

Zhao, H., (2013). *Inhibition of Fusarium Growth and Trichothecene Accumulation in Grain by Antifungal Compounds from Lactic Acid Bacteria*. North Dakota State University.

Zhao, Y., Selvaraj, J. N., Xing, F., Zhou, L., Wang, Y., Song, H., Tan, X., et al., (2014). Antagonistic action of *Bacillus subtilis* strain SG6 on *Fusarium graminearum*. *PloS One, 9*(3), e92486.

CHAPTER 15

Recent Advancement in Nanotechnology for Crop Disease Management

PRANJAL KR. KAMAN,[1] PRANAB DUTTA,[2] and ASHOK BHATTACHARYYA[1]

[1]*Department of Plant Pathology, College of Agriculture, Assam Agricultural University, Jorhat, Assam, India*

[2]*School of Crop Protection, College of Post-Graduate Studies in Agricultural Sciences, Central Agricultural University, Umiam, Meghalaya, India*

ABSTRACT

Recent disease management includes the use of chemicals, which leave a heavy amount of residue in the soil, causing significant soil pollution. However, the use of these agrochemicals is not recommended globally due to their hazardous effects on human beings, including livestock, high residual effect, toxicity to the environment, outbreaks of pests, and negative effects on beneficial insects. Approaches in nanotechnology include early detection of plant pathogens, disease management, nanosensors, etc., and many others. This chapter mainly deals with some achievements in better plant health management through the biologically synthesized nanoparticles.

15.1 INTRODUCTION

Nanotechnology can be defined as the design, characterization, production, and application of structures, devices, and systems by manipulating shape and size at the nano-level. The science mainly deals with materials at the nanoscale and their exploitation in agriculture for different applications. In the coming

New Advances in Crop Disease Management. Sangita Sahni, Ganesh Thapa, and Bishun Deo Prasad (Eds)

years, nanotechnology will emerge as a new science interconnecting with other disciplines like physics, chemistry, biology, material science, and medicine.

Nanomaterials is a branch of material science division that leads to nanotechnology. The science mainly studies the morphological characters at the nanolevel, and preference is mainly given at the nanoscale. The term nanoscale can be defined as one-tenth of a micrometer, which is much smaller in one dimension.

Nanoparticles are the basic building blocks of nanotechnology. Nanoparticles are usually in the range of 0.1 to 1,000 nm. The possible ways by which nanoparticles can be synthesized are chemical, physical, and biological. Though the synthesis of the chemical process takes a shorter period of time compared to other methods, it requires extra material like capping agents that are required for stabilization. Moreover, the chemicals used for the synthesis of nanoparticles are found to be toxic, and the end product formed will be non-eco-friendly. Thus, there is a need of the hour for approaches like biological synthesis for the formation of nanoparticles, which will use toxic-free chemicals and byproducts. Thus, there is an increasing demand for "Green Nanotechnology." Extracellular and intracellular nanoparticles synthesis are the approaches that have been used in the biological approaches with the help of microorganisms like fungi, bacteria, and viruses as well as plant extract.

Sondi & Salopek (2004) reported that silver ions and silver-based composites are highly detrimental for microorganisms. Therefore, silver ions have been used in numerous types of formulations (Aymonier et al., 2002) and lately, it was observed that combination of silver nanoparticle with amphiphilic hyper branched macromolecules having a surface coating proved to have effective antimicrobial activity (Retchkiman et al., 2006). Nanoparticles like copper, zinc, titanium (Gu et al., 2003), magnesium, gold, alginate (Ahmad et al., 2005) and silver have also been tested, but silver nanoparticles have an excellent antimicrobial effectiveness against bacteria, viruses, and fungi (Gong et al., 2007; Lead et al., 2006).

Environment friendly synthesis of nanoparticles include microorganisms and plants extract are used widely because of their availability in large quantity. Microbes are widely used for the synthesis purpose because they have the capability to incorporate inorganic material within or outside the cell leading to formation of nanoparticles. Microbe based synthesis of nanoparticle is a greener approach interlinking nanotechnology and microbial biotechnology. Biological synthesis of gold, silver, titanium nanoparticles by bacteria, actinomycetes, fungi, yeasts, and viruses have been recorded by various scientists and scholars. Out of different nanoparticles silver is the most preferred and widely used nanoparticles by the consumer. In the coming years, biological synthesis of nanoparticles is gaining more interest over physio-chemical methods. Biosynthesis of nanoparticles is an important aspect because it helps in synthesizing nanoparticle with pre-defined properties. Apart from this, toxic metals can be remediated through reduction of metal ions in which microorganisms play an important role. With regards to microbes' fungi proves to be widely used by the nanotechnologists because of its unique physio-chemical properties as compared to bacteria, yeast, plants (Dias et al., 2002). Moreover, fungi are easy to manage need simple nutrient, with high wall binding capacity, as well as intracellular metal uptake capabilities (Dutta et al., 2015). Some of the fungus which have been widely used for synthesis of nanoparticles include *Trichoderma reseei* (Vahabi et al., 2011), *T. viride* (Fayaz et al., 2010), *Phytophthora infestans* (Thirumurugan et al., 2009), *Aspergillus niger* (Jaidev & Narasimha, 2010), *A. flavus* (Jain et al., 2011), *A. clavatus* (Verma et al., 2011), *Fusarium oxysporum* (Duran et al., 2005), *Verticillium* spp. (Mukherjee et al., 2001), *Penicillium* spp. (Hemath et al., 2010), *Pleurotus sajor caju* (Nithya & Ragunathan, 2009).

15.2 GREEN SYNTHESIS OF NANOPARTICLES

Green synthesis is the process of synthesizing nanoparticles by the used of living organisms or materials of biological origin. In this approach living bacteria or fungi, or plant extracts are used for synthesis of nanoparticles. This method gives an advantage over conventional methods of synthesizing nanoparticles because they are eco-friendly, cheap, easy for synthesized with less intervention or input of energy. In these process, the biological origin materials like organisms can be easily cultured in readily available material and are also renewable. It is a known fact that microorganisms have the ability to reduce metal ions to metal because they can produce lager amount of enzymes. It has been demonstrated that bacteria have the ability to produce magnetic compounds by breaking the iron compounds by attaching the magnetic nanoparticles with the help of magnetosomes within their cells which can take place both.

15.2.1 NEED FOR GREEN SYNTHESIS

Synthesis of nanoparticles from biological sources is environmentally friendly approach that links between nanotechnology and biotechnology. Biosynthesis of Au, Ag, Au-Ag alloy, selenium, tellurium, platinum, palladium, silica, titania, zirconia, quantum dots (QDs), magnetite, and uraninite

(UO_2, a radioactive compound) nanoparticles by phototrophic eukaryotes such as plants (Herrera et al., 2003; Shankar et al., 2003) have been reported. The need for green synthesis is of great importance, despite of its disadvantages like low stability, polydispersity, and the rate of synthesis is slow owing to its toxic free in nature. The reason for polydispersity might be due to the long period of nucleation which occurs during the synthesis process as the concentration of component or macromolecule involved in the nucleation of particle differs in time. The remedy for this problem includes different methods like microbial cultivation methods and photobiological methods may be used. Mechanism at cellular, biochemical, and molecular level need to be studied for increasing the rate of synthesis as well as for improving the properties of nanoparticles. Plants and microbes having rich in biodiversity which have the potential for synthesis of nanoparticles are yet to be utilized and fully explored. In biological synthesis of nanoparticles bottom-up approach is used which involved the oxidation and reduction reactions. Need for greener approach of synthesis of nanoparticle is arose due to the cost expensive in chemical and physical methods. Moreover, chemical approach for synthesis of nanoparticles has a negative impact due to the production of toxic chemical which get absorbed on the surface that may be reluctant in the field of medical applications (Parasharu et al., 2009). Despite of its disadvantage, the only solutions is green synthesis by microbes and plant extracts (Begum et al., 2009) by the reduction of metal compounds into nanoparticles because of their antioxidant and reducing properties.

15.3 PRINCIPLE OF BIOSYNTHESIS OF NANOPARTICLES

Nanoparticles are considered to be building block of nanotechnology (Mansoori et al., 2007). In nanoscience and nano-engineering approaches for synthesis of nanoparticles is an important area where research has been done rapidly. In the current context there is a need of safe synthesis of nanoparticles without the outcome of hazardous material with good monodispersity along with good chemical composition which in turn favors the biological process for synthesis. In biological process mainly microorganisms like bacteria, fungi, herbal extracts, and yeasts are utilized. Microorganisms like bacteria, yeasts, and fungi helps in bioremediation of toxic metals through depletion of the metal ions in such a way that they are not toxic at all and can be used for synthesis process by forming a complex with metal ions that are insoluble in nature in the form of colloidal particles (Mehra & Winge, 1991). Microbes like, *Klebsiella pneumoniae*, *Bacillus licheniformis*, *Fusarium oxysporum*, *Aspergillus flavus*, *Cladosporium cladosporioides*, *Aspergillus clavatus*, and *Penicillium brevicompactum* (Ahmad et al., 2003; Shahverdi et al., 2007; Kalishwaralal et al., 2008; Balaji et al., 2009; Shaligram et al., 2009; Verma et al., 2010) have been reported for the reduction of Ag ions to Ag nanoparticles with the help of cell mass and extracellular components from microorganisms. The process of synthesis of silver nanoparticles by the used of microorganism is a bioreduction process. Some of the microorganisms from which nanoparticles can be synthesized are listed in Table 15.1. The main reason for the production of silver nanoparticles is the reduction of Ag ions by the extracellular reductase enzymes (Figure 15.1). The reaction involved in this process is as follows:

$$Ag^{+} + e^{-} Ag \xrightarrow[\text{Silver Nanoparticles}]{\text{Microbial Enzyme}} \text{Silver ion}$$

Reaction showing preparation of silver nanoparticles.

TABLE 15.1 List of Organism Synthesizing Nanoparticles (Dutta et al., 2015)

Microorganism	Extracellular/Intracellular Nanoparticle Synthesis	Metal
1. Actinomycetes		
Rhodococcus	Intracellular	Au
Streptomyces sp.	Extracellular	Cu/ZnO
2. Algae		
Plectonema boryanum	Intracellular	Au
Chlorella vulagaris	Extracellular	Ag
3. Bacteria		
Lactobacillus sp.	Intracellular	Au/Ag/alloy of Au-Ag
Magnetospirillum magnetotacticum	Intracellular	Magnetite/Greigite
P. stutzeri AG259	Intracellular	Ag
Bacillus subtilis	Extracellular	Ag
B. licheniformis	Extracellular	Ag
Escherichia coli	Extracellular	Ag
Morganella sp.	Extracellular	Ag
Pseudomonas aeruginosa	Extracellular	Au
Lactobacillus sp.	Intracellular	Au/Ag/Alloy of Au-Ag
Magnetospirillum magnetotacticum	Intracellular	Magnetite/Greigite
P. stutzeri AG259	Intracellular	Ag
Bacillus subtilis	Extracellular	Ag
B. licheniformis	Extracellular	Ag
Escherichia coli	Extracellular	Ag
Morganella sp.	Extracellular	Ag

TABLE 15.1 *(Continued)*

Microorganism	Extracellular/Intracellular Nanoparticle Synthesis	Metal
Pseudomonas aeruginosa	Extracellular	Au
Rhodopseudomonas capsulate	Extracellular	Au
Stenotrophomonas maltophilia	Extracellular	Au
4. Fungi		
Aspergillus flavus	Intracellular	Ag
Rhizopus oryzae	Intracellular	Au
Trichothecium sp.	Intracellular	Au
Verticillium sp.	Intracellular	Au/Ag
A. fumigates	Extracellular	Ag
Colletotrichum sp.	Extracellular	Au
F. solani	Extracellular	Ag
F. oxysporum	Extracellular	Magnetite/Ag/Au
Phanerochaete chrysporium	Extracellular	Ag
Trichoderma viride	Extracellular	Ag
5. Yeast		
Pichiajadinii	Intracellular	Au
Schizosaccharomyces pombe	Intracellular	CdS
Torulopsis sp.	Intracellular	PbS
Trichosporon ovoides	Extracellular	Ag
Candida glabrata	Intracellular	CdS

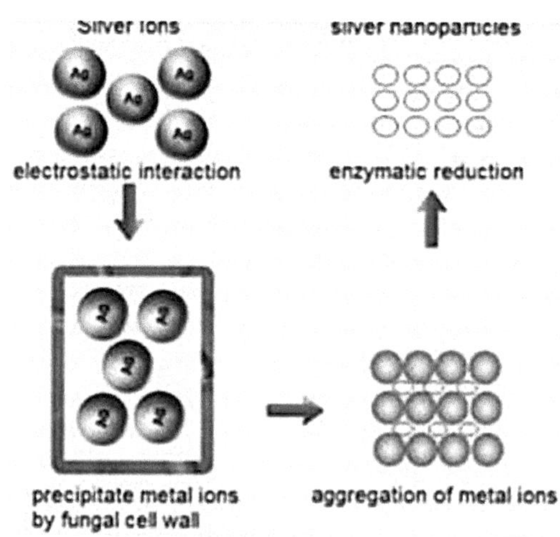

FIGURE 15.1 Mechanism of synthesis intracellular NPs in fungal species. Electrostatic interaction between the metal ion and the enzyme presents of the fungal cell wall.

Source: Reprinted with permission from Alghuthaymi et al., 2015. https://creativecommons.org/licenses/by/4.0/

15.4 NANO-SILVER

Nano-silver is the widely used nanopartiucle and studied for bio-system known to have antimicrobial activities from the immemorial times. As compared to bulk silver nano-form of silver has more antimicrobial activity as they possess high surface area and surface atoms. Silver nanoparticle solution that are found to be more adhesive on bacteria and fungus and are found to be more efficient than fungicide. Silver is known have antimicrobial properties and also have the capacity to act as disinfecting agent. It inhibits the growth of unicellular microorganisms by disrupting the enzyme metabolism by oligodynamic action because in ionic form of silver possess a high antimicrobial activity as compared to nonionic forms (Thomas & McCubin, 2003). Study revealed that, ionic silver is unstable because it gets easily oxidized or reduced into a metal depending on the surrounding media and high reactivity.

Antimicrobial action of silver nanoparticles is not clear and is still a debatable topic. Many researchers have given various mechanism of action of silver nanoparticles that cause microbicidal effect. It has been proved that silver nanoparticles penetrate the bacterial cell wall leading to structural changes in the cell permeability leading to death of the cell. Silver nanoparticles are known to produce some holes cell surface, creating a place for aggregation of the nanoparticles on the cell surface. Another mechanism by which the cells die is the formation of free radicals or ions formed by the silver nanoparticles. Electron spin resonance spectroscopy studies gives an idea about the free radicals' formations by the silver nanoparticles and when these free radicals come in contact with the bacteria, damaging the cell membrane and creating a porous leading to the death of the cells (Danilcauk et al., 2006; Kim et al., 2007). It has also been hypothesis that there is a possibility of release of silver ions by the nanoparticles (Feng et al., 2008), and these ions have the high potential to interact with the thiol groups of many vital enzymes and inactivate them (Matsumura et al., 2003). When the bacterial cells come in contact with silver ions causing damaged to the cells by producing reactive oxygen species (ROS) due to the restriction of respiratory enzyme. Disruption of DNA in bacteria occurs when silver nanoparticles interact with sulfur and phosphorus of the DNA.

Application of silver nanoparticles in disease management especially in plants has been limited till date (Lamsal et al., 2011). *Sclerotinia sclerotiorum*, *S. minor*, and *Rhizoctonia solani* belonging to ascomycetes and basidiomycetes group of plant pathogenic fungi causing diseases in plants by the production of sclerotia considered to have the largest host range with a wider adaptability. So far management is concerned different methods like chemical, physical, biological, cultural methods and genetic controls has been used but as they form sclerotia which can survive in soil for more than 20 years making it difficult to developed management strategies. Nel et al. (2003) lead the antifungal property of silver nanoparticles and the antifungal property might be due to suggested morphological and structural changes. Studies revealed that the minimal amount of nanoparticles is required for management of diseases in plants because it has high penetration power. Silver nanoparticle is found to be effective against antibiotic resistant microbes as they are unable to penetrate into the cells. Morones et al. (2005) reported that silver nanoparticles disrupt transport systems, including ion efflux. Silver nanoparticles can cause disruption of ion efflux leading to the rapid accumulation of silver ions and creating a barrier in the cellular process at very low concentration. Free radicals of silver or ions are capable of producing ROS leading to damage of proteins, lipids, and nucleic acids (NAs) (Hwang et al., 2008) (Figures 15.2 and 15.3; Tables 15.2–15.4).

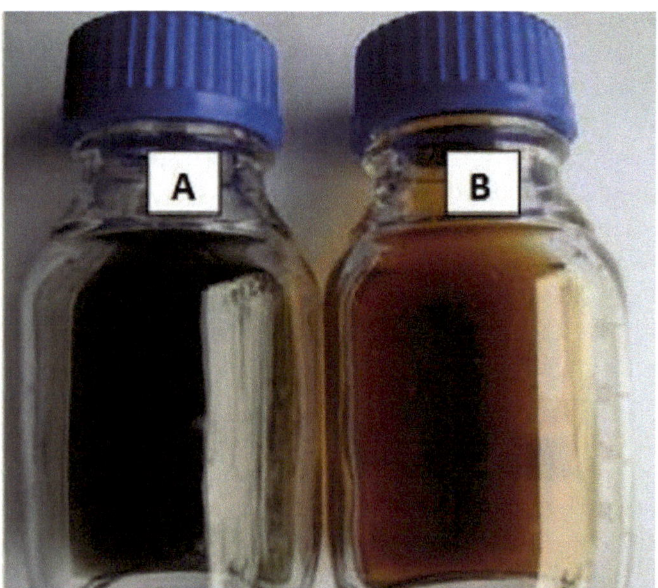

FIGURE 15.2 Bottle containing the extracellular filtrate of the *Trichoderma asperallum* biomass in aqueous solution of 1 m $MAgNO_3$ at the beginning of the reaction (A); and after 72 hrs. of reaction (B).

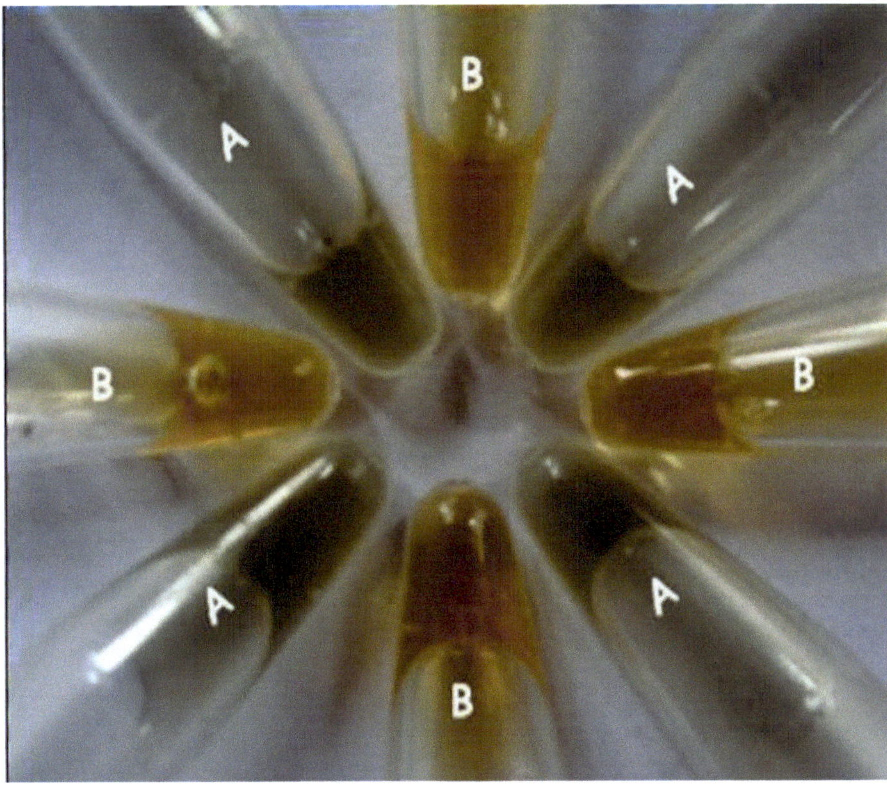

FIGURE 15.3 Centrifuge tube containing the extracellular filtrate of the *T. asperellum* in aqueous solution of 1 m $MAgNO_3$ at the beginning of the reaction (A); and after 72 hrs. of reaction (B).

TABLE 15.2 Fungi That Synthesize Silver Nanoparticles or Other Metallic Nanostructures (Dutta & Kaman, 2017)

Fungi	Mode of Synthesis	References
Fusarium oxysporum	Extracellular	Duran et al. (2005)
Aspergillus niger	Extracellular	Gade et al. (2008)
Fusarium acuminatum	Extracellular	Ingle et al. (2008)
Trichoderma asperellum	Extracellular	Mukherjee et al. (2008)
Penicillium sp.	Extracellular	Sadowski et al. (2008)
Phoma glomerata	Extracellular	Birla et al. (2009)
Fusarium solani	Extracellular	Ingle et al. (2009)
Fusarium semitectum	Extracellular	Basavaraja et al. (2007)
Phoma sp. 3.2883	Extracellular	Chen et al. (2003)

TABLE 15.3 Silver Nanoparticles Synthesized from Medicinal Plants and Their Reported Activity (Dutta & Kaman, 2017)

Sl. No.	Name of the Plants	AgNO3 Concentration Mode and Time of Plant Extraction	Part Used	Activity	References
1.	*Acalypha indica*	1 mM boiling 5 min	Leaf	Antifungal	Krishnaraj et al. (2012)
2.	*Allium sativum*	1 mM boiling 5 min	Clove	Antimicrobial	Ahmad et al. (2011)
3.	*Azadirachta indica*	10 mM 30–240 min	Leaf	Antimicrobial	Prathana et al. (2011)
4.	*Citrus sinensis*	1 mM boiling 2 min	Peel	Antibacterial	Kaviya et al. (2011)
5.	*Cynodon dactylon*	1 mM boiling 2–3 min	Leaf	Antibacterial	Sahu et al. (2012)
6.	*Mangifera indica*	0.1 mM boiling 1 min	Leaf	Antimicrobial	Philip (2011)
7.	*Medicago sativa*	1 mM soaking 10 mM	Seed	Antibacterial	Lukman et al. (2011)
8.	*Morinda pubescens*	1 mM boiling 2–3 min	Leaf	Antimicrobial	Inbathamizh et al. (2013)
9.	*Piper betle* L.	1 mM boiling 5 min	Leaf	Antimicrobial	Usha & Rajasekharreddy (2011)
10.	*Terminalia chebula*	10 mM 50°C, 2 min	Fruit	Antimicrobial	Jebakumar & Sethuraman (2012)

15.5 NANO-COPPER

Copper has been used for plant disease management from the ancient times because of its dual properties as micronutrient and protection from plant pathogens. The first record of using copper as fungicide is with the discovery of Bordeaux mixture by PMA Millardet against downy mildew of Grapes in the year 1885. The first evidence of using nano-copper was reported by Giannousi et al. (2013) who developed nano-CuO, Cu_2O, and Cu/Cu_2O composites and found to be effective against *Phytophthora infestans* in tomato plants when foliar application was done @ 150–340 μg/mL as compared to commercial available copper fungicide like Kocide & White (2016) reported that application of copper nanofertilizer increased the disease resistance against *Fusarium wilt* of tomato and *Verticillium* wilt of eggplant. Another experiment conducted by Elmer et al. (2018) found that foliar application of copper nanoparticles on watermelons reduced the infection of *F. oxysporum* f. sp. *niveum*. Ponmurugan et al. (2016) reported antifungal activity of green synthesized copper nanoparticles from *Streptomyces griseus* against root rot of tea caused by the fungus *Poria hypolateritia*.

Besides antifungal activity of nano-copper it has also antibacterial activity against plant-pathogenic Xanthomonads causing bacterial blight diseases in many crops (Oscoy et al., 2013; Strayer-Scherer et al., 2018). Application of copper nanoparticles is found to be effective against bacterial blight of

TABLE 15.4 Examples of Antimicrobial Activity of Silver Nanoparticle Against Various Plant Pathogens (Dutta & Kaman, 2017)

Source/Mode of Synthesis of Silver Nanoparticles Used in the Study	Target Pathogen	References
Nanosized silver–chitosan composite prepared using chitosan extracted from *Aspergillus niger* and silver nanoparticles (Sigma-Aldrich, St. Louis, MO, USA) according to the method described by Rhim et al. (2006).	Gray mold in strawberry	Moussa et al. (2013)
Colloidal solution of AgNPs provided by BioPlus Co. Ltd. (Pohang, Korea).	Various plant pathogenic fungi.	Kim et al. (2012)
Nanosized Ag–silica hybrid complex prepared by γ-irradiation.	*Pseudomonas syringae* pv. tomato	Chu et al. (2012)
Colloidal solution of AgNPs provided by BioPlus Co. (Pohang, Korea).	*Colletotrichum* sp.	Lamsal et al. (2011a)
AgNPs provided by BioPlus Co. Ltd. (Pohang, Korea).	Powdery mildews on cucumber and pumpkins.	Lamsal et al. (2011b)
Silver nanoparticles synthesized using high-voltage arc discharge method.	*Fusarium culmorum*	Kasprowicz et al. (2010)
AgNPs provided by Quantum Sphere Inc., Santa Ana, CA.	*Bipolaris sorokiniana* and *Magnaporthe grisea*	Jo et al. (2009)
AgNPs provided by BioPlus Co., Ltd.	Oak wilt pathogen *Raffaelea* sp.	Kim et al. (2007)
Nanosilver (Shanghai Huzheng Nano-Technology Co. Ltd., China).	Stem-end bacteria on cut gerbera.	Liu et al. (2009)
The nanosized silica–silver prepared by combining nanosilver with silica molecules and water-soluble polymer and exposing a solution including silver salt, silicate, and water-soluble polymer to radioactive rays.	Powdery mildews of pumpkin.	Park et al. (2006)

rice (*Xanthomonas oryzae* pv. *oryzae*) and leaf spot of mung (*X. campestris* pv. *phaseoli*) reported by Gogoi et al. (2009). Kala et al. (2016) biological synthesized copper nanoparticles *Datura innoxia* found to have antibacterial activity against bacterial blight of rice caused by a bacterium *Xanthomonas oryzae* pv. *oryzae* at a concentration of 100 ppm.

15.6 NANO-ZINC

Zinc is considered to be an important micro-nutrient for plant growth and in disease resistance against plant pathogens. Therefore, plant pathologist can utilize nano-zinc for plant disease management as well as for nutritional purpose. Workers like Graham et al. (2016); Hafez et al. (2014) and Indhumathy & Mala (2013) reported the antimicrobial activity of zinc-based nanoparticles for plant pathogens. Graham et al. (2016) study the antibacterial activity of Zinc Oxide nanoparticles against citrus canker *Xanthomonas citri* subsp. c*itri.* Hafez et al. (2014) synthesized ZnO nanoparticles from Neem leaves and found to have antibacterial activity against different plant pathogenic bacteria like *Pectobacterium carotovorum* subsp. *wasabiae*, *Pectobacterium atrosepticum*, *Dickeya chrysanthemi*, *Pectobacterium carotovorum*, and *Dickeya solani*.

Fungicidal property of ZnO nanoparticles was reported on two post-harvest fungal diseases *Botrytis cinerea* and *Penicillium expansum* when ZnO nanoparticles was applied at a dose of 12 mmol l^{-1}. Rosa-Garca et al. (2018) studied the antifungal activity of ZnO nanoparticles and found them to inhibit the *Colletotrichum gloesporoides* in papaya at a dose of 0.156 mg/mL.

15.7 NANOSENSORS

Currently nanosensors is required for precision farming as they are portable in size and gives a real picture of monitoring in large field areas. A sensor is a device which can detect a specific biological or chemical compound in the materials like seed and planting material which may contain pathogens and can be detected by electronic signals. Nanosensors are likely to use for detection of harmful or toxic chemical compounds at much lower concentrations (ultra-low). It is possible that seeds act as a carrier for transmission of various diseases from one place top another place either through transportations or through airports or in seaports and can be managed by nano-coating of seeds with the nanoparticles like Zn, Mn, Pa, Pt, Au, Ag. Su and Li (2004) developed QDs that can detect *E. coli* using fluorescence markers and immunomagnetic separation.

15.8 RNAI NANOTECHNOLOGY

RNA interference (RNAi) is a new technology that paves the way for plant disease management. RNAi interference can be defined as silencing a gene in a sequence-specific manner at a cellular level. RNAi interference technology was first discovered and observed in plants and subsequently in *Caenorhabditis elegans*. RNAi is highly conserved regions of eukaryotic organisms that restrict or interfere in the growth and development and also in the defense response of viruses, which can be utilized against pathogenic fungi, bacteria, and viruses. Therefore, it can be integrated with the science of nanotechnology, giving birth to a new branch called RNAi nanotechnology.

Mitter et al. (2017) synthesized layer double hydroxide nanoparticles loaded with dsRNA called BioClay. Its foliar application protects the plant against cucumber mosaic virus (CMV) and pepper mild mottle virus (PMMoV).

15.9 CONCLUSION

In conclusion, we can say that nanoparticles are useful in managing the fungal pathogens at much lower concentrations than the fungicide. Apart from the fungal pathogens, these nanoparticles can also be used for the management of bacterial and human pathogens. The use of nanoparticles as plant protection materials as fungicide, bactericide, or viricide will curb plant diseases by reducing the environmental hazards due to the non-judicious use of pesticides. But before its commercial use, ecological issues and other toxic conditions and their effects on the environment need to be studied.

KEYWORDS

- **biosynthesis**
- **disease management**
- **green synthesis**
- **nanoparticles**
- **nanosensors**
- **nano-zinc**
- **pathogens**

REFERENCES

Ahmad, M., Khana, M. A., Siddiqui, M. K. J., Salhi, M. S., & Alrokayan, S. A., (2011). Green synthesis, characterization, and evaluation of biocompatibility of silver nanoparticles. *Physica Electronica, 43,* 1266–1271.

Ahmad, Z., Pandey, R., Sharma, S., & Khuller, G. K., (2005). Alginate nanoparticles as antituberculosis drug carriers, formulation development, pharmacokinetics, and therapeutic potential. *Indian Journal of Chest Diseases and Allied Sciences, 48*, 171–176.

Alghuthaymi, M. A., Almoammar, H., Rai, M., Said-Galiev, E., & Abd-Elsalam, K. A., (2015). Myconanoparticles: Synthesis and their role in phytopathogens management. *Biotechnology & Biotechnological Equipment*, *29*(2), 221–236.

Aymonier, C., Scholotterbeck, U. L., Antonietti, P., Zacharias, R., Thomann, J. C., & Tiller, M. S., (2002). Hybrids of silver nanoparticles with amphiphilic hyper branched macromolecules exhibiting antimicrobial properties. *Chemical Communication, 24*, 3018, 3019.

Balaji, D. S., Basavaraja, S., Deshpande, R., Bedre, M. D., Prabhakara, B. K., & Venkataraman, A., (2009). Extracellular biosynthesis of functionalized silver nanoparticles by strains of *Cladosporium cladosporioides* fungus. *Colloids and Surfaces B: Biointerfaces, 68*, 88–92.

Basavaraja, S., Balaji, S. D., Lagashetty, A., Rajasa, H., & Venkataraman, A., (2008). Extracellular biosynthesis of silver nanoparticles using the fungus *Fusarium semitectum. Materials Research Bulletin, 43*, 1164–1170.

Begum, N. A., Mondal, S., Basu, S., Laskar, R. A., & Mandal, D., (2009). Biogenic synthesis of Au and Ag nanoparticles using aqueous solutions of black tea leaf extracts. *Colloids and Surfaces B: Biointerfaces, 71*, 113–118.

Birla, S. S., Tiwari, V. V., Gade, A. K., Ingle, A. P., Yadav, A. P., & Rai, M. K., (2009). Fabrication of silver nanoparticles by *Phoma glomerata* and its combined effect against *Escherichia coli, Pseudomonas aeruginosa* and *Staphylococcus aureus*. *Letters in Applied Microbiology*, *48*(2), 173–179.

Chen, J. C., Liu, Z. H., & Ma, X. X., (2003). Evidence of the production of silver nanoparticles via pretreatment of *Phoma* sp. 3.2882 with silver nitrate. *Letter of Applied Microbiology, 37*. 105–108.

Chu, H., Kim, H. J., Kim, J. S., Kim, M. S., Park, H. J., & Kim, C. Y., (2012). A nanosized Ag-silica hybrid complex prepared by γ-irradiation activates the defense response in *Arabidopsis*. *Radiation Physics Chemistry, 81*, 180–184.

Danilcauk, M., Lund, A., Saldo, J., Yamada, H., & Michalik, J., (2006). Conduction electron spin resonance of small silver particles. *Spectrochimaca. Acta. Part A, 63*, 189–191.

Dias, M. A., Lacerda, I. C. A., Pimental, P. F., Castro, H. F., & Rosa, C. A., (2002). Removal of heavy metals by an *Aspergillus terreus* strain immobilized in a polyurethane matrix. *Letters in Applied Microbiology, 34*(1), 46–49.

Duran, N., Marcato, P. D., Alves, O. L., De Souza, G., & Esposito, E., (2005). Mechanistic aspects of biosynthesis of silver nanoparticles by several *Fusarium oxysporum* strains. *Journal of Nanobiotechnology, 3*(8).

Dutta, P., & Kaman, P. K., (2017). Nanocentric plant health management with special reference to silver. *Int. J. Curr. Microbiol. Appl. Sci.*, *6*(6), 2821–2830.

Dutta, P., Kaman, K. P., Kaushik, H., & Boruah, S., (2015). Biotechnological and nanotechnological approaches for better plant health management. *Trends in Biosciences, 8*(22), 6051–6065.

Fayaz, A. M., Balaji, K., Girilal, M., Yadav, R., Kalaichelvan, P. T., & Venketesan, R., (2010). Biogenic synthesis of silver nanoparticles and their synergistic effect with antibiotics: A study against gram-positive and gram-negative bacteria. *Nanomedicine, 6*(1), 103–109.

Feng, Q., Wu, J., Chen, G. Q., Cui, F. Z., Kim, T. N., & Kim, J. O., (2008). A mechanistic study of the antibacterial effect of silver ions on *Escherichia coli* and *Staphylococcus aureus*. *Journal of Biomedical Material Research, 52*, 662–668.

Gade, A. K., Bonde, P. P., Ingle, A. P., Marcato, P. D., Duran, N., & Rai, M. K., (2008). Exploitation of *Aspergillus Niger* for synthesis of silver nanoparticles. *Journal of Biobased Materials and Bioenergy, 2*(3), 243–247.

Giannousi, K., Avramidis, I., & Dendrinou-Samara, C., (2013). Synthesis, characterization, and evaluation of copper-based nanoparticles as agrochemicals against *Phytophthora infestans*. *RSC Advances, 3*(44), 21743–21752.

Gogoi, R., Dureja, P., & Singh, P. K., (2009). Nanoformulations a safer and effective option for agrochemicals. *Indian Farming, 59*(8), 7–12.

Gong, P., Li, H., He, X., Wang, K., Hu, J., Tan, W., Zhang, S., & Yang, X., (2007). Preparation and antibacterial activity of Fe_3O_4@Ag nanoparticles. *Nanotechnology, 18*, 604–611.

Graham, J. H., Johnson, E. G., Myers, M. E., Young, M., Rajasekaran, P., Das, S., & Santra, S., (2016). Potential of nano-formulated zinc oxide for control of citrus canker on grapefruit trees. *Plant Disease, 100*(12), 2442–2447.

Gu, H., Ho, P. L., Tong, E., Wang, L., & Xu, B., (2003). Presenting vancomycin on nanoparticles to enhance antimicrobial activities. *Nano Letter, 3*(9), 1261–1263.

Hafez, E. E., Hassan, H. S., Elkady, M., & Salama, E., (2014). Assessment of antibacterial activity for synthesized zinc oxide nanorods against plant pathogenic strains. *International Journal of Science and Technology Research, 3*(9), 318–324.

He, L., Liu, Y., Mustapha, A., & Lin, M., (2011). Antifungal activity of zinc oxide nanoparticles against *Botrytis cinerea* and *Penicillium expansum*. *Microbiological Research, 166*(3), 207–215.

Hemath, K. S., Naveen, G., Kumar, L., Karthik, K. V. B., & Rao, E., (2010). Extracellular biosynthesis of silver nanoparticles using the filamentous fungus *Penicillium* sp. *Archives of Applied Science Research, 2*(6), 161–167.

Herrera, I., Gardea, T. J. L., Tiemann, K. J., Peralta, V. J. R., Armendariz, V., & Parsons, J. G., (2003). Binding of silver(I) ions by alfalfa biomass (*Medicago sativa*): Batch pH, time, temperature, and ionic strength studies. *Journal of Hazarduous Substance Research, 4*, 1–16.

Hwang, E. T., Lee, J. H., Chae, Y. J., Kim, Y. S., Kim, B. C., Sang, B. I., & Gu, M. B., (2008). Analysis of the toxic mode of action of silver nanoparticles using stress-specific bioluminescent bacteria. *Small. 4*(6), 746–750.

Inbathamizh, L., Mekalai, Ponnu, T., & Jancy, & Mary, E., (2013). *In vitro* evaluation of antioxidant and anticancer potential of *Morinda pubescens* synthesized silver nanoparticles. *Journal of Pharmacy Research, 1*, 32–38.

Indhumathy, M., & Mala, R., (2013). Photocatalytic activity of zinc sulphate nano material on phytopathogens. *International Journal of Agriculture, Environment and Biotechnology, 6*(4), 737–745.

Ingle, A., Gade, A., Pierrat, S., Sonnichsen, C., & Rai, M., (2008). Mycosynthesis of silver nanoparticles using the fungus *Fusarium acuminatum* and its activity against some human pathogenic bacteria. *Current Nanoscience, 4*, 141–144.

Ingle, A., Rai, M., Gade, A., & Bawaskar, M., (2009). *Fusarium solani*: A novel biological agent for the extracellular synthesis of silver nanoparticles. *Journal of Nanoparticle Research, 11*(8), 2079–2085.

Jaidev, L. R., & Narasimha, G., (2010). *Fungal Mediated Biosynthesis of Silver Nanoparticles, Characterization, and Antimicrobial Activity* (Vol. 81, No. 2, pp. 430–433). Epub.

Jain, N., Bhargava, A., Majumdar, S., Tarafdar, J. C., & Panwar, J., (2011). Extracellular biosynthesis and characterization of silver nanoparticles using *Aspergillus flavus* NJP08: A mechanism perspective. *Nanoscale, 3*, 635–641.

Jebakumar, I. E. T., & Sethuraman, M. G., (2012). Instant green synthesis of silver nanoparticles using *Terminalia chebula* fruit extract and evaluation of their catalytic activity on reduction of methylene blue. *Process Biochemistry, 47,* 1351–1357.

Jo, Y. K., Kim, B. H., & Jung, G., (2009). Antifungal activity of silver ions and nanoparticles on phytopathogenic fungi. *Plant Diseases, 93*, 1037–1043.

Kalishwaralal, K., Deepak, V., Ramkumarpandian, S., Nellaiah, H., & Sangiliyandi, G., (2008). Extracellular biosynthesis of silver nanoparticles by the culture supernatant of *Bacillus licheniformis*. *Materials Letters, 62*, 4411–4413.

Kasprowicz, M. J., Kozioł, M., & Gorczyca, A., (2010). The effect of silver nanoparticles on phytopathogenic spores of *Fusarium culmorum*. *Canadian Journal of Microbiology, 56*(3), 247–253.

Kaviya, S., Santhanalakshmia, J., Viswanathan, B., Muthumary, J., & Srinivasan, K., (2011). Biosynthesis of silver nanoparticles using *Citrus sinensis* peel extract and its antibacterial activity. *Spectrochimica Acta Part A: Molecular and Biomolecular Spectroscopy, 79,* 594–598.

Kim, S. J., Kuk, E., Yu, N. K., Kim, H. J., Park, J. S., Lee, J. H., Kim, H. S., et al., (2007). Antimicrobial effects of silver nanoparticles. *Nanomedicine: Nanotechnology, Biology and Medicine, 3*(1), 95–101.

Kim, W. S., Jung, H. J., Lamsal, K., Kim, S. Y., Min, S. J., & Lee, S. Y., (2012). Antifungal effects of silver nanoparticles (AgNPs) against various plant pathogenic fungi. *Mycobiology, 40*(1), 53–58.

Krishnaraj, C., Ramachandran, R., Mohan, K., & Kalaichelvan, P. T., (2012). Optimization for rapid synthesis of silver nanoparticles and its effect on phytopathogenic fungi. *Spectrochimica Acta Part A: Molecular and Biomolecular Spectroscopy, 93*, 95–99.

Lamsal, K., Kim, W. S., Jung, H. J., Kim, S. Y., Kim, S. K., & Lee, S. Y., (2011). Inhibition effects of silver nanoparticles against powdery mildews on cucumber and pumpkin. *Mycobiology, 39*(1), 26–32.

Lead, J. R., & Wilkison, K. J., (2006). Aquatic colloids and nanoparticles: Current knowledge and future trends. *Environmental Chemistry, 3*, 159–171.

Liu, J., He, S., Zhang, Z., Cao, J., Lv, P., He, S., & Joyce, D. C., (2009). Nano-silver pulse treatments inhibit stem-end bacteria on cut gerbera cv. Ruikou flowers. *Postharvest Biology and Technology, 54*(1), 59–62.

Lukman, A. I., Gong, B., Marjo, C. E., Roessner, U., & Harris, A. T., (2011). Facile synthesis, stabilization, and anti-bacterial performance of discrete Ag nanoparticles using *Medicago sativa* seed exudates. *Journal of Colloid and Interface Science, 353*(2), 433–444.

Mansoori, G. A., George, T. F., Zhang, G., & Assoufid, L., (2007). *Molecular Building Blocks for Nanotechnology.* Springer, New York.

Matsumura, Y., Yoshikata, K., Kunisaki, S., & Tsuchido, T., (2003). Mode of bacterial action of silver zeolite and its comparison with that of silver nitrate. *Applied Enviromental Microbiology, 69,* 4278–4281.

Mehra, R. K., & Winge, D. R., (1991). Metal ion resistance in fungi: Molecular mechanisms and their regulated expression. *Journal of Cellular Biochemistry, 45,* 30–40.

Morones, J., Elechiguerra, J. L., Camacho, A., Holt, K., Kouri, J. B., Ramirez, J. T., & Yacaman, M. J., (2005). The bactericidal effect of silver nanoparticles. *Nanotechnology, 16,* 2346–2353.

Moussa, S. H., Tayel, A. A., Alsohim, A. S., & Abdallah, R. R., (2013). Botryticidal activity of nanosized silver-chitosan composite and its application for the control of gray mold in strawberry. *Journal of Food Science, 78*(10), 589–594.

Mukherjee, P., Ahmad, A., Mandal, D., Senapati, S., Sainkar, S., Khan, M. I., Ramani, R., et al., (2001). Bioreduction of $AuCl_4$-ions by the fungus *Verticillium* sp. and surface trapping of the gold nanoparticles formed. *Angewante Chemie International Edition, 40*(19), 3585–3389.

Mukherjee, P., Roy, M., Mandal, B. P., Dey, G. K., Mukherjee, P. K., Ghatak, J., & Kale, S. P., (2008). Green synthesis of highly stabilized nanocrystalline silver particles by a non-pathogenic and agriculturally important fungus *T. asperellum*. *Nanotechnology*, *19*(7), 103–115.

Nel, A., Xia, T., Madler, L., & Li, N., (2003). Toxic potential of materials at the nano level. *Science, 311*, 622–627.

Nithya, R., & Ragunathan, R., (2009). Synthesis of silver nanoparticle using *Pleurotus sajor caju* and its antimicrobial study. *Digest Journal of Nanomaterials and Biostructures, 4*(4), 623–629.

Ocsoy, I., Paret, M. L., Ocsoy, M. A., Kunwar, S., Chen, T., You, M., & Tan, W., (2013). Nanotechnology in plant disease management: DNA-directed silver nanoparticles on graphene oxide as an antibacterial against *Xanthomonas perforans*. *ACS Nano*, *7*(10), 8972–8980.

Parasharu, K., Saxena, A., & Srivastava, A., (2009). Bioinspired synthesis of silver nanoparticles. *Digest Journal of Nanomaterials and Biostructures, 4*(1), 159–166.

Park, H. J., Kim, S. H., Kim, S. J., & Choi, S. H., (2006). A new composition of nanosized silica-silver for control of various plant diseases. *Plant Pathology Journal, 22,* 295–302.

Philip, D., (2011). *Mangifera indica* leaf-assisted biosynthesis of well-dispersed silver nanoparticles. *Spectrochimica Acta Part A: Molecular and Biomolecular Spectroscopy, 78,* 327–331.

Ponmurugan, P., Manjukarunambika, K., Elango, V., & Gnanamangai, B. M., (2016). Antifungal activity of biosynthesized copper nanoparticles evaluated against red root-rot disease in tea plants. *Journal of Experimental Nanoscience*, *11*(13), 1019–1031.

Prathna, T. C., Chandrasekaran, N., Raichur, A. M., & Mukherjee, A., (2011). Biomimetic synthesis of silver nanoparticles by *Citrus limon* (lemon) aqueous extract and theoretical prediction of particle size. *Colloids and Surfaces B: Biointerfaces*, *82*(1), 152–159.

Retchkiman, P. S., Canizal, G., Herrera-Becerra, R., Zorrilla, C., Liu, H. B., & Ascencio, J. A., (2006). Biosynthesis and characterization of Ti/Ni bimetallic nanoparticles. *Optical Materials*, *29*(1), 95–99.

Rosa-García, D., Susana, C., Martínez-Torres, P., Gómez-Cornelio, S., Corral-Aguado, M. A., Quintana, P., & Gómez-Ortíz, N. M., (2018). Antifungal activity of ZnO and MgO nanomaterials and their mixtures against *Colletotrichum gloeosporioides* strains from tropical fruit. *Journal of Nanomaterials, 55*(12), 220–226.

Sadowski, Z., Maliszewska, I. H., Grochowalska, B., Polowczyk, I., & Kozlecki, T., (2008). Synthesis of silver nanoparticles using microorganisms. *Materials Science-Poland, 26*(2), 419–424.

Sahu, N., Soni, D., Chandrashekhar, B., Sarangi, B. K., Satpute, D., & Pandey, R. A., (2013). Synthesis and characterization of silver nanoparticles using *Cynodon dactylon* leaves and assessment of their antibacterial activity. *Bioprocess and Biosystems Engineering, 36*(7), 999–1004.

Shahverdi, A. R., Fakhimi, A., Shahverdi, H. R., & Minaian, S. A., (2007). Synthesis and effect of silver nanoparticles on the antibacterial activity of different antibiotics against *Staphylococcus aureus* and *Escherichia coli. Nanomedicine and Nanotechnology, Biological Mededicine, 3*, 168–171.

Shaligram, N. S., Bule, M., Bhambure, R., Singhal, R. S., Singh, S. K., Szakacs, G., & Pandey, A., (2009). Biosynthesis of silver nanoparticles using aqueous extract from the compactin producing fungal strain. *Process Biochemistry, 44*, 939–943.

Shankar, S. S., Rai, A., Ahmad, A., & Sastry, M., (2003). Biosynthesis of silver and gold nanoparticles from extracts of different parts of the geranium plant. *Applied Nanoscience, 1*, 69–77.

Sondi, I., & Salopek, S. B., (2004). Silver nanoparticles as antimicrobial agent: A case study on *E. coli* as a model for gram-negative bacteria. *Journal of Colloid and Interface Science, 275*, 177–182.

Strayer-Scherer, A., Liao, Y. Y., Young, M., Ritchie, L., Vallad, G. E., Santra, S., & Paret, M. L., (2018). Advanced copper composites against copper-tolerant *Xanthomonas perforans* and tomato bacterial spot. *Phytopathology, 108*(2), 196–205.

Thirumurugan, G., Shaheedha, S. M., & Dhanaraju, M. D., (2009). *In vitro* evaluation of anti-bacterial activity of silver nanoparticles synthesized by using *Phytophthora infestans. International Journal of Chem Tech Research, 1*(3), 714–718.

Thomas, S., & Mc Cubbin, P., (2003). A comparison of the antimicrobial effects of four silver-containing dressings on three organisms. *Journal of Wound Care, 12*(3), 101–107.

Usha, R. P., & Rajasekharreddy, P., (2010). Green synthesis of silver-protein (core–shell) nanoparticles using *Piper betle* L. leaf extract and its ecotoxicological studies on *Daphnia magna. Colloids and Surfaces A: Physicochemical and Engineering Aspects, 389*, 188–194.

Vahabi, K., Mansoori, G. A., & Karimi, S., (2011). Biosynthesis of silver nanoparticles by fungus *Trichoderma reesei* (A route for large-scale production of Ag NPs). In: *Sciences Journals,* (Vol. 65, No. 1, pp. 543–547).

Verma, V. C., Kharwar, R. N., & Gange, A. C., (2010). Biosynthesis of antimicrobial silver nanoparticles by the endophytic fungus *Aspergillus clavatus. Nanomedicine, 5*, 33–40.

CHAPTER 16

Effectoromics in the Management of Crop Diseases

GANESH THAPA

Biohazards and Biosafety, Trinity College of Dublin, The University of Dublin, College Green, Dublin, Ireland

ABSTRACT

The use of plant disease resistance genes is an integral part of crop disease management. Since the discovery and sequence identification, validation of resistance plant genes is challenging. Hence, the characterization of the effector/avirulence products in pathogens recognized by host plant receptors is candid. Pathogens secrete small proteins, called 'effectors' that hijack, bypass, and modulate plant's defense upon attack. So, effectors can be milestones to direct breeding of resistance genes, to trigger defense responses, or to structure-driven decoy to inhibit effector production in pathogens, as strategies for crop protection. Given that multi-billion euros agriculture crops are at stake to increase yield, effector-receptor proteins interaction can be used as a biotechnology resource in agriculture. One way to develop novel control strategies based on the intricate mechanisms of pathogens' effectors-plant receptor interaction is to engineer plants to fight infection or restrict spread. This chapter covers the role of effector-driven biotechnology in controlling fungal plant diseases.

16.1 INTRODUCTION

Plants and pathogens during pathogenesis do use common cyto-areas, but with competitive imbalance, wherein pathogens fiercely cause disease and plants defends themselves. The pathogen uses an arsenal of biomolecules to

New Advances in Crop Disease Management. Sangita Sahni, Ganesh Thapa, and Bishun Deo Prasad (Eds)

evade the defense response of the host, which is called 'effector.' The host primarily harbors various defense responsive gene families and a particular gene class, called resistance (*R*) gene family. To overcome such defense barriers, the pathogen genome encodes 'effector' proteins. These fungal effectors are the prime arsenal to evade the host defense. The plant host generally uses a conserved 2-tier defense mechanism either by the pattern triggered immunity (PTI) process or by the more specialized mechanism of effector-triggered immunity (ETI) against the pathogens. This balance is most apparent where record-breaking crop yields losses were seen due to diseases, such as., *Zymoseptoria tritici*, causing Septoria tritici blotch (STB) in wheat (Fones & Gurr, 2015). Although the genome of *Z. tritici*, IPO323, 39.7 Mb, approximately 11,000 predicted genes dispersed over 21 chromosomes (Goodwin et al., 2011), disease cycle, pathogenesis, and mode already have been decoded (Eyal, 1987), But STB still remains a nemesis. The use of fungicides to control STB has led to fungicide resistance, rapidly evolution and diversity in natural isolates of *Z. tritici*, posing serious threat to global wheat production and food security (Hayes, Sackett, Anderson, Flowers, & Mundt, 2016). The sexual reproduction of *Z. tritici* facilitates the spread of newly acquired resistance traits across locations while the asexual reproductive cycle allows *Z. tritici* to propagate acquired resistant traits (Orton, Deller, & Brown, 2011). This enable *Z. tritici* with rapid gene flow and high genetic diversity to be a persistent uncontrollable pathogen of wheat (Talbot, 2015). The widespread monoculture of crop without the natural diversity of resistance against disease puts crop at high risk of an evolved pathogen to overcome resistance.

Plant disease resistance in plants may be quantitative and qualitative resistance. Quantitative resistance leads to a reduction in disease, rather than eradication of disease, and is generally thought to be achieved through the contribution of multiple genes (St. Clair, 2010). Whereas qualitative resistance is conferred usually by a single gene inherited in a typical Mendelian segregation pattern (Jones & Dangl, 2006). Furthermore, qualitative resistance sometimes results in a 'gene for-gene' interaction, whereby for each resistance gene in the plant there is a corresponding effector (termed avirulence) gene in the pathogen (Flor, 1956). Plant breeding strategies have traditionally focused on qualitative resistance, but recently breeding efforts have shifted towards quantitative resistance due to its durable nature of the resistance. In the 'gene-for-gene' interaction, a single avirulence (Avr) product from the fungus is recognized by a single resistance (R) product of the plant: this combination triggers a hypersensitive response (HR), i.e., a form of programmed cell death (PCD) that restricts subsequent spread of the pathogen. The plant protection is based on the recognition of molecules produced by the pathogen, known as pathogen-associated molecular patterns (PAMP), by receptor-like kinases (RLKs) that lead to downstream regulatory events giving rise to PTI. PTI can be evaded by the pathogen using effectors, that may manipulate host cell functions to prevent downstream signaling cascades, resulting in disease. However, in the presence of specific qualitative resistance proteins, a subset of effectors that are known as *Avr* proteins are detected, leading to ETI and a resistant phenotype via the expression of defense related genes and HR. To date, three different methods for ETI have been identified. First, the *Avr* effector protein is directly recognized by the resistance protein and triggers ETI (Dodd, Buffle, & Von, 2006). Second, the interaction between the *Avr* effector and resistance protein is indirect, whereby the *Avr* effector targets an alternative guard protein or 'guardee,' and it is this interaction that is detected by the R protein (Dangl & Jones, 2001). Third, in the decoy model, a decoy protein that mimics the effector target interacts with the *Avr* effector, which is then detected by the R-protein (Van & Kamoun, 2008). Despite the type of interaction between *Avr* effector and R protein, if the effector gene is mutated, rendering its protein undetectable by the plant, the pathogen goes undetected, and disease develops. Though it may be counter-intuitive to focus on the pathogen molecules instead of the plant genes, but there are two main reasons why work on effectors plays a prominent role in plant defense studies. Microbial genomics, genetics, and genetic manipulation are considerably simpler or easier to perform than in the host plants. Thus, generally more effectors can be identified and characterized than plant *R* genes. Furthermore, the 'breakdown' of resistance in crops is due to changes in the pathogen population. Cloning of pathogen *Avr* genes and plant *R* genes revealed several generalizations about their features. Fungal effector genes often encode small, secreted proteins, with a high proportion of cysteine residues and very little homology with other genes. However, there are always diverse exceptions to the rules, such as the Ecp6 effector of *Cladosporium fulvum* that prevents chitin-triggered immunity through the binding of conserved LysM domains with peptidoglycan or chitin (De Jonge et al., 2010; Sánchez-Vallet et al., 2013). Qualitative R proteins basically includes nucleotide binding-leucine-rich repeat domain (NB-LRRs) containing receptors and membrane localized RLKs and RLPs, which are involved in recognition of specific *Avr* effector proteins, domains for recognition and transmission of that signal (Jones & Dangl, 2006). Alternatively, wall-associated kinases (WAK), containing extracellular epidermal growth factor repeats domain confers quantitative resistance towards multiple pathogen species including *Fusarium oxysporum*, *Magnaporthe oryzae* and *Exserohilum turcicum* (Delteil et al., 2016; Hurni et al., 2015; Li, Zhou, Zhao, Su, & Peng, 2009; Zuo et al., 2015). The gene-for-gene system is the basis of the plant immune defense system that leads to ETI (Jones & Dangl, 2006). While the gene-for-gene concept still holds true not only to biotrophs but also to a subset of necrotrophic fungal pathogens like *Parastagono sporanodorum, Pyrenophora tritici-repentis*, and *Bipolaris sorokiniana*. These fungi produce host-specific proteinaceous toxins that are recognized by specific plant susceptibility receptors (Friesen, Meinhardt, & Faris, 2007; Friesen, Faris, Solomon, & Oliver, 2008; Lan, Liu, Song, Yin, Xiang, Qu, & Lu, 2019). Because of inverse relationship to Flor hypothesis, i.e., a pathogen virulence factor is perceived by a plant susceptibility receptor to promote disease progression rather than resistance, this opens a new scope for the implementation of effectors science in agriculture. Second, the *Avr*-R gene recognition process was traditionally thought to trigger an HR to provide plant protection, but the amplitude of defense arising from the recognition event can be variable. Some effector genes have a series of alleles encoding proteins with differing levels of impact on the host. For example, the different isoforms of the ToxA effector induce a range of levels of damage on wheat plants (Tan et al., 2012), while varied genetic background of host plants may lead to a range in severities to the same ToxA (or ToxB) isoform (See, Iagallo, Oliver, & Moffat, 2019). Third, the simplest model of one pathogen protein (an *Avr*) physically binding to a plant resistance receptor is not universally applicable for all situations (Van, Tian, Hoy, & Li, 2020). There may be case of multiple function like, more than one *Avr* is recognized by a single resistance protein or one *Avr* is recognized by more than one resistance protein (Maidment et al., 2021). Some '*R* resistance genes' are different alleles (variant forms) of the same gene; viz., LepR3 and Rlm2 from *Brassica napus* are alleles of the same gene with each allele recognizing different avirulence proteins from *Leptosphaeria maculans* (Larkan, Ma, & Borhan, 2015), and similarly YrSP and Yr5 are alleles from the same wheat that are involved in recognition of avirulence proteins from *Puccinia striiformis* f. sp. *tritici* (Marchal et al., 2018). There are complex interactions undergoing between the fungal *Avr* genes and their corresponding plant *R* genes, which results in diverse consequences to be considered for disease management (Petit-Houdenot & Fudal, 2017). Hence, the knowledge of how these systems work is important in terms of how to utilize *Avrs* in agriculture. Apart from *Avr*-R recognition leading to defense is not always true, plants may integrate multiple signals that they detect from a potentially dangerous microbe when initiating their defense machinery (Wu, Liu, Xu, Zhang, Shen, & Zhang, 2018). Though validated but these advances in effector science in terms of applications have not been translated into industry-related tools yet. The concepts and principles of effectors in agriculture can be used to control disease-causing organisms to drive the impetus to increase production.

16.2 EFFECTOR PROTEINS – THE DECEPTIVE ARSENAL OF BIOTROPHIC PATHOGENS

Pathogens must overcome the plant defense triggered by PAMPs and proteins secreted by it (Dangl & Jones, 2001) to cause disease. The pathogen evades host defense system by pathogen secreted specialized small-sized proteins (< 300 amino acids) called 'effectors' unique with specific potential to manipulate host machinery (Figure 16.1). The host species can defend the pathogen progression by *R* genes, which can detect secreted effector proteins inside the host cell that transforms virulent pathogen to avirulent one and ultimately providing protection (Jones & Dangl, 2006). Effectors generally have no or very little conservation but share common functions viz., cytotoxicity or attenuation of host cell machinery to facilitate infection and nutrition acquisition from the host (Divon & Fluhr, 2007; Sonah, Deshmukh, & Bélanger, 2016; Sperschneider, 2020). The effector must have a signal peptide, which is 18–30 amino acids at N-terminal (Sonah, Deshmukh, & Bélanger, 2016; Xu et al., 2020; Sperschneider et al., 2016). However, presence of the signal peptide alone in proteins does not classify them as an effector molecule, as many pathogen proteins may have a signal peptide, but essentially not extracellular. Apart from signal peptide, the effectors can be identified by: (i) small size, < 300 amino acids with repeat motif and Intrinsic domain; (ii) absence of transmembrane and Pfam domain; (iii) high cysteine content (Stergiopoulos & De Wit, 2009); (iv) high mRNA expression in the infection cycle (Djamei et al., 2011) or haustoria (Lo Presti et al., 2015); (v) high effector (Sperschneider, Dodds, Gardiner, Singh, & Taylor, 2018) and apoplast prediction (Sperschneider, Dodds, Singh, & Taylor, 2018) scores; and (vi) species or family specificity (Xu et al., 2020; De Torres et al., 2006) (Figure 16.1). The effector expression in plants is highly stage-dependent and varies with the type of host genotype (Haueisen et al., 2019; Sánchez-Vallet, Fouché, Fudal, Hartmann, Soyer, Tellier, & Croll, 2018). The effectors are highly expressed at the haustoria formed in case of obligate biotrophs, viz., *Puccinia* (Catanzariti, Dodds, Ve, Kobe, Ellis, & Staskawicz, 2010); powdery mildew infections (Ahmed, Pedersen, & Thordal-Christensen, 2016) and in the infection hyphae *in Ustilago maydis* (Doehlemann et al., 2009) or extracellular spaces in case of *Cladosporium fulvum* (Doehlemann et al., 2009). The effector may be cytoplasmic effectors (Selin, De Kievit, Belmonte, & Fernando, 2016) and apoplastic effectors (Sperschneider, Dodds, Singh, & Taylor, 2018) based on site of action. In case of biotrophic pathogens, Avr9, Avr4, Avr2, and ECP2 from *C. fulvum* are few apoplastic effectors that interacts with *R* gene in the apoplastic spaces (Franceschetti et al., 2017). The apoplastic effectors offers protection from chitinases and hydrolytic enzymes of the host and sequester chitin to strengthen fungal cell wall (Doehlemann & Hemetsberger, 2013; Villamil et al., 2019; Rocafort, Fudal, & Mesarich, 2020). Whereas most of the reported rust fungal pathogens have cytoplasmic effectors with few exception (Dagvadorj, Ozketen, Andac, Duggan, Bozkurt, & Akkaya, 2017; Nirmala et al., 2011). Bioinformatics analysis, viz., ApoplastP have also predicted the lowest probability of the apoplastic effectors in rust fungal pathogens (Sperschneider, Dodds, Singh, & Taylor, 2018). Recently several studies have reported non-proteinaceous molecules viz., RNA, specifically small RNA (sRNA) and secondary metabolite that can also act as potential effector molecules. The non-proteinaceous effectors (NPEs) can also hijack the host defense system and aid in the disease progression. The putative effector like sRNAs produced by the pathogens can translocate distantly and silenced the host genes (Hunt et al., 2019; Collemare, O'Connell, & Lebrun, 2019; Pradhan, Ghosh, Sahoo, & Jha, 2021). This has led to new unique mechanisms adapted by the pathogen for modulating diverse biological processes in host during pathogenesis. The apoplastic effectors are generally smaller compared to cytoplasmic effectors and have higher cysteine content. Effectors have also been known to contain a small stretch of amino acids imparting special functions. The RxLR and dEER motifs are present in numerous oomycetes effectors, specifically in *Phytophthora* species proteins (Van, Guo et al., 2008; Wang et al., 2011). These motifs facilitate translocation of these effectors by phospholipid-based endocytosis to host cells, even in the absence of the pathogen (Dou et al., 2008; Chen et al., 2019). However, the phospholipid-based entry mechanisms are not specific to all the RXLR proteins (Gan, Rafiqi, Ellis, Jones, Hardham, & Dodds, 2010; Wawra et al., 2012). In addition to RxLR, very few motifs have been reported in plant fungal pathogens that are widespread in effectors (Zhao et al., 2020). The Y/F/

WxC is one of the motifs reported in several effector proteins specific to haustoria forming pathogens such as powdery mildew and rust pathogens (Zhao et al., 2020; Godfrey et al., 2010).

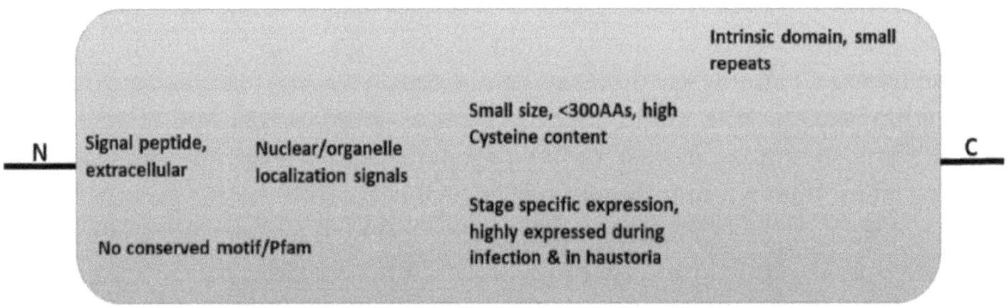

FIGURE 16.1 A characteristic representation of putative secretory effector proteins of plant fungi with high effector prediction score and positive evolution selection.

16.3 HOW EFFECTOR KNOWLEDGE CAN BE USED FOR PLANT DISEASE MANAGEMENT?

16.3.1 EFFECTOR-GUIDED BREEDING

The study of effectors, namely *Avr* genes and their interactions with R genes (or susceptibility genes in the necrotropic), and its use in crop breeding is known as effector-guided breeding. This technique can be utilized as a screening and selection process to create distinct sets of well-characterized pathogen isolates/races based on their virulence/avirulence phenotypes and/or genotypes. These isolates can then be used to screen plant cultivars to determine the underlying resistance genotype of those cultivars that can be commercially harvested. To evaluate the resistance genotype of all commercial *Brassica napus* (canola) cultivars, multiple sets of isolates of the canola pathogen *L. maculans* were used. However, due to the complexity of the Avr-R gene interaction and the pathogen's genome's close physical linkage of such genes, transgenic techniques should be used. Differential addition isolates, for example, are *L. maculans* isolates that have been specially engineered to express single avirulence genes. Using distinct sets of *L. maculans* isolates, this technique was utilized to find resistance genes in hitherto uncharacterized canola cultivars. The ToxA proteinaceous effector of the necrotrophic pathogen *Pyrenophora tritici-repentis* in wheat types can also be studied using a gene knock-out technique. Similarly, triple gene knockout strains of *Parastagonospora nodorum* were utilized to demonstrate that, whereas SnToxA, SnTox1, and SnTox3 effectors are the overwhelming casualties of the illness in wheat, certain other effectors are also involved. These triple knockout strains can be employed in breeding efforts to screen wheat germplasm for undiscovered susceptibility genes. One major limitation of this gene knockout strategy is that some pathogens have low homologous recombination rates, and in some species, such as *L. maculans* and *P. infestans*, Avr genes are embedded within highly repetitive DNA, limiting replacement using homologous recombination due to inaccessible target unique DNA. However, using the CRISPR/Cas9 system in fungal and oomycete diseases, which may exploit short target regions of 23 nucleotides, holds promise for the future development of such deletion strains. CRISPR/Cas9 was used to create a strain of *L. maculans* by interrupting various genes, including Avrones. Purified or heterologously produced individual effectors, in addition to microbial isolates, can be employed in effector-guided breeding to discover responses between each effector and R/susceptibility gene. This comprises effector agroinfiltration into plants, as well as the expression of the construct in plant cells. More than 50 unique *P. infestans* effectors were evaluated on potato clone SW93105, which shows novel resistance to *P. infestans* in the field, using this technique. Producing recombinant proteins for screening cultivars, such as screening wheat varieties in Australia for the presence of susceptibility genes to toxin effectors in both *P. tritici-repentis* and *P. nodorum*, is an alternate technique to agroinfiltration. Because of their modest size, the effectors should be reasonably easy to manufacture as recombinant proteins, giving valuable biotechnological tools. The aforementioned applications are concerned with qualitative resistance.

16.3.2 DECOY ENGINEERING

The plant immune system responds against pathogen invasion by detection of PAMPs and effectors through its receptors and co-receptor kinases. Engineering of these receptor variants that can recognize effectors efficiently could restrict pathogen from effector mediated manipulation and enhance disease resistance. One strategic 'decoy engineering' proof of concept includes, a modification of a decoy RLK protein PBS1 (protein kinase superfamily protein 1) with a protease cleavage site for the tobacco etch virus (TEV) protease that allowed activation of RPS5 (Resistance to

Pseudomonas syringae 5) leading to resistance against *P. syringae* and TEV (Pottinger & Innes, 2020). Similarly, recent crystal structures of the NLR Pik heavy metal associated (HMA) domain complexed with effector *Avr*-PikD and an unrecognized effector *Avr*-PikE showed that the C-terminus of the HMA domain played a key role in the interaction with effectors. The HMA domain represents the polymorphic regions of this NLR allowing recognition of multiple effectors (De La Concepcion et al., 2018). Therefore, these studies highlight the possibility of engineering decoys/guardees to provide broader spectrum resistance against pathogens. However, before proceeding it is central to establish whether such alteration within a decoy residue affects its structure and function. On the contrary, such modulation within the decoy protein might also allow further selection pressure in the pathogen effector repertoire and thus ultimately result in gain of virulence.

16.3.3 EFFECTOR-MEDIATED HOST DEFENSE RESISTANCE

The accessibility of genome sequences and functional effector assays has led to discovery of conserved effectors as well as effectors with unknown functions in inducing defense responses. A RxLR effector conserved between the oomycete species, *Phytophthora sojae* and *Hyaloperonospora arabidopsidis* has been identified, which contains a RxLR motif four amino acids, where x can be any amino acid residue (Deb et al., 2018). These RxLR effectors have a role in disease development with the RxLR part essential for translocation inside the plant cell (Gu et al., 2011). When transiently expressed in both *Nicotiana benthamiana* and soybean, the HasRxL23 and PsAvh73 effectors suppress PCD associated with PTI in *N. benthamiana* and ETI in soybean. Furthermore, when these genes were constitutively expressed in transgenic *Arabidopsis thaliana*, PTI was suppressed, and the lines showed increased susceptibility to both bacterial and oomycete attack (Deb et al., 2018). These conserved effectors suppress both PTI and ETI across diverse plant species, suggesting an essential role in virulence. Likewise, another RxLR effector, Cce1, has been identified in *Ustilago maydis* essential for virulence and triggers early plant defense responses (Seitner, Uhse, Gallei, & Djamei, 2018). Identification of conserved and potentially essential pathogen effectors along with corresponding host *R* genes for their recognition by plants could be exploited and may lead to more durable resistance (Zhao & Zhang, 2021). The identification of effector genes that are essential for virulence and trigger defense responses opens the door for control of these pathogens using spray-induced gene silencing (SIGS) where double stranded RNAs (dsRNAs) that target essential pathogen genes are sprayed onto plant surfaces, which leads to disease control (Koch et al., 2019). The SIGS technology relies on the processing of dsRNAs which have sequence homology to essential genes in the pathogen, into small RNAs by Dicer-Like (DCL) proteins (either by the host DCL proteins or the pathogen DCL proteins), which induces silencing of the corresponding gene target leading to control of the disease (Wang & Jin, 2017). The spraying of barley leaves with dsRNAs that target the three lanosterol C-14α-demethylase genes of *Fusarium graminearum* inhibited fungal growth, even in distal parts of the detached leaves that were not sprayed (Koch et al., 2019). One current limitation of SIGS is the stability of the dsRNAs when sprayed on the plants and the length of time for which the plant is protected. These technologies avoid the use of genetically modified (GM) organisms, such as plants engineered to express the dsRNAs, which are still a major concern to many consumers.

16.3.4 EFFECTOR-BASED ASSAYS TO AID INTEGRATED DISEASE MANAGEMENT (IDM) APPROACHES

Integrated disease management (IDM) involves manipulating different aspects of the pathogen, plant, or environment to reduce disease. An essential part of IDM is to deploy disease resistance genes to prolong their effectiveness in the field. Such management strategies for resistance gene deployment are influenced by the life traits of the pathogen. Pyramiding/stacking of *R* genes referred to the generation of a single host line with more than one qualitative *R*-gene, viz., the combination of *R*-genes Lr24, Lr28, and Lr9 to control leaf rust in wheat (Kim et al., 2020). More recently, stacking of *R*-genes has referred to the combination of one or more qualitative *R* genes with quantitative *R* genes, which leads to more durable resistance (Husain et al., 2021). Alternatively, rotation of cultivars with different single qualitative *R*-genes may be effective for sexually evolving pathogens (McDonald & Linde, 2002). Understanding how the pathogen evolves and information regarding what frequency of the population is virulent towards specific *R*-genes can aid in the deployment of *R*-genes across a landscape. However, by monitoring how the proportion of virulent isolates increases in a population, growers can be advised to adapt cultivars with a different *R* gene, to prevent probable yield losses related to 'breakdown' of resistance by altered pathogen population. Thus, the strategic resistance gene deployment is fully dependent on monitoring of virulence in the pathogen population, which can provide decisive information to determine whether specific resistance genes are at risk of being overcome. Alternatively, whole populations can be assessed for changes of avirulence towards specific *R* genes.

16.3.5 TAGGING EFFECTOR TARGETED HOST PROTEIN TO PREVENT MODULATION OF DEFENSE PATHWAYS

The plant pathogens secrete effectors which modulates the plant metabolism, signaling, and protein degradation pathways; thereby facilitating pathogen infection and spread in host. Hence, if the effector targeted host protein is identified and the pathogen's effector host protein is modulated, then the pathogen effector can bind with host target modified protein without disturbing the host mechanism and the pathogen infection can be inhibited. For instance, the *Ustilago maydis* effector proteins, Tin_2 and chorismite mutase (Cmu1) translocated into plant cells and the fungal Cmu1 binds with maize chorismate mutase, $ZmCm_2$ which is involved in salicylic acid (SA) production (activates plant immunity and induces cell death at high concentrations) (Djamei et al., 2011). This ultimately shifts the plant metabolism towards biosynthesis aromatic amino acids rather than SA accumulation. Similarly, the effector Tin2 has been shown to promote disease by diverting plant metabolism to anthocyanin biosynthesis rather than lignin biosynthesis though interaction with the host protein kinase ZmTTK1 (a positive regulator of anthocyanin biosynthesis) (Tanaka et al., 2019). The cytoplasmic effector of *Ustilago maydis*, Seedling efficient effector 1 (See1) interacts with the maize homolog of suppressor of G2 allele of skp1 (SGT1) and interferes with MAPK-induced phosphorylation of SGT1 resulting in maize leaves with tumor formation (Redkar et al., 2015). Some pathogen effectors are known to target this system to facilitate infection. The ubiquitin-proteasome system (UPS) is crucial in plant to maintain its proteome plasticity. Ubiquitin mediated degradation of stress responsive proteins until a stress signal is perceived prevents over accumulation or degradation of negative response regulators. *Magnaporthe oryzae* effector *Avr*Piz-t delivered into rice sheath epidermal cells during infection targets proteasome activity by interacting with the RING E3 ubiquitin ligase *Avr*Piz-t Interacting Protein 6 (APIP6). This interaction leads to degradation and suppression of ROS production during PTI in rice (Park et al., 2012). There are various cytoplasmic effectors which inhibit host PCD that restricts biotrophic pathogen growth. *Phytophthora* *infestans* effector *AVR*3a strongly suppresses infestin 1 (INF1)-triggered cell death (ICD) by binding and stabilizing a potato Ubox E3 ubiquitin ligase CMPG1 (Bos et al., 2010). Hence, a detail understanding of host proteins that are targeted by pathogen effectors will pave a way to engineer inhibitors to block effectors action mechanisms and restrict pathogens entry and spread in host cells.

16.4 FUTURE DIRECTIONS

There has been rapid development in effector research and progress of effector understanding and its interaction with host proteins with the advent of new omics technologies, availability of genome sequences, bioinformatics, and implementation of genetic manipulation technologies. This has led to the rapid identification of effectors in a vast number of pathogen species, though there is now a need of functional characterization of these candidate effector genes for a role in disease. CRISPR and SIGS provides the platform to accelerate the functional characterization needed to translate into improving agricultural practices. The identification of candidate plant *R* genes using pathogen effectors, provides high-throughput tools for screening germplasm and breeding material, though traditional breeding still requires rigorous generations of selection. The use of biotechnology approaches such as GM crops can accelerate this process. For instance, plant cultivars with *R* genes can be created by a transgenic approach faster rather than by a traditional crossing approach without linkage drag. This is because though *R* genes have been identified in hosts, the resistance is often negatively correlated with reduced yields and quality. *R* genes identified using knowledge about pathogen effectors can be introduced into susceptible cultivars, conferring resistance as reported in the Cavendish cultivar of banana for protection against a specific race of *F. oxysporum* (Dale et al., 2017). An alternative technology called host-induced gene silencing (HIGS) can be to control plant pathogens. The HIGS technology has been reported to offer various degrees of protection against a range of cereal pathogens (Qi, Guo, Peng, Liu, Kang, & Guo, 2019). Basically, HIGS is an RNAi based process whereby small RNAs made by the plant silence genes in the pathogen. The small RNAs are made from dsRNA molecules that target pathogen genes in transgenic plants and render inactivation. Anyhow, the bottleneck to overcome consumer concerns about GM organisms prevalent in today's world is always there. A reverse directional approach of using effectors in biotechnology to target pathogen populations using gene-drive elements can also be followed. Gene drives are genetic engineering approaches that aim to altered specific genes throughout a population that can used as self-sustaining technology in pest and vector management (Collins, 2018). The exposure to a plant resistance gene often leads to the emergence of populations predominated by strains with loss of the corresponding *Avr* gene whereas, in sexually propagating pathogens, the population tends to be 'driven' backwards to being predominantly avirulent. Hence, the risks they may impose, the more passive approach of monitoring *Avr*/virulence allele frequencies using molecular markers in populations without the presence of the R gene has the potential to provide insights into when cultivars of specific resistance could be reused.

16.5 CONCLUSIONS

The diverse mechanisms employed by effectors modulate or alter plant immunity, allowing for multiple approaches to harness and exploit their usefulness. The application-oriented work of effectors revolves around the subset involved in avirulence or acting as host-specific toxins, recombinant protein production, and purification to accelerate screening of plant lines for disease resistance/susceptibility. Throughout the whole process, the efficient delivery of the effector via agro-filtration to express a recombinant effector within plant cells has been aided by technology. The knowledge about the nature and function of effectors will facilitate the identification of specific strains of microbes with a known set of genes to screen plant cultivars for their resistance gene repertoire. Molecular marker screening using the detection of effector alleles in the population can be effective in selecting plant cultivars with the highest level of resistance and susceptibility. Each of these methods comes with caveats and limitations. With the continual improvements of genome technologies and functional genomics, the number of effectors identified has rapidly increased. The increase in gene discovery and predictions based on genome sequences of a fungus or an oomycete species has led to the identification of hundreds to over 1,000 such genes. However, many effectors have yet to be characterized and functionally validated. The increase in production and yield is always associated with the risk of disease and yield loss. The identification of thousands of such effector genes present in plant pathogenic fungi or oomycetes, the implementation of effector biotechnology holds promise to mitigate yield loss due to diseases and to meet the future agricultural production demand.

KEYWORDS

- **CRISPR/Cas9**
- **disease resistance genes**
- **effectoromics**
- **effectors**
- **effector-triggered immunity**
- **host-induced gene silencing**
- **pattern triggered immunity**
- **RNAi**
- **Septoria tritici blotch**

REFERENCES

Ahmed, A. A., Pedersen, C., & Thordal-Christensen, H., (2016). The barley powdery mildew effector candidates CSEP0081 and CSEP0254 promote fungal infection success. *PLoS One*, *11*(6), e0157586.

Bos, J. I., Armstrong, M. R., Gilroy, E. M., Boevink, P. C., Hein, I., Taylor, R. M., Zhendong, T., et al., (2010). Phytophthora infestans effector AVR3a is essential for virulence and manipulates plant immunity by stabilizing host E3 ligase CMPG1. *Proceedings of the National Academy of Sciences*, *107*(21), 9909–9914.

Catanzariti, A. M., Dodds, P. N., Ve, T., Kobe, B., Ellis, J. G., & Staskawicz, B. J., (2010). The AvrM effector from flax rust has a structured C-terminal domain and interacts directly with the M resistance protein. *Molecular Plant-Microbe Interactions*, *23*(1), 49–57.

Chen, X. R., Zhang, Y., Li, H. Y., Zhang, Z. H., Sheng, G. L., Li, Y. P., et al., (2019). The RXLR effector PcAvh1 is required for full virulence of *Phytophthora capsici*. *Molecular Plant-Microbe Interactions*, *32*(8), 986–1000.

Collemare, J., O'Connell, R., & Lebrun, M. H., (2019). Nonproteinaceous effectors: The terra incognita of plant–fungal interactions. *New Phytologist*, *223*(2), 590–596.

Collins, J. P., (2018). In *Gene Drives in Our Future: Challenges of and Opportunities for Using a Self-Sustaining Technology in Pest and Vector Management* (pp. 37–41). BMC proceedings, Springer.

Dagvadorj, B., Ozketen, A. C., Andac, A., Duggan, C., Bozkurt, T. O., & Akkaya, M. S., (2017). A *Puccinia striiformis* f. sp. *tritici* secreted protein activates plant immunity at the cell surface. *Scientific Reports*, *7*(1), 1–10.

Dale, J., James, A., Paul, J. Y., Khanna, H., Smith, M., Peraza-Echeverria, S., Garcia-Bastidas, F., et al., (2017). Transgenic cavendish bananas with resistance to *Fusarium* wilt tropical race 4. *Nature Communications*, *8*(1), 1–8.

Dangl, J. L., & Jones, J. D., (2001). Plant pathogens and integrated defense responses to infection. *Nature*, *411*(6839), 826–833.
De Jonge, R., Van, E. H. P., Kombrink, A., Shinya, T., Desaki, Y., Bours, R., Van, D. K. S., et al., (2010). Conserved fungal LysM effector Ecp6 prevents chitin-triggered immunity in plants. *Science*, *329*(5994), 953–955.
De La Concepcion, J. C., Franceschetti, M., Maqbool, A., Saitoh, H., Terauchi, R., Kamoun, S., & Banfield, M. J., (2018). Polymorphic residues in rice NLRs expand binding and response to effectors of the blast pathogen. *Nature Plants*, *4*(8), 576–585.
De Torres, M., Mansfield, J. W., Grabov, N., Brown, I. R., Ammouneh, H., Tsiamis, G., Forsyth, A., et al., (2006). *Pseudomonas syringae* effector AvrPtoB suppresses basal defense in *Arabidopsis*. *The Plant Journal*, *47*(3), 368–382.
Deb, D., Anderson, R. G., How-Yew-Kin, T., Tyler, B. M., & McDowell, J. M., (2018). Conserved RxLR effectors from oomycetes *Hyaloperonospora Arabidopsidis* and *Phytophthora sojae* suppress PAMP-and effector-triggered immunity in diverse plants. *Molecular Plant-Microbe Interactions*, *31*(3), 374–385.
Delteil, A., Gobbato, E., Cayrol, B., Estevan, J., Michel-Romiti, C., Dievart, A., Kroj, T., & Morel, J. B., (2016). Several wall-associated kinases participate positively and negatively in basal defense against rice blast fungus. *BMC Plant Biology*, *16*(1), 1–10.
Divon, H. H., & Fluhr, R., (2007). Nutrition acquisition strategies during fungal infection of plants. *FEMS Microbiology Letters*, *266*(1), 65–74.
Djamei, A., Schipper, K., Rabe, F., Ghosh, A., Vincon, V., Kahnt, J., Osorio, S., et al., (2011). Metabolic priming by a secreted fungal effector. *Nature*, *478*(7369), 395–398.
Dodd, M. C., Buffle, M. O., & Von, G. U., (2006). Oxidation of antibacterial molecules by aqueous ozone: Moiety-specific reaction kinetics and application to ozone-based wastewater treatment. *Environmental Science & Technology*, *40*(6), 1969–1977.
Doehlemann, G., & Hemetsberger, C., (2013). Apoplastic immunity and its suppression by filamentous plant pathogens. *New Phytologist*, *198*(4), 1001–1016.
Doehlemann, G., Van, D. L. K., Aßmann, D., Schwammbach, D., Hof, A., Mohanty, A., Jackson, D., & Kahmann, R., (2009). Pep1, a secreted effector protein of *Ustilago maydis*, is required for successful invasion of plant cells. *PLoS Pathogens*, *5*(2), e1000290.
Dou, D., Kale, S. D., Wang, X., Jiang, R. H., Bruce, N. A., Arredondo, F. D., Zhang, X., & Tyler, B. M., (2008). RXLR-mediated entry of *Phytophthora sojae* effector Avr1b into soybean cells does not require pathogen-encoded machinery. *The Plant Cell*, *20*(7), 1930–1947.
Eyal, Z., (1987). *The Septoria Diseases of Wheat: Concepts and Methods of Disease Management*. Cimmyt.
Flor, H., (1956). The complementary genic systems in flax and flax rust. In: *Advances in Genetics* (Vol. 8, pp. 29–54). Elsevier.
Fones, H., & Gurr, S., (2015). The impact of *Septoria tritici* blotch disease on wheat: An EU perspective. *Fungal Genetics and Biology*, *79*, 3–7.
Franceschetti, M., Maqbool, A., Jiménez-Dalmaroni, M. J., Pennington, H. G., Kamoun, S., & Banfield, M. J., (2017). Effectors of filamentous plant pathogens: Commonalities amid diversity. *Microbiology and Molecular Biology Reviews*, *81*(2), e00066–16.
Friesen, T. L., Faris, J. D., Solomon, P. S., & Oliver, R. P., (2008). Host-specific toxins: Effectors of necrotrophic pathogenicity. *Cellular Microbiology*, *10*(7), 1421–1428.
Friesen, T. L., Meinhardt, S. W., & Faris, J. D., (2007). The *Stagonospora nodorum*-wheat pathosystem involves multiple proteinaceous host-selective toxins and corresponding host

sensitivity genes that interact in an inverse gene-for-gene manner. *The Plant Journal*, *51*(4), 681–692.

Gan, P. H., Rafiqi, M., Ellis, J. G., Jones, D. A., Hardham, A. R., & Dodds, P. N., (2010). Lipid binding activities of flax rust AvrM and AvrL567 effectors. *Plant Signaling & Behavior*, *5*(10), 1272–1275.

Godfrey, D., Böhlenius, H., Pedersen, C., Zhang, Z., Emmersen, J., & Thordal-Christensen, H., (2010). Powdery mildew fungal effector candidates share N-terminal Y/F/WxC-motif. *BMC Genomics*, *11*(1), 1–13.

Goodwin, S. B., Ben M'Barek, S., Dhillon, B., Wittenberg, A. H., Crane, C. F., Hane, J. K., Foster, A. J., et al., (2011). Finished genome of the fungal wheat pathogen *Mycosphaerella graminicola* reveals dispensome structure, chromosome plasticity, and stealth pathogenesis. *PLoS Genetics*, *7*(6), e1002070.

Gu, B., Kale, S. D., Wang, Q., Wang, D., Pan, Q., Cao, H., Meng, Y., et al., (2011). Rust secreted protein Ps87 is conserved in diverse fungal pathogens and contains a RXLR-like motif sufficient for translocation into plant cells. *PLoS One*, *6*(11), e27217.

Haueisen, J., Möller, M., Eschenbrenner, C. J., Grandaubert, J., Seybold, H., Adamiak, H., & Stukenbrock, E. H., (2019). Highly flexible infection programs in a specialized wheat pathogen. *Ecology and Evolution*, *9*(1), 275–294.

Hayes, L. E., Sackett, K. E., Anderson, N. P., Flowers, M. D., & Mundt, C. C., (2016). Evidence of selection for fungicide resistance in *Zymoseptoria tritici* populations on wheat in Western Oregon. *Plant Disease*, *100*(2), 483–489.

Hunt, M., Banerjee, S., Surana, P., Liu, M., Fuerst, G., Mathioni, S., Meyers, B. C., et al., (2019). Small RNA discovery in the interaction between barley and the powdery mildew pathogen. *BMC Genomics*, *20*(1), 1–21.

Hurni, S., Scheuermann, D., Krattinger, S. G., Kessel, B., Wicker, T., Herren, G., Fitze, M. N., et al., (2015). The maize disease resistance gene Htn1 against northern corn leaf blight encodes a wall-associated receptor-like kinase. *Proceedings of the National Academy of Sciences*, *112*(28), 8780–8785.

Husain, R., Vikram, N., Gyanendra, K., Singh, V., Khan, N., Shamim, M., & Kumar, D., (2021). Molecular approaches for multiple genes stacking/pyramiding in tomato for major biotic stress management. *Biotic Stress Management in Tomato: Biotechnological Approaches*, 133.

Jones, J. D., & Dangl, J. L., (2006). The plant immune system. *Nature*, *444*(7117), 323–329.

Kim, K. M., Cho, S. W., Kang, C. S., Kim, K. H., Choi, C. H., Son, J. H., Park, C. S., & Mo, Y., (2020). Research advances in wheat breeding and genetics for leaf rust resistance. *Korean Society of Breeding Science*, *52*(3), 213–224.

Koch, A., Höfle, L., Werner, B. T., Imani, J., Schmidt, A., Jelonek, L., & Kogel, K. H., (2019). SIGS vs HIGS: A study on the efficacy of two dsRNA delivery strategies to silence *Fusarium* FgCYP51 genes in infected host and non-host plants. *Molecular Plant Pathology*, *20*(12), 1636–1644.

Lan, X., Liu, Y., Song, S., Yin, L., Xiang, J., Qu, J., & Lu, J., (2019). *Plasmopara viticola* effector PvRXLR131 suppresses plant immunity by targeting plant receptor-like kinase inhibitor BKI1. *Molecular Plant Pathology*, *20*(6), 765–783.

Larkan, N. J., Ma, L., & Borhan, M. H., (2015). The *Brassica napus* receptor-like protein RLM 2 is encoded by a second allele of the L ep R 3/R lm2 blackleg resistance locus. *Plant Biotechnology Journal*, *13*(7), 983–992.

Li, H., Zhou, S. Y., Zhao, W. S., Su, S. C., & Peng, Y. L., (2009). A novel wall-associated receptor-like protein kinase gene, OsWAK1, plays important roles in rice blast disease resistance. *Plant Molecular Biology*, *69*(3), 337–346.

Lo Presti, L., Lanver, D., Schweizer, G., Tanaka, S., Liang, L., Tollot, M., Zuccaro, A., et al., (2015). Fungal effectors and plant susceptibility. *Annual Review of Plant Biology*, *66*, 513–545.

Maidment, J. H., Franceschetti, M., Maqbool, A., Saitoh, H., Jantasuriyarat, C., Kamoun, S., Terauchi, R., & Banfield, M. J., (2021). Multiple variants of the fungal effector AVR-Pik bind the HMA domain of the rice protein OsHIPP19, providing a foundation to engineer plant defense. *Journal of Biological Chemistry*, *296*.

Marchal, C., Zhang, J., Zhang, P., Fenwick, P., Steuernagel, B., Adamski, N. M., Boyd, L., et al., (2018). BED-domain-containing immune receptors confer diverse resistance spectra to yellow rust. *Nature Plants*, *4*(9), 662–668.

McDonald, B. A., & Linde, C., (2002). Pathogen population genetics, evolutionary potential, and durable resistance. *Annual Review of Phytopathology*, *40*(1), 349–379.

Nirmala, J., Drader, T., Lawrence, P. K., Yin, C., Hulbert, S., Steber, C. M., Steffenson, B. J., et al., (2011). Concerted action of two avirulent spore effectors activates reaction to *Puccinia graminis* 1 (Rpg1)-mediated cereal stem rust resistance. *Proceedings of the National Academy of Sciences*, *108*(35), 14676–14681.

Orton, E. S., Deller, S., & Brown, J. K., (2011). *Mycosphaerella graminicola*: From genomics to disease control. *Molecular Plant Pathology*, *12*(5), 413–424.

Park, C. H., Chen, S., Shirsekar, G., Zhou, B., Khang, C. H., Songkumarn, P., Afzal, A. J., et al., (2012). The *Magnaporthe oryzae* effector AvrPiz-t targets the RING E3 ubiquitin ligase APIP6 to suppress pathogen-associated molecular pattern–triggered immunity in rice. *The Plant Cell*, *24*(11), 4748–4762.

Petit-Houdenot, Y., & Fudal, I., (2017). Complex interactions between fungal avirulence genes and their corresponding plant resistance genes and consequences for disease resistance management. *Frontiers in Plant Science*, *8*, 1072.

Pottinger, S. E., & Innes, R. W., (2020). RPS5-mediated disease resistance: Fundamental insights and translational applications. *Annual Review of Phytopathology*, *58*, 139–160.

Pradhan, A., Ghosh, S., Sahoo, D., & Jha, G., (2021). Fungal effectors, the double edge sword of phytopathogens. *Current Genetics*, *67*(1), 27–40.

Qi, T., Guo, J., Peng, H., Liu, P., Kang, Z., & Guo, J., (2019). Host-induced gene silencing: A powerful strategy to control diseases of wheat and barley. *International Journal of Molecular Sciences*, *20*(1), 206.

Redkar, A., Hoser, R., Schilling, L., Zechmann, B., Krzymowska, M., Walbot, V., & Doehlemann, G., (2015). A secreted effector protein of *Ustilago maydis* guides maize leaf cells to form tumors. *The Plant Cell*, *27*(4), 1332–1351.

Rocafort, M., Fudal, I., & Mesarich, C. H., (2020). Apoplastic effector proteins of plant-associated fungi and oomycetes. *Current Opinion in Plant Biology*, *56*, 9–19.

Sánchez-Vallet, A., Fouché, S., Fudal, I., Hartmann, F. E., Soyer, J. L., Tellier, A., & Croll, D., (2018). The genome biology of effector gene evolution in filamentous plant pathogens. *Annual Review of Phytopathology*, *56*, 21–40.

Sánchez-Vallet, A., Saleem-Batcha, R., Kombrink, A., Hansen, G., Valkenburg, D. J., Thomma, B. P., & Mesters, J. R., (2013). Fungal effector Ecp6 outcompetes host immune receptor for chitin binding through intrachain LysM dimerization. *eLife*, *2*, e00790.

See, P. T., Iagallo, E. M., Oliver, R. P., & Moffat, C. S., (2019). Heterologous expression of the *Pyrenophora tritici-repentis* effector proteins ToxA and ToxB, and the prevalence of effector sensitivity in Australian cereal crops. *Frontiers in Microbiology*, *10*, 182.

Seitner, D., Uhse, S., Gallei, M., & Djamei, A., (2018). The core effector Cce1 is required for early infection of maize by *Ustilago maydis*. *Molecular Plant Pathology*, *19*(10), 2277–2287.

Selin, C., De Kievit, T. R., Belmonte, M. F., & Fernando, W., (2016). Elucidating the role of effectors in plant-fungal interactions: Progress and challenges. *Frontiers in Microbiology*, *7*, 600.

Sonah, H., Deshmukh, R. K., & Bélanger, R. R., (2016). Computational prediction of effector proteins in fungi: Opportunities and challenges. *Frontiers in Plant Science*, *7*, 126.

Sperschneider, J., (2020). Machine learning in plant–pathogen interactions: Empowering biological predictions from field scale to genome scale. *New Phytologist*, *228*(1), 35–41.

Sperschneider, J., Dodds, P. N., Gardiner, D. M., Singh, K. B., & Taylor, J. M., (2018). Improved prediction of fungal effector proteins from secretomes with EffectorP 2.0. *Molecular Plant Pathology*, *19*(9), 2094–2110.

Sperschneider, J., Dodds, P. N., Singh, K. B., & Taylor, J. M., (2018). ApoplastP: Prediction of effectors and plant proteins in the apoplast using machine learning. *New Phytologist*, *217*(4), 1764–1778.

Sperschneider, J., Gardiner, D. M., Dodds, P. N., Tini, F., Covarelli, L., Singh, K. B., Manners, J. M., & Taylor, J. M., (2016). Effector P: Predicting fungal effector proteins from secretomes using machine learning. *New Phytologist*, *210*(2), 743–761.

St. Clair, D. A., (2010). Quantitative disease resistance and quantitative resistance loci in breeding. *Annual Review of Phytopathology*, *48*, 247–268.

Stergiopoulos, I., & De Wit, P. J., (2009). Fungal effector proteins. *Annual Review of Phytopathology*, *47*, 233–263.

Talbot, N. J., (2015). Taming a wild beast: Developing molecular tools and new methods to understand the biology of *Zymoseptoria tritici*. *Fungal Genetics and Biology*, *79*, 193–195.

Tan, K. C., Ferguson-Hunt, M., Rybak, K., Waters, O. D., Stanley, W. A., Bond, C. S., Stukenbrock, E. H., et al., (2012). Quantitative variation in effector activity of ToxA isoforms from *Stagonospora nodorum* and *Pyrenophora tritici-repentis*. *Molecular Plant-Microbe Interactions*, *25*(4), 515–522.

Tanaka, S., Schweizer, G., Rössel, N., Fukada, F., Thines, M., & Kahmann, R., (2019). Neofunctionalization of the secreted Tin2 effector in the fungal pathogen *Ustilago maydis*. *Nature Microbiology*, *4*(2), 251–257.

Van, D. H. R. A., & Kamoun, S., (2008). From guard to decoy: A new model for perception of plant pathogen effectors. *The Plant Cell*, *20*(8), 2009–2017.

Van, P. P. M., Guo, J., Van De, V. P. J., Jung, M. W., Birch, P. R., Whisson, S. C., & Govers, F., (2008). The *Phytophthora infestans* avirulence gene Avr4 encodes an RXLR-dEER effector. *Molecular Plant-Microbe Interactions*, *21*(11), 1460–1470.

Van, W. S., Tian, L., Hoy, R., & Li, X., (2020). Plant NLRs: The whistleblowers of plant immunity. *Plant Communications*, *1*(1), 100016.

Villamil, J. C. M., Mueller, A. N., Demir, F., Meyer, U., Ökmen, B., Hüynck, J. S., Breuer, M., et al., (2019). A fungal substrate mimicking molecule suppresses plant immunity via an inter-kingdom conserved motif. *Nature Communications*, *10*(1), 1–15.

Wang, M., & Jin, H., (2017). Spray-induced gene silencing: A powerful innovative strategy for crop protection. *Trends in Microbiology*, *25*(1), 4–6.

Wang, Q., Han, C., Ferreira, A. O., Yu, X., Ye, W., Tripathy, S., Kale, S. D., et al., (2011). Transcriptional programming and functional interactions within the *Phytophthora sojae* RXLR effector repertoire. *The Plant Cell*, *23*(6), 2064–2086.

Wawra, S., Agacan, M., Boddey, J. A., Davidson, I., Gachon, C. M., Zanda, M., Grouffaud, S., et al., (2012). Avirulence protein 3a (AVR3a) from the potato pathogen *Phytophthora infestans* forms homodimers through its predicted translocation region and does not specifically bind phospholipids. *Journal of Biological Chemistry*, *287*(45), 38101–38109.

Wu, G., Liu, Y., Xu, Y., Zhang, G., Shen, Q., & Zhang, R., (2018). Exploring elicitors of the beneficial rhizobacterium *Bacillus amyloliquefaciens* SQR9 to induce plant systemic resistance and their interactions with plant signaling pathways. *Molecular Plant-Microbe Interactions*, *31*(5), 560–567.

Xu, G., Zhong, X., Shi, Y., Liu, Z., Jiang, N., Liu, J., Ding, B., et al., (2020). A fungal effector targets a heat shock–dynamin protein complex to modulate mitochondrial dynamics and reduce plant immunity. *Science Advances*, *6*(48), eabb7719.

Zhao, K., & Zhang, Q., (2021). A climate-resilient R gene in rice traps two pathogen effectors for broad and durable resistance to bacterial blight. *Molecular Plant*, *14*(3), 366–368.

Zhao, S., Shang, X., Bi, W., Yu, X., Liu, D., Kang, Z., Wang, X., & Wang, X., (2020). Genome-wide identification of effector candidates with conserved motifs from the wheat leaf rust fungus *Puccinia triticina*. *Frontiers in Microbiology*, *11*, 1188.

Zuo, W., Chao, Q., Zhang, N., Ye, J., Tan, G., Li, B., Xing, Y., et al., (2015). A maize wall-associated kinase confers quantitative resistance to head smut. *Nature Genetics*, *47*(2), 151–157.

CHAPTER 17

Phytochemicals: An Alternate Approach Toward Various Disease Management

ANAND KUMAR,[1] RAVI RANJAN KUMAR,[2] TUSHAR RANJAN,[2] KUMARI RAJANI,[3] S. P. SINGH,[1] SATYENDRA,[1] MANKESH KUMAR,[1] SANGITA SAHNI,[4] BISHUN DEO PRASAD,[5] SANJAY KUMAR,[1] DINKAR,[1] and RAKESH DEO RANJAN[6]

[1]*Department of Plant Breeding and Genetics, Bihar Agricultural University, Sabour, Bhagalpur, Bihar, India*

[2]*Department of Molecular Biology and Genetic Engineering, Bihar Agricultural University, Sabour, Bhagalpur, Bihar, India*

[3]*Department of Seed Science Technology, Bihar Agricultural University, Sabour, Bhagalpur, Bihar, India*

[4]*Department of Plant Pathology, TCA, Dholi, Dr. Rajendra Prasad Central Agricultural University, Pusa, Samastipur, Bihar, India*

[5]*Department of AB&MB, CBS&H, Dr. Rajendra Prasad Central Agricultural University, Pusa, Samastipur, Bihar, India*

[6]*Department of Plant Physiology and Biochemistry, Bihar Agricultural University, Sabour, Bhagalpur, Bihar, India*

ABSTRACT

Plants contain a wide range of naturally occurring compounds that have biological relevance by assisting plants in defending themselves against various plant diseases by exhibiting antifungal, antimicrobial compounds through inhibition or killing processes. Synthetic antimicrobial chemicals have become widely employed in plants in preventing and curing various

New Advances in Crop Disease Management. Sangita Sahni, Ganesh Thapa, and Bishun Deo Prasad (Eds)

plant diseases after the discovery of different efficient and highly effective antimicrobial molecules. However, the emergence of resistant microbes, decreased antimicrobial efficiency and safety, and the search for new antimicrobial compounds to combat emerging noxious plant diseases caused by traditional antimicrobial compounds have reignited interest in different phytochemicals as a substitute for artificial antimicrobial chemical compounds. However, a lot of research on the efficiency of different phytochemicals as antimicrobial compounds has already been performed in vitro and in vivo in recent years. This chapter will give an overview of the different antifungal, antibacterial, and antiviral properties of some major groups of phytochemicals, namely flavonoids, phenols, alkaloids, saponins, steroids, glycosides, unsaturated lactones, tannins, terpenoids, sulfur compounds, essential oils, and glucosinolates. Exploration and investigation of the different phytochemicals already present inside different plants have hence come as a safer substitute for different disease management practices. This chapter aims to explore the current state of knowledge about phytochemicals as well as recent discoveries in this field to discover an alternate approach to various disease management.

17.1 INTRODUCTION

Phytochemicals, as the name suggests, are naturally occurring chemical compounds in plants that have biological relevance by assisting plants in defending themselves against various plant diseases by exhibiting anti-fungal, antimicrobial compounds through inhibition or killing processes. The amount of these substances secreted varies from plant to plant, with some producing more and others producing very little. They can be dangerous at times and extremely beneficial to others. Phytochemicals play a crucial part in their development and also in their growth. They defend plants against pathogens, insects, and microorganisms, as well as stressful events including ultraviolet (UV) irradiation and severe temperatures. Pollination, germination, and seed dissemination are all aided by the presence of helpful birds and insects. When consumed, phytochemicals give plants their colors and a variety of flavors, both pleasant and unpleasant. They are only found in certain plants and portions of plants, and their numbers tend to spike during stressful situations. When ingested, phytochemicals also have health benefits. They are made up of nutrients that are necessary for good health (e.g., proteins, carbs, vitamins, and minerals) as well as compounds (like flavonoids, phenolic acids, and also other phenolics) that play lesser-known roles in the body.

Several of these phytochemicals have been identified as biologically active components in traditional herbal drugs (like salicylates [aspirin]) that are present in willow bark, quinine in cinchona bark, and proanthocyanidins in cranberries, which are generally utilized to treat different infections of a urinary path). Polyphenols are the most common type of phytochemical and because of their numerous hydroxyl groups, they are potent antioxidants. Non-nutritive plant products with disease-protective natures are called phytochemicals. They are the compounds which are non-essential nutrients, which denote that they are not required by the human body for its survival. Plants synthesize such compounds to safeguard themselves, but a modern study has displayed that they can protect humans from a lot of diseases. A lot of phytochemicals have already been identified. Lycopene in tomatoes, isoflavones in soybean, and flavonoids in fruits are examples of well-known phytochemicals.

Phytoalexins are antimicrobial compounds with a low molecular weight that are synthesized by different plants in response to different abiotic and biotic stresses. The rapidity of phytoalexin accumulation inside the plant is related to the resistance of plants to different diseases caused by different fungi and bacteria, though all the genetic knowledge for phytoalexin production is present inside the susceptible as well as resistant plants. Phytoalexins are the only component of the complex process for disease resistance inside the plants. Phytochemicals are plant-specific defense systems. When plants photosynthesize, some of their sugars change to produce chemicals that protect them from the external environment (disease, insect pests, toxic substances, etc.) that attack them.

Phytochemicals are one of the biologically active non-nutrient plant chemical components that have a lot of curiosity as a photo-protective compound and also for their high antioxidant and free radical scavenging capacities. Polyphenols, isoflavonoids, phytoalexins, flavonoids, phenols, anthocyanidins, chlorophyll derivatives, nitrogen compounds (polyamines), beta carotene (pro-vitamin A), and other carotenoids like ascorbic acid (vitamin C), also folic acid, and tocopherol are among the different phytochemicals present in foods (fruits, vegetables, legumes, nuts, cereals (vitamin E) based on plants.

There are about 4,000 phytochemicals, with phenolics (one of the largest groups of phytochemicals) and carotenoids being the most studied, whereas other phytochemicals are only found in a few taxonomic groupings. Glucosinolates, for example, are only present in cruciferous vegetable crops, whereas sulfides are only found in the Liliaceae family. Furthermore, every fruit and vegetable species have a specific phytochemical profile that is present in a certain phytochemical category. Many phytochemicals, notably color compounds, are found in high concentrations in the outer layers of plant tissues. Lycopene comes from tomatoes, isoflavones come from soy, carotene comes from carrots, and anthocyanins come from blueberries and also from grapes are some examples of different phytochemicals. Lutein and zeaxanthin both are carotenoids that can be present inside spinach, kale, and turnip greens. One of the classes of phytochemicals called allyl sulfides is synthesized in garlic and onions.

17.2 CLASSIFICATION OF PHYTOCHEMICALS

Following are the classification of phytochemicals (Table 17.1):

- Alkaloids
- Anthocyanins
- Coumarins
- Flavonoids
- Fatty acids
- Lactones
- Polypeptides
- Polyphenols
- Tannins
- Terpenoids
- Saponins
- Sterols

17.3 WHAT ARE THE PHYTOCHEMICALS AND HOW DO THEY PERFORM THEIR FUNCTION?

Phytochemicals are a class of chemical compounds that are produced into plants to utilize in their mechanism of resistance to many infections caused by different fungi, bacteria, and also by plant viruses. There are a lot of phytochemicals present with a different mode of action like:

1. **Antioxidant:** A lot of phytochemicals have an antioxidant property, which protects our cells from the damage caused by oxidation and hence reduce the danger of some cancers. Antioxidant chemical compounds include allyl sulfides, carotenoids, flavonoids, and polyphenols.

TABLE 17.1 List of Phytochemicals

Sl. No.	Phytochemicals	Secondary Metabolites/Compounds
1.	*Alkaloids*	Caffeine, theobromine, theophylline
2.	Anthocyanins	Cyanidin, malvidin
3.	Carotenoids	Beta-carotene, lutein, lycopene
4.	Flavonoids	Proanthocyanidins, quercetin, rutin, tangeretin
5.	Hydroxycinnamic acids	Chicoric acid, coumarin, ferulic acid, scopoletin
6.	Isoflavones	Daidzein, genistein
7.	Lignans	Silymarin
8.	Monophenols	Hydroxytyrosol
9.	Monoterpenes	Geraniol, limonene
10.	Organosulfides	Allicin, glutathione, indole-3-carbinol, isothiocyanates, sulforaphane
11.	Other phytochemicals	Damnacanthal, digoxin, phytic acid
12.	Phenolic acids	Capsaicin, ellagic acid, gallic acid, rosmarinic acid, tannic acid
13.	Phytosterols	Beta-sitosterol
14.	Stilbenes	Pterostilbene, resveratrol
15.	Triterpenoids	Ursolic acid
16.	Xanthophylls	Astaxanthin, beta-cryptoxanthin

2. **Hormonal Action:** Soy isoflavones imitate the estrogens in human and guide in the reduction of the symptom of menopausal and also osteoporosis.
3. **Stimulator for a Lot of Enzymes:** Indoles, which are contained in cabbages, trigger enzymes that limit the effectiveness of estrogen, potentially reducing the danger of breast cancer. Protease inhibitors and terpenes are examples of phytochemicals that interfere with enzymes.
4. **Interference with DNA Replication:** Beans include saponins, which prevent cancer cells from multiplying by interfering with cell DNA replication. Capsaicin, a compound present in peppers, protect DNA from carcinogens.
5. **Anti-Bacterial Effect:** Garlic's phytochemical allicin has antibacterial effects.
6. **Physical Action:** Some of the phytochemicals get linked to the cell wall, and hence prevents adhering of the pathogen to human cell walls. Cranberry's anti-adhesion capacities are attributed to

proanthocyanidins. Consumption of cranberries reduces the risk of urinary path infections and also improves oral health.

17.4 SIGNIFICANCE AND IMPORTANCE OF PHYTOCHEMICALS

Plants with medicinal potential that are readily available in the area are a fantastic source of both food and effective medications. Plants are one of the greatest sources of different sorts of medications and naturally occurring goods because they tended to synthesize a wide spectrum of biologically active compounds. Because of this, they play an important role in the pharmaceutical industry's drug development programs. Exploration of biologically active chemicals derived from natural sources has always piqued researchers' curiosity as a potential source of novel infectious disease treatments.

The value of therapeutics of any plant is due to the presence of biologically active compounds or physiological components called "phytochemicals" that carries a particular mode of action on the human body. Chemicals like flavonoids, phenols, alkaloids, saponins, steroids, glycosides, unsaturated lactones, tannins, terpenoids, sulfur compounds, essential oils, and glucosinolates, are included in phytochemicals. Plants create these chemicals to protect themselves from microorganisms, insects, and herbivores. Plant stems, leaves, roots, twigs, fruit, flower, and modified plant organs are all employed for their therapeutic properties. Excessive usage of antibiotics can be damaging to the human body, and it could result in resistance to different diseases. As a result, the demand for natural chemicals with antimicrobial action is growing by the day. Even though phytotherapeutics is popular, still there is a need to research and develop cost-effective herbal medications with higher bioactive potential and fewer adverse effects. As a result, it is very critical to understand the potential of various indigenous plants and to validate such potentials using scientific methodologies. Many foods naturally contain phytochemicals.

Phytochemicals can be found in a variety of food crops including whole grains, vegetables, beans, fruits, and herbs. Eating more vegetables and fruits is the easiest approach to getting more phytochemicals (cauliflower, cabbage, carrots, broccoli, etc.) (Table 17.2). The consumption of 5 to 9 servings of fruits and vegetables is advisable every day. Minerals, vitamins, and fiber are abundant in fruits and vegetables, which are also low in saturated fat.

TABLE 17.2 List of Plants Containing Phytochemicals

Sl. No.	Plants	Plants Containing Phytochemicals
1.	Common herbs	Aloe vera, American ginseng, clary sage, common mallow, common yarrow, cornsilk, dandelion, ground ivy, hawthorn, hop, hyssop, Indian cress, Korean ginseng, lemon balm, lemon verbena, licorice, marigold, milk thistle, red clover, rooibos, rosemary, sage, schizandra, stinging nettle, sweet clover, tea, verbena, wild carrot, wild pansy, woodruff
2.	Medicinal plants	Comfrey, common broom, echinacea, ginkgo, goat's rue, lesser celandine, lungwort, opium poppy, passion fruit, periwinkle, red bryony, valerian, wintergreen
3.	Vegetables	Broccoli, fennel, garlic, tomato, wasabi
4.	Fruits and nuts	Acai, almond, bilberry, black raspberry, blackberry, blackcurrant, blueberry, cranberry, grape, guarana, hazelnut, mangosteen, maqui berry, noni, olive, orange, pomegranate, red raspberry, sea buckthorn, wild strawberry, wolfberry
5.	Beans and seeds	Cacao, flaxseed, soyabean

17.5 CHEMICAL DIVERSITY OF PHYTOALEXINS

Isoflavones, isoflavones, pterocarpans, Pterocarpus, isoflavones, and coumestans are six isoflavonoid category generated by Leguminosae (Table 17.3). Pisatin, phaseollin, glyceollin, medicarpin, and maackiain are few of the well-known pterocarpan phytoalexins. Worlds first phytoalexin was identified and explained in the garden pea, i.e., *Pisum sativum*, was known as pisatin. A very few numbers of pulses also produce some non-isoflavonoid phytoalexins like furanoacetylenes and stilbenes in addition to all these chemicals.

17.6 ROLE OF PHYTOCHEMICALS FOR DISEASE RESISTANCE

Those phytochemicals having antimicrobial characters have enormous capacity to reduce both plant as well as human diseases. Isolation, purification, and identification of these chemical compounds from a lot of different plant species is the initial step towards recognizing their different medicinal as well as agricultural applications (Table 17.4). In the sector of agriculture, the use of different antimicrobial phytochemicals like a significant part of an integrated disease management (IDM) system would scale back to the harmful remaining outcome of different manmade chemicals. The one and only aim of today's investigation in this sector has been the searching of additional control processes to reduce reliance on laboratory-made fungicides. Different

TABLE 17.3 Origin of Phytoalexin from Several Plant Families

Sl. No.	Plant Families	Types of Phytoalexins
1.	Amaryllidaceae	Flavans
2.	Brassicaceae	Indole phytoalexins/camalexin Sulfur-containing phytoalexins/brassinin
3.	Chenopodiaceae	Flavanones/betagarin isoflavones/betavulgarin
4.	Compositae	Polyacetylenes/safynol
5.	Convolvulaceae	Furanosesquiterpenes/Ipomeamarone
6.	Euphorbiaceae	Diterpenes/casbene
7.	Poaceae	Diterpenoids: momilactones; oryzalexins; zealexins; phytocassanes; kauralexins Deoxyanthocyanidins/luteolinidin and apigeninidin Flavanones/sakuranetin Phenylamides
8.	Leguminosae	Isoflavones, isoflavanones, isoflavans, coumestans Pterocarpans/pisatin, phaseollin, glyceollin, and maiackiain Furanoacetylenes/wyerone Stilbenes/resveratrol pterocarpens
9.	Linaceae	Phenylpropanoids/coniferyl alcohol
10.	Malvaceae	Terpenoids naphtaldehydes/gossypol
11.	Moraceae	Furanopterocarpans/moracins A-H
12.	Orchidaceae	Dihydrophenanthrenes/loroglossol
13.	Rutaceae	Methylated phenolic compounds/xanthoxylin
14.	Umbelliferae	Polyacetylenes/falcarinol Phenolics: xanthotoxin 6-methoxymellein
15.	Vitaceae	Stilbenes/resveratrol
16.	Rosaceae	Biphenyls/auarperin Dibenzofurans/cotonefurans
17.	Solanaceae	Phenylpropanoid related compounds Steroid glycoalkaloids Norsequi and sesquiterpenoids Coumarins Polyacetylenic derivatives

plants can synthesis a lot of different aromatic secondary metabolites for example, quinones, flavones, flavonoids, flavonols, phenols, phenolic acids, tannins, and coumarins. The components having the phenolic structures, like carvacrol, eugenol, and thymol, were noticed significantly active counter to different plant pathogens. These groups of chemical compounds execute antimicrobial activities and hence help in plant defense system in contrast to different plant pathogenic microorganisms. The volatile antimicrobial chemicals like allicin are produced in garlic when the tissues get injured and the substrate alliin gets intermixed with the enzyme allicin-lyase. Allicin is actively permeable in the membrane though it undergoes thiol-disulfide exchange reactions with free thiol groups in the proteins. Seed-borne pathogens are efficiently controlled by allicin. Use of different plant synthesized products specifically essential oils is one of the demanding attractive processes for protecting from different post-harvest diseases. The essential oil isolated from lemon grass controls post-harvest anthracnose of mango. The anti-viral protein isolated from *Bougainvillea spectabilis* and *Prosopis chilliness* was detected to be widely effective in reducing the infection of sunflower necrosis virus in cowpea and sunflower plants.

TABLE 17.4 List of Antimicrobial Phytochemicals

Antimicrobial Phytochemicals			
Phytoalexins	**Polyphenols**	**Essentials Oils**	**Other Compounds**
Pisatin	Eugenol	Caricacea	Ajoene
Phaseolin	Quercetin	Capparaceae	Capsanthin
Trifolirhizin	Ametoflavin	Terpineol	Capsaicin
Isocumarin	Ferulic acid	Thymol	S-methyl-1-cysteine sulfoxide
Orchinol	Catechol	P-cymene	Sinigrin

17.7 ANTIMICROBIAL SECONDARY METABOLITES

Plants have ultimate potential to synthesize a lot of aromatic secondary metabolites, mostly phenols or their oxygen-substituted derivatives. Most important sub-classes in this group of chemical compounds comprises of phenols, flavones, flavonoids, flavonols, phenolic acids, quinones, tannins, and coumarins. All these groups of chemical compounds execute an antimicrobial performance and also work in plant defense procedures against a lot of pathogenic microorganisms (Table 17.5). Basic phenols and phenolic acid are biologically vigorous phytochemicals consisting of a single substituted ring of phenol. The toxic effect of phenolic compounds to microorganisms is due to the site(s) and a lot of hydroxyl groups available inside the phenolic compound. Quinones is precisely vastly reactive chemical compound, colored chemical compounds which have two ketone substitutions inside the aromatic ring. Flavones, flavonoids, and flavonols are phenolic constructions having one carbonyl group. They are synthesized inside plants in response to a lot of different microbial infections and are commonly seen to be effective *in vitro* as an antimicrobial chemical compound against a wide array of a lot of different microorganisms. Polymeric phenolic chemical substances like tannins have the characteristic of astringent. These chemical compounds are soluble in water as well as in alcohol and also in acetone and give precipitates with proteins. Coumarins are also comes under phenolic compounds which is made up of fused benzene and α-pyrone rings. The crude liquid, volatile, and essential oil extracted from complete or specific plant organs like roots, stems, leaves, flowers, fruits, and seeds are widely used in manufacturing the antimicrobial chemical compounds which are precisely exploited against a lot of different plant diseases (Table 17.6).

Here under the list of six broad chemical groups like:

- Flavonoides and isoflavonoides;
- Saponins;
- Steroids;
- Tannins;
- Phenolic and phenolic acids—chlorogenic acid, protocatechuic acid, ferulic acid, caffeic acid;
- Coumarins and pyrones.

TABLE 17.5 Mode of Action of Phytochemicals

Sl. No.	Class	Sub-Class	Mechanism
1.	Phenolics	Simple phenols	Membrane disruption, substrate deprivation.
2.	Phenolic acids	Phenolic acids	Bind to adhesins, complex with the cell wall, inactivate enzymes.
3.	Terpenoids, essential oils	–	Membrane disruption.
4.	Alkaloids	–	Intercalate into the cell wall.
5.	Tannins	–	Bind to proteins, enzyme inhibition, substrate deprivation.
6.	Flavonoids	–	Bind to adhesins, complex with cell wall, Inactivate enzymes.
7.	Coumarins	–	Interaction with eukaryotic DNA.
8.	Lectins and polypeptides	–	Form disulfide bridges.

Source: Gurjar et al. (2012).

TABLE 17.6 Antimicrobial Compounds which are Used Against the Different Plant Pathogens/Diseases

Sl. No.	Compound	Class	Activity
1.	Tannin	Polyphenol	Fungi, bacteria, viruses
2.	Piperine	Alkaloid	Fungi
3.	Azadirachtin	Terpenoides	Fungi
4.	Allicin	Solfoxide	Fungi, bacteria
5.	Ricinine ricinoleic	Alkaloids	Fungi
6.	Curcumin	Terpenoids	Fungi, bacteria
7.	Caffeic acid	Terpenoid	Fungi, bacteria, viruses
8.	Withafarin A	Lactone	Bacteria, fungi
9.	Essential oil	Terpenoid	Fungi
10.	Hyoscymine scopolamine	Alkaloids	Fungi

17.8 ANTIMICROBIAL COMPOUNDS FROM PLANTS

Antimicrobial compounds from plants are widely classified into two groups:

1. **Phytoanticipins:** "Low molecular weight antimicrobial chemicals that are present in plants before the challenge by microorganisms, or are created primarily after infection from pre-existing precursors," according to the definition of phytoanticipins. Saponins, avenacin, and tomatine are the most common. One saponin, avenacin A-1, is found in the outer layer of oat roots, while another saponin, a-tomatine, is formed in tomatoes and has antifungal properties (Singh & Chandrawat, 2017).

2. **Phytoalexins:** These are very noxious antimicrobial chemical compounds synthesized in sufficient quantity inside the plants only because of stimulation by different various plant pathogenic bacteria or chemical and mechanical harm. Muller and Borger evolved it in 1940 during researching the association among potato and *Phytophthora infestans*. Majority of the known phytoalexins are poisonous to fungi, injurious to plants and also hinder their progress, but few among them are also dangerous for bacteria, nematodes, and other species. More than 350 phytoalexins extracted from roughly 30 different plants have been chemically characterized. From the pulse family, the largest count of 130 has been marked (Joseph, 1995). Phytoalexins synthesized by plants in the similar family generally have comparable chemical constructions; like, phytoalexins are flavonoids in most of

the legumes but terpenoids in most of the Solanaceae. Plants deploy a battery for initiation of defense approach, involving lytic enzymes for example glucanases and chitinases, oxidizing agents, cell wall lignifications, and a lot of proteins related to pathogenesis and transcripts with unclear activities (Dixon & Lamb, 1990). It's vital to recall that phytoalexin synthesis can be a part of a broader plant defense strategy, inside which any single factor cannot be abundant to keep a capable pathogen at bay (Mansfield, 1999).

17.7.1 ROLE OF PHYTOALEXIN IN DISEASE RESISTANCE

Phytoalexins gathered at the spots of infection and prevent the development of fungus and bacteria *in vitro*, hence they might be considered as a capable plant-defense chemicals in counter to fungi and bacteria-caused illnesses. Chemical fungicides are furthermore injurious than the phytoalexins. The suppression of extension of germ-tube, radial mycelial expansion, and rise in the dry weight of mycelia, as best illustrated by the effect of resveratrol on *B. cinerea*, the causal organism for gray mold in grapevine, is definite evidence of phytoalexin fungitoxicity. Fungal cells' cytological, morphological, and physiological characters may be effected by phytoalexins.

The mode of action of four different phytoalexins from the family Solanaceae on three different Phytophthora species concludes in zoospore motility loss, cell rounding with some type of swelling, and granulation in the cytoplasm and the rupturing of the membrane of cell (Harris & Dannis, 1997). The phytoalexins pisatin and phaseollin get gathered to fungi in lethal amounts in inoculum droplets placed on the opening pea or bean pods and inside the cells and tissues immediately below the inoculum droplets. In *B. cinerea* conidia treated by sub-lethal concentrations of resveratrol, asymmetric growth of the germ tube leads to the development of "curved-germ tubes." This cytological aberration displays that stilbenic chemical compounds may interact with tubulin polymerization, which is one of the common modes of action for artificial fungicides and anticancerous drugs.

Additionally, the correlation between phaseollin or kievitone and *Rhizoctonia solani* have showed that phytoalexins can have the capacity to alter glucose uptake procedure also by fungal cells. Conidia of *Botrytis cinerea* displayed complete disarray of mitochondria and also the fragmentation of the cellular membrane after getting treated with the stilbene phytoalexins resveratrol and pterostilbene. In *B. cinerea*, camalexin has newly been implicated in the acceleration of programmed cell death (PCD). Few phytoalexins, together with the coumarin phytoalexin scopoletin, were established to be effective *in vivo* in reducing green mold symptoms which are caused because of *Penicillium digitatum* on oranges. As like phenolic phytoalexins (resveratrol, scopoletin, scoparone, and umbelliferone), phenolic phytoalexins (resveratrol, scopoletin, scoparone, and umbelliferone) have been established to suppress *Penicillium expansum* development and patulin buildup inside apples. Phytoalexins also carry some antibacterial characteristic in addition to their antifungal characters. Rishitin, for example, suppress the viability of *Erwinia atroseptica* cells by around 100% at a dosage of 360 g/L (Lyon & Bayliss, 1975). Resveratrol also carries antibacterial characteristics against Chlamydia, Helicobacter, Staphylococcus, Enterococcus, Pseudomonas, and Neisseria gonorrhea. It is well known that phytoalexins are venomous to a broad range of organisms, both in prokaryotic and eukaryotic. In summary, phytoalexins are just only one portion of a complex system of disease resistance inside plants. Phyto biochemistry and molecular biology have helped greatly from studies on phytoalexins alone. Alarm ingredients synthesized and released by affected cells and spreading into nearby healthy cells promote phytoalexin synthesis and accumulation inside healthy plant cells surrounding all around the wounded or damaged cells.

Compatible biotrophic diseases never synthesize phytoalexins. Glucans, chitosan, glycoproteins, and polysaccharides are the examples of phytoalexin elicitors having a large molecular weight that are present inside the fungal cell wall. Host plant enzymes release the elicitor chemical compounds from the cell wall of fungi. Most of these elicitors are generic, which means that they are present in both compatible as well as in incompatible races of the plant pathogens and causes phytoalexin buildup irrespective of plant cultivar. Although, some of the phytoalexin elicitors have uniqueness in the phytoalexin accumulation that they produce on compatible as well as in incompatible cultivars is very similar to the phytoalexin buildup caused because of disease races. Phaseollin in beans, pisatin in peas, glyceollin in soybeans, alfalfa, and clover, rishitin in potatoes, gossypol in cotton, and capsidiol in pepper are few of the more well-known and well-studied phytoalexins.

17.7.2 THE ACTIVITY OF ANTIFUNGAL AND MODE OF ACTION OF METHANOL EXTRACTS FROM DIFFERENT PLANTS AGAINST PATHOGENS

The fungi are one of the major pathogens on plants and hence they can suppress agricultural productivity by even up to 90%. Plant diseases have been cured with a lot of systemic fungicides. Plants stock biologically active chemicals that can be utilized to fight against different infections. Though, a lot of key diseases have already developed resistance to a lot of presently available fungicides because of the indiscriminate application of synthetic fungicides (Gangawane, 1990). Additionally, these fungicides also pollute soil and waterbodies. Non-aimed organisms are also adversely affected by fungicides. Because of this, new anti-fungal chemicals should be developed as safe, ecofriendly, and economical substitute to synthetic fungicides extracted from different plants, because they produce a wide range of secondary metabolites that performs a defensive part in plants and prevent them from intruders. Essential oils and phyto-extracts have been found worldwide for their anti-fungal characteristic against a broad range of fungi (Ezzat, 2001; Abd-El-Khair & Hafez, 2006; Gupta et al., 2008).

A lot of plants have been already found to have anti-fungal mode of action against a lot of fungi (Mughal et al., 1996; Thirbhuvanmala & Doraisamy, 2004; Pandey et al., 2002; Sharma & Trivedi, 2002; Kumar, 2005). The mode of action of the plants are distinct against specific fungi, as these workers discovered through supervision of the results. The consequence of plant extracts on the character like percent seed germination and microbial infection was observed, and it was found that extracts of *Allium sativum* and *Einnicostema littorals* reduce microbial infection while enhancing germination of seed. Even 10% of extracts reduce germination of seed, extremely less concentrations of extracts can be used for treatment of seeds as seed protectants because high concentrations of the extract could alter the seed metabolism (Rathore, 1995). *Lawsonia inermis* extracts suppressed target fungal mycelial development the most (76.47%–87.77%), followed by extract of *Withania somnifera* (54.44%–78.88%). Antifungal mode of action was also seen in other plant extracts at various percentages. *In vivo*, a 1% extract of *Lawsonia inermis* enhanced germination of chickpea (*Cicer aeritinum* L.) seed by even up to 30% and protected against *Fusarium oxysporum*. The availability of a lot of different secondary metabolites was determined in each and every extracts. Alkaloids, flavonoids, saponins, steroids, and tannins were present in different plant extracts. Saponin was discovered in almost all of the extracts. On the cell membrane and enzyme synthesizing of fungi, the mechanism of action of extracts was determined. In *Aspergillus niger* and *Fusarium oxysporum*, *Lawsonia inermis* reduce catalase production.

Protein fractions of *Lawsonia inermis* and *Withania somnifera* suppress *Bipolaris oryzae* and *Colletotrichum lindemuthianum* mycelial development by four to five times more than non-protein fractions. The active chemical compound was proteinaceous or proteins in origin and hence were efficient against different plant diseases. The availability of these secondary metabolites can explain why all these plants have the property of antifungal. Pathogens can be opposed by the help of these substances in a variety of manners. Extract from *Lawsonia inermis* reduced the synthesis of catalase enzyme in *A. niger* and *F. oxysporum*. In fungi, catalase is utilized for the transformation of H_2O_2 to oxygen and water. If this enzyme cannot be produced because of any reason than H_2O_2 will get build up in higher concentration and hence become toxic to the cell. Rathore also reported on the effects of plant extracts on various enzyme secretion. Gupta & Bilgrami (1970) found that a decoction of bark from *Acacia arabica* and *Casuarina equisetifolia* suppresses cellulolytic enzyme production and activity in *Pythium aphanidermatum*, which is consistent with the current findings. The inability of plant extracts to inhibit enzymes could be related to a loss of adaptation or the presence of an inhibitor molecule. Khan & Nasreen (2010) reported that in comparison to non-protein fractions, protein fractions of *Lawsonia inermis* and *Withania somnifera* were reported to have stronger antifungal activity. It was confirmed that the active anti-fungal chemicals in both plants are proteinaceous in nature or proteins. The identification and elucidation of these chemicals will require more research. Fungi have distinct cell walls that are necessary for their survival. Any cell wall active substance that damages the cell wall component will cause the cell to lyse, but if there is sufficient osmo-protection in the medium, such as sorbitol, the cell will continue to proliferate.

17.7.3 USE OF PHYTOCHEMICALS AS BIO-FUNGICIDE

When compared to typical antifungal agents, the usage of natural formulations of antifungal compounds for safe, efficient, and natural-friendly characteristics against different fungus and mycotoxins is suitable for agricultural practices (Rai et al., 2020). The concept of fully understanding the exact principles underlying the antifungal and anti-mycotoxigenic mode of action of phytochemicals against mycotoxigenic fungi has been proposed in the literature (Chaudhari et al., 2020; Xu et al., 2020), facilitated by the following:

- Inhibition of ergosterol biosynthesis, a key sterol that governs plasma membrane biogenesis;
- Disruption of the fungal cell membrane;
- Generation of reactive oxygen species (ROS), which causes oxidative stress.

17.7.4 ROLE OF PHYTOCHEMICALS FOR CONTROLLING HUMAN DISEASES

17.7.4.1 ANTI-MICROBIAL PHYTOCHEMICALS

Plants having medicinal property and the phytochemical components present in their extracts can be a valuable therapeutic alternative, as well as a valuable tool in combating the threat of multi-drug-resistant pathogenic infections. To regulate, extracted crude of several well-known herbal medications are utilized. To combat microbial infections, crude extracts of various well-known herbal medications are employed. In the treatment of microbial infections that can arise in both acute and chronic wounds, antimicrobial components are utilized.

17.7.4.2 BIOMEDICINE FROM PHYTOCHEMICAL

A naturally occurring product can be a whole organism, like a plant, animal, or microorganism, or a component of it, such as a leaf of plant or flower, or an organ or isolated gland of animal. Humans have always been interested in the healing and toxic powers of plants. These plants' knowledge has been passed down through the generations. Herbal medicines and its formulations have been utilized in India, Japan, China, Korea, and other nations for over 1,000 years. The anti-inflammatory, anti-microbial, anti-oxidative, and wound-healing effects of different phytochemicals are related to their inhibition of platelet aggregation, anti-inflammatory, anti-microbial, and anti-oxidative activities, which reduce the risk of wound disorders. Antimicrobial, antioxidant, and wound-healing substances have been studied extensively to ease their application as phytomedicines in the pharmaceutical industry.

17.7.4.3 ANTIOXIDANT PHYTOCHEMICALS

Antioxidants are a type of defense system that protects the human body from damage caused by oxidation by free radicals. Excessive free radical production causes cellular stress, which can harm DNA, proteins, and other cellular structure and their functions. ROS are a cluster of free radicals, reactive particles, and ions synthesized from oxygen. Free radicals like superoxide anion (O_2^-), hydroxyl radical (OH^-), and non-radical compounds like hydrogen peroxide (H_2O_2) are examples of ROS. Medicinal plants have emerged as possible natural antioxidant sources, producing a variety of anti-oxidative chemicals with medicinal and therapeutic characteristics.

Antioxidant-containing drug formulations are widely used during treatment and protection of a broad range of infectious as well as complicated disorders. Antioxidants are said to prevent against oxidative stress exerted by reactive O_2 species (ROS) and free radicals. There are a lot of different kinds of naturally existing antioxidants in the nature, each with its composition, physical, and chemical characteristics, mode of action, and place of action. Antioxidant activity has been documented in medicinal plants high in flavonoids, vitamins, polyphenols, and anthocyanins. Antioxidant action is not limited to a single portion of the plant or specific groups of plants.

The plants are an important origin of medication with a wide range of biological characteristics, involving antioxidant character, antibacterial character, and also anti-fungal properties. Plants account for over a quarter of all conventional medications and primary health care for the majority of the world's population (Ekor, 2014). Natural antioxidants are broad-spectrum, safe, and effective in managing oxidative stress, which is caused by the overproduction of free radicals (Tan et al., 2018). Antibiotics, both synthetic and semi-synthetic, are commonly used in modern medicine to treat microbiological infections. Antibiotic resistance has emerged in numerous microorganisms, resulting in treatment failure. Furthermore, current antibiotics are prohibitively expensive. As a result, the utilization of plant extracts and chemicals generated from them has increased (Hassine et al., 2014). Several botanicals have been found to have curative potential during the course of treatment of different chronic diseases like cancer, diabetes, inflammation, stroke, and aging in earlier research (Kalita et al., 2012). According to these data, plants continue to be a valuable resource for the development of new medications and medicinal substances. Human cells, unlike plant cells, may not always produce sufficient levels of antioxidants (Pisoschi et al., 2009). The man-made antioxidants like butylated hydroxy toluene (BHT), ethoxquin, propyl gallate, tert-butylhydroquinone (TBHQ), and butylated hydroxy anisole (BHA) were commonly linked to poisonousness, carcinogenicity, and mutagenesis effects when utilized to battle oxidative stress (Bendiabdellah et al., 2012).

As a result, a natural and safer antioxidant source that can be used as a supplement is necessary. Vitamins, phenolics, flavonoids, tannins, and carotenes are plant-based antioxidants that are thought to play an important part in the oxidation process by interacting with free radicals and shielding cells from free radical-induced damage without generating any negative effects (Pisoschi et al., 2009; Prasad et al., 2010). Prior research has demonstrated a protective effect of high dietary natural antioxidants against chronic heart disease and cancer (Prasad et al., 2010). As a result, the demand for natural antioxidants in pharmaceuticals and nutraceuticals, as well as food additives, has skyrocketed in recent years. In agriculture, similar circumstances exist in terms of crop diseases. Major crop diseases and production losses are largely caused by fungi and bacteria (Kumar et al., 2020). As a rapid pest management solution, farmers typically use synthetic pesticides to prevent diseases and save yield. Synthetic pesticide residues, on the other hand, are very toxic and have been demonstrated to have harmful effects on non-targeted organisms, reducing biodiversity. Furthermore, indiscriminate, and haphazard use has resulted in the emergence of resistant pathogen strains and the recurrence of the disease. As a result, a better particular pesticide capable of controlling plant diseases must be created. T Plant-based eco-friendly greener botanical pesticides are an alternative to synthetic pesticides. Importantly, active plant chemicals are less harmful and biodegradable than inactive plant compounds (Fayaz et al., 2017). Organic foods are becoming more popular around the world. Consumer awareness of food safety has risen dramatically, leading to the prohibition of certain pesticides in agriculture. As a result of their efficacy, biodegradability, low toxicity, and vast availability of raw materials, botanical insecticides have gained traction (Campos et al., 2019). Several plant species are underutilized, and *L. camara* L. is one of them. Traditional healers have utilized *L. camara* extract to treat gastrointestinal, inflammatory, dermatological, rheumatism, fever, and headache for generations (Hernández et al., 2003; Mahdi-Pour et al., 2012). Interestingly, different *Lantana species* have been shown to have different types and levels of phytochemicals (Sharma et al., 1991).

17.7.4.4 ANTIVIRAL ACTIVITIES OF PHYTOCHEMICALS

Because of the unrivaled availability of chemical diversity and large-scale production capability, products with medicinal qualities derived from plants present limitless chances for new drug leads. Antivirals of chemical origin have often been linked to human toxicity, so finding alternatives has always been a must in antiviral therapy development. As a result, the need for medicinal medications derived from edible natural resources has increased over the world. Traditionally, a few countries with diverse flora have used plant-based treatments for the majority of their maladies, under various names and usually unverified wisdom. Such wisdom is known to exist in India, China, Africa, and several Sub-Saharan African countries. With herbal items and well-documented proof, Chinese medicine and Ayurveda from India are the main approaches.

Identifying active components from such plant-based medicines has become a vital part of the research with the introduction of modern allopathic practice and the science of drug discovery. The strategy is to find plants that are known to have antiviral activity against a specific virus and then isolate the different components to investigate which one is functioning on the virus and what mechanism it uses.

17.7.4.5 ANTI-BACTERIAL ACTIVITIES OF PHYTOCHEMICALS

Many diseases in humans and animals are caused by bacteria. Foodborne and waterborne infections caused by bacteria are important dangers in India. Staphylococcus aureus causes infections such as osteomyelitis, endocarditis, meningitis, pneumonia, and many others. They emit enterotoxins in food, causing food poisoning, and they also cause toxic shock syndrome in humans by releasing super antigens into the bloodstream. In humans, they make up the common microflora of the skin, nasal passage, and axillae. The existence of several bioactive chemicals or secondary metabolites in plant extracts has been associated with antibacterial activity. In India, traditional healers employ 2,500 types of plants, and 100 plant species are used daily as a source of medication. Chinese bell flowers have antimicrobial effects. Babul tree flowers, pods, and gum resins are used to alleviate diarrhea (Table 17.7).

TABLE 17.7 Anti-Mycobacterial Activities of Some Plants and Plant Parts

Plant Name	Plant Part Used	Active Constituent
Mallotus philippensis	Leaves	Ursolic acid, β-sitosterol
Piper nigrum	Seeds	Piperine
Alstonia scholaris	Bark, flower, fruit, and leaf	Not documented
Allium sativum	Bulb	Either fats and fixed oils or phenol and aryl amine derivative.
Acalypha indica	Leaves	
Adhatoda vasica	Leaves	
Leucas marrubioides	Roots	Not documented
Cassia fistula Linn	Roots	Alkaloids and tannins could be responsible.
Glycyrrhiza glabra L.	Rhizomes	Isoliquiritigenin and liquiritigenin
Celastrus vulcanicola	Dried leaves	Dihydro-β agarofuran sesquiterpenes.
Tiliacora triandra	Roots	Bisbenzylisoquinoline alkaloids, tiliacorinine, 2′-nortiliacorinine and tiliacorine

Source: Vijay et al. (2018).

Various plants such as *Acacia mearnsii, Acacia nilotica, Acalypa indica, Phyllanthus niruri, Allium rotundum,* and *Cinnamomum zeylanicum* were observed to possess antibacterial activities against *Klebsiella* spp. (Table 17.8).

TABLE 17.8 Anticancer Properties of Some Plants and Plant-Derived Bioactive Compounds

Plant Species Name	Bioactive Compound	Mode of Action
Murraya koenigii	Koenimbin	Induction of apoptosis and G0 arrest of cell cycle in breast cancer cells.
Catharanthus roseus	Vinblastine, vincristine	Disruption of microtubule function.
Taxus brevifolia	Paclitaxel	Interference with microtubule stabilization and inhibition of cancer cell propagation.
Camellia sinensis	Epigallocatechin-3-gallate	Inhibit cancer cell proliferation.
Curcuma longa	Curcumin	Inhibit cancer cell proliferation and inhibit NF-kB pathway that triggers the intracellular inflammatory response.
Podophyllum peltatum	Podophyllotoxin	Disruption of chromatin structure.
Moringa peregrina	Apigenin	Induces apoptosis, exhibit cytotoxic activities.
Saffron crocus	Crocetin	Inhibits the growth of cancer cells by inhibiting nucleic acid synthesis, inducing apoptosis, and enhancing antioxidative system.
Vitis vinifera	Resveratrol	Chemopreventive activity, anti-initiation, and anti-progression activities of cancer cells inhibit cancer metastasis by reducing hypoxia-inducible factor-1α and MMP-9 expression.
Glycine max	Genistein	Exhibits antioxidant, antihelmintic, antiangiogenic effects block the uncontrolled cell growth associated with cancer inhibits enzymes that regulate cell division and cell survival, functions as a tyrosine kinase inhibitor by inhibiting DNA topoisomerase II.
Zingiber officinale	Gingerol	Exhibits antioxidant, anti-inflammation, and antitumor promoting properties, further, decreases iNOS and TNF-alpha expression via suppression of IκBα phosphorylation and NF-κB nuclear translocation.
Solanum lycopersicum	Lycopene	Activates cancer preventive enzymes such as phase II detoxification enzymes and inhibits human cancer cell proliferation.
Rosmarinus officinalis	Rosmarinic acid	Dose dependently inhibits migration, adhesion, and invasion of cancer cells, induces apoptosis, suppresses NF-kB pathway through inhibition of phosphorylation and degradation of IκBα.

Source: Vijay et al. (2018).

17.8 CONCLUSION

Plants have a lot of different constituents and beneficial sources of new and biologically active chemical compounds with antimicrobial characteristics. The study on the basis of ethnobotanical of a plant is one of the most crucial for modern-day medicine but its benefits can't be overemphasized if procedures are not standardized to obtain similar and reproducible outcomes. These days, scientists are investigating for a lot of different plant products having antimicrobial characteristics. It would be really very much helpful to standardize procedures for extraction and antimicrobial efficacy testing in vitro, because of this the investigation for new bio-active plant products can be more organized and explanation of outcomes can be facilitated. A lot of different phytochemicals which have suppressive effects on each and every type of creature in vitro should be assigned to *in vivo* testing for approximation the efficiency in regulating the occurrence of a lot of different diseases in crops, plants, and also in humans. Effective associations with pharmacologists with doctors, plant pathologists with microbiologists are very much vital in searching the complete growth of an interesting lead compound into a usable product. Fungal diseases/infections are still one of the most predominant and serious difficulties that people face every day. A lot of antimycotics synthetic medicines are already present in the market; though, all these medicines have a wide range of ill effects, necessitating widespread treatment times, non-economical, and the occurrence of resistance in many fungal diseases. Itching, nausea, and abdominal distress are few of the disagreeable side effects of existing therapies and they can be risky at times. In many conditions, these side effects obstruct their therapeutic utilization. Herbal medications might be a best substitute to antibiotics in curing certain diseases because they are not or less harmful. Anti-fungal and antibacterial characteristics are present in a wide range of medicinal plants and their active ingredients, and they are commonly used in current drugs.

Traditional drugs from all around the world are being examined for antiviral activity, but we haven't been able to convert our discoveries into a ready-to-use medication. The high cost of clinical trials, product refinement and stabilization into a formulation, regulatory processes, unavailability of model animals, and monitoring resources are all contributing factors. If we want to explore more herbal remedies having well-documented antiviral characteristics on the market, we need to give such studies all the resources they need. A lot of government programs in a few nations have started to promote such works, but not to the amount that they should. The investigation is ongoing, with the hope of seeing novel chemical compounds with high efficiency in contrast to viral infections in the near future. In the similar vein as antimicrobial medications, the current scenario of cancer cells gradually developing resistance to chemotherapeutic treatments has made cancer treatment methods more difficult. Herbal medications are also in high demand for anti-diabetic therapy. Plants and their potential bioactive chemicals are thus the gold mines of the future medical world, which require more attention and focus.

KEYWORDS

- **antifungal action**
- **antimicrobial medications**
- **anti-mycotoxigenic mode**
- **antiviral characteristics**
- **disease management**
- **pathogens**
- **phytochemicals**

REFERENCES

Abd-El-Khair, H., & Hafiz, O. M., (2001). Effect of aqueous extract of some medicinal plants in controlling the green mould disease and improvement of stored "Washington" Novel orange quality. *Journal of Applied Science and Research, 2*(10), 664–674.

Bendiabdellah, A., El Amine, D. M., Djabou, N., Allali, H., Tabti, B., Muselli, A., et al., (2012). Biological activities and volatile constituents of *Daucus muricatus* L. from Algeria. *Chem. Cent. J., 6*, 48–58.

Campos, E. V., Proença, P. L., Oliveira, J. L., Bakshi, M., Abhilash, P. C., & Fraceto, L. F., (2019). Use of botanical insecticides for sustainable agriculture: Future perspectives. *Ecol. Indica, 105*, 483–495.

Chaudhari, A. K., Singh, V. K., Deepika, S. D., Prasad, J., Dwivedy, A. K., & Dubey, N. K., (2020). Improvement of *in vitro* and in situ antifungal, AFB1 inhibitory, and antioxidant activity of *Origanum majorana* L. essential oil through nanoemulsion and recommendation as a novel food preservative. *Food Chem. Toxicol.*, 111536.

Dixon, R. A., & Lamb, C. J., (1990). Molecular communication in the interaction between plants and microbial pathogens. *Ann. Rev. Plant Physiology and Plant Mol. Biol., 41*, 339–367.

Ekor, M., (2014). The growing use of herbal medicines: Issues relating to adverse reactions and challenges in monitoring safety. *Front Pharmacol., 4*, 177.

Ezzat, S. M., (2001). *In vitro* inhibition of *Candida albicans* growth by plant extracts and essential oil. *World Journal of Microbiology and Biotechnology, 17*, 757–759.

Fayaz, M., Bhat, M. H., Fayaz, M., Kumar, A., & Jain, A. K., (2017). Antifungal activity of *Lantana camara* L. Leaf extracts in different solvents against some pathogenic fungal strains. *Pharmacologia., 8*, 105–112.

Gangawane, L. V., (1990). Fungicide resistance in plant pathogens in India. *Indian Phytopathology, 40*, 551–553.

Gupta, C., Garg, A. P., Uniyal, R. C., & Kumari, A., (2008). Antimicrobial activity of some herbal oils against common food-borne pathogens. *African Journal of Microbiology Research, 2*, 254–261.

Gurjar, M. S., Ali, S., Akhtar, M., & Singh, K. S., (2012). Efficacy of plant extracts in plant disease management. *Agricultural Sciences, 3*(3), 425–433.

Harris, J. E., & Dennis, C., (1977). The effect of post- inflectional potato tuber metabolites and surfactants on zoospores of oomycetes. *Physiol. Plant Pathol., 11*, 163–169.

Hassine, M., Zardi-Berguaoui, A., Znati, M., Flamini, G., Jannet, H. B., & Hamza, M. A., (2014). Chemical composition, antibacterial and cytotoxic activities of the essential oil from the flowers of *Tunisian Convolvulus althaeoides* L. *Nat. Prod. Res., 28*, 769–775.

Hernández, T., Canales, M., Avila, J. G., Duran, A., Caballero., J., Romo De, V. A., et al., (2003). Ethnobotany and antibacterial activity of some plants used in traditional medicine of zapotitlán de las Salinas, Puebla (México). *J. Ethnopharmacol., 88*, 181–188.

Joseph, K., (1995). Phytoalexins, stress metabolism, and disease resistance in plants. *Annu. Rev. Phytopathol., 33*, 275–295.

Kalita, S., Kumar, G., Karthik, L., & Rao, K. V. B., (2012). A review on medicinal properties of lantana camara linn. *Res. J. Pharm. Tech., 5*, 711–715.

Kumar, A., Kumar, R., Sengupta, D., Das, S. N., Pandey, M. K., Bohra, A., et al., (2020). Deployment of genetic and genomic tools toward gaining a better understanding of rice-*Xanthomonas oryzae* pv. *oryzae* interactions for the development of durable bacterial blight-resistant rice. *Front. Plant Sci., 11*, 1152.

Lyon, G. D., & Bayliss, C. E., (1975). The effect of rishitin on *Erwinia carotovora* var. *atroseptica* and other bacteria. *Physiol. Plant Pathol., 6*, 177–186.

Mahdi-Pour, B., Jothy, S. L., Latha, L. Y., Chen, Y., & Sasidharan, S., (2012). Antioxidant activity of methanol extracts of different parts of *Lantana camara. Asian Pac. J. Trop Biomed., 2*, 960–965.

Mansfield, J. W., (1999). Antimicrobial compounds and resistance: The role of phytoalexins and anti-anticipins. In: Slusarenko, A. J., Fraser, R. S. S., & VanLoon, L. C., (eds.), *Mechanisms of Resistance to Plant Diseases*. Kluwer, Amsterdam.

Muller, K. O., & Borger, H., (1940). Experimentelle untersuchungen tuber die *Phytophthora* resistenz der *Kartoffel. Arbeit. Biol. Reichsant Land Forstwirtsch., 23*, 189–231.

Pisoschi, A. M., Cheregi, M. C., & Danet, A. F., (2009). Total antioxidant capacity of some commercial fruit juices: Electrochemical and spectrophotometrical approaches. *Molecules, 14*, 480–493.

Prasad, K. N., Chew, L. Y., Khoo, H. E., Kong, K. W., Azlan, A., & Ismail, A., (2010). Antioxidant capacities of peel, pulp, and seed fractions of *Canarium odontophyllum* Miq fruit. *J. Biomed. Biotech., 2010*, 871379.

Rai, M., Ingle, A. P., Pandit, R., Paralikar, P., Anasane, N., & Santos, C. A. D., (2020). Curcumin and curcumin-loaded nanoparticles: Antipathogenic and antiparasitic activities. *Expert Rev. Anti Infect. Ther., 18*(4), 367–379.

Rex, J. R. S., Muthukumar, N. M. S. A., & Selvakumar, P. M., (2018). Phytochemicals as a potential source for anti-microbial, anti-oxidant and wound healing: A review. *MOJ Biorg. Org. Chem., 2*(2), 61–70. doi: 10.15406/mojboc.2018.02.00058.

Roop, S., & Kuldeep, S. C., (2017). Role of phytoalexins in plant disease resistance. *Int. J. Curr. Microbiol. App. Sci., 6*(1), 125–129.

Sharma, O. P., Jagdish, V., & Sharma, P. D., (1991). Comparison of lantadenes content and toxicity of different varieties of the lantana plant. *J. Chem. Ecol., 17*, 2283–2291.

Tan, B. L., Norhaizan, M. E., Liew, W. P., & Sulaiman, R. H., (2018). Antioxidant and oxidative stress: A mutual interplay in age-related diseases. *Front. Pharmacol., 9*, 1162.

Vijay, N., Yogita, D., Joyita, B., & Neetu, M., (2018). Towards various disease management. In: Rani, V., & Yadav, U. C. S., (eds.), *Functional Food and Human Health*, https://doi.org/10.1007/978-981-13-1123-9_27.

Xu, J. L., Li, L. W., Luo, Y. X., Yuan, S. H., & Liu, N. N., (2020). *Antifungal Nanomaterials: Current Progress and Future Directions.* Innovations in Digital Health, Diagnostics, and Biomarkers 1.

CHAPTER 18

Recent Advances in the Development of Disease-Resistant Plants Utilizing Genetic Engineering Approaches

BISHUN DEO PRASAD,[1] SANGITA SAHNI,[2] DIKSHA KUMARI,[3] ANAND KUMAR,[4] TUSHAR RANJAN,[5] SUNAYANA RATHI,[6] and SUROJIT SEN[7]

[1]*Department of AB&MB, College of Basic Sciences and Humanities, Dr. Rajendra Prasad Central Agricultural University, Pusa, Samastipur, Bihar, India*

[2]*Department of Plant Pathology, TCA, Dholi, Dr. Rajendra Prasad Central Agricultural University, Pusa, Samastipur, Bihar, India*

[3]*Department of Plant Physiology, Institute of Agricultural Sciences, Banaras Hindu University, Varanasi, Uttar Pradesh, India*

[4]*Department of Plant Breeding and Genetics, Bihar Agricultural University, Sabour, Bhagalpur, Bihar, India*

[5]*Department of Molecular Biology and Genetic Engineering, Bihar Agricultural University, Sabour, Bhagalpur, Bihar, India*

[6]*Department of Biochemistry and Agricultural Chemistry, Assam Agricultural University, Jorhat, Assam, India*

[7]*Department of Zoology, Mariani College, Mariani, Jorhat, Assam, India*

ABSTRACT

Transgenic technologies are the only way to bridge the gap between an ever-increasing human population and their hunger, as well as diminishing

New Advances in Crop Disease Management. Sangita Sahni, Ganesh Thapa, and Bishun Deo Prasad (Eds)

crop productivity due to changing climatic conditions. Genetic engineering contributes directly to agricultural productivity by providing superior varieties with enhanced resistance to biotic stresses. Pathogen-derived resistance (PDR) refers to the resistance obtained from a pathogenic virus using viral protein like coat protein (CP), replicase protein (RP), and movement protein (MP). The biomolecules AMPs, PR proteins, RIPs, and enzymes involved in secondary metabolite biosynthesis (phytoalexins and phytoanticipins) also play a role in plant defense. Further, RNA interference (RNAi) is considered as a means of regulating immunity in GM disease-resistant plants. The development of the RNA silencing-based approach has been successfully employed as a potential tool for disease resistance in crop plants. This chapter deals with the production and status of transgenic plants engineered for disease resistance.

18.1 INTRODUCTION

Transgenic plants are those whose genome has been modified using recombinant DNA (rDNA) techniques. The purpose of making transgenics is to introduce a trait of interest to the plant. The gene(s) that have been incorporated into the host organism are known as the transgene, it may come from a different plant species. The purpose of developing transgenic plants is to produce desired and useful traits. Transgenic plants have been produced for improvement in versions of varieties that are already available or to develop new varieties for multiple traits.

Genetically modified (GM) crops are those that contain transgenes. Conventional plant breeding can produce cisgenic plants by using genes found within a species or closely related species. The use of cisgenic modification is considered beneficial for plants that are not developed conventionally (such as potatoes).

Transgenic approaches and the more recent genome-editing technologies are the most common forms of plant genetic engineering. The traditional transgenic method entails transforming plants at random locations with genes encoding desired agronomic traits (Lornce & Verpoorte, 2004). The result of these methods is usually varieties that contain foreign DNA. Gene editing, however, allows for modifications of endogenous plant DNA in a variety of ways, insertion/deletion/replacement of DNA at desired locations (Barrangou & Doudna, 2016). It depends on the editing process whether or not foreign DNA is introduced to the product. Many countries do not require additional regulatory measures for genome-edited plants that contain no foreign DNA, including the United States (USDA, 2018), Argentina (Whelan & Lema, 2015), and Brazil (CTNBio, Brazil, 2018). The use of traditional transgenic techniques and genome editing for crop improvement remains powerful, regardless of differences in regulatory policies.

Farmers face critical challenges to control plant diseases and pests, and this problem enhances major food security problems even after our best efforts are unable to control them. Approximately 11–30% of pests and diseases cause global loss to food (Savary et al., 2019). The hyper-virulent nature and short life cycle of pathogens have led the pathogens to overcome plant disease resistance quickly. The breakdown of resistance has led to the identification and introgression of new gene(s) in the crop plants. The limitation of natural sources of resistance in plants has argued for the use of GM technologies. Plant pathogens can be managed most effectively and sustainably through the use of GM and genome editing to expand the genetic tools available to breeders. There is great potential for farmers, consumers, and the agro-food industry to benefit from advances in GM technology. A growing number of pathogenic fungi, bacteria, and viruses are emerging as a consequence of global warming, making disease control a primary focus for sustained crop yield and quality in the coming years (Saharan et al., 2016). A description of some transgenic plants engineered for disease resistance is given in succeeding sections.

18.2 VIRUS RESISTANT TRANSGENIC PLANTS

Numerous viruses have been identified as causes of disease and crop losses. Sequencing, isolation, and cloning of key viruses' genes have revolutionized genetic engineering of virus-resistant crops. As a result of these advances, as well as advances associated with biotechnological transformation of several crops has develop opportunity towards controlling viruses.

18.2.1 PATHOGEN DERIVED RESISTANCE (PDR)

The resistance derived from pathogenic viruses is known as PDR. So, in this case the gene of interest (whole or part) is expressed in plants to obtain disease resists.

18.2.2 VIRAL PROTEIN MEDIATED RESISTANCE

This is one of the types of PDR in which transformed plants present several types of viral proteins such as movement protein (MP), replicase protein (RP), coat protein (CP). Viruses cannot replicate when these proteins are present. There are three types of resistance in this category – (i) coat protein mediated resistance (CPMR); (ii) replicase mediated resistance (Rep MR); and (iii) MP mediated resistance (MPMR) (Table 18.1).

TABLE 18.1 Viral Protein-Mediated Resistance in Transgenic Crop (Saharan et al., 2016)

Candidate PDR Genes	Host Plant	Viral Disease	Targeted Gene	References
CPMR	Papaya	Papaya ring spot virus (PRSV)	PRSV-CP	–
	Potato	Potato virus-X	PVX-CP	–
	Squash	Watermelon mosaic virus 2 and zucchini yellow mosaic virus.	CP	Tricoli et al. (1995)
	Squash	Cucumber mosaic virus, watermelon mosaic virus 2, and zucchini yellow mosaic virus.	CP	Tricoli et al. (1995)
	Sweet pepper	Cucumber mosaic virus.	CP	Zhu et al. (1996)
	Plum	Plum pox virus	CP	Scorza et al. (1994); Ilardi & Tavazza (2015)
	Tomato	Cucumber mosaic virus.	CP	Yang et al. (1995)
Rep MR	Rice	Rice yellow mottle virus (RYMV).	RdRp	–
	Potato	Potato leaf roll virus (PLRV).	PLRV-Rp	–
	Bean	Bean golden mosaic virus.	+, – RNA of viral replication protein.	Bonfim et al. (2007); Tollefson (2011)
MPMR	Tobacco	Tobacco mosaic virus (TMV).	MP-P^{30}	–

In CPMR, the unique sequences of viral CP were isolated and cloned in plant expression vector. The gene construct was transformed in to target plants for enhanced disease resistance. CPMR can be depicted in two ways (Beachy & Philos, 1999):

1. **Recoating of Invading Viruses:** The expressed CP submit in plant recoats the invading viral genetic material. The invading viruses cannot multiply and thus cannot infect the plants.
2. **Blocking of Receptors in Transgenic Cells:** CP subunit binds to receptors in transgenic plants and prevents viruses from attaching to them. Plants cannot be infected by viruses due to receptor binding. Transformation of tobacco to express TMV CP subunit induced CPMR expression for both TMV and closely related viruses. Perhaps this is the result of significant homology between the genes encoding each virus' CP subunit.

PRSV (papaya ring spot virus) is an extremely destructive virus that attacks papayas. It is necessary to spray aphicides on infected plants as part of the PRSV control strategy. When the disease has already spread, however, rouging will not be able to stop it. Additionally, aphidicides are often ineffective because the virus is transmitted to the plant before the pesticides kill the aphids. During 1992–1997, Hawaii, USA papaya industry was about to wipe out due to infection of PRSV. The transgenic papaya plants expressing CP gene of tobacco mosaic virus (TMV) prevent infection by PRSV (Azad et al., 2014). The transgenic papaya overexpressing PRSV CP genes was achieved using biolistic method (Fitch et al., 1992). For developing transgenic papaya, embryos of two commercial cultivars "Sunset" and "Kapoho" were bombarded with microprojectiles coated with CP gene construct (pGA48GG/cpPRV-4) (Fitch et al., 1992). Transgenic line 55-1 of Sunset was inbred to homozygosity for the single copy CP gene and named SunUp. However, no suitable transgenic lines were obtained in Kapoho cultivar. Further, hybrid papaya "Rainbow" was developed by crossing transgenic SunUp and nontransgenic Kapoho (Manshardt, 1998). The virus resistant cultivars of papaya were screened at field level and released in commercialized level by United State government. The cultivar SunUp and Rainbow released in USA in 1998, widely cultivated in Hawaii.

A replicase enzyme is required by viruses to replicate their genetic material within their host. In replicase mediated resistance (Rep MR), a mutated or truncated replicase enzyme is express in host plant, which hinder replication of virus genetic materials. Transgenic rice expressing the RdRp (RNA dependent RNA polymerase) of rice yellow mottle virus (RYMV) showed enhanced resistance against RYMV strains (Palukaitis & Zaitlin, 1997). The expression of complete sense PLRV replicase gene in potato showed higher resistance to potato leaf roll virus (PLRV) (Ehrifeild et al., 2004). The major disadvantages of Rep MR are its narrow spectrum range, i.e., effective against specific virus race. So, more experimental studies on Rep MR have not been widely explored. Viruses travel between cells via plasmodesmata and are assisted by MPs.

In order to prevent the movement of viruses from one cell to another, modified viral MP gene may be expressed in plant. The expressed modified MP and wild type viral MP competes, which hampers movement of viruses within cell. Transgenic tobacco plants reveal late symptoms of diseases when engineered with P^{30} MP. Mutated or dysfunctional MP genes have been reported to cause broader resistance than CP/Rep mediated resistance (Prins et al., 2008).

18.2.3 VIRAL RNA MEDIATED RESISTANCE

There are a number of plant viruses that cause diseases in plants that have RNA genomes that encode all the essential proteins, namely MPs, CPs, RPs, etc. Numerous reports indicate successful control of viral infection due to silencing of viral RNA in plants.

18.2.4 POST-TRANSCRIPTIONAL GENE SILENCING/RNA INTERFERENCE (PTGS/RNAI)

PTGS/RNAi an attractive approach for viral resistance in plants. This is based on the fact that small interfering RNAs (siRNAs) produced due to viral infection protect the further infection by same or related virus. SiRNAs have been synthesized by transgenic-mediated PTGS/RNAi technology (Leibman et al., 2011; Wang et al., 2010). The technique involves designing a hairpin RNA transformation vector with inverted repeats separated by non-coding sequences based on conserved sequences from the viral genome. These inverted repeats of the hairpin RNA transformation vector produce hairpin RNAs (dsRNA). In plant, the dsRNA was cleaved by RNAase III-like Dicer enzyme into small interfering RNAs (siRNAs) or miRNAs. The length of formed siRNA or miRNA is 21–25 nt in length. There are 2 to 3 nt overhang observed in siRNA to miRNA. These small RNAs are first incorporated into a multi-subunit ribonuclease complex, known as RISC (RNA-induced silencing complex). Sense strand (passenger strand) of siRNA get degraded. The siRNAs guide the RISC to homologous RNA sequences, promoting a sequence-specific RNA degradation. One of the important components of

RISC is the Argonaute (Ago) protein, which cleaves single-stranded (ss) RNA complementary to siRNAs into small RNA fragments. This gene inactivation process protects plants from invading viral RNA.

PTGS/RNAi technology has been successfully applied to several species of economically important plant viruses, including transgenic tomato plants exhibited resistance against potato spindle viroid through siRNA (Schwind et al., 2009), transgenic cassava plants showed resistance to African cassava mosaic virus (ACMV) by expressing dsRNAs (Vanderschuren et al., 2009), transgenic *N. benthamiana* showed resistance to a dicot-infecting mastrevirus (Nahid et al., 2011), transgenic soybean showed resistance against soybean mosaic virus (SMV) (Luan et al., 2020), etc. Some of the commercially available crops based on PTGS/RNAi approach in the U.S. include pumpkins resistant to watermelon mosaic virus (WMV), Zucchini mosaic virus (ZYMV), and cucumber mosaic virus (CMV), and virus-resistant potatoes.

This strategy has created the majority of GMOs with viral pathogen resistance that cultivated and consumed according to approved standards. There has been commercial cultivation of transgenic squash (*Cucurbita sp.*) using RNAi technology in the United States for over 20 years. Fuchs & Gonsalves (2007) report that disease resistance achieved by using RNAi-mediated strategies is highly durable. Furthermore, squash varieties that are resistant to several viral species have been developed using the RNAi strategy (Table 18.1). Rather than using individual RNAi constructs, these varieties are produced by gene stacking (Tricoli et al., 1995).

As a result of the demonstration that dsRNA induces RNAi (Lindbo & Dougherty, 2005), transgenes were designed encoding inverted repeat sequences, the transcripts of which form dsRNA (Waterhouse et al., 1998). Bonfim et al. (2007) employed this strategy to create a transgenic common bean variety that was resistant to the DNA virus bean golden mosaic virus (BGMV). BGMV is made up of single-stranded DNA, which is translated into double-stranded DNA and transcribed in the host to produce proteins necessary for its replication (Hanley-Bowdoin et al., 2013). A spliced hairpin RNA expression cassette containing sense and antisense sequences of BGMV replication gene AC1 was used to generate small interfering RNA targeting viral transcripts (Bonfim et al., 2007). The genetically engineered bean (Bonfim et al., 2007) was shown to be resistant in greenhouse and field conditions by Arago & Faria (2009). Brazilian farmers are now allowed to grow BGMV-resistant beans since 2011 (Table 18.1). The crop is the only deregulated genetically engineered crop to demonstrate virus resistance to date. It is relatively simple to generate heterogeneous small interfering

RNAs targeting multiple transcripts by transgenic expression of an artificially designed inverted repeat, as opposed to co-suppression.

18.2.5 NON-PATHOGEN DERIVED RESISTANCE (NPDR)

The resistance obtained from a nonvirus origin, i.e., gene(s) derived either from plants or any other non-pathogenic sources is known as non-pathogen derived resistance (NPDR).

18.2.6 RIBOSOME INACTIVATING PROTEINS (RIPS)

Ribosome-inactivating proteins (RIPs) inhibit the translation process in eukaryotes and prokaryotes. RIP is toxic N-glycosidases which depurinate rRNA component of protein synthesis in prokaryotes and eukaryotes. Diverse plant species and tissues contain RIPs. Transgenic plants producing RIPs inhibits virus protein synthesis in the host plant cells and Gutierrez therefore the infected virions are not able to generate the protein for their multiplication. The pokeweed antiviral protein (PAP; RIP Type-1) coding gene expressed in transgenic tobacco shows a low level of resistance against many unrelated viruses.

18.2.7 VIRAL PROTEASE INHIBITORS FROM PLANTS

Viral protease inhibitors from plants inhibits the viral protease enzymes, which are essential for survival in host cells. The tobacco etch virus (TEV) is successfully controlled with transgenic tobacco expressing cysteine protease inhibitors from rice (Gutierrez-Campos et al., 1999). However, certain viruses didn't require protein in lifecycle completion. In addition, viral protease inhibitor produces deleterious effect on plants.

18.2.8 PLANT ANTIBODIES

To protect plants from pathogens, plant antibodies (Av-plantibodies) offer an alternative approach. The heterologous expression of antibodies has been shown to provide resistance against invading viruses. These plant antibodies are known as plantibodies. The first demonstration of antibody-mediated resistance to a plant virus was observed in transgenic *N. benthamiana* plants expressing a scFv against Artichoke mottled crinkle virus (AMCV) (Tavladoraki et al., 1993). The transgenic plants showed delay producing diseases (Tavladoraki et al., 1993). A list of plantibodies producing transgenic plants were listed in Table 18.2.

TABLE 18.2 Plantibody-Mediated Resistance to Viruses in Various Transgenic Plant Species

Virus Species	Plant Species
Artichoke mottled crinkle virus (AMCV)	*N. benthamiana*
Arabis mosaic virus (ArMV)	*N. benthamiana*
Plum pox virus (PPV)	*N. benthamiana*
Potato leafroll virus (PLRV)	Potato
Potato virus Y (PVY)	Potato
Tomato spotted wilt virus (TSWV)	*N. benthamiana*
Turnip mosaic virus (TuMV)	Chinese cabbage

18.3 FUNGAL RESISTANT TRANSGENIC PLANTS

There are a variety of plant pathogenic fungi that have the ability to adapt to diverse environment and destroy the growth of their host plants. Pesticides and breeding programs have been commonly used to control pathogens. The naturally available R-gene resistance has incorporated major resistance in many cultivars against several pathogens, mainly biotrophs. Furthermore, pesticides used to kill pathogens are harmful to beneficial microbes as well as causing resistance to develop. By using a wide range of genes in a carefully designed manner, genetic engineering can overcome many of the shortcomings of traditional breeding.

18.3.1 USING GENES FOR CHITINASES AND GLUCANASES

In many fungi cell wall components chitin and chitin degraded by, chitinase and glucanase hydrolyze Plants and microorganisms have been found to contain chitinase genes. They are classified as protein classes PR-3. A number of reports of obtaining fungal-disease resistance through transformation of chitinase genes in many crops are available. The other enzyme glucanase is classified as a PR-2 class of proteins. Rhizoctonia solani, for instance, contains significant amounts of chitin, which can be hydrolyzed by chitinase. β-1,3-glucanase is known to degrade glucans which are also present in the fungal cell walls. Several laboratories have transformed plants with genes encoding β-1,3-glucanase and chitinase in order to develop transgenic crops with enhanced resistance to fungal diseases. The genes for chitinase from varied sources have been used to generate transgenics in grapevine (Yamamoto et al., 2000), rice (Takakura et al., 2000), peanut (Rohini & Rao, 2001). Further, pyramiding of these two genes in transgenic crops promotes higher levels of resistance against fungal pathogen (Ram & Mohandas, 2003).

18.3.2 USING ANTIFUNGAL GENES

Apart from chitinases and glucanases, many other antifungal/antimicrobial proteins (APs) and peptides such as osmotin, thaumatin-like protein (TLP), ribosome-inactivating protein (RIP), *Allium cepa* Anti-Microbial Protein (Ace-AMP1b), *Raphanus sativus* anti-fungal protein (Rs-AFP2), *Dahlia merckii* Anti-Microbial Protein (Dm-AMP1), *Mirabilis jalapa* AP (Mj-AMP2), etc. Plants transgenic with these genes have been very effective at resisting fungal diseases. Osmotin and thaumatin-like proteins (OLP and TLP) are important anti-fungal compounds. These proteins have molecular masses of 22–26 kDa and are classified in the PR-5 family of pathogenesis-related (PR) proteins. These TLPs get induced upon microbial infection, oxidative stress, ABA, salicylic acid (SA), methyl jasmonate, ethylene, and certain wounding (Saharan et al., 2016). Transgenic potato plants expressing the tobacco osmotin (similar to TLP) showed delayed development of disease symptoms against *Phytophthora infestans* (Liu et al., 1994). Overexpression of rice TLP in rice itself, American ginseng, carrot, and tobacco enhanced the resistance to various fungal diseases (Saharan et al., 2016).

18.3.2.1 SMALL CYSTEINE RICH PROTEINS

Small cysteine rich proteins are usually small, cationic, and amphipathic proteins having open-chain forms. The cationic hydrophobic residues are organized as segregate patches, resulting in a structure that is capable of forming ion channels through membrane bilayers. Furthermore, the compact and rigid structure is maintained by three or four disulfide bonds through cysteine residues. Following are two important small cysteine rich proteins which have an immense role in antifungal activities:

1. **Plant Defensins:** A plant AP, are small sized (45 to 54 aa), cystine-rich and cationic peptide. Defensins were discovered for the first time in wheat and barley, and most plants likely express them in a variety of organs. The best characterized defensins from radish Rs-AFP2 peptide shows enhanced resistance against the fungus *Alternaria longipipes* and *Magnaporthe oryzae* in transgenic tobacco and rice, respectively. Dahlia merckii antimicrobial protein 1 (Dm-AMP 1) was also widely used for genetic engineering in the plant for enhancing fungal diseases resistance.
2. **Hevein-Type Peptide:** It is a small cysteine and glycine-rich chitin-binding peptide of 43 aa residues. It was first discovered in the latex of rubber trees. Since then, a number of hevein-like antimicrobial peptides (AMPs) such as *Pn*-AMPs from morning glory (*Pharbitis nil*), GAFP from Ginkgo biloba, Ee-CBP from the bark of spindle tree (*Euonymus europaeus* L.), and *Wj*-AMP-1 from the leaves of *Wasabia japonica* L. Hevein-type peptides have shown a wide range of antifungal activity. The transgenic tomato plants showed enhanced resistance against both the non-chitinous fungus *Phytophthora capsici*.
3. **Knottin-Type Peptides:** Knottins are a group of structurally related molecules with six disulfide linked cysteines with a consensus pairing pattern (C1–C4, C2–C5, C3–C6). Knottins fold into a triple-stranded β-sheet and form a "knot-like" feature. Few knottin-like AMPs have been reported so far. Examples include two highly homologous AMPs *Mj*-AMP1 and *Mj*-AMP2 (36 and 37 residues, respectively) from *Mirabilis jalapa* L. seed. As compared to the untransformed control plant, transgenic rice expressing Mj-AMP2 inhibited the growth of M oryzae by 63% (Prasad et al., 2008).

18.4 PLANT RIBOSOME INACTIVATING PROTEINS

Plant ribosome inactivating proteins (RIPs) are RNA N-glycosidases that cleave a specific adenine residue in highly conserved sequence of 28S rRNA and inhibit the elongation factor eF-la to bind with ribosome. RIPs do not act on their own ribosome because they are targeted to vacuoles that sequester a certain development process. A RIP isolated from barley was shown to exhibit *in vitro* antifungal activity against a number of plant pathogenic fungi. Transgenic tobacco plant expressing isolated barley RIP gene under the control of inducible promoter showed increased resistance to *Rhizoctonia solani* (Logemann et al., 1992). An effective resistance was recorded in tobacco transgenics expressing a combination of RIP and chitinase (Chi-a) gene against *R. solani* (Jach et al., 1995).

18.5 PHYTOALEXINS

Phytoalexins are low molecular mass secondary metabolites with antimicrobial activity, which are induced by pathogens. In 1940, Müller & Börger proposed the concept of phytoalexins after finding that the effect of an infection of a potato tuber with Phytophthora infestans strain capable of causing hypersensitive reactions. It was significantly inhibited by an infection with another strain of this strain. This inhibition was linked to a "principle" produced by the plant cells reacting hypersensitively that they named phytoalexin. The expression of grapevine stilbene synthase gene in rice plants has been shown to enhance disease resistance (Stark-Lorenzen et al., 1997). Similarly, resveratrol synthase and isoflavone methyltransferase genes have been proved to enhance disease resistance in transgenic alfalfa (Hipskind & Paiva, 2000). A list of plant species overexpression phytoalexins for enhanced disease resistance is given in Table 18.3.

TABLE 18.3 List of Plant Species Genetically Engineered to Enhance Fungal Disease Resistance

Name of Plant Species Engineered	Expressed Gene Product	Disease Resistance
Alfalfa (*M. sativa*)	Alfalfa isoflavone o'methyltransferase.	Reduced lesion size due to *Phoma medicaginis*.
–	Peanut resveratrol synthase.	Reduced lesion size and sporulation of *Phoma medicaginis*.
Barley (*H. vulgare*)	Grape stilbene (resveratrol) synthase.	Reduced colonization by *Botrytis cinerea.*
Rice (*O. sativa*)	Grape stilbene (resveratrol) synthase.	Reduced lesion development due to rice blast.
Tobacco (*N. tabacum*)	Grape stilbene (resveratrol) synthase.	Reduced colonization by *Botrytis cinerea.*
Tomato (*L. esculentum*)	Grape stilbene (resveratrol) synthase.	Reduced lesion development by Phytophthora infestans; no effect on *Alternaria solani* or *Botrytis cinerea.*
Arabidopsis	*Polygonum cuspidatum* resveratrol synthase gene (*PcRS*).	Enhanced resistance against the fungal pathogen *C. higginsianum.*

18.6 R-GENES

Resistance genes (R-genes) act as an immune system in plants by recognizing pathogens and inducing defensive pathways. Many R-gene has been identified and introgressed for disease resistance breeding program. In order to transfer the R-gene from unrelated plant species, transgenic technology has been utilized. The R-gene Rxo1 from maize was successfully introduced into rice and conferred resistance against bacterial streak disease caused by *Xanthomonas oryzae* pv. *oryzicola*. R-gene RCT1 from *Medicago truncatula* that was expressed in alfalfa and conferred resistance to *Colletotrichum trifolii*, and RPI-BLB2 from wild potato *Solanum bulbocastanum* conferring resistance to *Phytophtohora infestans* in cultivated potato. The *LpiO* gene, one of the tested effectors from *Solanum* species, when co-expressed along with *Rpi-blb1* (as resistance gene) in *Nicotiana benthamiana*, it led to rapid identification of *Rpi-sto1* and *Rpi-pta1* as resistance genes to late blight (Vleeshouwers et al., 2008). Stacking of three broad-spectrum potato *R* genes (*Rpi*), *Rpi-sto1* (*Solanum stoloniferum*), *Rpi-vnt1.1* (*S. venturii*) and *Rpi-blb3* (*S. bulbocastanum*) in potato showed HR against pathogenic effects of *Phytophtora* (Zhu et al., 2012).

18.7 BACTERIAL DISEASE RESISTANT TRANSGENIC PLANTS

Crop plants are affected by a number of diseases caused by bacterial pathogens. The development of bacterial disease resistant transgenic plants holds considerable promise to combat these pathogens.

18.7.1 ANTI-MICROBIAL PROTEIN

Numerous AMPs have been identified and used for enhancing disease resistance in crop plants against bacterial pathogens using genetic engineering technologies. Among them, cationic antimicrobial peptides (CAPs) are the foremost active peptides among the AMPs. These peptides fall in two classes: α-helical peptides, such as cecropines and maganins and β-sheet peptides, such as defensins, protegrins, and lactoferrin (Huang et al., 2010).

Cecropins are potent antimicrobial linear amphipathic peptides consisting of 31–39 amino acids. Transgenic tomato plant expressing cationic peptide cecropin B (CB) enhanced resistance against bacterial diseases against

Xanthomonas campestris pv. *vesicatoria* and *Ralstonia solanacearum* (Jan et al., 2010).

Lactoferrin glycoprotein found to play important role in several defense related function and it was obtained from bovine. Transgenic tobacco over-expressing lactoferricin gene showed enhanced resistance to *Pseudomonas syringae* pv. *tabaci* (Fukuta et al., 2012).

18.7.2 TRANSGENIC EXPRESSION OF LYSOZYMES

Lysozymes are known antimicrobial agent, which cleaves the peptidoglycan component of bacterial cell. The human and egg white lysozyme have been expressed in crop plants to enhance bacterial disease resistance. The trans-genic potato overexpressing T4 bacteriophage lysozyme gene prevent the spread of the bacterium *Erwinia carotovora* (Rivero et al., 2012). Transgenic tobacco overexpressing human lysozyme gene exhibited slightly fewer symptoms against the fungus *Erysiphe cichoracearum* and the bacterium *P. syringae* pv. *tabaci* (Nakajima et al., 1997).

18.7.3 TRANSGENIC EXPRESSION OF BACTERIAL VIRULENCE FACTORS

Bacterial virulence factors are important for developing diseases in plant. Virulence entities includes locomotory structural components (flagella and pili) that allow for attachment and secretion of toxic compound upon host plants and degrade the host tissue. The suppression of bacterial virulence factors can lead to resistance in plants. The wildfire disease of tobacco is caused by *P. syringae* pv. *tabaci* which produces tabotoxin, a dipeptide toxin containing an uncommon β-lactum amino acid causing the chlorotic symp-toms. The overexpression of tabotoxin resistance gene (*ttr*) in transgenic tobacco showed enhanced resistance against *P. syringae* pv. Tabaci (Anzai et al., 1989).

18.7.4 ETHYLENE RESPONSIVE TRANSCRIPTION FACTORS (ERF)

The ethylene responsive transcription factors (ERF) have been demonstrated to have a role in controlling the expression of PR genes. The expression of cotton ERF in tobacco showed exhibition of greater level of resistance to *Xanthomonas* (Champion et al., 2009). It is to be noted that bacterial blight is a destructive disease of domesticated rice caused by the pathogen *Xanthomonas oryzae* pv. *oryzae.*

18.7.5 HAIRPIN GENES

The Harpins (*hrp*) genes encode type III secretory pathways and are required by many phytopathogenic bacteria for pathogenesis on susceptible hosts and to elicit a hypersensitive response (HR) on non-host or resistant host plants. Several studies indicated that enhanced HrpNEa levels in transgenic plants have effectively increased resistance to bacteria. Harpin NEa (HrpNEa) is encoded by the gene hrpN located on the chromosome of *Erwinia* causing the fire blight disease of apple. HrpNEa is a known inducer of systemic acquired resistance (SAR) in plants. Transgenic plants resistant to bacterial pathogens have been produced, making use of this property.

18.8 GENETIC ENGINEERING APPLICATIONS TO CONTROL DISEASES AND PARASITE

Aspergillus flavus and *Aspergillus parasiticus* produce aflatoxins, which are secondary metabolites of fungi. Aflatoxin poisons are known to cause teratogenic, mutagenic, and metabolic abnormalities in humans, including liver and kidney cancer and weakened immune systems. They can also contribute to infants failing to gain weight and height as expected (Kumar et al., 2017). As a result of their contamination of agricultural crops, aflatoxins pose a worldwide threat, such as cotton, maize, peanut (*Arachis hypogaea* L.) and some tree nuts. The exposure to contaminated crops also affects meat and dairy products from animals fed those crops (Kumar et al., 2017). Food contamination by aflatoxin occurs in both high- and low-income countries, but people in developing countries tend to be more exposed to it, usually as a result of inadequate processing, storage, and inspection in addition to lack of information regarding the problem (Darwish et al., 2014; Mupunga et al., 2017). Crop production, processing, and storage must be optimized to control Aspergillus spp. For reducing contamination, aspergillus-resistant cultivars are desirable, although conventional breeding has not produced varieties that are resistant enough for aflatoxins to be reduced to an acceptable level (Sharma et al., 2018). Therefore, genetic engineering may be necessary to develop the required varieties. Pathogens can be controlled via RNAi by downregulating genes in the critical pathways for establishing infection or spreading (Majumdar et al., 2017). According to Arias et al. (2015), peanut resistance was gained through silencing and siRNA-mediated suppression of five genes (*aflS, alfR, aftC, pes1*, and *aflep*), all involved in aflatoxin synthesis and accumulation. Overexpression of the lipoxygenase (*LOX*) genes, which can provide resistance to *Aspergillus* spp. (Müller et al., 2014; Song et al., 2016), has been proposed by Bhatnagar-Mathur et al. (2015). The overexpression of two plant-derived antifungal genes resulted in *A. flavus*-resistant peanut (Sundaresha et al., 2016). *BjNPRI* from brown mustard (*Brassica juncea*) and *TGFd* from fenugreek (*Trigonella foenum-graecum*) were found to be responsible. Sharma et al. (2018) overexpressed antifungal defensin genes from *Medicago sativa* (*MSDef1*) and *Medicago truncatula* (*MSDef4.2*) to reduce initial infections, and HIGS of the *aflM* and *aflP* genes responsible for aflatoxin biosynthesis to reduce aflatoxin levels in peanut. Some products only had aflatoxin levels well below US and European Union (EU) limits.

Many resource-poor farmers depend on maize, sorghum, pearl millet, and rice, which are all staples. Striga plants are parasitic plants in the Orobanchaceae family that attacks these crops. In Striga spp., controlling the seeds is a challenge since they are dormant until a host contact them and exudes a germination stimulant. When Striga germinates, it attaches itself to the root of its host and forms a haustorium, allowing it to ingest nutrients and water. Striga control cannot be achieved with traditional breeding or weed management methods. In order to control Striga, genetic engineering may be one solution, but multiple approaches will likely be needed since *Striga spp.* A mutant low-germination stimulant allele (lgs) in sorghum confers resistance to Striga (Ramaiah et al., 1990). Striga germination stimulant strigol (Gobena et al., 2017) is reduced by the mutant allele altering the strigolactone biosynthesis pathway. Mutations do not harm a host but alter biosynthesis so that a nongermination stimulant strigolactone is released by the roots. Other host crops (e.g., maize) might be able to develop resistant varieties by editing the LGS gene. We have inhibited haustorial growth on tobacco (Alakonya et al., 2012) and lettuce (Tomilov et al., 2008) by silencing RNAi genes in *Cuscuta pentagona* and *Triphysaria versicolor*. Striga was not found to be providing sufficient resistance by RNAi in maize but using viral-induced gene silencing (VIGS) in *Striga hermonthica*, effectively silenced a gene (Kirigia et al., 2014).

KEYWORDS

- **anti-microbial protein**
- **ethylene responsive transcription factors**
- **lysozymes**
- **phytoalexins**
- **plant antibodies**
- **R-genes**
- **RNAi**
- **transgenics**

REFERENCES

Alakonya, A., Kumar, R., Koenig, D., Kimura, S., Townsley, B., et al., (2012). Interspecific RNA interference of shoot meristemless-like disrupts *Cuscuta pentagona* plant parasitism. *Plant Cell, 24*(7), 3153–3166.

Anzai, H., Yoneyama, K., & Yamaguchi, I., (1989). Transgenic tobacco resistance to a bacterial disease by detoxification of a pathogenic toxin. *Mol. Gen. Genet., 219*, 492–494.

Aragão, F. J. L., & Faria, J. C., (2009). First transgenic geminivirus-resistant plant in the field. *Nat. Biotechnol., 27*, 1086–1089.

Arias, R. S., Dang, P. M., & Sobolev, V. S., (2015). RNAi-mediated control of aflatoxins in peanut: Method to analyze mycotoxin production and transgene expression in the peanut/*Aspergillus* pathosystem. *J. Vis. Exp., 106*, e53398.

Azad, A. K., Amin, L., & Nik, M. S. N. M., (2014). *Gene Technology for Papaya Ringspot Virus Disease Management*. Article ID 768038.

Barrangou, R., & Doudna, J. A., (2016). Applications of CRISPR technologies in research and beyond. *Nat. Biotechnol., 34*, 933–941.

Beachy, R. N., (1999). Coat-protein-mediated resistance to tobacco mosaic virus: Discovery mechanism and exploitation. *Philos. Trans. R Soc. Lond. Biol. Sci., 35*, 659–664.

Bhatnagar-Mathur, P., Sunkara, S., Bhatnagar-Panwar, M., Waliyar, F., & Sharma, K. K., (2015). Biotechnological advances for combating *Aspergillus flavus* and aflatoxin contamination in crops. *Plant Sci., 234*, 119–132.

Bonfim, K., Faria, J. C., Nogueira, E. O. P. L., Mendes, E. A., & Aragão, F. J. L., (2007). RNAi-mediated resistance to bean golden mosaic virus in genetically engineered common bean (*Phaseolus vulgaris*). *Mol. Plant Microbe. Interact., 20*, 717–726.

Champion, A., Hebrard, E., Parra, B., et al., (2009). Molecular diversity and gene expression of cotton ERF transcription factors reveal that group IXa members are responsive to jasmonate, ethylene and *Xanthomonas*. *Mol. Plant Pathol., 10*(4), 471-485.

CTNBio, Brazil (2018). *National Biosafety Technical Commission Normative Resolution No. 16*. https://agrobiobrasil.org.br/wp-content/uploads/2018/05/Normative-Resolution-16-of-January-15-2018.pdf (accessed on 12 December 2023).

Darwish, W. S., Ikenaka, Y., Nakayama, S. M. M., & Ishizuka, M., (2014). An overview on mycotoxin contamination of foods in Africa. *J. Vet. Med. Sci., 76*(6), 789–797.

Ehrifeild, N., Romano, E., Serrano, C., & Arce-Johnson, P., (2004). Replicase mediated resistance against potato leaf roll virus disease plants. *Biol. Res., 37*(1), 71–82.

Fitch, M. M. M., Manshardt, R. M., Gonsalves, D. D., et al., (1992). Virus resistant papaya plants derived from tissues bombarded with the coat protein gene of papaya ringspot virus. *Nature Biotechnology, 10*(11), 1466–1472.

Fuchs, M., & Gonsalves, D., (2007). Safety of virus-resistant transgenic plants two decades after their introduction: Lessons from realistic field risk assessment studies. *Annu. Rev. Phytopathol., 45*, 173–202.

Fukuta, S., Kawamoto, K. I., Mizukami, Y., Yoshimura, Y., Ueda, J. I., & Kanbe, M., (2012). Transgenic tobacco plants expressing antimicrobial peptide bovine lactoferricin show enhanced resistance to phytopathogens. *Plant Biotechnol., 29*, 383–389.

Gobena, D., Shimels, M., Rich, P. J., Ruyter-Spira, C., Bouwmeester, H., et al., (2017). Mutation in sorghum low germination stimulant 1 alters strigolactones and causes striga resistance. *PNAS, 114*(17), 4471–4476.

Gutierrez-Campos, R., Torres-Acosta, J. A., Saucedo-Arias, A. J., & Gomez-Lim, M., (1999). The use of cysteine proteinase inhibitors to engineer resistance against potyviruses in transgenic tobacco plants. *Nat. Biotechnol., 17*(12), 1223–1226.

Hanley-Bowdoin, L., Bejarano, E. R., Robertson, D., & Mansoor, S., (2013). Geminiviruses: Masters at redirecting and reprogramming plant processes. *Nat. Rev. Microbiol., 11*, 777–788.

Hipskind, J. D., & Paiva, N. L., (2000). Constitutive accumulation of a resveratrol glucoside in transgenic alfalfa increases resistance to *Phoma medicaginis*. *Mol. Plant Microbe. Interact., 13*, 551–562.

Huang, Y., Huang, J., & Chen, Y., (2010). Alpha-helical cationic antimicrobial peptides: Relationships of structure and function. *Prot. Cell, 1*, 143–152.

Ilardi, V., & Tavazza, M., (2015). Biotechnological strategies and tools for plum pox virus resistance: Trans-, intra-, cis-genesis, and beyond. *Front Plant Sci., 6*, 379.

Jach, G., Gornhardt, B., Mundy, J., et al., (1995). Enhanced quantitative resistance against fungal disease by combinational expression of different barley antifungal proteins in transgenic tobacco. *Plant J., 8*, 97–109.

Jan, P. S., Huang, H. Y., & Chen, H. M., (2010). Expression of a synthesized gene encoding cationic peptide cecropin B in transgenic tomato plants protects against bacterial diseases. *Applied and Environmental Microbiology, 76*(3), 769–775.

Kirigia, D., Runo, S., & Alakonya, A., (2014). A virus-induced gene silencing (VIGS) system for functional genomics in the parasitic plant *Striga hermonthica*. *Plant Methods, 10*, 16.

Kumar, P., Mahato, D. K., Kamle, M., Mohanta, T. K., & Kang, S. G., (2017). Aflatoxins: A global concern for food safety, human health, and their management. *Front. Microbiol., 7*, 2170.

Lehmann, P., Jenner, C. E., Kozubek, E., et al., (2003). Coat protein-mediated resistance to turnip mosaic virus in oilseed rape (*Brassica napus*). *Mol. Breed., 11*, 83–94.

Leibman, D., Wolf, D., Saharan, V., Zelcer, A., Arazi, T., Yoel, S., Gaba, V., & Gal-On, A., (2011). A high level of transgenic viral small RNA is associated with broad potyvirus resistance in cucurbits. *Molecular Plant-Microbe Interactions, 24*(10), 1220–1238.

Lindbo, J. A., & Dougherty, W. G., (2005). Plant pathology and RNAi: A brief history. *Annu. Rev. Phytopathol., 43*, 191–204.

Liu, D., Raghothama, K. G., Hasegawa, P. M., & Bressan, R. A., (1994). Osmotin overexpression in potato delays development of disease symptoms. *Proc. Natl. Acad. Sci. USA, 91*, 1888–1892.

Liu, Z. Y., Zhuang, C. X., Sheng, S. J., Shao, L., Zhao, W., & Zhao, S. J., (2011). Overexpression of a resveratrol synthase gene (PcRS) from *Polygonum cuspidatum* in transgenic *Arabidopsis* causes the accumulation of trans-piceid with antifungal activity. *Plant Cell Rep., 30*, 2027–2036.

Logemann, J., Jach, G., Tommerup, H., et al., (1992). Expression of a barley ribosome-inactivating protein leads to increased fungal protection in transgenic tobacco plants. *Nat. Biotechnol., 10*, 305–308.

Lorence, A., & Verpoorte, R., (2004). Gene transfer and expression in plants. *Methods Mol. Biol., 267*, 329–350.

Luan, H., Liao, W., Song, Y., et al., (2020). Transgenic plant generated by RNAi-mediated knocking down of soybean *Vma12* and *soybean mosaic virus* resistance evaluation. *AMB Expr., 10*, 62. https://doi.org/10.1186/s13568-020-00997-6.

Majumdar, R., Rajasekaran, K., & Cary, J. W., (2017). RNA interference (RNAi) as a potential tool for control of mycotoxin contamination in crop plants: Concepts and considerations. *Front. Plant Sci., 8*, 200.

Manshardt, R. M., (1998). *'UH Rainbow' Papaya* (pp. 1, 2). University of Hawaii College of Tropical Agriculture and Human Resources New Plants for Hawaii.

Müller, K. O., & Börger, H., (1940). Experimentelle untersuchungen über die phytophthora: Resistenz der kartoffel. *Arb. Biol. Reichsanst. Land Forstwirtsch, 23*, 189–231.

Müller, V., Amé, M. V., Carrari, V., Gieco, J., & Asis, R., (2014). Lipoxygenase activation in peanut seed cultivars resistant and susceptible to *Aspergillus parasiticus* colonization. *Phytopathology, 104*(12), 1340–1348.

Mupunga, I., Mngqawa, P., & Katerere, D. R., (2017). Peanuts, aflatoxins, and undernutrition in children in Sub Saharan Africa. *Nutrients, 9*(12), e1287.

Nahid, N., Amin, I., Briddon, R. W., et al., (2011). RNA interference-based resistance against a legume mastrevirus. *Virol J., 8*, 499. https://doi.org/10.1186/1743-422X-8-499.

Nakajima, H., Muranaka, T., Ishige, F., et al., (1997). Fungal and bacterial resistance in transgenic plants expressing human lysozyme. *Plant Cell Rep., 16*, 674–679.

Palukaitis, P., & Zaitlin, M., (1997). Replicase-mediated resistance to plant virus disease. *Adv. Virus Res., 48*, 349–377.

Prasad, B. D., Jha, S., & Chattoo, B. B., (2008). Transgenic indica rice expressing *Mirabilis jalapa* antimicrobial protein (Mj-AMP2) shows enhanced resistance to the rice blast fungus *Magnaporthe oryzae*. *Plant Science, 175*, 364–371.

Prins, M., Laimer, M., Noris, E., et al., (2008). Strategies for antiviral resistance in transgenic plants. *Mol. Plant. Pathol., 9*(1), 73–83.

Ram, M. S. N., & Mohandas, S., (2003). Transformation of African violet (*Saintpaulia ionantha*) with glucanase-chitinase genes using agrobacterium tumefacien. *Acta Horticult., 624*, 471–478.

Ramaiah, K. V., Chidley, V. L., & House, L. R., (1990). Inheritance of striga seed-germination stimulant in sorghum. *Euphytica, 45*(1), 33–38.

Rivero, M., Funrman, N., Mencacci, N., et al., (2012). Stacking of antimicrobial in potato transgenic plants conferring resistance to bacterial and fungal pathogen. *J. Biotechnol., 157*(2), 334–343.

Rohini, V., & Rao, S., (2001). Transformation of peanut (*Arachis hypogaea* L.) with tobacco chitinase gene: Variable response of transformants to leaf spot disease. *Plant Science: An International Journal of Experimental Plant Biology, 160*, 889–898.

Saharan, G. S., Mehta, N., & Meena, P. D., (2016). *Alternaria Diseases of Crucifers: Biology, Ecology and Disease Management.* Singapore: Springer.

Saharan, V., Jain, D., Pareek, S., et al., (2016). Viral, fungal, and bacterial disease resistance in transgenic plants. *Advances in Plant Breeding Strategies: Agronomic, Abiotic and Biotic Stress Traits,* 627–656.

Savary, S., Willocquet, L., Pethybridge, S. J., Esker, P., McRoberts, N., & Nelson, A., (2019). The global burden of pathogens and pests on major food crops. *Nature Ecology and Evolution, 3*, 430–439.

Schwind, N., Zwiebel, M., Itaya, A., et al., (2009). RNAi-mediated resistance to potato spindle tuber viroid in transgenic tomato expressing a viroid hairpin RNA construct. *Mol. Plant. Pathol., 10*(4), 459–469.

Scorza, R., Ravelonandro, M., Callahan, A. M., Cordts, J. M., Fuchs, M., Dunez, J., & Gonsalves, D., (1994). Transgenic plums (*Prunus domestica* L.) express the plum pox virus coat protein gene. *Plant Cell Rep., 14*, 18–22.

Sharma, K. K., Pothana, A., Prasad, K., Shah, D., Kaur, J., et al., (2018). Peanuts that keep aflatoxin at bay: A threshold that matters. *Plant Biotechnol. J., 16*(5), 1024–1033.

Song, H., Wang, P., Li, C., Han, S., Lopez-Baltazar, J., et al., (2016). Identification of lipoxygenase (LOX) genes from legumes and their responses in wild type and cultivated peanut upon *Aspergillus flavus* infection. *Sci. Rep., 6*, 35245.

Stark-Lorenzen, P., Nelke, B., Hänßler, G., Mühlbach, H. P., & Thomzik, J. E., (1997). Transfer of a grapevine stilbene synthase gene to rice (*Oryza sativa* L.). *Plant Cell Reports, 16*(10), 668–673.

Sundaresha, S., Rohini, S., Appanna, V. K., Arthikala, M. K., Shanmugam, N. B., et al., (2016). Co-overexpression of *Brassica juncea* NPR1 (BjNPR1) and *Trigonella foenum-graecum* defensin (Tfgd) in transgenic peanut provides comprehensive but varied protection against *Aspergillus flavus* and *Cercospora arachidicola. Plant Cell Rep., 35*(5), 1189–1203.

Takakura, Y., Ito, T., Saito, H., Inoue, T., Komari, T., & Kuwata, S., (2000). Flower-predominant expression of a gene encoding a novel class I chitinase in rice (*Oryza sativa* L.). *Plant Mol Biol., 42*, 883–897.

Tavladoraki, P., Benvenuto, E., Trinca, S., et al., (1993). Transgenic plants expressing a functional single-chain Fv antibody are specifically protected from virus attack. *Nature, 366*, 469–472.

Tollefson, J., (2011). Brazil cooks up transgenic bean. *Nature, 478*, 168.

Tomilov, A. A., Tomilova, N. B., Wroblewski, T., Michelmore, R., & Yoder, J. I., (2008). Trans-specific gene silencing between host and parasitic plants. *Plant J., 56*(3), 389–397.

Tricoli, D. M., Carney, K. J., Russell, P. F., Mcmaster, J. R., Groff, D. W., Hadden, K. C., Himmel, P. T., et al., (1995). Field evaluation of transgenic squash containing single or multiple virus coat protein gene constructs for resistance to cucumber mosaic virus, watermelon mosaic virus 2, and zucchini yellow mosaic virus. *Bio-technology (N Y), 13*, 1458–1465.

USDA, (2018). *Secretary Perdue Issues USDA Statement on Plant Breeding Innovation.* USDA press release no. 0070.18. https://www.usda.gov/media/press-releases/2018/03/28/secretary-perdue-issues-usda-statement-plant-breeding-innovation (accessed on 12 December 2023).

Van, E. H. P., Reuber, T. L., & Van, D. D. D., (2020). Genetic modification to improve disease resistance in crops. *New Phytologist, 225*(1), 70–86.

Vanderschuren, H., Alder, A., Zhang, P., & Gruissem, W., (2009). Dose dependent RNAi-mediated Geminivirus resistance in the tropical root crop cassava. *Plant Mol. Biol., 70*, 265–272.

Vleeshouwers, V. G. A. A., Van, D. W., Keizer, L. C. P., et al., (1999). A laboratory assay for *Phytophthora infestans* resistance in various *Solanum* species reflects the field situation. *Eur. J. Plant Pathol., 105*, 241–250.

Wang, X. B., Wu, Q., Ito, T., et al., (2010). RNAi-mediated viral immunity requires amplification of virus derived siR-NAs in *Arabidopsis thaliana. Proc. Natl. Acad. Sci USA, 107*, 484–489.

Waterhouse, P. M., Graham, M. W., & Wang, M. B., (1998). Virus resistance and gene silencing in plants can be induced by simultaneous expression of sense and antisense RNA. *Proc. Natl. Acad. Sci. USA, 95*, 13959–13964.

Whelan, A., & Lema, M., (2015). Regulatory framework for gene editing and other new breeding techniques (NBTs) in Argentina. *GM Crops Food, 6*, 253–265.

Yamamoto, T., Iketani, H., Leki, H., et al., (2000). Transgenic grapevine plants expressing a rice chitinase with enhance resistance to fungal pathogens. *Plant Cell Rep., 19*, 639–646.

Yang, R. C., Xu, H. L., Yu, W. G., Lu, C. G., Long, M. S., Liu, C. Q., Pan, N. S., & Chen, Z. L., (1995). Transgenic tomato plants expressing cucumber mosaic virus coat protein (CMV-cp) and their resistance to cucumber mosaic virus (CMV). *Acta Agric, Jiangsu, 11*, 42–46.

Zhu, S., Li, Y., Vossen, J. H., Visser, R. G., & Jacobsen, E., (2012). Functional stacking of three resistance genes against *Phytophthora infestans* in potato. *Transgenic Res., 21*(1), 89–99.

Zhu, Y. X., Ou-Yang, W. J., Zhang, Y. F., & Chen, Z. L., (1996). Transgenic sweet pepper plants from *Agrobacterium* mediated transformation. *Plant Cell Rep., 16*, 71–75.

CHAPTER 19

Editing Genomes for Disease Resistance via CRISPR/Cas9 Holds Endless Opportunities

DIKSHA KUMARI,[1] BISHUN DEO PRASAD,[2] PADMANABH DWIVEDI,[1] SANGITA SAHNI,[3] and SUHAIL MUZAFFAR[4]

[1]*Department of Plant Physiology, Institute of Agricultural Sciences, Banaras Hindu University, Varanasi, Uttar Pradesh, India*

[2]*Department of Agricultural Biology and Molecular Biology, College of Basic Sciences and Humanities, Dr. Rajendra Prasad Central Agricultural University, Pusa, Samastipur, Bihar, India*

[3]*Department of Plant Pathology, TCA, Dholi, Dr. Rajendra Prasad Central Agricultural University, Pusa, Samastipur, Bihar, India*

[4]*UAB Research Center of Excellence in Arsenicals and Department of Dermatology, University of Alabama at Birmingham, Birmingham, Alabama, USA*

ABSTRACT

Phytopathogens can adversely affect crop production and productivity, eventually threatening food security worldwide. For accurate management of phytopathogens, it is imperative to study their genetic mechanisms of virulence. The latest generation of genome editing tools, with site-specific nucleases (SSNs) such as ZNF and TALEN and clustered regularly interspaced short palindromic repeats (CRISPR)/CRISPR-associated proteins, are shaping our future. With the discovery of CRISPR/Cas9, we have been able to gain a fundamental understanding of how plants resist and respond

New Advances in Crop Disease Management. Sangita Sahni, Ganesh Thapa, and Bishun Deo Prasad (Eds)

to phytopathogens. Scientists can use this tool to develop genetic resistance in crop plants with smaller off-target effects by performing precise genome manipulation, genetic screening, regulation, and correction. Besides providing a new avenue for disease improvement, it also strengthens the breeding program for resistant variety. Here we discuss the challenges and potential applications of CRISPR/Cas9-based targeted gene modification as well as developing disease-resistant crop plants without introducing transgenes.

19.1 INTRODUCTION

A plant and microbe live together in nature. Their interaction is critical to their survival. This interaction (plant-microbe interaction) is classified as either beneficial or harmful based on the effects it has on the host, especially plants. Plants are vulnerable to many pathogens, including bacteria, fungi, and viruses, which cause continuous financial losses for farmers. The most significant challenge facing the world is feeding its continuously growing population, which will double by 2050 as a result of this yield loss. Even with the depletion of agricultural land and water, the global food demand is projected to rise by 25 to 70% above current production levels (Hunter et al., 2017). According to a statistical survey of major staple crops, yield losses in rice, wheat, maize, potatoes, and soybean were 30.0%, 21.5%, 22.5%, 17.2%, and 21.4% respectively from 2010 to 2014 (Savary et al., 2019). Moreover, major losses have been reported among the expanding population in areas of greater food scarcity. To control biotic stress, chemical pesticides are applied. These pesticides can harm the environment and food chain. Due to the regular application of the same pesticides, phytopathogens are becoming increasingly resistant to a variety of pesticides. On the other hand, the constitutive ON defense mechanism causes dwarfism in plants (Gao et al., 2020; van Butselaar & van den Ackerveken, 2020). In order to balance defense growth and yield, an efficient and fast resistance mechanism is required. Using conventional breeding, mutation breeding, transgenic or editing methods, plants which are naturally resistant to pests and diseases can help to minimize pesticide and yield loss (Figure 19.1). In conventional breeding, disease resistance is achieved through backcrossing of various varieties, however, this technology is unsuitable due to its limitations in terms of the introduction of resistance genes (R-genes) and the breakage of resistance after some time. A major limitation of this approach is the lack of genetic variation in a specific genotype, long duration, and labor requirements of the crossing program (Gao et al., 2018). Consequently, conventional breeding faces a major challenge to handle the growing threat of pathogen armies towards plants while simultaneously meeting the rising population requirements for food (Figure 19.1). Even though mutational breeding is a widely used approach, there is less demand due to rigorous strategies for selecting and evaluating particular phenotypes and genotypes for specific characteristics. The problem of accidental mutation in the genome is one of the most important limitations (Figure 19.1). A transgenic method of crop improvement is robust, less time-consuming, but cannot be used in the public domain because of the label describing it as a GMO (Figure 19.1). Using genome editing technology, we can increase the demand for food and increase disease resistance using faster and newer methods that are site-specific (Figure 19.1).

A variety of genomic changes can be made using nucleases, including small insertions and deletions (indels) as well as detection of useful or deleterious mutations. Numerous plants, animals, and cultures of cells have been engineered with sequence-specific nucleases (Mali et al., 2013). Plant genomes have been modified using ZFNs, TALENs, and CRISPR/Cas9 programmable nucleases (SSNs). A major advantage of the CRISPR/Cas9 technique is its widespread use and ease of implementation (Schiml et al., 2016) that has a magnificent role in disease resistance (Kumar & Jain, 2015; Paul & Qi, 2016; Zhang et al., 2016). CRISPR-based technologies are being used and performed more efficiently due to technological advancements.

19.2 PROGRAMMABLE SITE-SPECIFIC NUCLEASES (SSNS)

Nucleases that target specific genes can alter the genome in a highly precise manner. The nuclease family consists of proteins with DNA-binding domains that can hydrolyze phosphodiester bonds between nucleotides and cause single- or double-strand breaks in DNA (Yang et al., 2011). A site-specific nuclease is a modified protein that targets specific DNA sequences in the genome.

Meganucleases are naturally occurring endonucleases that bind to large DNA segments and recognize large (> 14 bp) recognition sites in the genomes. These DNA fragments cause precise breaks in DNA strands, which activate regeneration by homology-directed repair (HDR) (Molina et al., 2011). ARCUS is a new class of meganucleases developed to solve the

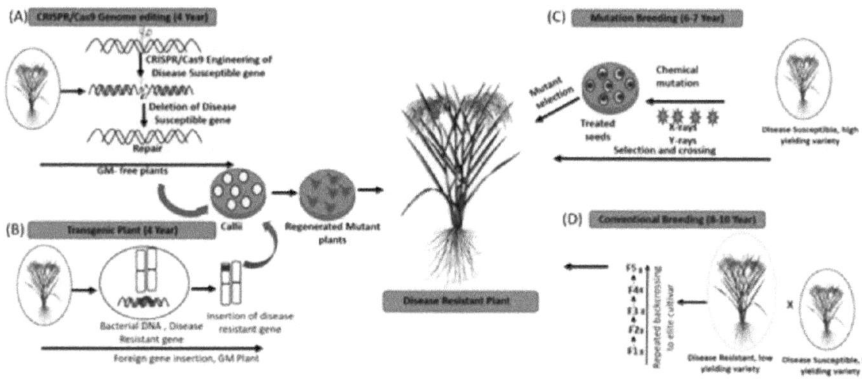

FIGURE 19.1 Comparative representation of methods used in crops.

Note: (A) CRISPR/Cas9 is an efficient robust system that require less time than any other method. Furthermore, because no foreign DNA is present this procedure may allow the plants to avoid GMO legislation and be classified as non-GMOs. (B) Chemical or physical mutagenesis is referred to as mutation breeding. Typically planting materials are treated with mutation causing agents (chemical and physical mutations) to generate mutation in genomes of plants. Method of selection desired mutant and entire process is hectic but helpful in identifying and mapping new gene(s) in the genome. Mutation breeding taking around 6–7 yrs. to achieve desired outcomes which makes this process exhaustive. Random changes in the genome which can be difficult to identify and anticipate are one of the technique's key limits and drawbacks. (C) Transgenic breeding involves transfer of gene with desired characteristic to another organism. The intended trait(s) is improved by inserting a gene of interest into the genome of the host elite variety with remarkable accuracy. However, because foreign DNA/elements are inserted into one's genome these are genetically engineered plants. Despite number of challenges the major issue with acceptance of GMO in public domain remain. And (D) Conventional breeding method involves crossing of two selected parent one with donor of desired trait and another with good recipient. This technique needed several backcrossing which is laborious time consuming and limited to some of traits.

Source: This figure is adapted from: Chen et al. (2019).

challenges of engineering sequence specificity by detecting differences in one base pair between target sites (Bartsevich et al., 2016).

ZFNs revolutionized genome editing and enabled molecular biologists to study the functions of genes for the first time in history. Zinc finger proteins (ZFPs) are attached to the FokI restriction enzyme's domain to form ZFNs. The zinc finger motif binds the DNA double helix's main groove and recognizes a 3 bp length. To produce a DSB by ZFNs, two monomers of zinc finger nucleases must bind to a complementary 12 bp site each. A double-strand break is generated when two FokI nuclease domains dimerize to form an active catalytic site (Kim et al., 1996) (Figure 19.2 and Table 19.1). In addition, four-finger zinc finger nucleases are able to identify segments of the genome by recognizing 24 base pairs in their recognition sites (Ramalingam et al., 2011).

For genome targeting, TALEN nucleases are a viable alternative to ZFN and Meganucleases of the second generation (2010). A DNA-binding domain is found in transcription activator-like effectors (TALEs) that are found in a genus of plant pathogenic bacteria called *Xanthomonas*. The core DNA-binding region of TALEs consists of 13–28, 34-amino-acid repeat monomers. Except for positions 12 and 13, which contain hypervariable amino acids (for example: HD recognizes C, NI recognizes A, NN recognizes G and NG or HG recognizes T), amino acids in each monomer are substantially conserved (Boch et al., 2009). While ZF motifs recognize more than one nucleotide, TALE motifs recognize only one. TALENs are created by attaching the *FokI* RE domain to a TALE DNA binding domain. A dimer of FokI recognizes the spacer region and generates DSBs when the targets are separated by a 12–20 bp spacer between two TALE-binding sites (Figure 19.2 and Table 19.1) (Christian et al., 2010).

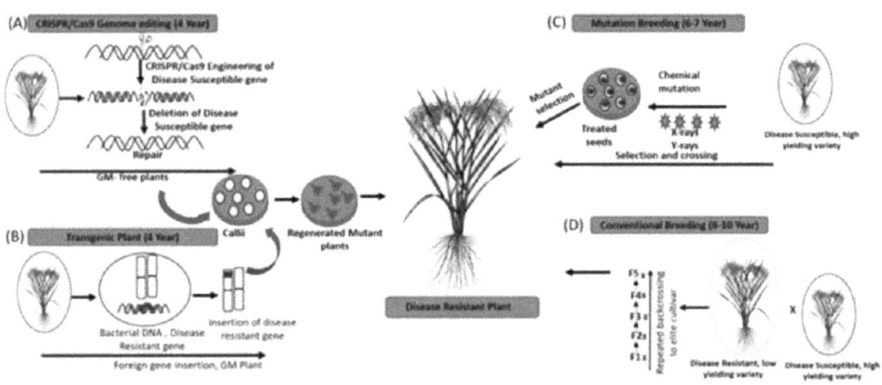

FIGURE 19.2 Schematic representations of programmable site-specific nucleases (SSNs).

19.3 CLUSTERED REGULARLY INTERSPACED SHORT PALINDROMIC REPEATS (CRISPR)-ASSOCIATED PROTEIN

19.3.1 CRISPR/CAS9

CRISPR-Cas9 is a tool developed recently for targeting a region of the genome. The unique genomic locus of this system was first discovered in strains of *E. coli* in 1987 and later discovery revealed that it is hallmark

TABLE 19.1 Comparison Among ZFNs, TALENs, and CRISPR/Cas9

Features	ZFNs	TALENs	CRISPR/Cas9
Essential components	Zinc finger proteins and *FokI*.	Transcription activator-like effector protein and *FokI*.	Synthetic guide RNA and *Cas9*.
Length of binding site.	2 sites consist of 12–15 bp each.	2 sites consist of 13–18 bp each).	1 site consist of 20 bp and 3 bp PAM sequence.
Feasibility of library construction and selection.	Technically challenging require protein engineering.	Technically challenging require molecular cloning.	Simple
Challenging	Multiplexing	Multiplexing	Feasible multiplexing
Efficiency of target.	Limiting factors.	Average	Good
Off-target efficiency.	Average	Low	Variable

of acquired immunity against bacterial infections. There are three components of the CRISPR/Cas9 system: crRNAs, tracrRNAs (gRNA), and Cas proteins (Ishino et al., 1987). CrRNA and tracrRNA (gRNA) were combined to produce Guide RNA, a single RNA chimera. A complex of crRNAs and gRNA binds to the Cas9 nuclease's recognition site in the PAM region. A DNA binding system that uses RNA guided DNA binding, in which CrRNA matches A DNA sequence made up of 20 nucleotides, then Cas9 makes a blunt-ended double strand break situated 3 bases upstream of the upstream region of PAM (Table 19.1). The Cas9 protein contains multiple domains (Ran et al., 2015). It was the REC (recognition) lobe that was responsible for identifying the complex, while the PI (PAM interaction) domain deals with PAM recognition. Both HNH and RuvC nucleases cut complementary and target strands during the DSB process, respectively (Figure 19.3) (Karvelis et al., 2013; Ran et al., 2015).

19.3.2 CRISPR/CAS9 AS A NICKASE

To create the Cas9 nickase system, Cas9's nuclease domain (HNH/RuvC) is mutated. To solve the off-target problem in genomes, a single-strand break is produced instead of a double-strand break, and then two adjacent gRNAs are used to reduce the off-target problem (Jinek et al., 2012). Cas9 gets the name "deadCas9" (dCas9) after the RuvC and HNH catalytic domains are mutated. Using this protein, multiple effector proteins can be recruited

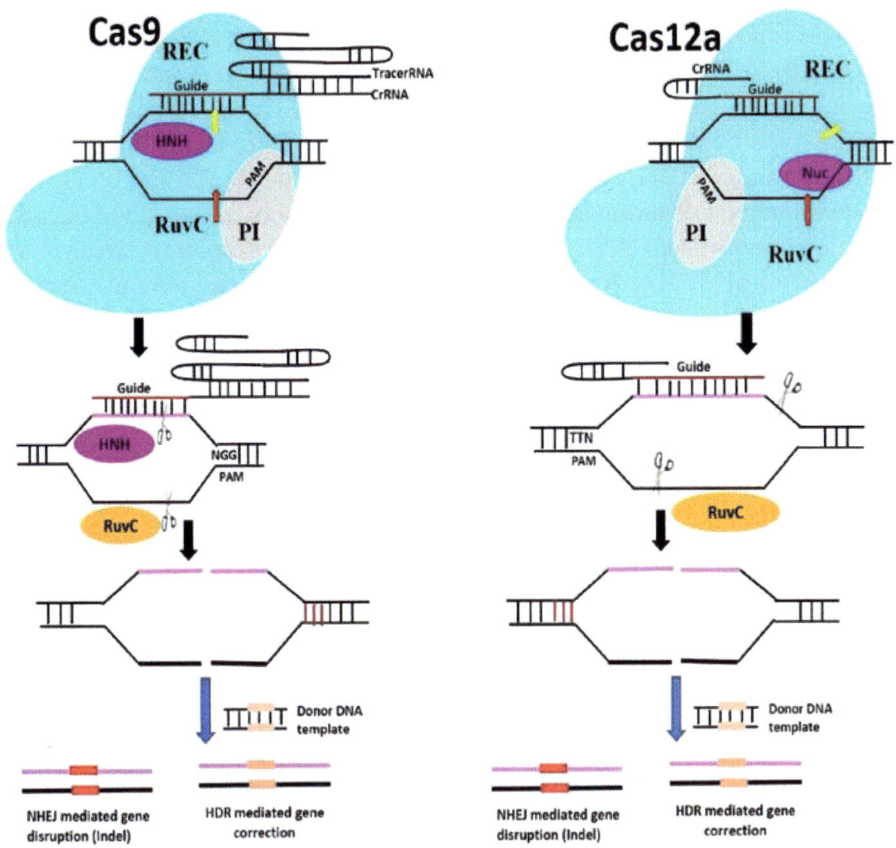

FIGURE 19.3 An illustration of the native forms of comparative Cas proteins.

Note: (a) The Crispr/Cas9 is worked with the help of important Cas9 protein along with crRNA and tracrRNA. At the time of hybridization these RNA binds to Cas9 nucleases followed by binding to PAM upstream region. This system is RNA guided DNA binding, for that CrRNA matches 20-nt sequence of DNA then Cas9 makes blunt ended double strand break which located 3 bp of PAM upstream region. Rec-lobe and PI (PAM interacting) domain recognizes crRNA-tracrRNA-target complex and PAM, respectively. spCas9 nuclease uses two domain HNH and RuvC to cleave target and non-target DNA strands, respectively. (b) The CRISPR/Cas12a system functioning is mediated by Cas12a protein and crRNA. In case of Cas12a nuclease, crRNA bind to 23–25-nucleotide sequence of DNA makes staggered-ended DSB downstream from the PAM sequence. The target site is cleaved by Nuc domain and nontarget site is cleaved by RuvC domain which is 18 bp and 23 bp downstream of the PAM, respectively (FnCas12a).

Source: This figure is adapted from: Schindele et al. (2018).

to specific areas. A Cas9 nickase-mediated mutagenesis revealed that the nickase is not able to perform NHEJ-based mutagenesis, as Cas9 can (Fauser et al., 2014). By using SpCas9 variant or paired nickases, Cas9 activity can be improved, and off-target mutations reduced (Kleinstiver et al., 2016). In this type Cas9 nickase requires two sgRNA each of which binds to one of the strands of DNA closely which results in a double strand break with 5'-overhangs. Off-target activity is sharply reduced by this method because single-strand breaks (SSBs) don't activate the NHEJ repairing mechanism (Schiml et al., 2014; Ran et al., 2013). The newest technology involves targeting a genome region using single-base conversion instead of double-strand breaks. When the Cas9 nickase is attached to a cytidine deaminase or an adenosine deaminase enzyme, then C/G to T/A or A/T to G/C may change. When a nickase is used instead of Cas9 which is inactive catalytically, base-editing is more effective because the nick stimulates mismatch repair by using the unchanged strand as a template (Komor et al., 2016; Gaudelli et al., 2017; Zong et al., 2017; Ren et al., 2018). Both adenosine deaminase and cytidine deaminase have lately been used as bases-editing tools in plants (Yan et al., 2018).

19.3.3 CRISPR/CAS12A (CPF1) SYSTEM IN PLANT

There are some similarities to the CRISPR/Cas9 system, but there are some important differences that provide an alternative in the genome-editing area of the CRISPR/Cas12a system (Figure 19.3) (Zetsche et al., 2015; Shmakov et al., 2017). Cas12a's activity is primarily determined by the presence of a short crRNA, whereas specificity to targets is determined by the presence of an extra component that must be not less than 22–23 nt long for optimum efficiency (Zetsche et al., 2015; Lei et al., 2017). As a result, Cas12a can identify T-rich PAMs in places where Cas9 targets G-rich PAMs, substantially increasing the number of target sites Cas12a can bind to. As a result of Cas12a's location upstream of the guide sequence, it generates staggered DSBs distal to the PAM instead of Cas9, which generates blunt-ended DSBs proximal to the PAM (Figure 19.3). Complimentary and non-complementary strands of a DNA molecule are cut after 5 nucleotides and 23 nucleotides, respectively, resulting in a 5'-overhang of 5 nucleotides. In eukaryotic cells, Cas12a orthologs also exhibit only minimal activity similar to Cas9. The ortholog of Cas12a, i.e., FnCas12a and AsCas12a identifies 5'-TTV (A/C/G)-3' and 5'-TTTV-3' sequences, respectively (Kim et al., 2017). Recent research suggests that

Cas12a in yeast (5'-TYCV-3) and in rice (5'-TATV-3') (Swiat et al., 2017; Zhong et al., 2018) engineered for efficiency as well as specificity for wide range of targeting (Gao et al., 2017).

19.4 DEFENSE SYSTEM MEDIATED VIA CRISPR

The CRISPR/Cas9 and CRISPR/Cas12a editing mechanisms evolved for the purpose of protecting cells from foreign DNA. We describe the three phases of the defensive process in detail, the first being spacer acquisition, the second being RNA synthesis, and the third being target interference (Figure 19.4).

19.4.1 ADAPTATION/SPACER ACQUISITION

To adapt or acquire a spacer by CRISPR, foreign genetic material is selected and kept as bacterial immunological memory on CRISPR arrays with unique sequences specific to each bacteriophage (Figure 19.4(a) and 19.4(b)). CRISPR-mediated immunity is triggered when foreign genetic material is

(a)

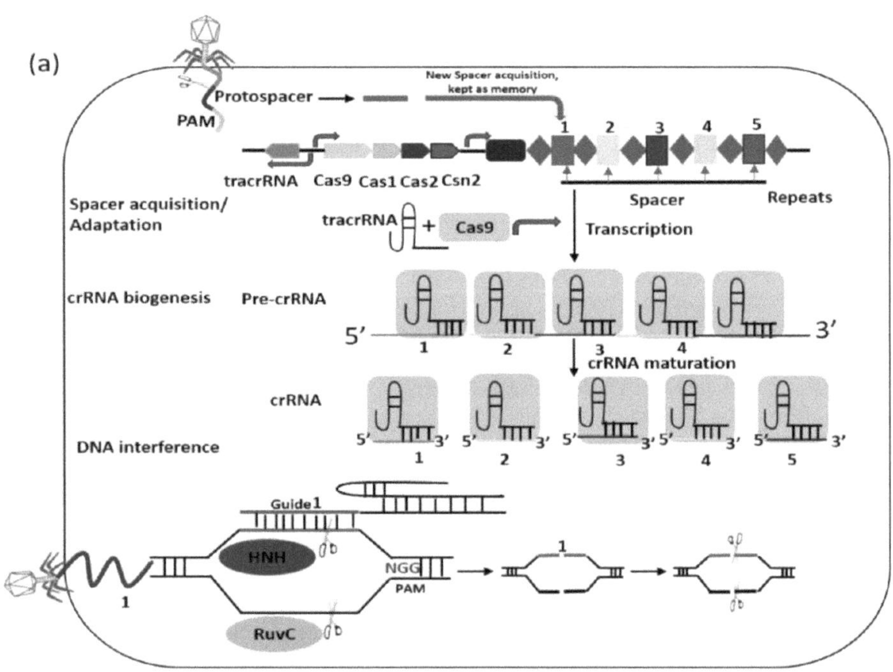

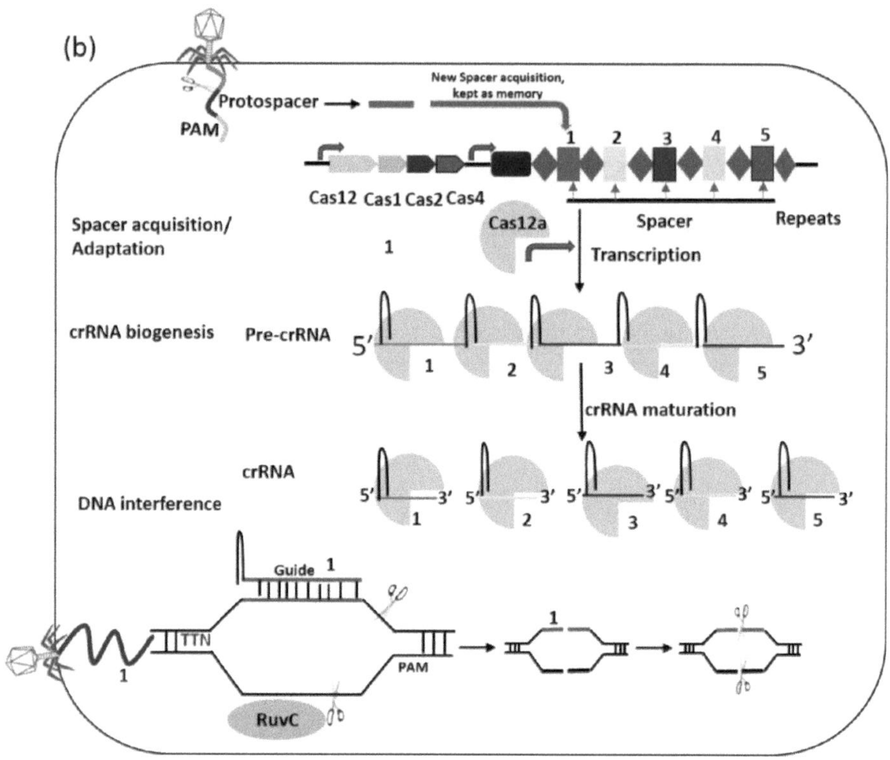

FIGURE 19.4 The CRISPR/Cas adaptive mechanism. (a) Type-II CRISPR/Cas9; and (b) type V CRISPR/Cas12a.

Note: The three stages of CRISPR/Cas presented in upper three stages: Spacer acquisition/ adaptation CrRNA biogenesis and DNA interference. In CRISPR adaption process the phage DNA injected into bacterial cell (illustrated in upper left) activates the Cas1–Cas2 adaptive proteins which facilitate excision of phage DNA and channels then for incorporation into CRISPR array/spacer as a memory. In next step of CRISPR RNA biogenesis CRISPR array is transcribed into a long precursor CRISPR RNA (pre-crRNA) and then processed into mature CRISPR RNAs to guard against DNA invasion (crRNA) afterwards each crRNAs are bounded by Cas protein effectors. In DNA interference phage when phage again attack then phage DNA sequence matches with CRISPR spacer appeared in colored cubical cell (upper middle) effectors programmed by appropriate crRNA bind to it and destroyed by Cas executor nuclease. spCas9 nuclease uses two domain HNH and RuvC to cleave target and non-target strands of DNA, respectively while whereas fnCas12a has only RuvC nuclease domain to cleave both the strand of DNA.
Source: This figure is adapted from: Yao et al. (2018).

detected and incorporated into the CRISPR array, allowing the host organisms to memorize invaders (Jackson et al., 2017). There are two conserved proteins called Cas1 and Cas2, present in virtually all CRISPRs. They help integrate pre-spacers into arrays. Each CRISPR-Cas type has its own mechanism to acquire a spacer. For the type II-A system, Cas9, Cas1, Cas2, and Csn2 are needed along with TracRNA (Heler et al., 2015; Wei et al., 2015) (Figure 19.4). Cas12a, Cas1, and Cas2 are the four Cas proteins used in the Type V-A system to acquire spacers (Stella et al., 2017a) (Figure 19.4).

Generally, new protospacers are not selected at random. Before DNA interrogation can take place, a short PAM-proximal segment (1 or 2 bp) within the protospacer must be unpaired (Deveau et al., 2008; Mojica et al., 2009; Zetsche et al., 2015; Fonfara et al., 2016; Swarts et al., 2018). After CRISPR is triggered, Cas9 makes a selection of a protospacer next to a PAM, activating the acquisition proteins (Cas1, Cas2, and Casn2) for the integration of the new spacer into the array (Wei et al., 2015). Cas1–Cas2 insert new DNA sequences into the leader end of the array to produce new spacers (Figure 19.4). In order to create a new spacer, the Cas1–Cas2 complex inserts a sequence of invading DNA into the array's leader end before duplicating the original repeat of the array in order to maintain this structure repeat-spacer-repeat (Barrangou et al., 2007). CRISPR does not contain the PAM site and but in targets sites, preventing self-targeting (Anders et al., 2014; Jiang et al., 2015).

19.4.2 CRRNA BIOGENESIS

This step involves translating CRISPR RNA into precursor CRISPR RNA, which, when further processed, becomes mature CRISPR RNAs that guard against DNA invasion (crRNA) (Haurwitz et al., 2010). The maturation of pre-crRNA requires a tracrRNA that possesses a sequence corresponding to the pre-crRNA (Figure 19.4). An endogenous RNase III breaks the duplex, which Cas9 recognizes and stabilizes (Deltcheva et al., 2011). By trimming the crRNA's 5'end by an unidentified nuclease, the mature crRNA is formed (Deltcheva et al., 2011; Gasiunas et al., 2012). Cas12a, a type V-A Cas system, is also capable of maturing crRNA without the use of tracrRNA or RNase III (Zetsche et al., 2015; Fonfara et al., 2016).

19.4.3 INTERFERENCE

At the end of the CRISPR defense process, the interference stage occurs when the crRNA recognizes the effector molecules and cleaves them away from the invading DNA (Gasiunas et al., 2012; Jinek et al., 2014; Zetsche et al., 2015). A crucial part of the interference process is played by Cas9 and Cas12a nucleases. In its crystal structure, Cs9 has two lobes: a nuclease (NUC) lobe and a helical recognition (REC) lobe (Jiang et al., 2015; Nishimasu et al., 2014) (Figure 19.3) where, DNA and sgRNA are bound by the REC lobe. HNH and RuvC, both nuclease domains, are located in the NUC lobe. These two domains cleave targeted complementary and noncomplementary regions of the DNA strand. The REC and NUC lobes of Cas12a are bi-lobed, similar to Cas9 (detailed in Yamano et al., 2016) (Figure 19.3). Cleavage of Cas12a strands is performed using the RuvC nuclease domain, along with its own crRNA and RNase WED III, whereas Cas9 requires two domains (Gao et al., 2016).

In the type II-A system, Cas9 recognizes a target DNA sequence through its PI domain by recognizing an NGG sequence located immediately downstream of a protospacer (e.g., 5′-NGG-3′ in *S. pyogenes* spCas9) (Anders et al., 2014; Jiang et al., 2015). There is a 10–12-nt seed region requirement for interference (Sternberg et al., 2015) (Figure 19.4). Cas9 requires tracrRNA whereas Cas12a requires only crRNA. In *Francisella novicida fnCas12a,* the PAM sequences are found directly upstream, but in Cas9, the PAM sequences are located downstream (Zetsche et al., 2015; Fonfara et al., 2016).

19.5 DEFENSE MECHANISM SYSTEM IN PLANTS

Plants establish defense mechanisms through communication with pathogens via cell-to-cell interaction. A series of interactions between pathogens and plants occur during this interaction process (Silva et al., 2018). Biomolecules needed for defense are largely determined by the nature of the pathogen (bacteria, fungi, viruses), such as bacteria disseminating virulence-associated molecules via type I, type II, and type III secretion systems (T3SSs) (Kamber et al., 2017). To hijack the first line of defense, fungi secrete enzymes that smash cell walls and recruit fungi for further infection processes, or they build haustoria to steal the entire machinery of a plant (Kamoun, 2006). Plants become infected by viral particles instead of through mechanical damage, inserting their genetic material and causing infection directly.

Kachroo et al. (2017) describe how plants possess trans-membrane protein receptors (PRRs) that recognize pathogen/microbe associations (PAMPs or MAPS) that will initiate highly specific PAMP-triggered immunity (PTI) against infections (Boutrot et al., 2017; Andersen et al., 2018; Tyagi et al., 2018). Upon contacting the plant, some pathogens produce effector proteins that fight with the plant's defense mechanisms and defeat PTI by causing the plant to produce susceptibility (S) proteins results effector-triggered susceptibility (ETS) (Kachroo et al., 2017).

As in response to ETS plants develop their second line of defense, i.e., use of resistance genes (R) after recognition of avirulence (AVR) proteins or effector molecules. This is called effector-triggered immunity (ETI). Plants' ETI is a type of extremely specific immunity which is only triggered when pathogens are spotted, then proceeds to a hypersensitive response (HR) through programmed cell death (PCD) and monitors overall pathogen surveillance in the plant (Musidlak et al., 2014). In addition to target identification in plants, pathogen effector research is encouraged as a way to develop new biological tools for disrupting plant-pathogen interactions, either by mutation or manipulation of the relationship (Doron et al., 2018).

19.6 EFFICIENT ROLE OF CRISPR/CAS9 IN RESISTANCE

Even though bacteria are simpler in structure and smaller in size, they are very diverse organisms. Since they are diverse, they can survive in extreme conditions and multiply rapidly in different ways. Hence, bacterial diseases are difficult to control, and they can lead to epidemics. Furthermore, because of their potential to hijack host systems, susceptible genes have been targeted via genome editing, which is found to be more durable under field conditions (Chen, 2014; Bezrutczyk et al., 2017). A detailed list of genes edited for disease resistance in crop has been shown in Table 19.2.

Plants' resistance to disease can be enhanced by editing the genes encoding susceptibility factors (Zaidi et al., 2018). *Xanthomonas oryzae* pv. *oryzae* (*Xoo*) is the causal organism of bacterial blight of rice and results in up to 10%–20% yield losses, but it can also cause up to 50% yield losses under favorable growing conditions. A humid agroclimatic condition sometimes makes the problem worse.

There are many varieties of resistance R-genes or susceptible S-genes in plants, and these can be grouped according to their functions – (i) nucleotide-binding leucine-rich repeat receptor (NB-LRR) genes (*Xa21 Xa3/Xa26*, and *Xa4*); (ii) sugars transporter (*SWEET*) genes (*Xa13/SWEET11, Xa25/SWEET13,* and *Xa41/SWEET14*); (iii) executor R genes (*Xa10, Xa23*, and *Xa27*); and (iv) others (*Xa1* and *Xa5*). Among them *Xa1 Xa5, Xa10, Xa13, Xa23, Xa25, Xa27*, and *Xa41* have been significantly involved in interaction of pathogen like effector protein called TALEs during the host cell's defense response (TALEs) (Jiang et al., 2020).

A pathogen's nutrition may be improved by triggering the *SWEET* gene expression, which transports sugar in cells (Chen, 2014; Bezrutczyk et al., 2017). A *Xoo* strain lacks SWEET gene expression is virtually non-pathogenic, and every Xoo strain studied thus far contains at least one gene for a major TALe (Oliva et al., 2019). *Xoo* secretes the TALE protein inside the plant, which activates the *SWEET* gene, and the *SWEET* gene provides an infection favorable environment (Doyle et al., 2013). Developing a bacterial blight infection relies heavily on the SWEET gene and its interaction with TALE. A few TALE effectors such as PthXo1, PthXo2, and AvrXa7, PthXo3, TalC, and TalF (originally Tal5) in bacteria can cause sugar transporter genes *SWEET11, SWEET13*, and *SWEET14* to be expressed, as well as hijack the genome (Streubel et al., 2013; Tran et al., 2018). Researchers used CRISPR/Cas9 editing tools to edit effector-binding elements (EBEs). The bacterial blight resistance could be achieved without compromising plant health in this way by destroying TALE protein binding (Jiang et al., 2013). *Rice genes SWEET11, SWEET13*, and *SWEET14* have been edited with CRISPR/Cas9 to target disease resistance (Oliva et al., 2019). In addition to editing the coding region, CRISPR can be used to edit the promoter region, preventing the gene from fully expressing or interrupting an EBS in an attempt to interfere with the binding of pathogenic effectors. Rice sugar transporter's promoter has been altered by this technique (Zaidi et al., 2020). Using CRISPR/Cas9 genome editing, Olivia et al. (2019) added mutations to all three *SWEET* genes, *SWEET11, SWEET13*, and *SWEET14*. In order to test if resistance could be developed to bacterial blight, the mutations were then transferred to the rice line Kitaake and to the elite mega varieties IR64 and Ciherang-Sub1. Rice cv. Kitaake, which carries the recessive resistance allele of *SWEET13*, showed similar responses after disruption of the EBEs of *SWEET11* and *SWEET14*. Based on testing against a variety of *Xoo* strains, the modified MS14K line appeared to be highly resistant (Xu et al., 2019).

SWEET14 contains the EBEs for four different TALEs, PthXo3, AvrXa7 TalC and TalF (formerly identified as Tal5) (Antony et al., 2010; Yu et al., 2011; Streubel et al., 2013; Oliva et al., 2019). According to Zafar et al. (2020), TALEs targeting *SWEET* gene promoter regions are six of the best known. To achieve broad-spectrum resistance, it is critical to focus on the *SWEET14* promoter. By targeting four EBEs within *SWEET14* promoter, CRISPR-Cas9 modified rice genomes to make them resistant to bacterial blight. When deletions were made in the EBE of AvrXa7, the lines became resistant to *Xoo* (Zafar et al., 2020). This notwithstanding, the *Xa13* gene is increasing in popularity in practical breeding due to its unique function and disease resistance mechanism (Hajira et al., 2016), as well as its promoter being investigated for resistance to disease. The *Xa13* gene promoter has been altered to reduce the promoter's ability to stimulate expression without affecting another gene's function and expression (Li et al., 2020). In this way, rice fertility is not altered by the deletion method, but the resistance of rice is improved (Li et al., 2020). Furthermore, these results show CRISPR/Cas9 multiplex targeting can be used to edit several EBEs simultaneously. Consequently, scientists have been able to develop BB resistant variants under different environmental conditions (Kim et al., 2019; Li et al., 2019). As well, using the CRISPR/Cas9 technology, a targeted mutation in the *OsMPK5* gene, which encodes a mitogen-activated protein kinase for rice, significantly increased resistance to *Burkholderia glumae* (Xie et al., 2013).

In Arabidopsis thaliana, it has been found that a single gene knockout of DMR6 (downy mildew resistance 6) confers resistance to a broad range of pathogens. SlDMR6-1, the tomato ortholog of this gene, displays broad spectrum resistance to *Pseudomonas syringae* pv. tomato and *Phytophthora capsica* (de Toledo Thomazella et al., 2016). Further, genetic modification of defense pathways can result in development of resistance to specific kinds of pathogens. In spite of this, limited success has been found due to the antagonistic relationship of salicylic acid (SA) and jasmonic acid (JA) defense pathways. SA-based defenses result in a reduction of JA-mediated resistance. A spatially uncoupled interaction between SA and JA was used to overcome this trade-off problem. By doing so, we have created a tomato that is resistant to the bacterial speck disease caused by *Pseudomonas syringae* pv. tomato (Pto) DC3000, which produces JA which mimics the bioactive hormone coronatine (COR). Recently, tomato's *SlJAZ2* ortholog of the COR stomatal co-receptor *AtJAZ2* was edited with CRISPR/Cas9 technology. By inhibiting stomatic reopening by COR, the edited SlJAZ2 mutant also inhibited bacterial invasion through stomata (Ortigosa et al., 2019). In addition, the multiplication of citrus canker disease-causing bacterial *Xanthomonas citri* sp. *citri* (Xcc) was stimulated by a transcription factor called *CsLOB1*, so using CRISPR/Csa9 this gene was knocking outs in the coding region (Jia et al., 2017) and promoter region (Peng et al., 2017) of *CsLOB1*.

Among the list of plant disease pathogens, fungi are ranked first, which are the cause of numerous plant diseases that affect both the quantity and quality of agricultural products. According to studies, fungi are responsible for nearly 30% of emerging diseases (Giraud et al., 2010), and necessitating new management strategies. Mutations in the host S gene/s have been used to confer broad-spectrum disease resistance to plants using the CRISPR/Cas9 system. *Magnaporthe oryzae* causes rice blast, which is the second-most important rice disease, and its prevalence varies depending on the agroclimatic region (Dean et al., 2012). By mutating the negative regulator *OsERF922* through targeted loss of function, rice blast resistance could be improved (Wang et al., 2016). In addition, mutations of the Pi21 gene resulting from CRISPR/Cas9 enhanced rice plant resistance to *M. oryzae* (Li et al., 2019). Further, it was discovered that the coding area of OsMAPK5 targeted to increase resistance to *Magnaporthe grisea* (Xie et al., 2013).

Through the use of advanced CRISPR/Cas9 and TALEN technologies, it has been possible to introduce targeted mutations in the three homoeoalleles that encode the MILDEW RESISTANCE LOCUS (MLO) proteins in wheat, which encode a protein that can defend plants against powdery mildew diseases (Büschges et al., 1997). A mutation of *TaMLO-A1* by CRISPR/Cas9 is used for durable resistance to powdery mildew caused in wheat by *Blumeria graminis* f. sp. *tritici* (Bgt) (Wang et al., 2014). Further, enhanced disease resistance 1 (EDR1) acts as a positive regulator of powdery mildew susceptibility, resulting in Taedr1 mutant wheat plants being protected from *Bgt* by CRISPR/Cas9 (Zhang et al., 2017). *SlMlo1* is the most important gene associated with powdery mildew susceptibility in tomato, and CRISPR/Cas9 mediated knockouts of this gene provide resistance to *Oidium neolycopersici* (Nekrasov et al., 2017). CRISPR/Cas9 mutations result in resistance to powdery mildew disease in tomato by changing the callose synthase gene (*SlPMR4*) (Koseoglou et al., 2017). Further, mutations in CRISPR/Cas9 genes in tomato *Solyc08g075770* (Prihatna et al., 2018) and mitogen-activated protein kinase 3 (MAPK3) (Zhang et al., 2017) indicate their role in the development of disease. The CRISPR/Cas9 gene editing technology targets the transcription factor WRKY52 (VvWRKY52) which increases resistance to Botrytis cinerea in grapes (Wang et al., 2018).

19.7 CHALLENGES SOLUTIONS AND FUTURE PROSPECTS

A combination of robustness, simplicity, and multiplexing make CRISPR/Cas9 systems attractive to agriculture, but they also pose challenges in terms

TABLE 19.2 List of Genes Edited for Disease Resistance

Protein/Region/ Mutation	Targeted Gene in Rice	Findings	Targeted Disease	Transgene Free	References
		Rice			
Cas9/promoter/ deletion	*Xa13*	Enhanced disease resistance to Xoo.	Bacterial blight	Yes	Li et al. (2020)
Cas9/promoter/ deletion	*SWEET14*	Enhanced disease resistance to Xoo.	Bacterial blight	Yes	Zafar et al. (2020)
Cas9/CDS/In-dels	*Xa13*	Enhanced disease resistance to Xoo.	Bacterial blight	n.d.	Li et al. (2019)
Cas9/CDS/In-dels	*OsSWEET11*	Enhanced resistance to Xoo	Bacterial blight	Yes	Kim et al. (2019)
Cas9/CDS and promoter/In-dels	TALE-binding elements (EBEs) in *OsSWEET13* promoter *OsSWEETT11* and *OsSWEEt14* genes.	Broad spectrum resistance against multiple Xoo strains.	Bacterial blight	No	Xu et al. (2019)
Cas9/promoter/ In-dels	*OsSWEET11*, *OsSWEET13*, *OsSWEET14*	Enhanced resistance to Xoo.	Bacterial blight	Yes	Oliva et al. (2019)
Cas9/CDS/deletion	*OsSWEET13*	Resistance not confirmed.	Bacterial blight	No	Zhou et al. (2015)
Cas9/promoter/ In-dels	*OsSWEET11*, *OsSWEET14*	Disease resistance not confirmed.	Bacterial blight	No	Jiang et al. (2013)
Cas9/CDS/In-dels	*OsMPK5*	Disease resistance not confirmed.	*Burkholderia glumae*	No	Xie et al. (2013)
Cas9/CDS/In-dels	*OsERF922*	Enhanced disease resistance.	Rice blast	Yes	Wang et al. (2016)
Cas9/CDS/In-dels	*Pi21*	Enhanced disease resistance.	Rice blast	Yes	Li et al. (2019)
Cas9/CDS/In-dels	*OsMPK5*	Disease resistance not confirmed.	Rice blast	No	Xie et al. (2013)
		Tomato			
Cas9/CDS/deletion	*SlDMR6–1*	Enhanced disease resistance.	Bacterial speck	No	de Toledo Thomazella et al. (2016)

TABLE 19.2 *(Continued)*

Protein/Region/ Mutation	Targeted Gene in Rice	Findings	Targeted Disease	Transgene Free	References
Cas9/CDS/deletion	*SlDMR6–1*	Enhanced disease resistance.	Phytophthora blight	No	de Toledo Thomazella et al. (2016)
Cas9/CDS/deletion	*SlDMR6–1*	Enhanced disease resistance.	Bacterial spot	No	de Toledo Thomazella et al. (2016)
Cas9/CDS	*SlJAZ2*	Enhanced disease resistance.	Bacterial speck	n.d.	Ortigosa et al. (2019)
Cas9/CDS/deletion	*SlMlo1*	Enhanced disease resistance.	Powdery mildew	Yes	Nekrasov et al. (2017)
Cas9/CDS/deletion inversion.	*SlPMR4*	Enhanced disease resistance.	Powdery mildew	No	Koseoglou et al. (2017)
Cas9/genomic region	*Solyc08g075770*	Enhanced susceptibility to disease.	*Fusarium wilt*	No	Prihatna et al. (2018)
Cas9	*SlMAPK3*	Enhanced susceptibility to disease.	Gray mold	No	Zhang et al. (2018)
		Citrus			
Cas9/CDS/In-dels	*CsLOB1*	Enhanced disease resistance.	Citrus canker	–	Jia et al. (2017)
Cas9/promoter/ deletion	*CsLOB1*	Enhanced disease resistance (83.2–98.3%).	Citrus canker	No	Peng et al. (2017)
Wheat					
TALEN and Cas9/ cds/In-dels.	*TaMLO-A1*	Enhanced disease resistance.	Powdery mildew	Yes	Wang et al. (2014)
Cas9/CDS/In-dels	*TaEDR1*	Enhanced disease resistance.	Powdery mildew	–	Zhang et al. (2017)
Grapes					
Cas9/CDS/In-dels	*MLO-7*	Resistance to disease developed.	Powdery mildew	Yes	Malnoy et al. (2016)
Cas9/CDS/In-dels	*VvWRKY52*	Enhanced disease resistance.	Gray mold	No	Wang et al. (2018)

of developing resistance. The high efficiency and precision of CRISPR-Cas9 have great potential to overcome the limitations of traditional breeding for resistance development in crop plants. Though CRISPR/Cas9 has many benefits and a wide range of applications, there are a few disadvantages.

Despite this, targeting S genes may result in an adverse fitness penalty due to a few major roadblocks that could hinder the evolution of disease resistance given its links to various desired genes that control plant growth. A targeted mutation of an S gene may disrupt a product's route and, in turn, many other products along that route may be adversely affected as a result of that mutation. While this fitness cost is not as devastating as the first, nutritional deficiencies and phenotypic abnormalities may result. Fitting cost, however, may change when an S gene is targeted and is affected by a number of factors, such as those that increase disease susceptibility, those that suppress defense, and those that affect replication (van Schie & Takken, 2014). The use of promoter editing to generate new variations (Rodriguez-Lea et al., 2017) to knockdown the desired qualities of the S gene (Bastet et al., 2018) or the use of the base editing method to control these phenomena are all viable approaches (Yan et al., 2018). During the last two decades, research has focused on the promoters or the regulatory regions of pathogen-induced genes (Rushton et al., 2002). Through CRISPR/Cas9, the regulatory/promoter regions of S/S genes could be knocked out to avoid fitness implications related to sugar transportation and help prevent pathogen colonization of the plant (Kim et al., 2019; Zafar et al., 2020).

Its use has been a source of considerable concern due to the potential for off-target mutation, also known as nonspecific or unwanted alterations in the genome, such as point mutations and insertions (Wilson et al., 2018). Regarding the specificity of CRISPR/Cas9, CRISPR offers a distinct advantage over traditional Y-ray mutations or chemically induced mutations. In this regard, gRNA complements genomic DNA to what extent it does so without binding to another region (Off-target region) of genomic DNA which is then potentially cleaved by Cas9 protein and causes undesirable mutations. Researchers have generally found two types of off-target mutations: those that are expected to occur as a result of the sequence's high similarity to the target sequence and those that are unexpected and occur as a result of an unrelated sequence. In order to detect off-target mutations expected to be present within the targeted region, PCR amplification and sequencing of the targeted region are used, while whole-genome sequencing is required when there are off-target mutations that are not expected (Feng et al., 2016).

Sequencing the genome is an essential tool for predicting awaited off-target mutations. Searching sequences is done based on 20 bp pairings, excluding PAM-5'-NGG-3' that is necessary for target recognition (Shah et al., 2013). The CRISPR/Cas9 system accepts a minimum of three base pair mismatches in the 20-bp target DNA sequence. In order to minimize mismatches, CRISPR/Cas9 tools suggest only specific gRNA designs that don't bind to off-target regions. When a gene has numerous paralogs, it is more challenging to design a specific gRNA, but it is easier when the gene is unique within the genome. According to this, designing gRNAs for diploid genomes is quite simpler than for polyploid genomes (Bortesi et al., 2016). A high level of specificity of 87.3 to 94.3% was seen in the relatively simple genomes of Arabidopsis, rice, tomato, and soybean, but in maize only 29.5% was observed due to its repetitive DNA (Bortesi et al., 2016).

KEYWORDS

- **CRISPR/Cas9**
- **defense system**
- **disease resistant transgene-free**
- **programmable site-specific nucleases**
- **spacer acquisition**
- **TALEN**
- **ZNF**

REFERENCES

Anders, C., Niewoehner, O., Duerst, A., & Jinek, M., (2009). Structural basis of PAM-dependent target DNA recognition by the Cas9 endonuclease. *Nature, 513*(7519), 569–573. doi: 10.1038/nature13579.

Andersen, E. J., Ali, S., Byamukama, E., Yen, Y., & Nepal, M. P., (2018). Disease resistance mechanisms in plants. *Genes, 9*(7), 339. doi: 10.3390/genes9070339.

Antony, G., Zhou, J., Huang, S., Li, T., Liu, B., White, F., & Yang, B., (2010). Rice xa13 recessive resistance to bacterial blight is defeated by induction of the disease susceptibility gene Os-11N3. *Plant Cell, 22*(11), 3864–3876. doi: 10.1105/tpc.110.078964.

Barrangou, R., Fremaux, C., Deveau, H., Richards, M., Boyaval, P., Moineau, S., Romero, D. A., & Horvath, P., (2007). CRISPR provides acquired resistance against viruses in prokaryotes. *Science, 315*(5819), 1709–1712. doi: 10.1126/science.1138140.

Bartsevich, V. V., Morris, J., Tomberlin, G., Turner, C., Sharer, W., Nicholson, M. G., Smith, J., et al., (2016). 579. Meganucleases as an efficient tool for genome editing. *Mol. Ther., 24*, S230. doi: 10.1016/S1525-0016(16)33387-1.

Bastet, A., Lederer, B., Giovinazzo, N., Arnoux, X., German-Retana, S., Reinbold, C., Brault, V., et al., (2018). Trans-species synthetic gene design allows resistance pyramiding and broad-spectrum engineering of virus resistance in plants. *Plant Biotechnol. J., 16*(9), 1569–1581. doi: 10.1111/pbi.12896.

Bezrutczyk, M., Yang, J., Eom, J. S., Prior, M., Sosso, D., Hartwig, T., Szurek, B., et al., (2018). Sugar flux and signaling in plant–microbe interactions. *Plant J., 93*(4), 675–685. doi: 10.1111/tpj.13775.

Boch, J., Scholze, H., Schornack, S., Landgraf, A., Hahn, S., Kay, S., Lahaye, T., et al., (2009). Breaking the code of DNA binding specificity of TAL-type III effectors. *Science, 326*(5959), 1509–1512. doi: 10.1126/science.1178811.

Bortesi, L., Zhu, C., Zischewski, J., Perez, L., Bassié, L., Nadi, R., Forni, G., et al., (2016). Patterns of CRISPR/Cas9 activity in plants, animals, and microbes. *Plant Biotechnol. J., 14*(12), 2203–2216. doi: 10.1111/pbi.12634.

Boutrot, F., & Zipfel, C., (2017). Function, discovery, and exploitation of plant pattern recognition receptors for broad-spectrum disease resistance. *Ann. Rev. Phytopathol., 55*, 257–286. doi: 10.1146/annurev-phyto-080614–120106.

Büschges, R., Hollricher, K., Panstruga, R., Simons, G., Wolter, M., Frijters, A., Van, D. R., et al., (1997). The barley Mlo gene: A novel control element of plant pathogen resistance. *Cell, 88*(5), 695–705. doi: 10.1016/S0092-8674(00)81912-1.

Chen, K., Wang, Y., Zhang, R., Zhang, H., & Gao, C., (2019). CRISPR/Cas genome editing and precision plant breeding in agriculture. *Ann. Rev. Plant Biol., 70*, 667–697. doi: 10.1146/annurev-arplant-050718-100049.

Chen, L. Q., (2014). SWEET sugar transporters for phloem transport and pathogen nutrition. *New Phytol., 201*(4), 1150–1155. doi: 10.1111/nph.12445.

Christian, M., Cermak, T., Doyle, E. L., Schmidt, C., Zhang, F., Hummel, A., Bogdanove, A. J., & Voytas, D. F., (2010). Targeting DNA double-strand breaks with TAL effector nucleases. *Genetics, 186*(2), 757–761. doi: 10.1534/genetics.110.120717.

De Toledo, T. D. P., Brail, Q., Dahlbeck, D., & Staskawicz, B., (2016). *CRISPR-Cas9 Mediated Mutagenesis of a DMR6 Ortholog in Tomato Confers Broad-Spectrum Disease Resistance* (Vol. 1, p. 064824). BioRxiv. doi: 10.1101/064824.

Dean, R., Van, K. J. A., Pretorius, Z. A., Hammond-Kosack, K. E., Di Pietro, A., Spanu, P. D., Rudd, J. J., et al., (2012). The top 10 fungal pathogens in molecular plant pathology. *Mol. Plant Pathol., 13*(4), 414–430. doi: 10.1111/j.1364-3703.2011.00783.x.

Deltcheva, E., Chylinski, K., Sharma, C. M., Gonzales, K., Chao, Y., Pirzada, Z. A., Eckert, M. R., et al., (2011). CRISPR RNA maturation by trans-encoded small RNA and host factor RNase III. *Nature, 471*(7340), 602–607. doi: 10.1038/nature09886.

Deveau, H., Barrangou, R., Garneau, J. E., Labonté, J., Fremaux, C., Boyaval, P., Romero, D. A., et al., (2008). Phage response to CRISPR-encoded resistance in *Streptococcus thermophilus*. *J. Bacteriol., 190*(4), 1390–1400. doi: 10.1128/JB.01412-07.

Doron, S., Melamed, S., Ofir, G., Leavitt, A., Lopatina, A., Keren, M., Amitai, G., & Sorek, R., (2018). Systematic discovery of antiphage defense systems in the microbial pangenome. *Science, 359*(6379). doi: 10.1126/science.aar4120.

Doyle, E. L., Stoddard, B. L., Voytas, D. F., & Bogdanove, A. J., (2013). TAL effectors: Highly adaptable phytobacterial virulence factors and readily engineered DNA-targeting proteins. *Trends Cell Biol., 23*(8), 390–398. doi: 10.1016/j.tcb.2013.04.003.

FAO, (2015). Keeping plant pests and diseases at bay: experts focus on global measures. Online. Available at: http://www.fao.org/news/story/en/item/280489/icode (accessed on 12 December 2023).

Fauser, F., Schiml, S., & Puchta, H., (2014). Both CRISPR/C as-based nucleases and nickases can be used efficiently for genome engineering in *Arabidopsis thaliana. Plant J., 79*(2), 348–359. doi: 10.1111/tpj.12554.

Feng, C., Yuan, J., Wang, R., Liu, Y., Birchler, J. A., & Han, F., (2016). Efficient targeted genome modification in maize using CRISPR/Cas9 system. *Journal of Genetics and Genomics, 43*(1), 37–43. doi: 10.1016/j.jgg.2015.10.002.

Fonfara, I., Richter, H., Bratovič, M., Le Rhun, A., & Charpentier, E., (2016). The CRISPR-associated DNA-cleaving enzyme Cpf1 also processes precursor CRISPR RNA. *Nature, 532*(7600), 517–521. doi: 10.1038/nature17945.

Gao, C., (2018). The future of CRISPR technologies in agriculture. *Nature Reviews Mol. Cell Biol., 19*(5), 275–276. doi: 10.1038/nrm.2018.2.

Gao, L., Cox, D. B., Yan, W. X., Manteiga, J. C., Schneider, M. W., Yamano, T., Nishimasu, H., et al., (2017). Engineered Cpf1 variants with altered PAM specificities. *Nature Biotechnol., 35*(8), 789–792. doi: 10.1038/nbt.3900.

Gao, P., Yang, H., Rajashankar, K. R., Huang, Z., & Patel, D. J., (2016). Type V CRISPR-Cas Cpf1 endonuclease employs a unique mechanism for crRNA-mediated target DNA recognition. *Cell Res., 26*(8), 901–913. doi: 10.1038/cr.2016.88.

Gao, Z., Liu, Q., Zhang, Y., Chen, D., Zhan, X., Deng, C., Cheng, S., & Cao, L., (2020). OsCUL3a-associated molecular switches have functions in cell metabolism, cell death, and disease resistance. *J. Agricul. Food Chemist., 68*(19), 5471–5482. doi: 10.1021/acs.jafc.9b07426.

Gasiunas, G., Barrangou, R., Horvath, P., & Siksnys, V., (2012). Cas9–crRNA ribonucleoprotein complex mediates specific DNA cleavage for adaptive immunity in bacteria. *Proceed. Nat. Acad. Sci., 109*(39), E2579–2586. doi: 10.1073/pnas.1208507109.

Gaudelli, N. M., Komor, A. C., Rees, H. A., Packer, M. S., Badran, A. H., Bryson, D. I., & Liu, D. R., (2017). Programmable base editing of A• T to G• C in genomic DNA without DNA cleavage. *Nature, 551*(7681), 464–471. doi: 10.1038/nature24644.

Giraud, T., Gladieux, P., & Gavrilets, S., (2010). Linking the emergence of fungal plant diseases with ecological speciation. *Trends Ecol. Evol., 25*(7), 387–395. doi: 10.1016/j.tree.2010.03.006.

Hajira, S. K., Sundaram, R. M., Laha, G. S., Yugander, A., Balachandran, S. M., Viraktamath, B. C., Sujatha, K., et al., (2016). A single-tube, functional marker-based multiplex PCR assay for simultaneous detection of major bacterial blight resistance genes Xa21, xa13 and xa5 in rice. *Rice Science, 23*(3), 144–151. doi: 10.1016/j.rsci.2015.11.004.

Haurwitz, R. E., Jinek, M., Wiedenheft, B., Zhou, K., & Doudna, J. A., (2010). Sequence-and structure-specific RNA processing by a CRISPR endonuclease. *Science, 329*(5997), 1355–1358. doi: 10.1126/science.1192272.

Heler, R., Samai, P., Modell, J. W., Weiner, C., Goldberg, G. W., Bikard, D., & Marraffini, L. A., (2015). Cas9 specifies functional viral targets during CRISPR–Cas adaptation. *Nature, 519*(7542), 199–202. doi: 10.1038/nature14245.

Hunter, M. C., Smith, R. G., Schipanski, M. E., Atwood, L. W., & Mortensen, D. A., (2017). Agriculture in 2050: Recalibrating targets for sustainable intensification. *Bioscience, 67*(4), 386–391. doi: 10.1093/biosci/bix010.

Ishino, Y., Shinagawa, H., Makino, K., Amemura, M., & Nakata, A., (1987). Nucleotide sequence of the iap gene, responsible for alkaline phosphatase isozyme conversion in *Escherichia coli*, and identification of the gene product. *J. Bacteriol., 169*(12), 5429–5433. doi: 10.1128/jb.169.12.5429-5433.1987.

Jackson, S. A., McKenzie, R. E., Fagerlund, R. D., Kieper, S. N., Fineran, P. C., & Brouns, S. J., (2017). CRISPR-Cas: Adapting to change. *Science, 356*(6333). doi: 10.1126/science.aal5056.

Jia, H., Zhang, Y., Orbović, V., Xu, J., White, F. F., Jones, J. B., & Wang, N., (2017). Genome editing of the disease susceptibility gene Cs LOB 1 in citrus confers resistance to citrus canker. *Plant Biotechnol. J., 15*(7), 817–823. doi: 10.1111/pbi.12677.

Jiang, F., Zhou, K., Ma, L., Gressel, S., & Doudna, J. A., (2015). A Cas9–guide RNA complex preorganized for target DNA recognition. *Science, 348*(6242), 1477–1481. doi: 10.1126/science.aab1452.

Jiang, N., Yan, J., Liang, Y., Shi, Y., He, Z., Wu, Y., Zeng, Q., et al., (2020). Resistance genes and their interactions with bacterial blight/leaf streak pathogens (*Xanthomonas oryzae*) in rice (*Oryza sativa* L.)—An updated review. *Rice, 13*(1), 1, 2. doi: 10.1186/s12284-019-0358-y.

Jiang, W., Zhou, H., Bi, H., Fromm, M., Yang, B., & Weeks, D. P., (2013). Demonstration of CRISPR/Cas9/sgRNA-mediated targeted gene modification in *Arabidopsis*, tobacco, sorghum, and rice. *Nuc. Acid. Res., 41*(20), e188. doi: 10.1093/nar/gkt780.

Jinek, M., Chylinski, K., Fonfara, I., Hauer, M., Doudna, J. A., & Charpentier, E., (2012). A programmable dual-RNA–guided DNA endonuclease in adaptive bacterial immunity. *Science*. *337*(6096), 816–821. doi: 10.1126/science.1225829.

Jinek, M., Jiang, F., Taylor, D. W., Sternberg, S. H., Kaya, E., Ma, E., Anders, C., et al., (2014). Structures of Cas9 endonucleases reveal RNA-mediated conformational activation. *Science 343*(6176). doi: 10.1126/science.1247997.

Kachroo, A., Vincelli, P., & Kachroo, P., (2017). Signaling mechanisms underlying resistance responses: What have we learned, and how is it being applied? *Phytopathology 107*(12), 1452–1461. doi: 10.1094/PHYTO-04-17-0130-RVW.

Kamber, T., Pothier, J. F., Pelludat, C., Rezzonico, F., Duffy, B., & Smits, T. H., (2017). Role of the type VI secretion systems during disease interactions of *Erwinia amylovora* with its plant host. *BMC Genom., 18*(1), 1, 2. doi: 10.1186/s12864-017-4010-1.

Kamoun, S., (2006). A catalogue of the effector secretome of plant pathogenic oomycetes. *Annu. Rev. Phytopathol., 44*, 41–60. doi: 10.1146/annurev.phyto.44.070505.143436.

Karvelis, T., Gasiunas, G., Miksys, A., Barrangou, R., Horvath, P., & Siksnys, V., (2013). crRNA and tracrRNA guide Cas9-mediated DNA interference in *Streptococcus thermophilus*. *RNA Biol., 10*(5), 841–851. doi: 10.4161/rna.24203.

Kim, H. K., Song, M., Lee, J., Menon, A. V., Jung, S., Kang, Y. M., Choi, J. W., et al., (2017). *In vivo* high-throughput profiling of CRISPR–Cpf1 activity. *Nature Meth., 14*(2), 153–159. doi: 10.1038/nmeth.4104.

Kim, Y. A., Moon, H., & Park, C. J., (2019). CRISPR/Cas9-targeted mutagenesis of Os8N3 in rice to confer resistance to *Xanthomonas oryzae* pv. oryzae. *Rice, 12*(1), 1–3. doi: 10.1186/s12284-019-0325-7.

Kim, Y. G., Cha, J., & Chandrasegaran, S., (1996). Hybrid restriction enzymes: Zinc finger fusions to Fok I cleavage domain. *Proceed Nat. Acad. Sci., 93*(3), 1156–1160. doi: 10.1073/pnas.93.3.1156.

Kleinstiver, B. P., Pattanayak, V., Prew, M. S., Tsai, S. Q., Nguyen, N. T., Zheng, Z., & Joung, J. K., (2016). High-fidelity CRISPR–Cas9 nucleases with no detectable genome-wide off-target effects. *Nature, 529*(7587), 490–495. doi: 10.1038/nature16526.

Komor, A. C., Kim, Y. B., Packer, M. S., Zuris, J. A., & Liu, D. R., (2016). Programmable editing of a target base in genomic DNA without double-stranded DNA cleavage. *Nature, 533*(7603), 420–424. doi: 10.1038/nature17946.

Koseoglou, E., (2017). *The Study of SlPMR4 CRISPR/Cas9-Mediated Tomato Allelic Series for Resistance Against Powdery Mildew* (Vol. 34). Wageningen: Department of Plant Breeding Wageningen University and Research.

Kumar, V., & Jain, M., (2015). The CRISPR–Cas system for plant genome editing: Advances and opportunities. *J. Exp. Bot., 66*(1), 47–57. doi: 10.1093/jxb/eru429.

Lei, C., Li, S. Y., Liu, J. K., Zheng, X., Zhao, G. P., & Wang, J., (2017). The CCTL (Cpf1-assisted cutting and Taq DNA ligase-assisted ligation) method for efficient editing of large DNA constructs *in vitro*. *Nuc. Acids Res., 45*(9), e74. doi: 10.1093/nar/gkx018.

Li, C., Li, W., Zhou, Z., Chen, H., Xie, C., & Lin, Y., (2020). A new rice breeding method: CRISPR/Cas9 system editing of the Xa13 promoter to cultivate transgene-free bacterial blight-resistant rice. *Plant Biotechnol. J., 18*(2), 313. doi: 10.1111/pbi.13217.

Li, S., Shen, L., Hu, P., Liu, Q., Zhu, X., Qian, Q., Wang, K., & Wang, Y., (2019). Developing disease-resistant thermosensitive male sterile rice by multiplex gene editing. *J. Integ. Plant Biol., 61*(12), 1201–1205. doi: 10.1111/jipb.12774.

Malnoy, M., Viola, R., Jung, M. H., Koo, O. J., Kim, S., Kim, J. S., Velasco, R., & Nagamangala, K. C., (2016). DNA-free genetically edited grapevine and apple protoplast using CRISPR/Cas9 ribonucleoproteins. *Front Plant Sci., 7*, 1904. doi: 10.3389/fpls.2016.01904.

Mojica, F. J., Díez-Villaseñor, C., García-Martínez, J., & Almendros, C., (2009). Short motif sequences determine the targets of the prokaryotic CRISPR defense system. *Microbiol., 155*(3), 733–740. doi: 10.1099/mic.0.023960-0.

Molina, R., Montoya, G., & Prieto, J., (2011). Meganucleases and their biomedical applications. *eLS*. doi: 10.1002/9780470015902.a0023179.

Musidlak, O., Buchwald, W., & Nawrot, R., (2014). Plant defense responses against viral and bacterial pathogen infections. Focus on RNA-binding proteins (RBPs). *Herba Polonica, 60*(4). doi: 10.1515/hepo-2015-0005.

Nekrasov, V., Wang, C., Win, J., Lanz, C., Weigel, D., & Kamoun, S., (2017). Rapid generation of a transgene-free powdery mildew resistant tomato by genome deletion. *Sci. Rep., 7*(1), 1–6. doi: 10.1038/s41598-017-00578-x.

Nishimasu, H., Ran, F. A., Hsu, P. D., Konermann, S., Shehata, S. I., Dohmae, N., Ishitani, R., et al., (2014). Crystal structure of Cas9 in complex with guide RNA and target DNA. *Cell, 156*(5), 935–949. doi: 10.1016/j.cell.2014.02.001.

Oliva, R., Ji, C., Atienza-Grande, G., Huguet-Tapia, J. C., Perez-Quintero, A., Li, T., Eom, J. S., et al., (2019). Broad-spectrum resistance to bacterial blight in rice using genome editing. *Nat. Biotechnol., 37*(11), 1344–1350. doi: 10.1038/s41587-019-0267-z.

Ortigosa, A., Gimenez-Ibanez, S., Leonhardt, N., & Solano, R., (2019). Design of a bacterial speck resistant tomato by CRISPR/Cas9-mediated editing of Sl JAZ 2. *Plant Biotechnol. J., 17*(3), 665–673. doi: 10.1111/pbi.13006.

Paul, J. W., & Qi, Y., (2016). CRISPR/Cas9 for plant genome editing: Accomplishments, problems, and prospects. *Plant Cell Rep., 35*(7), 1417–1427. doi: 10.1007/s00299-016-1985-z.

Peng, A., Chen, S., Lei, T., Xu, L., He, Y., Wu, L., Yao, L., & Zou, X., (2017). Engineering canker-resistant plants through CRISPR/Cas9-targeted editing of the susceptibility gene Cs LOB 1 promoter in citrus. *Plant Biotechnol. J., 15*(12), 1509–1519. doi: 10.1111/pbi.12733.

Prihatna, C., Barbetti, M. J., & Barker, S. J., (2018). A novel tomato *fusarium* wilt tolerance gene. *Front Microbiol., 9*, 1226. doi: 10.3389/fmicb.2018.01226.

Ramalingam, S., Kandavelou, K., Rajenderan, R., & Chandrasegaran, S., (2011). Creating designed zinc-finger nucleases with minimal cytotoxicity. *J. Mol. Biol., 405*(3), 630–641. doi: 10.1016/j.jmb.2010.10.043.

Ran, F. A., Cong, L., Yan, W. X., Scott, D. A., Gootenberg, J. S., Kriz, A. J., Zetsche, B., et al., (2015). *In vivo* genome editing using *Staphylococcus aureus* Cas9. *Nature, 520*(7546), 186–191. doi: 10.1038/nature14299.

Ran, F. A., Hsu, P. D., Lin, C. Y., Gootenberg, J. S., Konermann, S., Trevino, A. E., Scott, D. A., et al., (2013). Double nicking by RNA-guided CRISPR Cas9 for enhanced genome editing specificity. *Cell, 154*(6), 1380–1389. doi: 10.1016/j.cell.2013.08.021.

Ren, B., Yan, F., Kuang, Y., Li, N., Zhang, D., Zhou, X., Lin, H., & Zhou, H., (2018). Improved base editor for efficiently inducing genetic variations in rice with CRISPR/Cas9-guided hyperactive hAID mutant. *Mol. Plant, 11*(4), 623–626. doi: 10.1016/j.molp.2018.01.005.

Rodríguez-Leal, D., Lemmon, Z. H., Man, J., Bartlett, M. E., & Lippman, Z. B., (2017). Engineering quantitative trait variation for crop improvement by genome editing. *Cell, 171*(2), 470–480. doi: 10.1016/j.cell.2017.08.030.

Rushton, P. J., Reinstadler, A., Lipka, V., Lippok, B., & Somssich, I. E., (2002). Synthetic plant promoters containing defined regulatory elements provide novel insights into pathogen-and wound-induced signaling. *Plant Cell, 14*(4), 749–762. doi: 10.1105/tpc.010412.

Savary, S., Willocquet, L., Pethybridge, S. J., Esker, P., McRoberts, N., & Nelson, A., (2019). The global burden of pathogens and pests on major food crops. *Nat. Ecol. Evol., 3*(3), 430–439. doi: 10.1038/s41559-018-0793-y.

Schiml, S., Fauser, F., & Puchta, H., (2014). The CRISPR/C as system can be used as nuclease for in planta gene targeting and as paired nickases for directed mutagenesis in *Arabidopsis* resulting in heritable progeny. *Plant J., 80*(6), 1139–1150. doi: 10.1111/tpj.12704.

Schiml, S., Fauser, F., & Puchta, H., (2016). Repair of adjacent single-strand breaks is often accompanied by the formation of tandem sequence duplications in plant genomes. *Proceed Nat. Acad. Sci., 113*(26), 7266–7271. doi: 10.1073/pnas.1603823113.

Schindele, P., Wolter, F., & Puchta, H., (2018). Transforming plant biology and breeding with CRISPR/Cas9, Cas12 and Cas13. *FEBS Letters, 592*(12), 1954–1967. doi: 10.1002/1873-3468.13073.

Shah, S. A., Erdmann, S., Mojica, F. J., & Garrett, R. A., (2013). Protospacer recognition motifs: Mixed identities and functional diversity. *RNA Biol., 10*(5), 891–899. doi: 10.4161/rna.23764.

Shmakov, S., Smargon, A., Scott, D., Cox, D., Pyzocha, N., Yan, W., Abudayyeh, O. O., et al., (2017). Diversity and evolution of class 2 CRISPR–Cas systems. *Nat. Rev. Microbiol., 15*(3), 169–182. doi: 10.1038/nrmicro.2016.184.

Silva, M. S., Arraes, F. B., De Araújo, C. M., Grossi-de-Sa, M., Fernandez, D., De Souza, C. E., Cardoso, M. H., et al., (2018). Potential biotechnological assets related to plant immunity modulation applicable in engineering disease-resistant crops. *Plant Sci., 270*, 72–84. doi: 10.1016/j.plantsci.2018.02.013.

Stella, S., Alcón, P., & Montoya, G., (2017). Structure of the Cpf1 endonuclease R-loop complex after target DNA cleavage. *Nature, 546*(7659), 559–563. doi: 10.1038/nature22398.

Sternberg, S. H., LaFrance, B., Kaplan, M., & Doudna, J. A., (2015). Conformational control of DNA target cleavage by CRISPR–Cas9. *Nature, 527*(7576), 110–113. doi: 10.1038/nature15544.

Streubel, J., Pesce, C., Hutin, M., Koebnik, R., Boch, J., & Szurek, B., (2013). Five phylogenetically close rice SWEET genes confer TAL effector-mediated susceptibility to *Xanthomonas oryzae* pv. *oryzae*. *New Phytol., 200*(3), 808–819. doi: 10.1111/nph.12411.

Swarts, D. C., & Jinek, M., (2018). Cas9 versus Cas12a/Cpf1: Structure–function comparisons and implications for genome editing. *Wiley Interdisciplinary Reviews: RNA, 9*(5), e1481. doi: 10.1002/wrna.1481.

Świat, M. A., Dashko, S., Den, R. M., Wijsman, M., Van Der, O. J., Daran, J. M., & Daran-Lapujade, P., (2017). Fn Cpf1: A novel and efficient genome editing tool for *Saccharomyces cerevisiae*. *Nuc. Acid. Res., 45*(21), 12585–12598. doi: 10.1093/nar/gkx1007.

Tran, T. T., Pérez-Quintero, A. L., Wonni, I., Carpenter, S. C., Yu, Y., Wang, L., Leach, J. E., et al., (2018). Functional analysis of African *Xanthomonas oryzae* pv. oryzae TALomes reveals a new susceptibility gene in bacterial leaf blight of rice. *PLoS Pathog., 14*(6), e1007092. doi: 10.1371/journal.ppat.1007092.

Tyagi, S., Mulla, S. I., Lee, K. J., Chae, J. C., & Shukla, P., (2018). VOCs-mediated hormonal signaling and crosstalk with plant growth promoting microbes *Critical Rev. Biotechnol., 38*(8), 1277–1296. doi: 10.1080/07388551.2018.1472551.

Van, B. T., & Van, D. A. G., (2020). Salicylic acid steers the growth–immunity tradeoff. *Trends Plant Sci., 25*(6), 566–576. doi: 10.1016/j.tplants.2020.02.002.

Van, S. C. C., & Takken, F. L., (2014). Susceptibility genes 101: How to be a good host. *Ann. Rev. Phytopathol., 52*, 551–581. doi: 10.1146/annurev-phyto-102313-045854.

Wang, F., Wang, C., Liu, P., Lei, C., Hao, W., Gao, Y., Liu, Y. G., & Zhao, K., (2016). Enhanced rice blast resistance by CRISPR/Cas9-targeted mutagenesis of the ERF transcription factor gene OsERF922. *PloS One, 11*(4), e0154027. doi: 10.1371/journal.pone.0154027.

Wang, X., Tu, M., Wang, D., Liu, J., Li, Y., Li, Z., Wang, Y., & Wang, X., (2018). CRISPR/Cas9-mediated efficient targeted mutagenesis in grape in the first generation. *Plant Biotechnol. J., 16*(4), 844–855. doi: 10.1111/pbi.12832.

Wang, Y., Cheng, X., Shan, Q., Zhang, Y., Liu, J., Gao, C., & Qiu, J. L., (2014). Simultaneous editing of three homoeoalleles in hexaploid bread wheat confers heritable resistance to powdery mildew. *Nat. Biotechnol., 32*(9), 947–951. doi: 10.1038/nbt.2969.

Wei, Y., Terns, R. M., & Terns, M. P., (2015). Cas9 function and host genome sampling in type II-A CRISPR–Cas adaptation. *Gene Develop., 29*(4), 356–361. doi: 10.1101/gad.257550.114.

Wilson, L. O., O'Brien, A. R., & Bauer, D. C., (2018). The current state and future of CRISPR-Cas9 gRNA design tools. *Front. Pharmacol., 9*, 749. doi: 10.3389/fphar.2018.00749.

Xie, K., & Yang, Y., (2013). RNA-guided genome editing in plants using a CRISPR–Cas system. *Molecul. Plant, 6*(6), 1975–1983. doi: 10.1093/mp/sst119.

Xu, Z., Xu, X., Gong, Q., Li, Z., Li, Y., Wang, S., Yang, Y., et al., (2019). Engineering broad-spectrum bacterial blight resistance by simultaneously disrupting variable TALE-binding elements of multiple susceptibility genes in rice. *Mol. Plant., 12*(11), 1434–1446. doi: 10.1016/j.molp.2019.08.006.

Yamano, T., Nishimasu, H., Zetsche, B., Hirano, H., Slaymaker, I. M., Li, Y., Fedorova, I., et al., (2016). Crystal structure of Cpf1 in complex with guide RNA and target DNA. *Cell, 165*(4), 949–962. doi: 10.1016/j.cell.2016.04.003.

Yan, F., Kuang, Y., Ren, B., Wang, J., Zhang, D., Lin, H., Yang, B., et al., (2018). Highly efficient A·T to G·C base editing by Cas9n-guided tRNA adenosine deaminase in rice. *Molecul. Plant, 11*(4), 631–634. doi: 10.1016/j.molp.2018.02.008.

Yang, W., (2011). Nucleases: Diversity of structure, function, and mechanism. *Quarterly Rev. Biophy., 44*(1), 1–93. doi: 10.1017/S0033583510000181.

Yao, R., Liu, D., Jia, X., Zheng, Y., Liu, W., & Xiao, Y., (2018). CRISPR-Cas9/Cas12a biotechnology and application in bacteria. *Synt. Sys. Biotechnol., 3*(3), 135–149. doi: 10.1016/j.synbio.2018.09.004.

Zafar, K., Khan, M. Z., Amin, I., Mukhtar, Z., Yasmin, S., Arif, M., Ejaz, K., & Mansoor, S., (2020). Precise CRISPR-Cas9 mediated genome editing in super basmati rice for resistance against bacterial blight by targeting the major susceptibility gene. *Front Plant Sci., 11*, 575. doi: 10.3389/fpls.2020.00575.

Zaidi, S. S., Mahas, A., Vanderschuren, H., & Mahfouz, M. M., (2020). Engineering crops of the future: CRISPR approaches to develop climate-resilient and disease-resistant plants. *Genome Biology, 21*(1), 1–9. doi: 10.1186/s13059-020-02204-y.

Zaidi, S. S., Mukhtar, M. S., & Mansoor, S., (2018). Genome editing: Targeting susceptibility genes for plant disease resistance. *Trends in Biotechnology, 36*(9), 898–906. doi: 10.1016/j.tibtech.2018.04.005.

Zetsche, B., Gootenberg, J. S., Abudayyeh, O. O., Slaymaker, I. M., Makarova, K. S., Essletzbichler, P., Volz, S. E., et al., (2015). Cpf1 is a single RNA-guided endonuclease of a class 2 CRISPR-Cas system. *Cell, 163*(3), 759–771. doi: 10.1016/j.cell.2015.09.038.

Zhang, D., Li, Z., & Li, J. F., (2016). Targeted gene manipulation in plants using the CRISPR/Cas technology. *Journal of Genetics and Genomics, 43*(5), 251–262. doi: 10.1016/j.jgg.2016.03.001.

Zhang, S., Wang, L., Zhao, R., Yu, W., Li, R., Li, Y., Sheng, J., & Shen, L., (2018). Knockout of SlMAPK3 reduced disease resistance to *Botrytis cinerea* in tomato plants. *Journal of Agricultural and Food Chemistry, 66*(34), 8949–8956. doi: 10.1021/acs.jafc.8b02191.

Zhang, Y., Bai, Y., Wu, G., Zou, S., Chen, Y., Gao, C., & Tang, D., (2017). Simultaneous modification of three homoeologs of Ta EDR 1 by genome editing enhances powdery mildew resistance in wheat. *The Plant Journal, 91*(4), 714–724. doi: 10.1111/tpj.13599.

Zhong, Z., Zhang, Y., You, Q., Tang, X., Ren, Q., Liu, S., Yang, L., et al., (2018). Plant genome editing using FnCpf1 and LbCpf1 nucleases at redefined and altered PAM sites. *Molecular Plant, 11*(7), 999–1002. doi: 10.1016/j.molp.2018.03.008.

Zhou, J., Peng, Z., Long, J., Sosso, D., Liu, B., Eom, J. S., Huang, S., et al., (2015). Gene targeting by the TAL effector PthXo2 reveals cryptic resistance gene for bacterial blight of rice. *The Plant Journal, 82*(4), 632–643. doi: 10.1111/tpj.12838.

Zong, Y., Wang, Y., Li, C., Zhang, R., Chen, K., Ran, Y., Qiu, J. L., et al., (2017). Precise base editing in rice, wheat, and maize with a Cas9-cytidine deaminase fusion. *Nature Biotechnology, 35*(5), 438–440. doi: 10.1038/nbt.3811.

CHAPTER 20

Robotics in Plant Diseases Diagnosis and Management

S. AMEER BASHA, V. RAMYA, G. UMA DEVI, and A. U. AKASH

Department of Plant Pathology, College of Agriculture, Professor Jayashankar Telangana State Agricultural University, Hyderabad, Telangana, India

ABSTRACT

Agricultural robotics is an automation technology involving development, fabrication, functioning, and application of robots as a cost-effective solution in the fields of agriculture, horticulture, forestry, fisheries, and livestock. Robots and machine learning are increasingly being used across different disciplines in agriculture including plant protection. In the field of plant pathology, robots can be employed for crop health monitoring, diagnosis of plant diseases, detection of plant pathogens, assessing disease severity, and to carry out precision management practices. A robot typically comprises of sensors, manipulators, control system, EOAT/end effectors, and actuators. Developing an accurate and reliable sensor is the first real step in the development of a robot plant doctor. Spectral sensors like visual (RGB), multi-spectral, hyper-spectral, thermal chlorophyll fluorescence, and volatile organic compound (VOCs) sensors are major ones exploited for plant disease detection and monitoring. Robots have not yet been used extensively for agriculture in India, let alone in pathogen management. Still, most of the potential robots are in the prototype stage, major advances are seen in the detection of leaf diseases of crop plants and automated spraying of pesticides. In this chapter, we discussed how robots are employed in agriculture in general and plant pathology in particular.

New Advances in Crop Disease Management. Sangita Sahni, Ganesh Thapa, and Bishun Deo Prasad (Eds)

20.1 INTRODUCTION

World population is growing; we need food to feed them. Agricultural yields are not increasing over the years. Across the world, labor crisis is prominent and labor wages are increasing. Conventional steps taken to increase the crop yields have an impact on our environment. Heavy machinery like tractors have reached their limit; we cannot push them further sustainably. Robots are the future of farming.

More than 820 million people, i.e., 1 in every 9 people are hungry globally. By the year 2050, the world's population is estimated to be 9.7 billion (UNWPP, 2019), and in order to feed them we need to produce more food. By 2027, India will be the most populated country (UNWPP, 2019). Though the global hunger index (GHI) is a moderate 18.2, from 28.2 (serious) during the year 2000, India is still at a "serious" level of hunger with an index of 27.2 occupying the 94th position out of a total of 107 countries. Some of the other countries featuring behind India are Indonesia, Nepal, Bangladesh, and Pakistan at 70, 73, 75, and 88 ranks (GHI, 2020) (Figure 20.1).

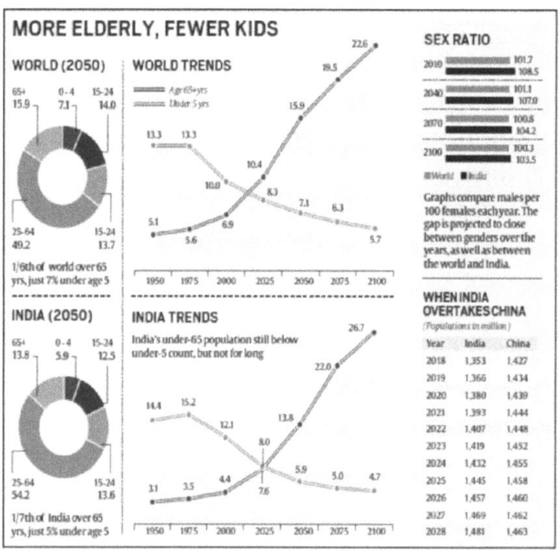

World and India population trends as given in UN 'World Population Projections 2019'.

FIGURE 20.1 World and India population trends as given in UN 'World Population Projections 2019.'

Over the last decade, there has been a decline in agricultural laborers in terms of workforce and total workers. As a percentage of workforce, they declined from 81.0% to 48.2% in the developing countries and from 35.0% to 4.2% in the developed ones (ILO, 2014). The percentage of agricultural laborers among total workers also decreased from 59.1% (1991) to 54.6% (2011). It is projected to come further down to 40.6% by 2020, constituted by women workers at 45% (Mehta et al., 2019). Farmers around the world are aging and we need to have our younger generations involved in agriculture.

How do we find enough food for these 9 billion people without overwhelming the planet? The answer to this question is sustainable agriculture, which means practicing farming in sustainable and viable ways to meet the food and textile demands of society while maintaining the resources for future generations to meet their needs. It involves a number of eco-friendly agricultural practices that maintain an ecological balance by avoiding diminution of natural resources. Robots are helping farmers in practicing sustainable agriculture as they can do away with the excessive use of resources and inputs and supply only what is needed to the crops, thus reducing the impact on the environment. Furthermore, robots can be a possible alternative to labor, especially during shortages and crisis situations. They have the potential to attract the younger generation towards agriculture. The use of robots and machine learning in different disciplines of agriculture, including plant health management, has been increasing since the past few decades (Prabha, 2020).

20.2 SMART AND DIGITAL FARMING

Precision agriculture is a modern management strategy in farming to monitor, process, and optimize agricultural production processes using digital technique to support management decisions in agricultural production. The core point here is optimization. For instance, if one were to apply fertilizers, in precision agriculture, the fertilizer application strategy would involve measuring the soil variations within the field and adapting accordingly instead of uniformly applying the fertilizers in the entire field without taking into account the variations within-field. This strategy not to leads to optimizing the fertilizer usage but is also cost saving and reduces the impact on environment.

20.2.1 SMART FARMING

Smart farming is using modern information and data for managing complex farming systems efficiently by optimizing the labor requirements by accessing the data and using it in a smart way.

20.2.2 DIGITAL FARMING

Digital farming is an amalgamation of smart farming and precision farming with the core idea of creating value from data. Digital farming refers to regular application of the methods of precision and smart farming, internal, and external networking of the farm and application of web-based platforms coupled with big data analyzes. It is not just the availability or presence of data but creating meaningful and intelligent added value from such type of data.

20.3 ARTIFICIAL INTELLIGENCE

Artificial intelligence or AI is the intelligence exhibited by machines and the capability of the machine to imitate human behavior. AI mimics the cognitive functions exhibited by humans and is based on the premise that intelligence is not "real" or "human" (Table 20.1).

20.3.1 ARTIFICIAL INTELLIGENCE AND MACHINE LEARNING – APPLICATIONS

20.3.1.1 APPLICATIONS OF AI

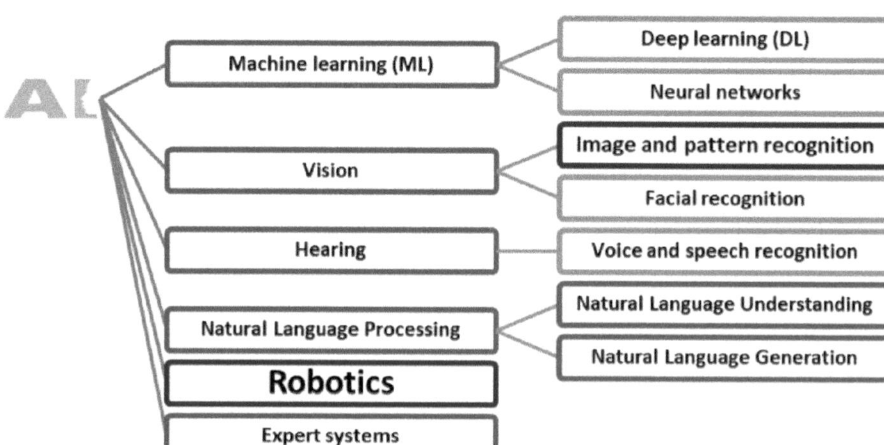

TABLE 20.1 Three Stages of Artificial Intelligence

Type of Artificial Intelligence	**Narrow**	**General**	**Super**
Year	By 2015	By 2020	By 2050
Features	• User-driven big data systems for machine learning. • Limited to one or two functional areas. • Are not self-aware or self-conscious.	• Machine intelligence. • Advanced networks trained to build ad-hoc systems and improve themselves using data. • Human like intelligence and reasoning.	• Machine consciousness. • System characterizing cognitive self-learning. • Surpasses human intelligence.

Machine Learning: It is a type of AI that allows software applications to learn and improve from experience and improve from experience (without being programmed explicitly) to predict the outcomes by finding patterns in a data (Table 20.2).

Types of machine learning:

1. **Supervised learning**
   i. Machine is provided with an input, an output, and feedback to build a model.

      Labeled Training Data (Input Data + Expected Output) → Machine Learning Algorithm → learned model

   ii. The ML program:

      a. is provided with training data along with the expected output or rules to categorize this data also known as labels.
      b. uses the set of inputs and outputs to predict the output for future unseen inputs.

   iii. Used in classification.

2. **Unsupervised Learning:**

   i. Deep learning is used to derive patterns and conclusions from unlabeled dataset.

Training Data (Input Data) → Machine Learning Algorithm → Learned Model

   ii. The ML algorithm:

      a. learns from an unlabeled dataset.
      b. uses input data to train the model.
      c. is expected to find patterns and anomalies.

   iii. Used in image detection, customer segmentation, MRI, etc.

3. **Semi-Supervised Learning:**

   i. It uses both labeled and unlabeled data.

4. **Reinforcement Learning:** It is a blended approach using a combination of supervised and unsupervised.

   i. Learning is through self-interpretation based on the method of trial and error with an aim to seek maximum reward.

ii. The learning system observes the environment to learn the ideal behavior.
iii. Selects and takes certain actions and receives rewards in return.
iv. Receives feedback in a loop.
v. Learns the tactic that maximizes rewards.

20.4 ROBOTS IN AGRICULTURE

20.4.1 ROBOTS

Term "robot" meaning servitude or forced labor, was first used in a play entitled "Rossum's Universal Robots" in the year 1920 by Karel Capek, a Czech writer by profession. The word "robotics" was accidentally coined by Isaac Asimov, a scientist and science fiction writer in 1940s. Initially, he proposed the three "Laws of Robotics," and added the "zeroth law" later on.

1. **Zeroth Law:** A robot may not injure humanity, or, through inaction, allow humanity to come to harm.
2. **First Law:** A robot may not injure a human being, or, through inaction, allow a human being to come to harm, unless this would violate a higher order law.
3. **Second Law:** A robot must obey orders given it by human beings, except where such orders would conflict with a higher order law.
4. **Third Law:** A robot must protect its own existence as long as such protection does not conflict with a higher order law.

   After technological revolutions during World War II, engineer and inventor, George C. Devol, CEO of Condep Corp., Norman Schafler and physicist and engineer, Joseph F. Engelberger launched a robotics-based firm "Unimation" and manufactured the first robot named "Unimate." Joseph F. Engelberger is known as the 'Father of Robotics.'

20.4.2 AGRICULTURAL ROBOTICS

It is use of automation in agriculture and allied fields as a cost-effective and efficient alternative for carrying out various operations and processes. The role played by robots in agriculture is becoming prominent with each day especially

TABLE 20.2 Common Algorithms in ML

Type of Machine Learning	Supervised	Unsupervised	Semi-Supervised	Reinforcement
Features and applications	• Support vector machine. • Image classification. • Diagnostics.	• Dimensional reduction. • Image recognition. • Big data visualization.	• Generalized adversarial networks.	Q learning
	• Regression. • Risk assessment. • Prediction • Decision trees.	• Clustering. • Biology. • Plant pathology.	• Self-trained naive bayes classifier. • Natural language processing (NLP)	• Model based value estimation. • Used mostly in robotics.

in agricultural operations including plant protection. They are focused in improving automation of traditional and regular agricultural machines and processes from sowing to harvesting. Agricultural tasks which are systematic, repetitive, and time-dependent are best-suited for robotic applications (Ampatzidis et al., 2017) (Figure 20.2). Following are some of the prominent agricultural robots (Kushwaha et al., 2016; Shamshiri et al., 2018):

1. **MF-Scamp Robots:** MF-Scamp robot is developed by Blackmore for cutting and removing the weeds.
2. **Lady Bird, University of Sydney:** Ladybird designed by the University of Sydney is a multi-use farm robot used in surveillance, mapping, classification, and identification of vegetables.
3. **Vitirover Solar Robots, New Zealand:** Vitrover is an autonomous robot fitted with sensors and global positioning system (GPS) and works at 500 m/hr. for cutting grass and weeds in grape vineyards.
4. **Image Processing Design, CROPS, European Union:** "Clever Robots for Crops" is a CROPS research project to facilitate automation in harvesting.
5. **AgBot II, Australia:** AgBot II is a robot designed at the Queensland University of Technology (QUT) in Australia to assist agricultural producers and farmers on operations like watering and application of chemicals including fertilizers and pesticides, etc., help farmers to take decisions on the use of herbicides, pesticides, fertilizers, and watering.
6. **Sub-Canopy Robot ISAAC 2:** Designed at Hohenheim University, Germany, ISAAC 2 is fitted with multiple sensors for timely collection of data and processing of information to assess the crop health.
7. **BoniRob Farming Robot:** BoniRob, designed by Deep field robotics, Germany is a multi-use robotic system for applications in remote farming.
8. **Lettuce Bot, California:** This was designed by Heraurd and Lee Redden at Stanford University. It is also referred as "Rambo of Weeds."
9. **AgriBot, India:** It is an agricultural robot developed by the Birla Institute of Technology Pilani – Hyderabad campus for carrying out plowing, sowing, and covering the seeds with soil.

10. **Autonomous Plant Inspection (API) Research Platform:** This by the Danish Institute of Agricultural Science is designed and has a GPS (real-time) for agricultural operations like spraying and weeding.
11. **HortiBot, Denmark:** It is a robust tool carrier developed for weeding in vegetables and outdoor gardens. The steering of the bot is carried out by a computer-aided navigation system.
12. **RIPPA:** It stands for robot for intelligent perception and precision application, which is a solar powered robot designed by the Australian Center for Field Robotics at Sydney University.

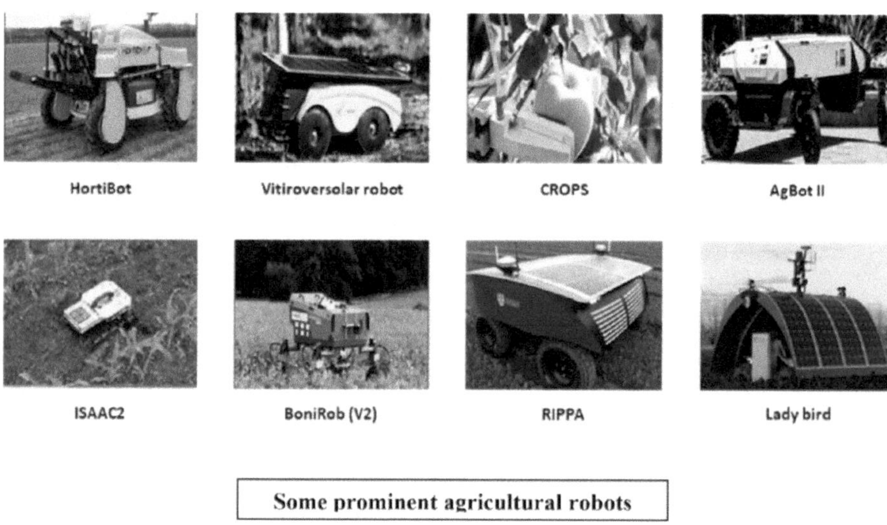

FIGURE 20.2 Some prominent agricultural robots.

20.4.3 ROBOTICS IN PLANT PATHOLOGY

Robots and machine learning are increasingly being used in various fields and disciplines of agriculture including plant protection. In plant pathology, robots can be employed for crop health monitoring, diagnosis of plant diseases, detection of plant pathogens, assessing disease severity and to carry out precision management practices. Robots used in plant pathology can be broadly grouped into two:

- Unmanned ground vehicles (UGV) – on-field robots;
- Unmanned aerial vehicles (UAV) – drones.

The recent advancements in AI and machine learning with robots, drones, and sensors have a wide array of applications in plant pathology. Over the years, robotics are used in disease forecasting, diagnosis, early detection before the onset of visual symptoms, understanding host-pathogen interactions, physiological changes during disease infection, assessment of disease severity in field and horticultural crops have all gained momentum through internet of things (IoT) based applications (Prabha, 2020).

UAVs are used for phenotyping fields and orchards for accurate application of chemicals including fertilizers and pesticides. Remote diagnostics of plant diseases in combination with precision plant protection can be made possible by assimilation of AI and augmented reality. Even though application of AI in plant protection is still in the nascent stage, the core concept of "self-sufficient, disease free, perfect plant" can soon become a reality using plant-robot bio-hybrids (Prabha, 2020).

20.4.4 MAIN PARTS OF A ROBOT

A robot mainly consists of the following parts (Celen et al., 2015) (Figure 20.3):

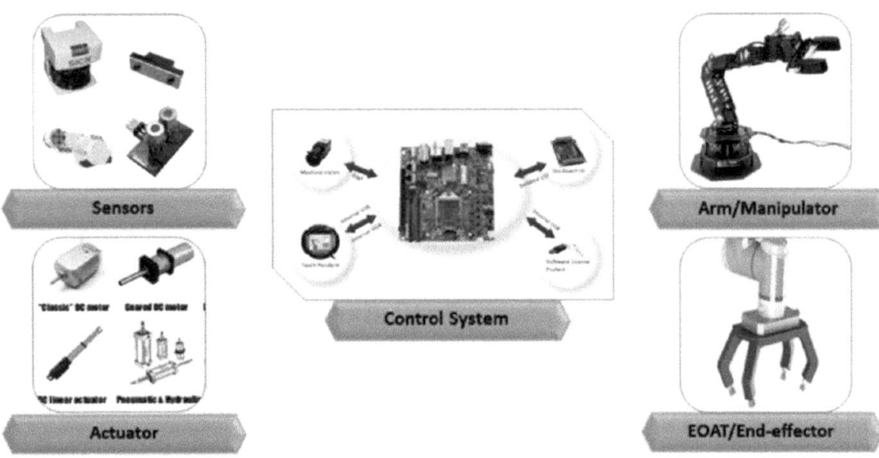

FIGURE 20.3 Main parts of a robot.

1. **Sensors:** These are input devices for robots to determine robot's environment and positioning. The sensors respond to various stimuli such as sound, light, heat, motion, magnetism, and pressure. They

transmit the signal and/or data for operating a control, providing a measurement or both.

2. **Control System:** The control system of a robot is a typically a computer to store data pertaining to robot itself and the work environment. It has programs, data, algorithms, logic analysis and other processing activities, which enable the robot to perform the intended tasks.
3. **Actuator:** It is a power mechanism for maintaining the position of a robot or to effect motion. The actuator works in response to the signal from the central control system.
4. **Arm/Manipulator:** It is a series of interconnected joints and links powered to support and move the hand or wrist through space.
5. **End of Arm Tooling (EOAT)/End-Effector:** Device fitted as an accessory to the hand of the robot. It can also be a mounting plate for tools enabling the robot to do the required task.

20.4.5 HOW DOES A ROBOT WORK?

1. **Gathers Information from Sensors:** Active sensors like laser system using LIDAR and passive sensors like digital cameras, photometric sensors, etc., are mainly used.
2. **Processes the Input and Makes Decision:** Various algorithms like decision making, operation planning, navigation, spectroscopic or image analysis techniques and IoT based techniques are employed to arrive at a result to carry out activities.
3. **Perform Desired Actions:** Actions like actuating other devices or operators, alert operators, spraying pesticides or performing other management actions will be the output.

20.5 SENSORS USED IN PLANT DISEASE DIAGNOSIS

The spectral signature of the plant or plant canopy is defined and calculated as the reflected radiation divided by the incident radiation for each wavelength and multiplied by 100. The nature of interaction varies with the wavelength of the radiation (Figures 20.4 and 20.5).

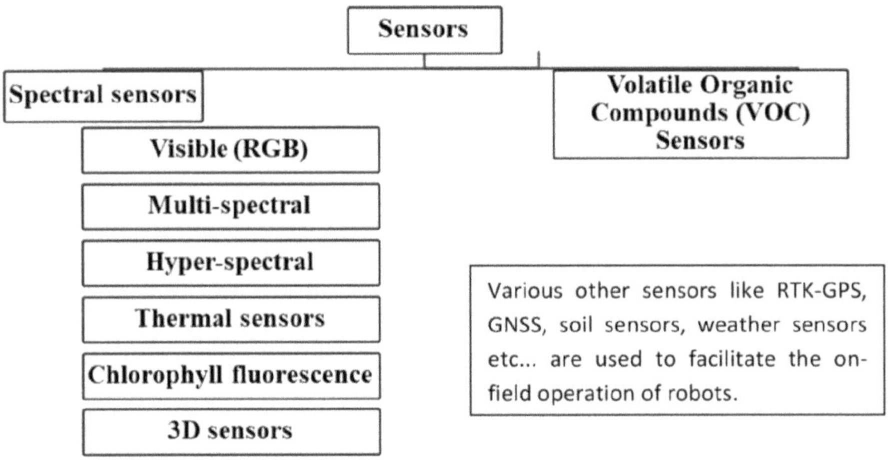

FIGURE 20.4 Principle behind spectral sensors.

Source: Adapted from Madden et al., 2007; Bock & Nutter, 2011

- Visible (VIS in the wavelength range of 400–700 nm); divided into:
- Violet-blue (~380–450 nm);
- Green (~550 nm);
- Red (~650–700 nm).
- Near-infrared (NIR in the wavelength of range 700–1,200 nm);
- Shortwave infrared (SWIR in the wavelength range 1,200–2,400 nm); and
- Thermal (TIR with wavelength ~10 μm).

Reflectance characteristics of a healthy plant (canopy):

- Low reflectance in the visible and shortwave infrared regions;
- High reflectance in the near-infrared regions;
- Within the visible spectrum, green region has relatively high reflectance.

Reflectance characteristics of a diseased plant (canopy):

- Overall increase in the reflectance in visible region;
- Low reflectance in green region due to reduction in chlorophyll;
- High reflectance in red and blue regions;
- Reduced reflectance in near infrared region due to defoliation.

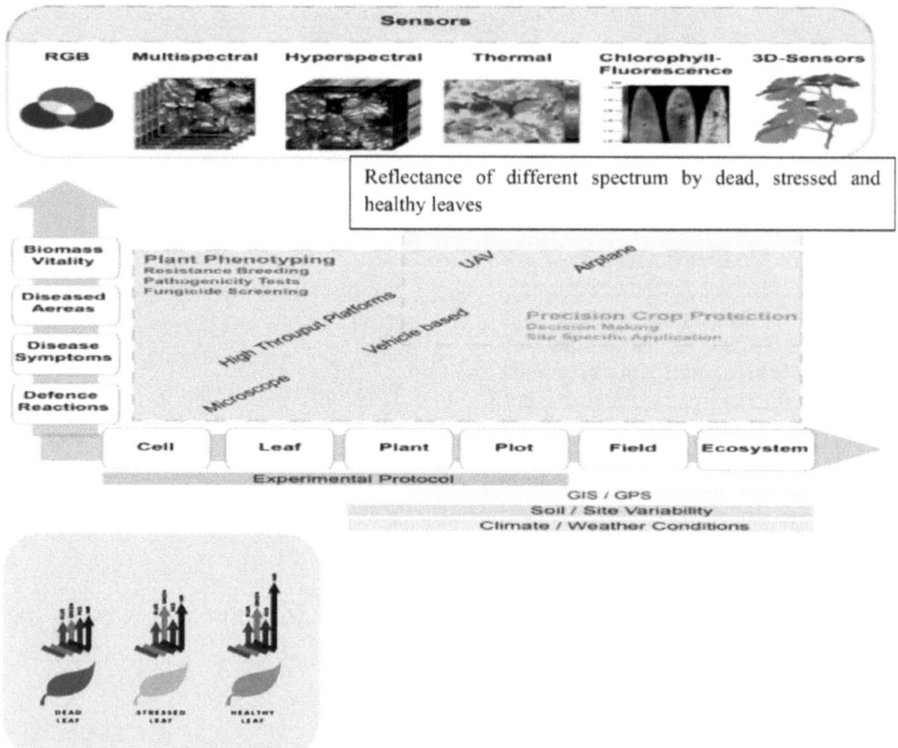

FIGURE 20.5 Reflectance of different spectrum by dead, stressed and healthy leaves.

20.5.1 FIELD APPLICATION WITH EXAMPLES

In citrus, different imaging methods were used for detection of greening (huanglongbing) and canker. A leaf recognition system detected citrus greening with a high detection accuracy of 92%. Elevated levels of starch in infected leaves detected by polarized imaging identified greening even before the symptoms were visual to the naked eye. Hyperspectral (400–1,000 nm) imaging of leaves and immature fruits at different developmental stages *viz.*, asymptomatic, early and late was used to detect canker on the leaves and fruits of citrus using a UAV with cent per cent accuracy and differentiate canker-infected trees from healthy ones.

Integration of thermal imaging and chlorophyll fluorescence can generate specific fingerprints for different host-pathogen interactions to identify early detection of plant diseases. For example, electronic nose, Cyranose 320 and artificial neural networks (ANN) not only detected the disease early in oil palm trees early but also were able to discriminate the different stages of symptoms caused by *Ganoderma boninense.*

Multispectral (R-G-NIR) camera to CROPS – Clever Robot for crOPS, a multipurpose manipulator, enabling it for disease detection and selective spraying of grapevine diseases, majorly Powdery mildew. Robotic system may produce reliable warnings of the outbreak of *Botrytis* and anthracnose in strawberry by integrating the phenology of the susceptible host and the epidemiological factors influencing the disease development.

Robots have not been yet used extensively for agriculture in India, let alone in pathogen management. Still, most of the potential robots are in the prototype stage, major advances are seen in the detection of leaf diseases of crop plants and automated spraying of pesticides. The use of robotic systems in plant pathology carries high accuracy, sensitivity, and early detection as major advantages while the inability to recognize certain variations and difficulty to manage them on field are the main disadvantages. Several challenges are there in developing and adopting robotic systems in the field of plant health management. Though a huge amount of money, time, and effort would be required to harness these systems into the field of plant pathology, it is worth trying owing to the wide array of advantages offered by this technology. Hopefully, we can see more of robotic plant doctors in near future (Tables 20.3 and 20.4; Figure 20.6).

20.6 CASE STUDIES ON ROBOTS IN INDIA

Shaikh et al. (2016) have developed an agri-Robo system for monitoring crops, identifying leaf diseases, and spraying pesticides. This agri-Robo system was able to detect cotton diseases by capturing the images in real time. The system detected and classified infection in a particular field and sprayed pesticides when required. The system is placed on robotic car which can move around the space in between the crop rows in the farm. A camera placed on the car captured the images of the plant and it is connected to the PC wirelessly, on which the main process of disease detection algorithm is run using the MATLAB software. The robot has a spraying system which automatically sprays the pesticides in proper amount automatically upon disease detection. Nireeshma et al. (2020) have demonstrated a novel prototype to monitor the crops in a timely manner using agricultural robot and image processing technique. The system detects the disease and sprays pesticides on the crops. Kumar & Vani (2018) used image processing coupled with

TABLE 20.3 Case Studies on Application of Robotic Systems in Plant Pathology

Disease	Type of Sensor	Procedure	Key Finding	References
Tobacco mosaic in tobacco. Cercospora leaf spot in sugarbeet.	Thermal and chlorophyll fluorescence imaging (Chl-FI) sensor.	Chl-FL and thermography were carried out in parallel by visualizing photosynthetic efficiency and transpiration, respectively to supervise the changes in the physiological functioning of the plants due to attack by the pathogens. Under controlled conditions, the robotic set up captured series of images of the infected leaves *viz.*, thermal, visual, chlorophyll fluorescence, etc. While thermal imaging detected local change in temperature, chlorophyll fluorescence monitored co-localized changes in the intensity of fluorescence.	The detection system has the potential for early identification of disease by generating specific patterns and signatures for host-pathogen interactions using combination of both thermal imaging and chlorophyll fluorescence imaging.	Chaerle et al. (2004)
Basal stem rot in oil palm.	Volatile organic compound sensor.	The detection system consists of two components *viz.*, handheld electronic nose at the front end and a pattern recognition tool. In this system, Cyranose 320 is the electronic nose fitted with various attachments for different functions such as – (i) sampling; (ii) sensory and signal processing; and (iii) algorithm for pattern recognition. In this case, the tool for pattern recognition is artificial neural networks (ANN), which is widely used for data processing.	Successful differentiation of healthy and infected trunk in basal stem rot.	Markom et al. (2009)
Grape powdery mildew.	Optical sensor.	The robotic system set up integrates three different units *viz.*, manipulator, an optical sensor, and a precision spraying actuator. Based on the feedback received by the optical sensor, the manipulator positions the spraying actuator to spray the pesticides solely on the infected areas with accuracy. The sprayer produces a circular spraying pattern of 0.15 m diameter with adjustable velocity covering a wide range of distances. The sprayer can deliver an air-carrier flow with an adjustable velocity, producing a circular spraying pattern of a constant diameter of 0.15 m over a wide range of spraying distances.	CROPS–clever robot for cOPS, a multipurpose manipulator provided with multispectral (R-G-NIR) camera demonstrated the potential for detection of diseases in grapevines, majorly powdery mildew, and subsequent need-based spraying to manage the disease.	Oberti et al. (2013)

TABLE 20.3 *(Continued)*

Disease	Type of Sensor	Procedure	Key Finding	References
Citrus greening (Huanglongbing).	Visual sensor	Citrus leaves were categorized into 10 different classes data collection using Nikon D600 camera. Image pre-processing and feature extraction for background removal, gray level imaging and texture feature extraction using local binary pattern. HLB classification using supervised binary classification method viz., cost-support vector classification (C-SVC) and principal component analysis (PCA). Validation using k-fold cross-validation.	A method for leaf symptom recognition using image processing based on visible spectrum for successful detection of citrus greening, also known as Huanglongbing or HLB with high accuracy of 92%.	Deng et al. (2016)
Citrus greening (Huanglongbing)	Visual sensor	Graft inoculation where a cut-disc of infected leaf is placed in a similar sized healthy leaf and after 8 weeks, grafting evaluation was conducted. Using a vision-based sensor that had a sensitive camera at 591 nm and following the polarized imaging procedure, series of polarized images of citrus leaves were captured. Symptomatic leaves with HLB infection had increased levels of starch and were detected at 7 days after grafting. Results were confirmed by polymerase chain reaction.	Early detection of Huanglongbing even before the visual onset of symptoms can be achieved by using polarized imaging to monitor hyper accumulation of starch in HLB infected leaves.	Pourreza et al. (2016)
Citrus canker	Hyper-spectral sensor.	The hyperspectral data (indoor and outdoor) were collected using *Resonon pika* L. imaging system and an unmanned aerial vehicle Pro Hexacopter covering a spectral range of 400–100 nm. The spectral fingerprints of four disease development stages were generated *viz.*, healthy, diseased, and symptomatic, early disease development and late disease development were generated, and classification methods were determined.	Successful identification and differentiation of healthy and canker-infected leaves. Within the infected leaves, various disease development stages were differentiated using 100% accuracy *viz.*, diseased, and asymptomatic stage, and early stage.	Abdulridha et al. (2019)

TABLE 20.4 Some Other Algorithms Developed in Disease Detection

Crop	Objective	Technique	References
Multiple crops	Anthracnose detection.	Spectral vegetation indices.	Pujari et al. (2013)
Sugar beet	Cercospora leaf spot detection.	Robust template matching.	Zhou et al. (2014)
Pomegranate	Bacterial blight control.	Image processing approach.	Bhange & Hingoliwala (2015)
Tomato	Powdery mildew and tomato spotted wilt disease.	PCA and CV	Schor et al. (2016)
Betel vine	Leaf decay ailment.	Machine vision and image processing.	Dey et al. (2016)
Sugar beet	Control of powdery mildew, downy mildew and cercospora leaf spot diseases.	Spectral vegetation indices.	Mahlein et al. (2013)
Citrus	Detect diseased regions.	Imaging sensors.	Garcia-Ruiz et al. (2013)

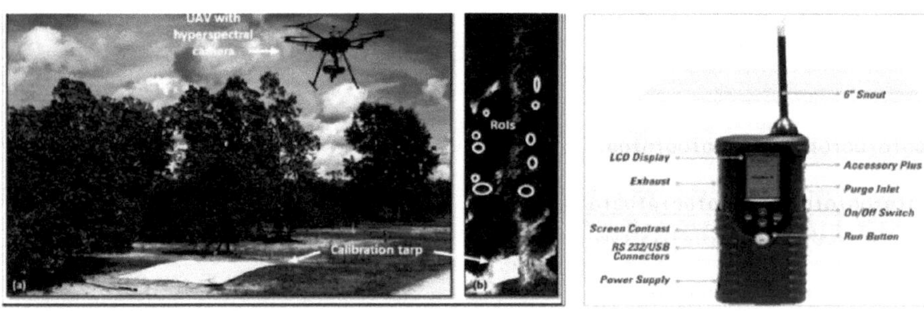

UAV with hyperspectral camera e-nose, Cyranose 320

FIGURE 20.6 Images captured by (a) UAV with hyperspectral camera; and (b) e-nose, Cyranose 320.

machine learning to develop a for detection of diseases in plants. The robot, which is designed using Lattepanda, an advanced processor is integrated with machine learning model. It has image processing for feature extraction, mean shift algorithm for segmentation and an SVM classifier for classification of disease. The robot is controlled by android application for performing all the operations. The robot monitors the soil moisture and other field conditions and takes up need-based irrigation and application of chemicals (pesticides, fertilizers, etc.) for obtaining good yields. The information on readings and operations are sent to the farmers and producers as SMS alerts (Figure 20.7).

Agri-robo system

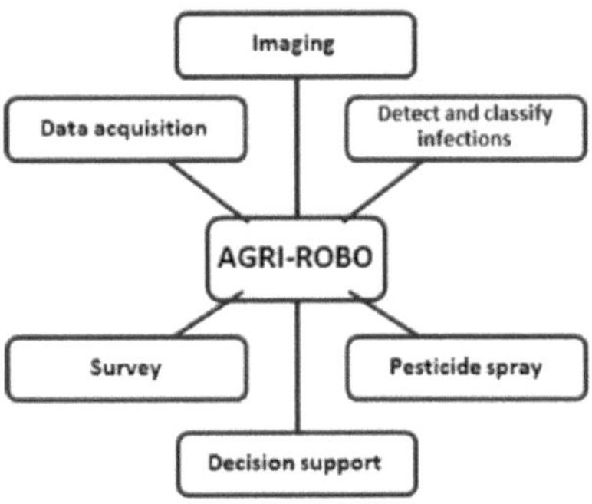

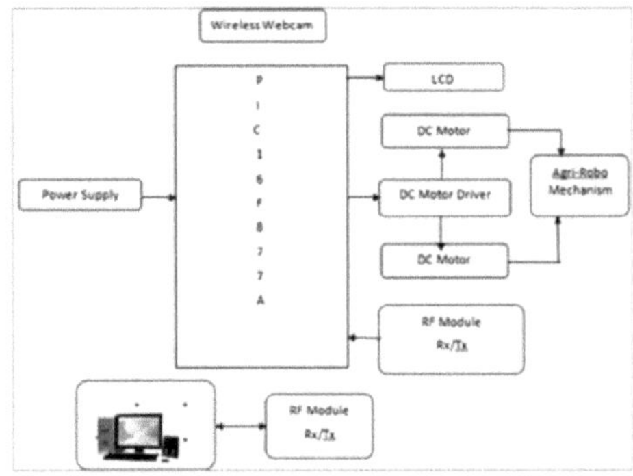

System architecture of Agri-robo system

Flow of work

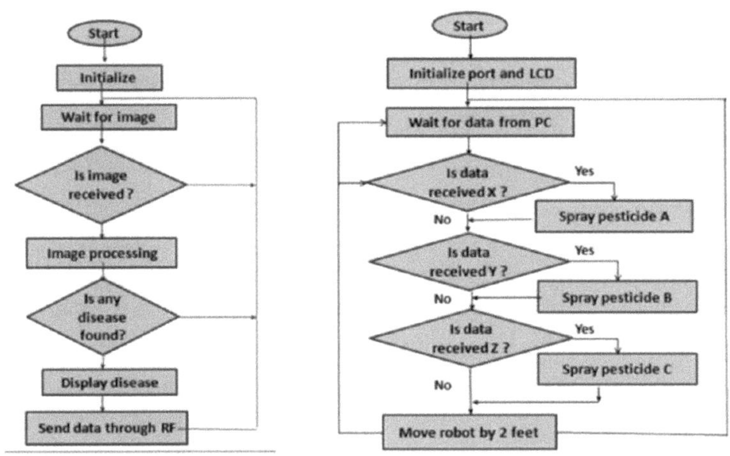

Image Processing Spraying operations

FIGURE 20.7 Agri-robo system flow of work.

20.7 PROS AND CONS OF ROBOTS IN PLANT PATHOLOGY

20.7.1 ROBOTS IN PLANT PATHOLOGY: INDIAN SCENARIO

Robots have not been yet used extensively for agriculture in India let alone in pathogen management. Still, most of the potential robots are in prototype stage. Even use of UAVs for plant health monitoring has also not reached farmers' level. Major advances are seen in the detection of foliar diseases of crop plants and automation in pesticide spraying of pesticides (Shaikh et al., 2016; Kumar & Vani, 2018; Nireeshma et al., 2020). A lot of factors provide hurdles in the use of robotic technology for plant disease management in India. These hurdles range from social stigmas like use of robots will give way to increased unemployment to technical difficulties like lack of skill, small farm size, etc. Also, negligence of farmers to adopt robotic technologies due to its high initial cost is also a factor. But we can hope its development in the near future.

20.7.2 PROS AND CONS OF ROBOTS IN PLANT PATHOLOGY

- **Pros:** Robots can work every day for 24 hours without rest and also don't pose problem of drudgery. It can continuously monitor the crop health status. It is extremely accurate, more efficient, and faster compared to humans. Some of them can detect crop diseases, even before visible symptoms appear. It can perform precise application of pesticides in the field thus reducing amount of pesticide used and impact on environment.
- **Cons:** Development of robots with accuracy for disease detection in open field condition is costly and tedious approach. Robots are very expensive in short run although it is profitable in long run. It cannot accustom with unusual conditions as humans, it can perform well only in places for which it is trained for. As of now, robots don't perform well in open field conditions as various factors such as weather can affect its detection accuracy. It can give false positive results, if programmed wrong. It cannot differentiate between some symptoms or signs which a human eye can.

20.7.3 PROS AND CONS OF UAVS IN PLANT PATHOLOGY

- **Pros:** This technology is effective and adaptive. It provides greater safety of farmers as it can help reduce contact with pesticides when we employ UAVs for spraying operations. It can provide faster data which may be of help for taking quick decision thus minimizing losses. Due its ability to provide real time insight of crop health status and perform precise management practices it helps reduce wastage of resources. It does not require any dedicated airport and navigation station. It is useful for insurance claim such as PMFBY.

- **Cons:** Flight time and range of UAVs is a constraint as it has to be stopped in between operations. We can't load large quantity of spray fluid at a time. Although operation costs are not so high, initial cost is very high. Farmers have to ace the bottlenecks in National Laws. Connectivity issues in rural areas may pose hurdle in its operations. It cannot be operated in harsh weather conditions. Knowledge and skill required for its operation. Misuse of drones for illegal activities also is a disadvantage.

20.8 LEGALITIES AND REGULATIONS ON UAV USE IN INDIA

In India, use of UAV is regulated by the Directorate General of Civil Aviation, Ministry of Civil Aviation, New Delhi. The drone regulations through a Civil Aviation Requirement (CAR) came into effect on 1st December 2018. The process for registering and operating drones in India is an all-digital process. Some of the important features of the drone policy are as follows:

- Operations are enabled through digital sky platform, a first of its kind national unmanned traffic management platform.
- Strict implementation of no permission no takes off (NPNT).
- Mandatory requirement of unmanned aircraft operator's permit (UAOP) for piloting UAV.
- Clearly defined no drone zones.
- Suspension/cancellation of permits in case of regulatory violations subjected to penalties.
- Firms registered in the DigitalSky platform of DGCA can only manufacture drones.

20.8.1 CHALLENGES

Like any other detection techniques in plant pathology, the accuracy of detection using agricultural robotics is influenced by factors like technique for sampling, size and duration of the sample, weather conditions, stage of the crop, background noise, etc.

In machine learning, though the use of large datasets can annul sampling issues, handling such a huge amount of data incurs cost. However, it is important to note that as the usage of such data increases over time, the cost of technology will lessen. In diagnostics based on vision, sometimes the same pathogen causes varying symptoms and different pathogens can cause similar looking symptom. Under such cases, it is important that interpretation is not only standardized but also validated to minimize the errors. In image-based diagnostics, the success is determined by the variations in plant density, architecture, canopy, foliage shapes, etc. The limited skills in information technology in case of botanists and plant pathologists and the limited skills in biology in case of engineers (data or AI) is another challenge in agricultural robotics. One strategy to overcome this is through proper interactions and combined trainings cum workshops on the basic concepts across the disciplines involving the researchers.

On field, the management of robots is challenging in comparison to that of laboratory environment and controlled conditions. Hence it is imperative that on-field robotics is focused on agronomic operations. In plant protection related operations, especially using computer vision technology, an important concern is the development of symptoms to the level of detection which usually happens after the pathogen has colonized or established in the host. However, by that time, delayed identification can make management difficult. In e-nose based diagnostics, one of the major concerns is the sensor drift affecting the accuracy of detection. This pre-requisite frequent calibration and drift correction which makes the technique costlier.

20.9 CONCLUSION

Agricultural robotics is an emerging market valued at $3.42 billion in 2017 with an expected compound annual growth rate of 21.1% by the year 2023. There is immense scope to use robotics in plant health management. Extensive research should be done in the field of sensors, algorithm development and UGVs and UAVs to exploit this potential. An interdisciplinary approach is needed in order to achieve better outputs by removing the knowledge constraints of data scientists/engineers and plant pathologists. Research should have an ultimate aim of developing robots that can be adapted to a broad spectrum of conditions. Commercialization is the next big step wherein the robots developed should reach the farmers' field. Robots can be helpful for plant pathologists and farmers in detecting and managing plant diseases.

Robotic systems with the ability to forecast, early detect, and precisely manage plant diseases will help in reducing yield losses, thus solving a lot of other associated problems. The development of reliable and user-friendly robots for plant health management is the need of the moment. However, there are several challenges in developing and adopting robotic systems in the field of plant health management. A huge amount of money, time, and effort would be required to harness these systems into the field of plant pathology. But the try is worth owing to the array of applications it offers and hopefully, we can see more robotic plant doctors in the near future.

KEYWORDS

- **agricultural robotics**
- **artificial intelligence**
- **digital farming**
- **machine learning**
- **plant disease diagnosis**
- **plant pathology**
- **robots**

REFERENCES

Abdulridha, J., Batuman, O., & Ampatzidis, Y., (2019). UAV-based remote sensing technique to detect citrus canker disease utilizing hyperspectral imaging and machine learning. *Remote Sensing*, *11*(1373), 1–22.

Agro Flow System, (2019). *What is the Difference Between Precision, Digital and Smart Farming?* https://agroflowsystem.com/blog/what-is-the-difference-between-precision-digital-and-smart-farming/#:~:text=According%20to%20a%20paper%20on%20Digital%20Agriculture%20by,webbased%20data%20platforms%20together%20with%20Big%20Data%20analyses%E2%80%9D (accessed on 12 December 2023).

Bhange, M., & Hingoliwala, H. A., (2015). Smart farming: Pomegranate disease detection using image processing. *Procedia Computer Science*, *58*, 280–288.

Bock, C. H., & Nutter, Jr. F. W., (2011). Detection and measurement of plant disease symptoms using visible-wavelength photography and image analysis. *CAB Reviews: Perspectives in Agriculture, Veterinary Science, Nutrition and Natural Resources*, *6*(27), 1–7.

Celen, I. H., Onler, E., & Kilic, E., (2015). A design of an autonomous agricultural robot to navigate between rows. *Proceedings of the 2015 International Conference on Electrical,*

Automation and Mechanical Engineering (pp. 349–352). Phuket Island, Thailand. Atlantis Press.

Chaerle, L., Hagenbeek, D., Bruyne, E. D., Valcke, R., & Der, S. D. V., (2004). Thermal and chlorophyll–fluorescence imaging distinguish plant–pathogen interactions at an early stage. *Plant Cell Physiology*, *45*(7), 887–896.

Deng, X., Lan, Y., Hong, T., & Chen, J., (2016). Citrus greening detection using visible spectrum imaging and C-SVC. *Computers and Electronics in Agriculture, 130*, 177–183.

Dey, A. K., Sharma, M., & Meshram, M. R., (2016). Image processing-based leaf rot disease, detection of betel vine (*Piper betle*L.). *Procedia Computer Science*, *85*, 748–754.

Garcia-Ruiz, F., Sankaran, S., Maja, J. M., Lee, W. S., Rasmussen, J., & Ehsani, R., (2013). Comparison of two aerial imaging platforms for identification of huanglongbing-infected citrus trees. *Computers and Electronics in Agriculture, 91*, 106–115.

GHI: Global Hunger Index, (2020). *2020 Global Hunger Index by Severity.* https://www.globalhungerindex.org/ (accessed on 12 December 2023).

ILO: International Labor Organization, (2014). *World of Work Report 2014.* https://www.ilo.org/global/research/global-reports/world-of-work/2014/lang--en/index.htm (accessed on 12 December 2023).

Kumar, V. V., & Vani, K. S., (2018). Agricultural robot- leaf disease detection and monitoring the field condition using machine learning and image processing. *International Journal of Computational Intelligence Research*, *14*(7), 551–561.

Kushwaha, H. L., Sinha, J. P., Khura, T. K., Kushwaha, D. K., Ekka, U., Purushottam, M., & Singh, N., (2016). Status and scope of robotics in agriculture. *International Conference on Emerging Technologies in Agricultural and Food Engineering* (pp. 264–277). IIT, Kharagpur, India. Agricultural and Food Engineering Department, IIT Kharagpur.

Madden, L. V., Hughes, G., & Bosch, F. V., (2007). Measuring plant diseases. In: *The Study of Plant Disease Epidemics* (pp. 11–31). USA: The American Phytopathological Society.

Mahlein, A. K., (2016). *Plant Disease Detection by Imaging Sensors – Parallels and Specific Demands for Precision Agriculture and Plant Phenotyping* (pp. 241–251). *Plant Disease*. APS Press.

Mahlein, A. K., Rumpf, T., Welke, P., Dehne, H. W., Plümer, L., Steiner, U., & Oerke, E. C., (2013). Development of spectral indices for detecting and identifying plant diseases. *Remote Sensing of Environment, 128*, 21–30.

Markom, M. A., Md Shakaff, A. Y., Adom, A. H., Ahmad, M. N., Hidayat, W., Abdullah, A. H., & Fikri, N. A., (2009). Intelligent electronic nose system for basal stem rot disease detection. *Computers and Electronics in Agriculture, 66*, 140–146.

Mehta, C. R., Chandel, N. S., Jena, P. C., & Jha, A., (2019). Indian agriculture counting on farm mechanization. *Agricultural Mechanization in Asia, Africa and Latin America*, *50*(1), 84–89.

Nireeshma, R., Rashmi, R., Sangeetha, M., Spoorthi, M. E., & Shilpa, B. L., (2020). Plant leaf disease detection and automated medicine using IoT. *International Research Journal of Engineering and Technology*, *7*(4), 1729–1731.

Oberti, R., Marchi, M., Tirelli, P., Calcante, A., Iriti, M., Hočevar, M., Baur, J., et al., (2014). The CROPS agricultural robot: Application to selective spraying of grapevine's diseases. *Proceedings International Conference of Agricultural Engineering* (pp. 1–8). Zurich. European Society for Agricultural Engineers.

Pourreza, A., Lee, W. S., Etxeberria, E., & Zhang, Y., (2016). Identification of citrus huanglongbing disease at the pre-symptomatic stage using polarized imaging technique.

In: 5th *IFAC Conference on Sensing, Control and Automation Technologies for Agriculture AGRICONTROL 2016* (Vol. 49, No. 16, pp. 110–115). Seattle, USA. Elsevier.

Prabha, K., (2020). Disease sniffing robots to apps fixing plant diseases applications of artificial intelligence in plant pathology—A mini review. *Indian Phytopathology*, 1–8.

Pujari, J. D., Yakkundimath, R., & Byadgi, A. S., (2013). Grading and classification of anthracnose fungal disease of fruits based on statistical texture features. *International Journal of Advanced Science and Technology*, *52*(1), 121–132.

Rieder, R., Pavan, W., Maciel, J. M., Fernandes, J. M., & Pinho, M. S., (2014). A virtual reality system to monitor and control diseases in strawberry with drones: A project. In: 7th *International Congress on Environmental Modelling and Software* (pp. 919–926). San Diego, California, USA. Brigham Young University.

Schor, N., Bechar, A., Ignat, T., Dombrovsky, A., Elad, Y., & Berman, S., (2016). Robotic disease detection in greenhouses: Combined detection of powdery mildew and tomato spotted wilt virus. *IEEE Robotics and Automation Letters*, *1*(1), 354–360.

Shaikh, D. A., Ghorale, A. G., Chaudhari, P. A., & Kale, P. L., (2016). Intelligent autonomous farming robot with plant disease detection using image processing. *International Journal of Advanced Research in Computer and Communication Engineering*, *5*(4), 1012–1016.

Shamshiri, R. R., Weltzien, C., Hameed, I. A., Yule, I. J., Grift, T. E., Balasundram, S. K., Pitonakova, L., et al., (2018). Research and development in agricultural robotics: A perspective of digital farming. *International Journal of Agricultural and Biological Engineering*, *11*(4), 1–14.

United Nations, Department of Economic and Social Affairs, Population Division. (2019). *World Population Prospectus 2019: Highlights.* ST/ESA/SER.A/423.

Zhou, R., Kaneko, S. I., Tanaka, F., Kayamori, M., & Shimizu, M., (2014). Disease detection of *Cercospora* leaf spot in sugar beet by robust template matching. *Computers and Electronics in Agriculture*, *108*, 58–70.

Index

D

N

O

P

Q

T

U

V

W

X

Y

Z